Seismic Assessment and Buildings

NATO Science Series

A Series presenting the results of scientific meetings supported under the NATO Science Programme.

The Series is published by IOS Press, Amsterdam, and Kluwer Academic Publishers in conjunction with the NATO Scientific Affairs Division

Sub-Series

I. Life and Behavioural Sciences	IOS Press
II. Mathematics, Physics and Chemistry	Kluwer Academic Publishers
III. Computer and Systems Science	IOS Press
IV. Earth and Environmental Sciences	Kluwer Academic Publishers
V. Science and Technology Policy	IOS Press

The NATO Science Series continues the series of books published formerly as the NATO ASI Series.

The NATO Science Programme offers support for collaboration in civil science between scientists of countries of the Euro-Atlantic Partnership Council. The types of scientific meeting generally supported are "Advanced Study Institutes" and "Advanced Research Workshops", although other types of meeting are supported from time to time. The NATO Science Series collects together the results of these meetings. The meetings are co-organized bij scientists from NATO countries and scientists from NATO's Partner countries – countries of the CIS and Central and Eastern Europe.

Advanced Study Institutes are high-level tutorial courses offering in-depth study of latest advances in a field.
Advanced Research Workshops are expert meetings aimed at critical assessment of a field, and identification of directions for future action.

As a consequence of the restructuring of the NATO Science Programme in 1999, the NATO Science Series has been re-organised and there are currently five sub-series as noted above. Please consult the following web sites for information on previous volumes published in the Series, as well as details of earlier sub-series.

http://www.nato.int/science
http://www.wkap.nl
http://www.iospress.nl
http://www.wtv-books.de/nato-pco.htm

Series IV: Earth and Environmental Sciences – Vol. 29

Seismic Assessment and Rehabilitation of Existing Buildings

edited by

S. Tanvir Wasti
Department of Civil Engineering,
Middle East Technical University, Ankara, Turkey

and

Guney Ozcebe
Department of Civil Engineering,
Middle East Technical University, Ankara, Turkey

Kluwer Academic Publishers

Dordrecht / Boston / London

Published in cooperation with NATO Scientific Affairs Division

Proceedings of the NATO Science for Peace Workshop on
Seismic Assessment and Rehabilitation of Existing Buildings
Izmir, Turkey
13–14 May 2003

A C.I.P. Catalogue record for this book is available from the Library of Congress.

ISBN 1-4020-1624-7 (HB)
ISBN 1-4020-1625-5 (PB)

Published by Kluwer Academic Publishers,
P.O. Box 17, 3300 AA Dordrecht, The Netherlands.

Sold and distributed in North, Central and South America
by Kluwer Academic Publishers,
101 Philip Drive, Norwell, MA 02061, U.S.A.

In all other countries, sold and distributed
by Kluwer Academic Publishers,
P.O. Box 322, 3300 AH Dordrecht, The Netherlands.

Printed on acid-free paper

Printed in the Netherlands.

Courtesy: METU GISAM ARCHIVE

Contents

Contributing Authors

Umut AKGUZEL
Research Assistant
Bogazici University
Department of Civil Engineering
80815 Bebek, Istanbul, Turkey

Sinan AKKAR
Assistant Professor
Middle East Technical University
Department of Civil Engineering
06531 Ankara, Turkey

Ugurhan AKYUZ
Assistant Professor
Middle East Technical University
Department of Civil Engineering
06531 Ankara, Turkey

Aysegul ASKAN
Research Assistant
Carnegie Mellon University
Department of Civil and Environmental Engineering
Computational Mechanics Laboratory
Pittsburgh, 15213, PA, USA

Volkan AYDOGAN
Research Assistant
Middle East Technical University
Department of Civil Engineering
06531 Ankara, Turkey

B. Sadik BAKIR
Assistant Professor
Middle East Technical University
Department of Civil Engineering
06531 Ankara, Turkey

Mehmet BARAN
Research Assistant
Middle East Technical University
Department of Civil Engineering
06531 Ankara, Turkey

Dionysis BISKINIS
c/o University of Patras
Department of Civil Engineering
Structures Laboratory
Greece

Stathis N. BOUSIAS
Assistant Professor
University of Patras
Department of Civil Engineering
Structures Laboratory
Greece

Murat DUVARCI
Middle East Technical University
Department of Civil Engineering
06531 Ankara, Turkey

Ibrahim ERDEM
Research Assistant
Middle East Technical University
Department of Civil Engineering
06531 Ankara, Turkey

Emrah ERDURAN
Research Assistant
Middle East Technical University
Department of Civil Engineering
06531 Ankara, Turkey

Ugur ERSOY
Professor
Middle East Technical University
Department of Civil Engineering
06531 Ankara, Turkey

Michael N. FARDIS
Professor
University of Patras
Department of Civil Engineering
Structures Laboratory
Greece

Robert J. FROSCH
Associate Professor
Purdue University
School of Civil Engineering
West Lafayette, Indiana, USA

Luis E. GARCIA
Professor
Universidad de los Andes
Bogota, Colombia

Mihail GAREVSKI
Professor
St. Cyril and Methodius University
Institute of Earthquake Engineering and
Engineering Seismology
Salvador Aljende 73, 1000 Skopje
Republic of Macedonia

Polat GULKAN
Professor
Middle East Technical University
Department of Civil Engineering
06531 Ankara, Turkey

Viktor HRISTOVSKI
Assistant Professor
St. Cyril and Methodius University
Institute of Earthquake Engineering and
Engineering Seismology
Salvador Aljende 73, 1000 Skopje
Republic of Macedonia

Alper ILKI
Assistant Professor
Istanbul Technical University
Civil Engineering Department
Structural and Earthquake Engineering
Laboratory 34469, Maslak, Istanbul, Turkey

James O. JIRSA
Professor
University of Texas
Department of Civil Engineering
78712 Austin, Texas, USA

Faruk KARADOGAN
Professor
Istanbul Technical University
Civil Engineering Department
Structural and Earthquake Engineering
Laboratory 34469, Maslak, Istanbul, Turkey

Antonis KOSMOPOULOS
c/o University of Patras
Department of Civil Engineering
Structures Laboratory
Greece

Michael E. KREGER
Professor
University of Texas
Department of Civil Engineering
78712 Austin, Texas, USA

Guney OZCEBE
Professor
Middle East Technical University
Department of Civil Engineering
06531 Ankara, Turkey

Sevket OZDEN
Assistant Professor
Kocaeli University
Department of Civil Engineering
Kocaeli, Turkey

Turan OZTURAN
Professor
Bogazici University
Department of Civil Engineering
80815 Bebek, Istanbul, Turkey

Telemachos B. PANAGIOTAKOS
c/o University of Patras
Department of Civil Engineering
Structures Laboratory
Greece

Aleksandar PASKALOV
Dr., Researcher
St. Cyril and Methodius University
Institute of Earthquake Engineering and
Engineering Seismology
Salvador Aljende 73, 1000 Skopje
Republic of Macedonia

Kyriazis PITILAKIS
Professor
Aristotle University of Thessaloniki
Department of Civil Engineering
Greece

Murat SAATCIOGLU
Professor
University of Ottawa
Department of Civil Engineering
Ottawa, CANADA

Mete A. SOZEN
Professor
Purdue University
School of Civil Engineering
West Lafayette, Indiana, USA

Haluk SUCUOGLU
Professor
Middle East Technical University
Department of Civil Engineering
06531 Ankara, Turkey

Kosta TALAGANOV
Professor
St. Cyril and Methodius University
Institute of Earthquake Engineering and
Engineering Seismology
Salvador Aljende 73, 1000 Skopje
Republic of Macedonia

Tugrul TANKUT
Professor
Middle East Technical University
Department of Civil Engineering
06531 Ankara, Turkey

Ahmet TURER
Assistant Professor
Middle East Technical University
Department of Civil Engineering
06531 Ankara, Turkey

S. M. UZUMERI
Professor
Middle East Technical University
Department of Civil Engineering
06531 Ankara, Turkey

Yaman UZUMERI
Chief Building Official (Ret.)
City of Toronto, Canada

S. Tanvir WASTI
Professor
Middle East Technical University
Department of Civil Engineering
06531 Ankara, Turkey

Ahmet YAKUT
Assistant Professor
Middle East Technical University
Department of Civil Engineering
06531 Ankara, Turkey

Ufuk YAZGAN
Research Assistant
Middle East Technical University
Department of Civil Engineering
06531 Ankara, Turkey

M. T. YILMAZ
Middle East Technical University
Department of Civil Engineering
06531 Ankara, Turkey

M. Semih YUCEMEN
Professor
Middle East Technical University
Department of Civil Engineering
06531 Ankara, Turkey

Ercan YUKSEL
Assistant Professor
Istanbul Technical University
Civil Engineering Department
Structural and Earthquake Engineering
Laboratory 34469, Maslak, Istanbul, Turkey

Preface

The present volume contains a total of 23 papers centred on the research area of Seismic Assessment and Rehabilitation of Existing Buildings. This subject also forms the core of Project SfP977231, sponsored by the NATO Science for Peace Office and supported by the Scientific and Technical Research Council of Turkey [TUBITAK]. Most of these papers were presented by the authors at a NATO Science for Peace Workshop held in Izmir on 13 – 14 May, 2003 and reflect a part of their latest work conducted within the general confines of the title of the NATO Project. Middle East Technical University, Ankara, Turkey serves as the hub of Project SfP977231 and coordinates research under the project with universities within Turkey, e.g. Istanbul Technical University and Kocaeli University, and with partner institutions in Greece and the Former Yugoslav Republic of Macedonia.* A few articles have also been contributed by invited experts, who are all noted researchers in the field. Altogether, the contents of the volume deal with a vast array of problems in Seismic Assessment and Rehabilitation and cover a wide range of possible solutions, techniques and proposals. It is intended to touch upon many of these aspects separately below.

Earthquakes constitute possibly the most widely spread and also the most feared of natural hazards. Recent earthquakes within the first six months of 2003, such as the Bingol Earthquake in Turkey and the Algerian earthquake, have caused both loss of life and severe damage to property. Most earthquake prone countries are taking steps to ensure that new housing

* Turkey recognizes the Republic of Macedonia with its constitutional name

conforms to the requirements of well developed earthquake codes and specifications which have the seal of international approval. This is by no means an easy task, as much building construction in many areas is non-engineered and poorly if not dangerously detailed. Supervision and inspection of construction sometimes exist in name only.

A far greater problem is the unsatisfactory performance exhibited in any earthquake by the large stock of already existing buildings. What complicates the problem is that in many earthquake regions, some of these buildings may be fairly sound, and may well satisfy the requirements of the seismic codes that were valid when they were built, whereas others may have been deficiently engineered from the very start. However, by the demanding standards of today's codes and specifications, which are periodically updated and upgraded, such buildings have ceased to be acceptable. And yet, so impressive has been the accumulation of analytical and experimental knowledge, so swift the development in new materials and the improvements in techniques, that the gap between challenge and response in this area continues to narrow.

The health of buildings, like that of people, requires monitoring, diagnosis and treatment. Health is by no means a question of the survival of the fittest; more attention and care often need to be lavished on the weakest individual. Health is just the means; the end and aim is welfare. Structures have to be rendered fit for earthquakes, whether they are old or new, resistant or decrepit – if for nothing other than that people do and must live in them. *Mutatis mutandis*, diagnosis and treatment, examination and prescription, give way to assessment and rehabilitation.

The contents of the book reflect many – though mainly structural – aspects of the earthquake problem and, from the work presented herein, solutions of varying efficacy may be distilled. It is not a coincidence that one of the articles is pointedly titled "Bringing to buildings the healing touch". Several of the most comprehensive papers, e.g. those by Baran et al., Fardis and Bousias, Ozden et al. and Saatcioglu deal with the seismic 'retrofit' of structures, indicating the measures that can be taken to prepare a structure for its seismic encounter after construction is complete but before an earthquake occurs. Closely related strengthening procedures are discussed by Erdem et al., Ersoy et al. and Karadogan et al., who also present the results of much arduous and well-planned experimental work. Their experimentation is supplemented by the valuable work of Frosch, Jirsa and Kreger, who validate a sound and economically feasible retrofit method involving precast infill walls. In addition to conventional strengthening techniques, the use of carbon fibre reinforced polymers – CFRP to insiders – begins to gain prominence in these applications. Allied to all experimental research is the advantage provided by tests on reinforced concrete models, some of which, though scaled to a third of the original size, enable the simulated representation of actual buildings of several storeys. Garevski et al. provide useful initial information on the shaking table tests of such a large model subjected to horizontal and vertical base motions.

Another term, loaded with uncertainty, is that of 'seismic vulnerability', which features prominently in the papers by Garcia, and Ozcebe et al. Risk assessment and damage probabilities are non-deterministic concepts that fit in well with the unannounced nature of the earthquake event, and are treated in the papers by Sucuoglu and Yazgan, and Yucemen and Askan. Such papers also indicate the need for new methods of structural evaluation and point towards the direction in which both codes and practice might be heading. Pitilakis reminds us that earthquakes emanate from the soil and that microtremors may be used for site characterization and microzonation. Akkar and Sucuoglu start with the analysis of earthquake motions and relate them to procedures for the assessment of subsequent damage. Separately, Gulkan et al. suggest that building damage prediction in its current form still needs to be taken with an occasional pinch of salt.

A touch of colour is added by Turer whose paper deals with the condition assessment of bridges as well as historical and monumental structures that go back many centuries. The history of structural drift from its role as a humble bystander into a central character in seismic codes has been traced by Sozen with inimitable style. Uzumeri and Uzumeri draw attention to prevention rather than cure by reiterating the importance of enforcing seismic codes that embody available expertise on the subject, thereby complementing the paper by Wasti wherein the educational spin-offs accruing from the project are emphasized, with the message that formation is more important than information, even in the seismic world.

Syed Tanvir Wasti
Guney Ozcebe

Ankara, July 2003

Foreword

At the turn of the last century, builders pondering the havoc wreaked by earthquakes had scant resources to guide their reconstruction efforts. A century later, earthquakes continue tragic devastations of our cities, with one important difference – today's builders can turn to a rich literature in earthquake engineering to guide their reconstruction efforts. This volume on *Seismic Assessment and Rehabilitation of Buildings* is bound to become one of the valued resources of earthquake engineers who are building to reduce the tragedies of future earthquakes.

The volume is filled with brilliant contributions from renowned experts in earthquake engineering. Topics include seismic earthquake strong ground motion, dynamic response of structures, seismic assessment of older hazardous construction, and practical seismic rehabilitation methods derived from extensive laboratory and field research. Many of the contributions have been motivated by the tragic consequences, yet real-world experiences, of the 1999 Kocaeli and Düzce earthquakes in Turkey, and demonstrate the rapid progress that can be inspired by necessity. While the topic of this volume is especially important for Turkey and her neighbors, the lessons learned and clearly presented are universal and should be studied by serious earthquake engineers worldwide.

Among the volume's contents is a quote from W. Shakespeare, *Julius Caesar*, Act I, Scene III:

"Are you not mov'd, when all the sway of earth
Shakes like a thing unfirm?"

The earthquake engineering community can be thankful that Professors S. Tanvir Wasti and Güney Özcebe, Middle East Technical University, were moved to organize the workshop and associated papers that form the contents of this memorable volume. Their effort, documented in this volume, is one of great and lasting value.

Jack Moehle
Professor and Director
Pacific Earthquake Engineering Research Center
University of California, Berkeley
July 2003

BRINGING TO BUILDINGS THE HEALING TOUCH
A Challenging Task for Engineers

Syed Tanvir Wasti and Ugur Ersoy
Structural Mechanics Division, Department of Civil Engineering, Middle East Technical University, Ankara, 06531, Turkey

Are not you mov'd, when all the sway of earth
Shakes like a thing unfirm?
W. Shakespeare, Julius Caesar, Act I, Scene III.

Abstract: Civilization and civil engineering are intimately connected, and few animate – inanimate relationships are more enduring than those between human beings and their built environment. This interactive relationship goes back thousands of years. The assessment of how a building will react to a natural disaster like an earthquake requires a detailed examination of the architectural features and engineering design. Structural damage can always be investigated after it has occurred, but the challenge is to pre-empt such damage by suitably equipping a building in advance to resist the onslaught of disaster.

Keywords: Structural rehabilitation, seismic behaviour, technical insight, assessment, appraisal, system amelioration

1. BRIEF HISTORICAL BACKGROUND

Of the seven wonders of the ancient world, only the Pyramids of Egypt have survived. Most of the other massive structures, built to last forever,

S.T. Wasti and G. Ozcebe (eds.), Seismic Assessment and Rehabilitation of Existing Buildings, 1–10.

happened to be located near the shores of the Mediterranean which, then as now, are prone to earthquakes. As with the statues of Ozymandias of Egypt, despair surrounds their ruins.[1] Over the centuries, monumental structures continued to be constructed, all over the world, by builders with imagination, talent and ingenuity whose names have mostly been forgotten.[2] Where in-built safety or brilliant craftsmanship enabled buildings to sustain the ravages of earthquakes and similar natural disasters, the great buildings of the past have remained standing, to teach and inspire us.[3]

More interestingly, it may be noted that the history of structural repair and strengthening is about as old as that of building construction itself. The famous code of Hammurabi, formulated *circa* 2200 B.C. by the great king of Mesopotamia, contains several vital clauses relating to the building of houses, and even a clause that is most pertinent to structural rehabilitation.[4]

In contrast, the theory of the elastic bending of beams is not yet two hundred years old.[5] It is only within the last century that large numbers of buildings have been engineered, i.e., subjected to the twin constraints of formal analysis and design. Furthermore, the importance of lateral loads such as those engendered by earthquakes crept into the realm of structural

1 The reference is to the famous poem by Percy Bysshe Shelley (1792 – 1822) . Ozymandias is the Greek version of one of the names of Ramesses II, Pharaoh of Egypt, who died at the age of 92 in 1213 B.C. Accounts suggest that the destruction of his statues and mortuary temple in Thebes was probably due to an earthquake.

2 An exception is that of Imhotep, the architect / engineer of the great stepped pyramid at Saqqara, Egypt, who was one of two commoners deified during the long history of ancient Egypt. Imhotep flourished between 2667 B.C. and 2648 B.C., and this pyramid was the first to be built in Egypt.

3 In the case of Turkey, such structures would include the grand mosques in Istanbul, such as the Blue Mosque and the Süleymaniye. A special example is that of the monumental structure of Santa Sophia which still stands, although it has seen rehabilitation once every few centuries after suffering fire and earthquake damage. William Butler Yeats has rightly drawn attention to the seeming permanence of Santa Sophia in his well-known lines: 'A starlit or moonlit dome disdains / All that man is, / All mere complexities / The fury and the mire of human veins.'

4 "If a builder build a house for a man and do not make its construction meet the requirements and a wall fall in, that builder shall strengthen the wall at his own expense." [From the translation by R. F. Harper]

5 Although attempts at fixing the location of the neutral axis of a beam cross-section occupied Coulomb and others, the solution of the elastic bending problem was first published by Navier in 1826. See Jacques Heyman, *Coulomb's Memoir on Statics*, Cambridge University Press, Cambridge, 1972, p. 101.

analysis less than a hundred years ago. [6] Prior to this period, it was tacitly assumed by engineers and builders that a design that had proper provisions for gravity loads with reasonably large safety factors would prevent buildings from toppling over in an earthquake.

During the last fifty years, however, enormous progress has been achieved in the area of seismic analysis and design because of the advent of powerful computers, sophisticated experimentation and the introduction of new techniques such as seismic control and base isolation. Furthermore, research input from the disciplines of seismology, geotechnical engineering and statistics has materially modified the manner in which earthquakes are treated.

It is estimated that the population of the whole world in 1800 was one billion, whereas just the increase in the world population during the decade 1990 – 2000 was also one billion. This extraordinary rise in population, combined with technological development, has resulted in a housing stock measured in the hundreds of millions. To this must be added the vast building infrastructure available worldwide for the transportation of people and goods. It can therefore be concluded that while measures need to be taken to ensure that all new construction satisfies the latest version of the seismic code in every country, there is bound to remain a huge backlog of existing structures that do not satisfy the latest or even any seismic code. The assessment of the seismic vulnerability of these existing structures before and after an earthquake is therefore an urgent task of great magnitude.

All constructive physical modification carried out to buildings and structures may be classified as **rehabilitation**. Structural rehabilitation of a building is intended to repair damage if any and/or to improve structural behaviour [often in addition to providing enhanced utility and aesthetics] and thereby may be said to lead to structural **upgrading**. Other expressions used in this connection, occasionally with small differences in implication, are *renovation*, *refurbishment* and *restoration* – the last being reserved usually for buildings of historical importance.

Repair usually applies to construction methods aimed at bringing a damaged structure back to its original or higher level of use, from the viewpoint of both strength and stiffness. As the name indicates, *strengthening* incorporates the provision of extra strength and stiffness to structures that may or may not have been damaged. *Augmentation* is an expression for strengthening that usually includes the provision of extra

6 Glen V. Berg, *Elements of Structural Dynamics*, Prentice Hall, Englewood Cliffs, N.J., 1989, pp. 216 *et seqq*. Thus, the first edition of the Uniform Building Code appeared in 1927.

dimensions for structural members, e.g. by jacketing. Measures taken to strengthen structures in advance to minimize possible damage and danger in the event of a disaster come under the operation of *retrofitting* or simply strengthening. All the above terms are occasionally applied without rigour to various rehabilitation procedures.

Ivo Andric's remarkable saga, in which the main 'character' is a stone bridge spanning the centuries, which was built as a pious bequest by an Ottoman Grand Vizier,[7] is not devoid of a serious reference to repair and strengthening:

The indefatigable lifts lowered cement and gravel, load after load, and the three strong piers which were the most exposed . . . and the most corroded were filled in at the bases as a rotten tooth is filled at its root.[8]

2. ASSESSMENT AND REHABILITATION OVERVIEW

Reinforced concrete structures comprise a large percentage of buildings in all parts of the world. Concrete structures are generally sturdy, with a long useful life measured in terms of several decades. However, a variety of factors may cause deterioration or damage in concrete structures, or the undamaged structure may lack the strength and stiffness envisaged or desired, in which case remedial constructive measures are needed. Accepting that the annual production of cement in the world is currently about 1.5 billion tonnes [9] it is observed that the task of replacing old RC structures and of improving existing structures is enormous indeed.

Naturally, the decision as to whether a building that has suffered damage, say in an earthquake, is to be rehabilitated or demolished is not one that can be taken lightly. While financial considerations are always important, economics can never be the sole arbiter. The engineer cannot be callous about ordering the demolition of buildings that have structural 'life' left in them. Although one must not stretch the medical analogy too far, a detailed and professional assessment and diagnosis will be required and various tests conducted in order to decide not only if rehabilitation is feasible but also to

7 The Grand Vizier in question was Sokullu Mehmed Pasha (1506 – 1579), who originated from the Bosnian village of Sokolovici. Several accounts suggest that the architect of the famous bridge on the Drina was the great architect Koca Sinan (1490 – 1588).

8 Ivo Andric, *The Bridge on the Drina*, The University of Chicago Press, Chicago, 1977, p.205.

9 Estimates range from 1.3 to 2 billion metric tonnes.

propose the method and technique to be employed for the case in hand. Such appraisal becomes especially acute after a disastrous earthquake when large numbers of damaged structures – some of which may still have occupants within them – have to be assessed within a highly limited time period.

An important reason that necessitates the upgrading of existing structures is that by way of incorporation of the results of continuing research, building codes are frequently revised to allow new structures to withstand higher lateral earthquake loads or, in the case of bridges, heavier wheel loads. Existing structures designed to earlier codes and specifications perforce become delinquent or deficient and out of date.

It should be pointed out at once that the dual tasks of assessing all kinds of structures from the viewpoint of performance in a future earthquake and of equipping them in advance by suitable methods of rehabilitation are being actively pursued in all countries and regions which are susceptible to moderate or severe ground motions. Academic institutions in relatively earthquake-free countries are also developing methods for structural assessment and rehabilitation. In particular, extensive research in this area is being conducted in California, Illinois and Texas in addition to other states of the United States, in Turkey, in Japan and in the Balkan countries. A simple computer search reveals the extraordinary number of publications dealing with all aspects of structural rehabilitation, many of which also promote new materials and techniques. Valuable review publications on the subject are also available.[10]

During the last 25 years, pre- or post-disaster strengthening of concrete structures was often conducted by concrete and steel jacketing and bonding of steel plates to the surface of the structural members. Every method has some drawbacks and, depending on the strengthening needs for each particular case, recourse was had to the provision of reinforced infills and / or cast-in-place shear walls. Composite materials and fibre reinforced plastics are currently replacing steel for strengthening purposes because of a very high strength / weight ratio and good corrosion resistance. A recent development in this area is that of using liquidized plastic sprayed along with fibres to form an external strengthening layer for concrete members, which provides both protection as well as confinement.

10 J. P. Moehle, J. P. Nicoletti and D. E. Lehman [Editors], *Review of Seismic Research Results on Existing Buildings*, Report No. SSC 94-03, Seismic Safety Commission, State of California, Fall 1994, 428 pp.

3. TURKEY – A NATURAL EARTHQUAKE LABORATORY

The Turkish peninsula[11] possesses a history and geography each of which is unique and rich. Geologically very active, Turkey is frequently rocked by large earthquakes. Descriptions of many destructive earthquakes punctuate Turkish history during the past thousand years. Six major earthquakes in the last decade[12] have resulted in the deaths of thousands of people and caused huge economic losses. A further legacy of such earthquakes is many thousands of damaged buildings. After each of these earthquakes, the Turkish Government funded projects to rehabilitate buildings which were moderately damaged. Middle East Technical University [METU] was involved in most of these projects and its contribution is the subject of a separate section below.

Surveys made on damaged buildings have revealed that the majority of existing buildings do not meet the current code requirements. Also it was found out that most RC buildings in Turkey comprise framed structures with inadequate lateral stiffness. It would be possible to rehabilitate the damaged buildings by improving the strength and/or deformability of the deficient members. However, almost all of the members of the building would usually have to be upgraded in order to accomplish life safety. Needless to say, such rehabilitation would not be feasible. For this reason, the METU group involved in structural rehabilitation have preferred to conduct what they have called 'system improvement'.

In system improvement the aim is to replace the existing lateral load resisting system, which consists of frames having various deficiencies, with a new lateral load resisting system consisting of structural walls. Structural walls are made by filling the selected bays of the frame structure with reinforced concrete walls properly connected to the frame members. In the system improvement approach the objective is to introduce an adequate number of infilled frames (structural walls) in each principal direction of the structure *in-situ* and these infilled frames are envisaged to carry the total lateral load. These infilled frames not only provide strength but also increase the lateral stiffness of the structure significantly.

Rehabilitation of damaged reinforced concrete frame structures by introducing infilled frames was found to be a very feasible method for Turkey because it is both simple and easy to apply. The connection between

11 "From far-off Asia at full gallop, like a mare's head stretching into the Mediterranean sea" in the words of the poet Nazım Hikmet (1902 – 1963).

12 The 1992 Erzincan, 1995 Dinar, 1998 Ceyhan, 1999 Marmara, 1999 Düzce and 2002 Afyon – Sultandağı earthquakes.

the infill and frame members is provided by dowels consisting of steel bars, bonded by using epoxy resin into the holes drilled in the frame members.

4. THE METU EXPERIENCE

Active and sustained interest in the seismic behaviour of RC structures began at METU with the 1967 Mudurnu valley earthquake.[14] Seismic rehabilitation by using infilled frames was first applied in Turkey after the 1968 Bartin earthquake.[15] The first experimental research was initiated at METU after this application to see the effectiveness of the method used. Since then, several experimental research projects have been conducted at METU to understand the behaviour of infilled frames and to produce criteria for design. During this period research results were used in various applications, and problems encountered at the design and construction stages were brought to the laboratory. The criteria and details used in the design and construction of infilled frames by the METU group have been developed as a result of this interaction between research and application. For more comprehensive documentation on the experimental research of the METU group from the inception of the METU Structural Laboratory till the year 1999, the survey by Tankut[15] may be consulted.

As mentioned earlier, surveys made in Turkey have revealed that a great majority of the existing residential buildings do not meet the requirements of the current seismic code and are found to be vulnerable to damage and even collapse in a major earthquake. Needless to say, similar problems exist in all countries located in seismic regions. The most common deficiencies in these buildings are found to be:

- inadequate confinement at the end of columns and beams
- absence of ties in beam-column joints
- low concrete strength
- lapped splices in column longitudinal bars made at the floor levels, usually with inadequate lap length and

13 Ş. Z. Uzsoy and U. Ersoy, "Damage to Reinforced Concrete Buildings caused by the July 22, 1967 Earthquake in Turkey", *Bulletin of the Seismological Society of America*, Vol. 59, No. 2, April 1969, pp. 631-650.

14 U. Ersoy and Ş. Z. Uzsoy, "The Behavior of Infilled Frames" [in Turkish], TÜBİTAK Report MAG-205, Ankara, 1971.

15 T. Tankut, "Forty Years of Experimental Research in METU Structural Mechanics Laboratory", *Proceedings of the Ugur Ersoy Symposium on Structural Engineering*, Ankara, July 1999, pp. 3-36.

– improperly anchored ties (90^0 hooks).

In addition to these drawbacks, most of the framed structures have inadequate lateral stiffness. Furthermore, irregularities in plan and elevation, strong beam – weak column connections, soft stories and short columns are quite common in all these buildings.

Existing buildings having the deficiencies mentioned above can all be rehabilitated by using the "infilled frame" technology. The results of experimental research at METU in this area have been documented in the literature.[16 17] Work on the retrofitting of RC frames with RC infill walls has also reached a level of completion.[18] Canbay *et al.* have recently produced a comprehensive paper on the effect of RC infills on the seismic response of structures.[19]

One important aspect dealing with the rehabilitation of RC structures using infills needs to be pointed out here. Virtually all deficient buildings are found to be inhabited even after an earthquake. The buildings need to be evacuated for several months in order to construct and install the infilled frames. Of course, this turns out to be practically impossible. For such buildings alternative technologies have to be developed, using conventional or innovative materials which will enable seismic rehabilitation with minimum disturbance to the occupants.

The first step in the rehabilitation of existing undamaged structures is to determine the vulnerability of these buildings. This is not an easy task since usually thousands of buildings have to be assessed, and both time and funds are limited. Therefore the practical approach would be to carry on the assessment in several steps. The methodologies to be used in the early steps should be as simple as possible and should take minimum amount of time. The first step should aim to screen the most vulnerable and least vulnerable buildings. The method to be used in this step should not require the services of any specialists. The main objective in the first – and perhaps the second – step should be to reduce the number of buildings for which more detailed analyses are needed for the final decision. If quick and practical screening

16 S. Altın, U. Ersoy and T. Tankut, "Hysteretic Response of RC Infilled Frames", *ASCE Structural Journal*, Vol. 118, No. 8, August 1992, pp. 2133-2150 .

17 U. Ersoy, "Seismic Rehabilitation – Application, Research and Current Needs", 11th World Conference on Earthquake Engineering, Acapulco, Mexico, 1996 [Invited Paper].

18 A. M. Türk, U. Ersoy and G. Özcebe, "Retrofitting of RC Frames with RC Infill Walls", FIP Symposium on Concrete Structures in Seismic Regions, Athens, Greece, May 2003.

19 E. Canbay, U. Ersoy and G. Özcebe, "Contribution of RC Infills to the Seismic Behaviour of Structural Systems" ACI Structural Journal [to be published within 2003].

methods are not applied, the problem of assessment of existing buildings cannot be solved within a reasonable time period.

It is very essential to calibrate the screening methods developed using the data obtained from the damaged structures. In addition, using existing data, criteria should be established for the final assessment stage.

To develop methodologies for the seismic assessment of existing buildings and techniques which will enable seismic rehabilitation with minimum disturbance to the occupants appear to be challenging tasks for the engineers involved. Both assessment methods and rehabilitation techniques require extensive research. The methods developed should result in safe, economical and practical solutions if they are to find wide application.

Current research at METU, sponsored in part by NATO and TUBITAK[20] aims at continuing to develop suitable seismic assessment methodologies and rehabilitation techniques for existing buildings.[21] [22] The assessment methods developed are being calibrated using the data bank obtained from the six major earthquakes in Turkey during the last decade. Strengthening techniques using high-strength innovative materials such as CFRP [carbon fibre reinforcement polymers] and conventional materials such as precast panels seem to show much promise. Tests made at METU indicate that seismic strengthening technologies developed using these materials result in significant strength increase. The techniques developed seem to be both economical and practical and will result in minimum disturbance to the occupants of a building being rehabilitated. Past experimental experience at METU over more than three decades was very helpful in developing these new technologies.

5. SUMMING UP

From the technical viewpoint as well as from a consideration of economics, it is surprising that a systematic and research-based approach to the question of pre- or post-disaster structural rehabilitation has been so late

20 The Turkish acronym for the The Scientific and Technical Research Council of Turkey.

21 An overview is to be found in U. Ersoy, "The METU Approach to Structural Repair and Strengthening – Experimental Research and Applications" [in Turkish], *Symposium in Memory of Prof. Kemal Özden*, Istanbul University Faculty of Civil Engineering Publication, Istanbul, 2002, pp. 1-26.

22 G. Özcebe, U. Ersoy, T. Tankut, E. Erduran, R.S.O. Keskin and H.C. Mertol, "Strengthening of Brick-Infilled RC Frames with CFRP", SERU Report No. 2003/1, Middle East Technical University, Ankara, March 2003.

in arriving upon the scene. Although methods of structural rehabilitation may have lacked the *cachet* of more abstract research in the past, innovative and incisive engineering approaches have now made their presence felt in this vital area. Methods for enhancing the safety of all buildings minimize the loss of life and national wealth in the event of a disaster.

Research for developing and improving effective, economical and swift assessment methods and rehabilitation techniques for existing buildings with minimum disturbance to the occupants should therefore be encouraged and, where necessary, suitable funding should be made available. As a result of research, overall assessment and rehabilitation guidelines should be established and their success in the wake of a large earthquake should be critically examined. Furthermore, unless programs are designed to train and educate practising engineers, disaster relief attempts which incorporate structural amelioration will continue to remain seriously flawed.

THE VELOCITY OF DISPLACEMENT

Mete A. Sozen
Purdue University, School of Civil Engineering, West Lafayette, IN

Abstract: During the twentieth century, consideration in design of drift response has moved with speed from a minor serviceability criterion to a central proportioning issue. It is suggested that the time is ripe for uncoupling the drift computation from calculated and often unrealistic lateral design forces and relating it directly to a design displacement spectrum. It is also argued that the "cracked-section" concept for determining stiffness of reinforced concrete structures may be realistic for gravity loading but is not for earthquake effects.

Keywords: Drift, Earthquake-resistant design, Displacement spectrum, Code, Lateral stiffness, Lateral displacement, Cracked section, Reinforced concrete

1. INTRODUCTION

Both the perceived impact and the anticipated magnitude of lateral displacement of buildings in earthquakes (drift) have changed with time since the 1930's when the first movements occurred toward the assembly of detailed professional canons for earthquake-resistant design. This note summarizes some of the highlights that pertain to drift in the development of model codes for earthquake-resistant design in North America to trace the velocity of its travel from an optional check to a central design issue.

S.T. Wasti and G. Ozcebe (eds.), Seismic Assessment and Rehabilitation of Existing Buildings, 11–28.

2. DRIFT REQUIREMENTS

In organizing the available experience and science with respect to earthquake resistant design, the initial focus of the profession was exclusively on strength. The interested reader is referred to two publications that capture the perspectives of the period 1930 to 1960: [AND 1951] and [BIN 1960]. Drift was, besides being negligibly small, a concern related to preserving the investment and possibly to reducing the likelihood of pounding. But it was not a consideration for safety. This was made abundantly clear in the book by Blume et al. [BLU 1961] that contained the statement, well in keeping with the spirit of the times, "*...lateral displacement is seldom critical in a multi-story reinforced concrete building* [p 200]" despite the farseeing suggestion made earlier in the book in reference to determining the likelihood of pounding, "*A less rigorous appearing rule, but one which may in fact be both more accurate and more rational, is to compute the required separation as the sum of deflections computed for each building separately on the basis of an increment in deflection for each story equal to the yield-point deflection of that story, arbitrarily increasing the yield deflections of the two lowest stories by multiplying them by a factor of 2.*"

Perhaps the most candid expression of professional attitudes toward drift was contained in a committee report in 1959 of the Structural Engineers Association of Southern California reproduced in full as Appendix A to this note because it is brief and because the writer does not wish to distort the authors' perspective. The reader will note that the concerns with respect to drift were (1) debris ejection leading to a safety problem, (2) damage to nonstructural items, and (3) comfort of occupants. The members of the committee were quick to dismiss items (2) and (3) as issues inviting paternalism. They did observe that "...deflections of buildings subjected to earthquake computed as a statical force are not necessarily a true measure of the actual deflections of these buildings under earthquake shock" but went on to concede the field to drift related to wind effects.

Despite the sensitivity to the contradiction of using a static analysis for a dynamic effect, professional documents have continued to use, in general, static equivalent lateral forces to calculate drift, although this process has taken different turns and magnitude as summarized briefly in Table 1.

Table 1 provides a narrow perspective of professional opinion in North America with respect to drift. It is compiled in reference to a specific structure, one that may be considered to be a seven-story frame with an infinite number of spans so that its response can be understood in terms of a "tree" as shown in Figure 1. It is assumed to have the appropriate details to qualify as a special moment-resisting frame. Its calculated period is a little less than 0.7 sec. All requirements included in Table 1 refer to a frame of seven stories with a calculated period barely less than 0.7 sec.

The report by Anderson et al. [AND 1951] implied that the drift should be calculated by using the lateral forces selected for design but provided no guidance as to what to do with the results. The decision was left to opinion of the engineer.

The 1959 issue of the "Blue Book" [SEI 1959] introduced the sensitivity of the design shear force to the type of framing. For the selected frame, the coefficient was 0.67. The nominal period for a frame was set simply at 0.1N, where N is the number of stories. Distribution of the lateral forces over the height of the building was made linearly proportional to mass and height, the "linear distribution."

In 1972, the code developed by the Veterans Administration [VET 1972], being a "landlord's code," set high standards in new directions. The design base shear coefficient was increased. It was made a function of the specific site. By itself, that is not so significant, because it can be reduced in the next step. However, the VA Code also prescribed a modest "response reduction factor" of 4 for "ductile" frames and an amplifier of 3 for the deflection obtained using the design shear. In addition, the designer was asked to use "cracked sections" in determining the stiffness of the structural elements. The drift requirement was increased substantially.

In 1974, the "Blue Book" definition of the base shear coefficient [SEI 1974] was changed as was the distribution of lateral forces over the height of the building. If the calculated period was 0.7 seconds or less, the distribution conformed to that for the "linear mode shape."

Report ATC-3 by the Applied Technology Council [APP 1978], a product of a nation-wide collection of engineers familiar with earthquake-resistant design, introduced a different approach to period determination. Unless it was determined by dynamic analysis, the period was determined as a function of the total height of the frame. For reinforced concrete framing, a coefficient of 0.025 was recommended. The definition of the base shear coefficient was based on two different approaches. For low periods, it was set as a constant in recognition of the "nearly constant acceleration response" range identified by N. M. Newmark [BLU 1961]. For higher periods, shown in Table 1, it was set as an inverse function of the period ("nearly constant velocity response" range). For reinforced concrete frames, the "response reduction factor" was 7. The amplifier for drift was less. It was set at 6.

The 1994 edition of the Uniform Building Code [INT 1994] used the ATC-3 definition for the period with a different constant and raised the "response reduction factor" for frames to a generous 12.

After the Northridge 1994 and Kobe 1995 events, the Uniform Building Code [INT 1997] was modified as recorded in Table 1. The critical change in the draft requirement was that the designer was asked to assume cracked sections to determine stiffness but was not told how to determine the cracked section stiffness.

From the narrow perspective of a specific seven-story frame, the events over the years are summarized in Figure 2 and 3 showing the changes in required base shear coefficients and related maximum story-drift ratios. (The tree structure for which the story drift was calculated was assigned the following properties: Young's Modulus, 30,000 MPa, moment of inertia 0.067m^4 columns and 0.043m^4 girders, mass 500 kN/joint).

The data presented in Figure 2 are distorted by the fact that required nominal strength is also a function of the load factors used for strength as well as of the base shear force so that the UBC 1959 demand is not necessarily less than the UBC 1997 demand. However, with respect to determination of drift, the base shear coefficients are comparable. The salient and true change over the years occurred in the drift demand. The calculated drift in 1959 was indeed small enough to provide a foundation for the general opinion that it was not an important issue. The drift ratio determined by the method included in Veterans Administration Code of 1972 [VET 1972] changed the scene. While the calculated story drift ratio for the specific frame considered was 3%, the allowable was 0.8%. if there were no brittle elements involved. Otherwise, the limit to drift was reduced to approximately 0.25%. It is also seen in Figure 3 that the drifts calculated according to the 1994 and 1997 versions of UBC were comparable and in both cases they were admissible. The last bar shown in Figure 3 represents an estimate of the maximum story drift ratio, for the assumed frame based on the method by LePage [LEP 1997] which provides a frame of reference for the drift such a frame might experience subjected to strong ground motion.

3. WHY CRACKED SECTION?

One of the latest developments in the UBC-specified determination of drift was the demand for cracked section.

Figure 4 show an idealized load-displacement relationship for a reinforced concrete element loaded transversely. The initial response is quite stiff. It relates to uncracked section. The onset of cracking results in a rapid change in the stiffness of the beam. The slope of the unbroken line represents the "cracked section stiffness."

Under gravity loading, the beam is expected to work at approximately half its capacity. For moderately reinforced beams, the beam is likely to have cracked before the service load is reached. There is a logical argument for determining the short-time deflection for a given service load recognizing the stiffness of the beam after cracking. This is easier said than done because of random distribution of discrete cracks. The stiffness of the beam along its span is not easy to define. Nevertheless, the idea is defensible and leads to a reasonably good estimate of the short time deflection. At service load, the deflection of the beam can be estimated using the cracked-section stiffness.

Figure 5 shows four cycles in the lateral loading sequence of a test frame [GUL 1971] that experienced a displacement history simulating the one it would experience in response to a particular ground motion. The cracked-section stiffness for the frame is indicated in the plots for cycles 1 and 4. Cycle 4 represents the maximum response of the frame. That is the cycle in which we would be interested in determining the drift if we wish to limit the drift. The cracked-section stiffness is quite different from the effective stiffness at the time of attainment of the maximum drift.

Used as a scheme to increase the calculated drift, the cracked-section idea is plausible. But it should not be presented as being "realistic." It may be more effective to amplify the drift arbitrarily by a plausible factor rather than asking the designer to go on a chase to determine how much the columns would crack vs. how much the girders or the walls would crack.

4. DRIFT DETERMINATION

The main issue in process of determining drift is whether it should be related to static forces. It has been well established that the drift of a low to moderate rise building is dominated primarily by Mode 1 in a given plane. This is certainly acceptable if the building is reasonably uniform in distributions of mass, stiffness, and story height. To handle this problem simply and specifically (the generalization of nonlinear drift being equal to linear drift is not specific enough to be used in design as the linear drift may vary by 100% depending on the assumed damping factor), LePage proposed a very simple procedure. In the nearly acceleration and nearly constant velocity ranges as identified by Newmark, LePage [LEP 1997] proposed that the displacement spectrum representing the reasonable upper bound to what would be expected in a strong earthquake for a building with nonlinear response would be expressed by Eq. 1.

$$S_{da} = KT\sqrt{2} \tag{1}$$

K = a constant, with the dimension length/time, determined from displacement spectra (damping factor = 2%) suitable for the site. For stiff soil, the constant was proposed as 250 mm/sec for ground motions indexed by an effective peak ground acceleration of 0.5G.

T = calculated period (uncracked section) in sec.

LePage specified that the period T in this equation be the period calculated for the uncracked reinforced concrete structure amplified by $\sqrt{2}$. It is important to note that this was not in expectation of a cracked section. It was used because it had been found convenient in an earlier study by Shimazaki [SHI 1984]. Shimazaki had also defined the spectral displacement response by determining the envelope to specific displacement response spectra calculated using a damping factor of 2% for a collection of ground motions assumed to represent motions from comparable site conditions. Shimazaki normalized the response spectra in deference to the energy spectrum, on the premise that the drift increase caused by nonlinear response would be less if the energy response did not increase or if it increased at a low rate with increase in period. Using the basic concept by Shimazaki and LePage of normalizing the spectrum in relation to velocity, the LePage spectrum was restated in terms of the peak ground velocity.

$$S_{dv} = \frac{V_g}{\sqrt{2}} \cdot T \quad (2)$$

S_{dv} = spectral displacement
V_g = peak ground velocity
T = calculated first mode period

In a study of the effects of the ground motions measured in Anatolia during the two earthquakes of 1999, Ozturk noted that for structural systems with low base shear strengths, the drift response tended to exceed the limit set by LePage. He also noted that the difference increased with peak ground velocity. To generalize what he had observed, he devised Eq. 3.

$$S_{de} = \frac{V_g^2}{G\pi c_y}(1+T) \quad (3)$$

S_{de} = spectral displacement
V_g = peak ground velocity
c_y = base shear strength coefficient
G = acceleration of gravity

Combining Eq. 2 and 3,.

$$S_d = \max(S_{da}, S_{de}) \quad (4)$$

The results of Eq. 4 are shown in Figure 6 and 7 for base shear strength coefficients of 0.15 and 0.3. It is seen that as the base shear strength coefficient increases, the maximum response spectrum tends to conform to the LePage spectrum over a larger range of periods.

5. CONCLUDING REMARKS

The recent Bingöl tragedy has once again emphasized that frills in analysis for design represent no more than "playing in the sand." To protect lives, the essential structural concerns are detail and framing. Once the elementary and, by now, self-evident requirements for detailing are understood and the simple solutions are implemented in the field as well as in the drawings, there is room for thinking about drift because it impacts detail as well as framing. In the calculus of syllogisms, zero drift is tantamount to no need for detail. But the physical environment does not encourage thinking in the abstract. There will be drift and there will be need for proper framing and detail. The understanding of drift should improve the implementation of detail.

The concept of drift has moved with speed from an optional check to compete with strength as a driving design criterion. The time may have come to uncouple its determination from design forces and approach it directly using calculated period and a simple design spectrum. Equation 6 may provide an initial platform.

REFERENCES

1. [AND 1951] Anderson, Arthur W., et al. 1951, *Lateral Forces of Earthquake and Wind,* Proceedings Vol. 77, American Society of Civil Engineers, April, 1951.
2. [SEI 1959] Seismology Committee, Structural Engineers Association of California, 1959, *Recommended Lateral Force Requirements and Commentary*, 171 2nd Street, San Francisco, CA.
3. [BIN 1960] Binder, R. W. and W. T. Wheeler, *Building Code Provisions for Aseismic Design,* World Congress of Earthquake Engineering, Tokyo, 1960, pp. 1843-1857.
4. [SEI 1960] Seismology Committee, Structural Engineers Association of California, 1960, *Recommended Lateral Force Requirements and Commentary*, 171 2nd Street, San Francisco, CA.
5. [BLU 1961] Blume, John E., N. M Newmark, and Leo H. Corning, 1961, *Design of Multi-Story Reinforced Concrete Buildings for Earthquake Motions,* Portland Cement Association, Skokie, IL.
6. [SEI 1967] Seismology Committee, Structural Engineers Association of California, 1967, *Recommended Lateral Force Requirements and Commentary*, 171 2nd Street, San Francisco, CA.

7. [GUL 1971] Gulkan, P., *Response and Energy Dissipation of Reinforced Concrete Frames Subjected to Strong Base Motion,* thesis submitted to the Graduate College of the University of Illinois, Urbana, IL, June 1971.
8. [SEI 1971] Seismology Committee, Structural Engineers Association of California, 1971, *Recommended Lateral Force Requirements and Commentary*, 171 2nd Street, San Francisco, CA.
9. [VET 1972] Veterans Administration Office of Construction, Report of the Earthquake and Wind Forces Committee, *Earthquake Resistant Design Requirements for VA Hospital Facilities,* Washington, DC, 1972.
10. [SEI 1974] Seismology Committee, Structural Engineers Association of California, 1974, *Recommended Lateral Force Requirements and Commentary*, 171 2nd Street, San Francisco, CA.
11. [SEI 1975] Seismology Committee, Structural Engineers Association of California, 1975, *Recommended Lateral Force Requirements and Commentary*, 171 2nd Street, San Francisco, CA.
12. [APP 1978] Applied Technology Council, Tentative Provisions for the Development of Seismic Regulations for Buildings, ATC 3-06, 1978.
13. [SEI 1980] Seismology Committee, Structural Engineers Association of California, 1980, *Recommended Lateral Force Requirements and Commentary*, 171 2nd Street, San Francisco, CA.
14. [SHI 1984] Shimazaki, K. and M. A. Sozen, *Seismic Drift of Reinforced Concrete Structures,* Technical Report, Hazama-Gumi, Tokyo, 1984, pp. 145-165.
15. [INT 1985] International Conference of Building Officials, 1985, *Uniform Building Code, 1985 ed.,* 5360 S. Workman Mill Rd., Whittier, CA.
16. [INT 1994] International Conference of Building Officials, 1994, *Uniform Building Code, 1994, Vol. 2,* 5360 S. Workman Mill Rd., Whittier, CA.
17. [INT 1997] International Conference of Building Officials, 1997, *Uniform Building Code, 1997, Vol. 2,* 5360 S. Workman Mill Rd., Whittier, CA.
18. [LEP 1997] LePage, A., 1997, *A Method for Drift Control in Earthquake-Resistant Design of Reinforced Concrete Building Structures,* thesis submitted to the Graduate College of the University of Illinois, Urbana, IL, June 1997.
19. [OZT 2003] Ozturk, B. O., 2003, *Seismic Drift Response of Building Structures in Seismically Active and Near Fault Regions,* thesis submitted to the faculty of Purdue University for the degree of Doctor of Philosophy, in May 2003

Table 1

Year	Source	Nominal Period	Base Shear Coeff.	Base Shear	Story-Force Distribution	Amplifier for Soil	Cracked Section	Limits for Drift
1951	[AND 1972]	$T = \frac{0.05H}{\sqrt{b}}$	C=0.015/T	V=CW	Option A	No	No	No
1959	[BLU 1963]	T=N/10	$C = 0.05/\sqrt[3]{T}$	V=0.67CW	Option B	No	No	No
1967	[SEI 1967]	T=N/10	$C = 0.05/\sqrt[3]{T}$	V=0.67CW	Option B	No	No	No
1972	[VET 1972]	T=0.08N	C=5Amax/4T<3	V=CW/4	Option B	Yes	Yes	See Note D1
1974	[SEI 1974]	T=N/10	$C = 1/15\sqrt{T} < 0.12$	V=0.67CSW	Option C	Yes	No	See Note D2
1978	[APP 1978]	$T = 0.025h_n^{3/4}$	C=1.2AvST2/3	C=CW/7	Option C	Yes	No	See Note D3
1994	[INT 1994]	$T = 0.03h_n^{3/4}$	C=1.5S/4T2/3	V=CW/12	Option C	Yes	No	See Note D4
1997	[INT 1997]	$T = 0.03h_n^{3/4}$	C=Cv /T	V=CW/8.5	Option C	Yes	Yes	See Note D5

NOTATIONS FOR TABLE 1:

T: Period
H: Total height
H_n: Total height
W: Total weight
N: Number of stories
A_v: Coefficient reflecting response in the range of nearly constant velocity response
C_v: Coefficient reflecting response in the range of nearly constant velocity response and varying with site characteristics
S: Coefficient reflecting site characteristics

NOTES ON REQUIREMENTS FOR DRIFT CONTROL

D1

The VA Code [VET 1972] used different factors for force and drift. For example, while the force reduction factor was ¼ for "ductile moment resisting" frames with light and flexible walls, the amplifier for the calculate drift using the reduced force was set at 3. In addition, the VA Code required that the stiffness of reinforced concrete frames be based on "cracked sections." The drift-ratio limit was set at 0.8%. The limit was reduced to 0.26% for frames encasing brittle glass windows.

D2

In 1974, the limiting story-drift ratio was set at 0.5%. In addition, the lateral force was amplified by 1/K, where K=0.67 for properly detailed frames.

D3

Similarly to what was done in the VA Code, the ATC-3 Model Code recommended a force-reduction factor of 7 for shear in reinforced concrete frames and an amplification factor of 6 for deflection. It is interesting to note that these factors were set at 8 and 5.5 for steel frames.

D4

Calculated story drift ratio (for the reduced force) for a "ductile" frame shall not exceed 0.33% if the period is less than 0.7 sec and 0.25% otherwise.

D5

Story drift ratio to be determined for the reduced force but then amplified by 70% of the force-reduction factor and to be limited by 2.5% for frames with calculated periods less than 0.7 sec and by 2% for frames with higher periods.

Option A

$$F_x = V\frac{w_x h_x}{\Sigma(wh)}$$

F_x = lateral force at level x
w_x = weight at level x
h_x = height of level x above base
V = design base shear

Option B

$$F_x = \frac{(V - F_t)w_x h_x}{\sum_{i=1}^{n} w_i h_i}$$

$$F_t = 0.004V\left(\frac{h_n}{D_s}\right)^2$$

$$\text{if } \left(\frac{h_n}{D_s}\right) > 3, \text{otherwise } F_t = 0.$$

F_i , F_x = Lateral force at level i, x
w_i, w_x = Lateral force at level i, x
h_n, h_i, w_x = height to level n, i, x
V = Design base shear

Option C

$$F_x = \frac{(V - F_t)w_x h_x}{\sum_{i=1}^{n} w_i h_i}$$

$$F_t = 0.07TV < 0.25V$$

and F_t=0 if $T \leq 0.7$ sec
F_i , F_x = Lateral force at level i, x
w_i, w_x = Lateral force at level i, x
h_n, h_i, w_x = height to level n, i, x
V = Design base shear

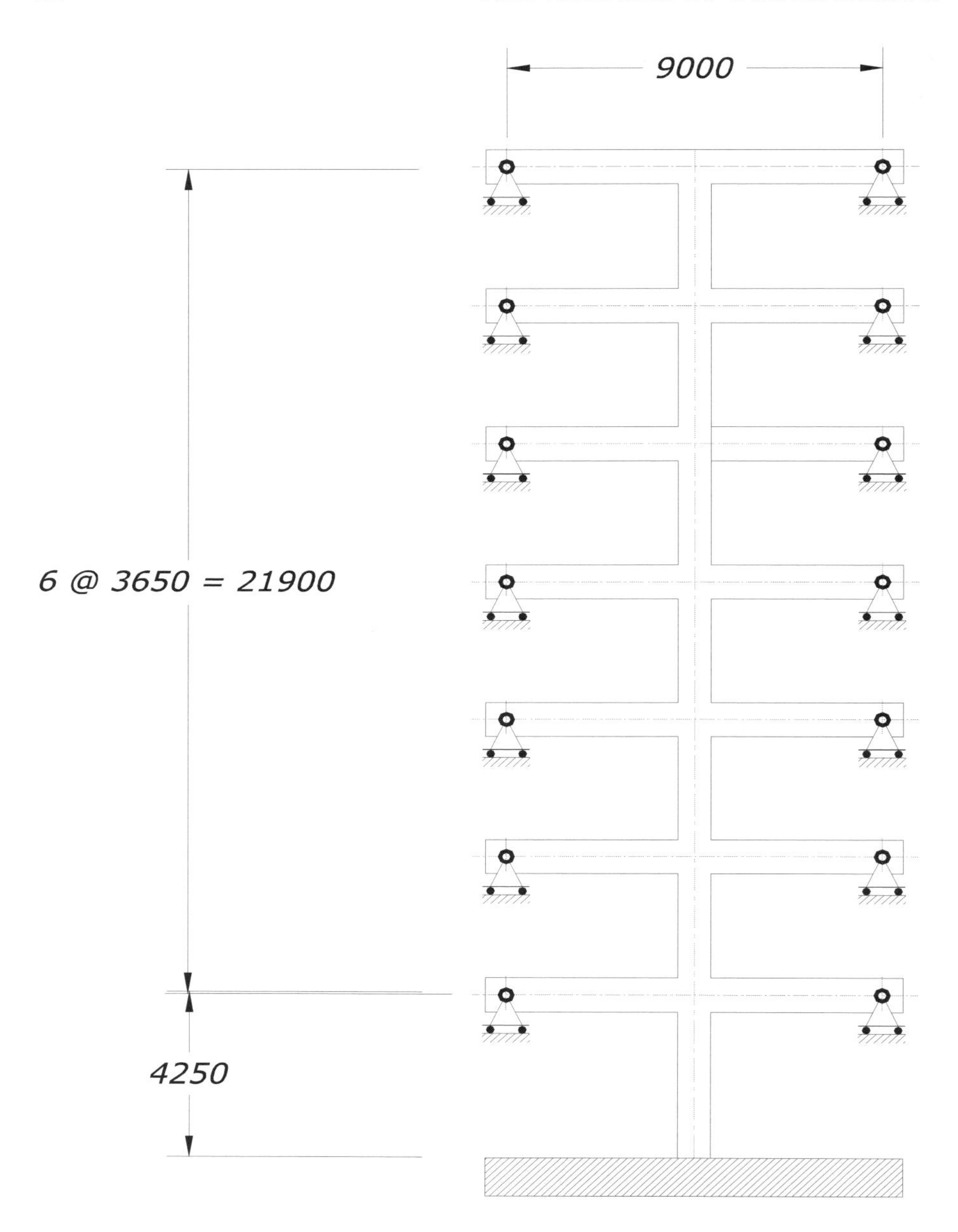

Figure 1

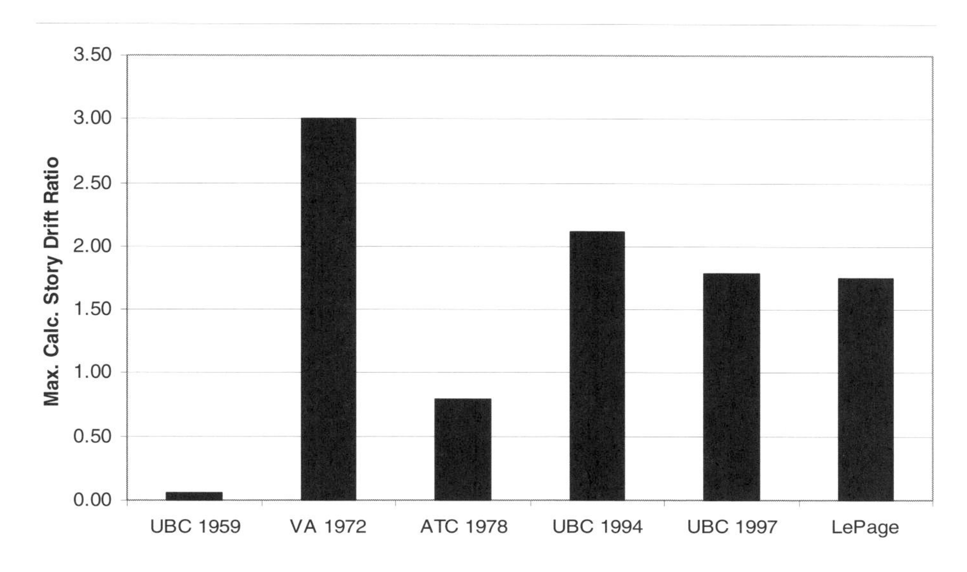

Figure 2

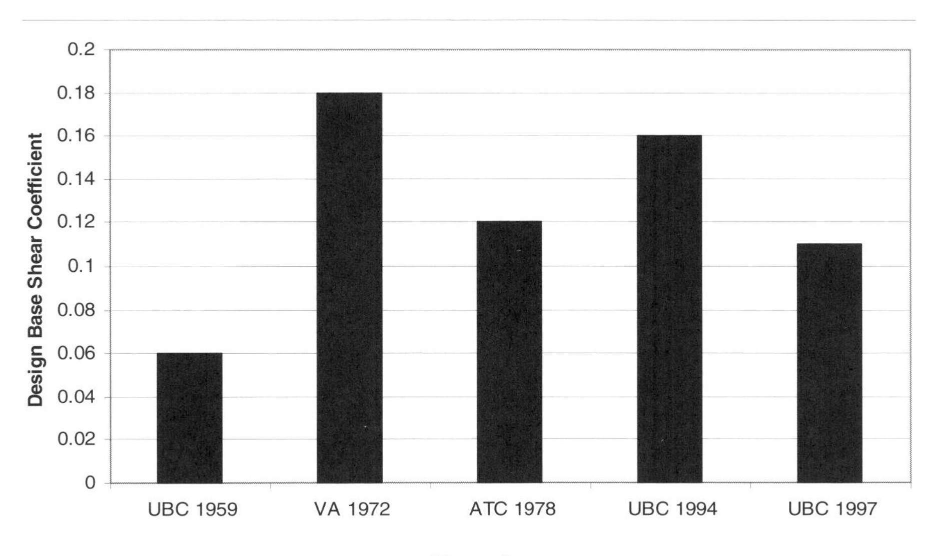

Figure 3

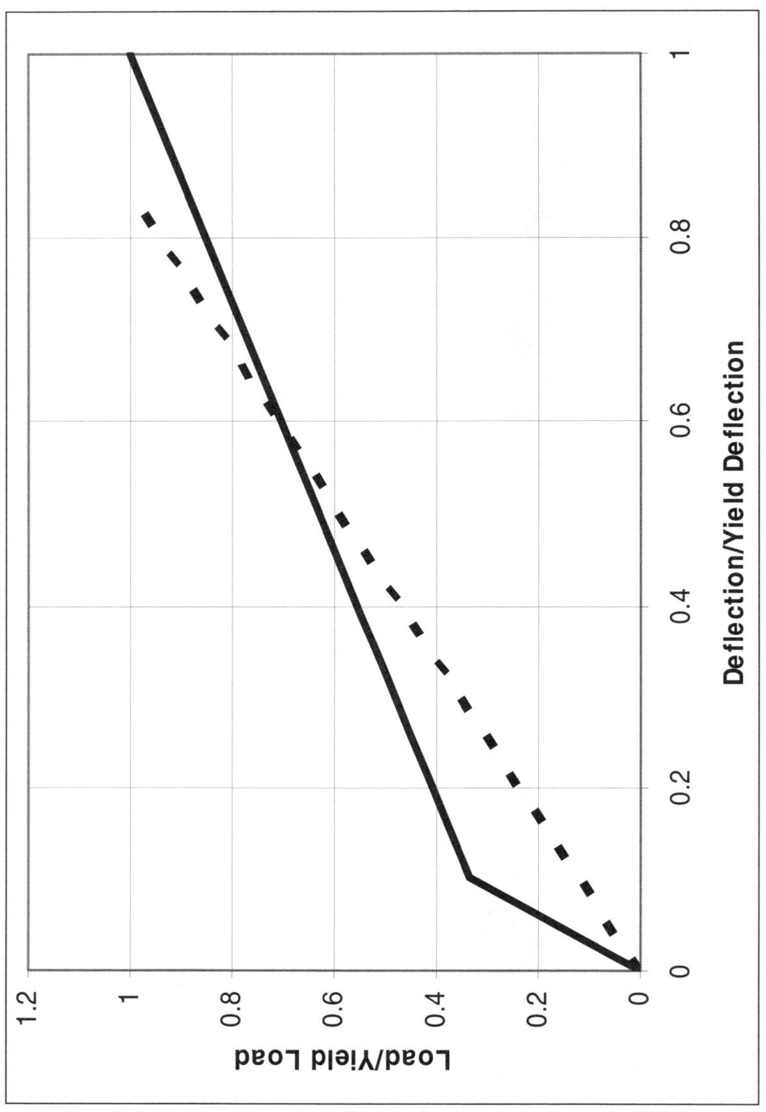
Load/Yield Load
Deflection/Yield Deflection
0
0.2
0.4
0.6
0.8
1
1.2

Figure 4

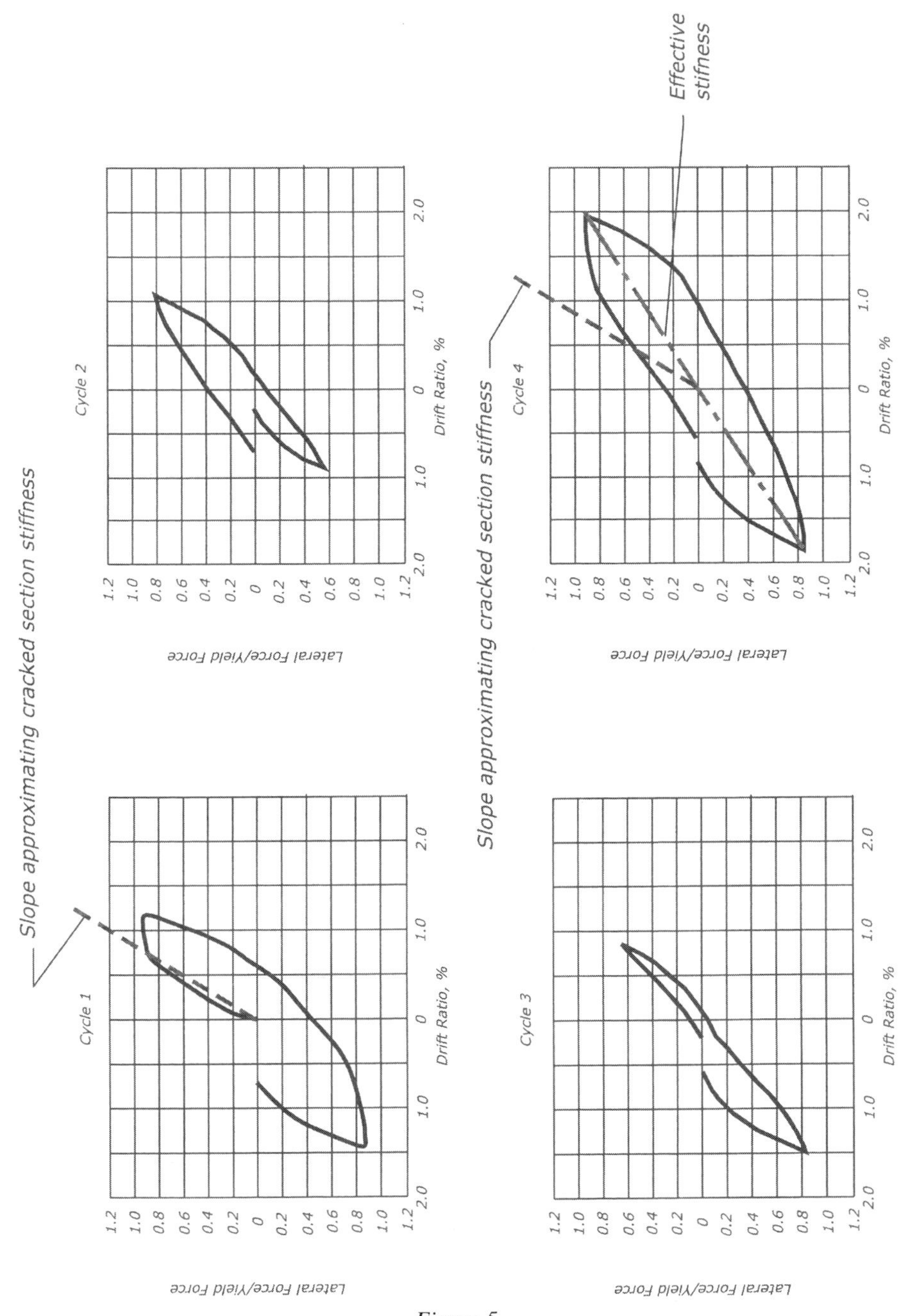
Slope approximating cracked section stiffness
Cycle 1
Cycle 2
Cycle 3
Cycle 4
Effective stiffness
Lateral Force/Yield Force
Drift Ratio, %
1.2
1.0
0.8
0.6
0.4
0.2
0
2.0

Figure 5

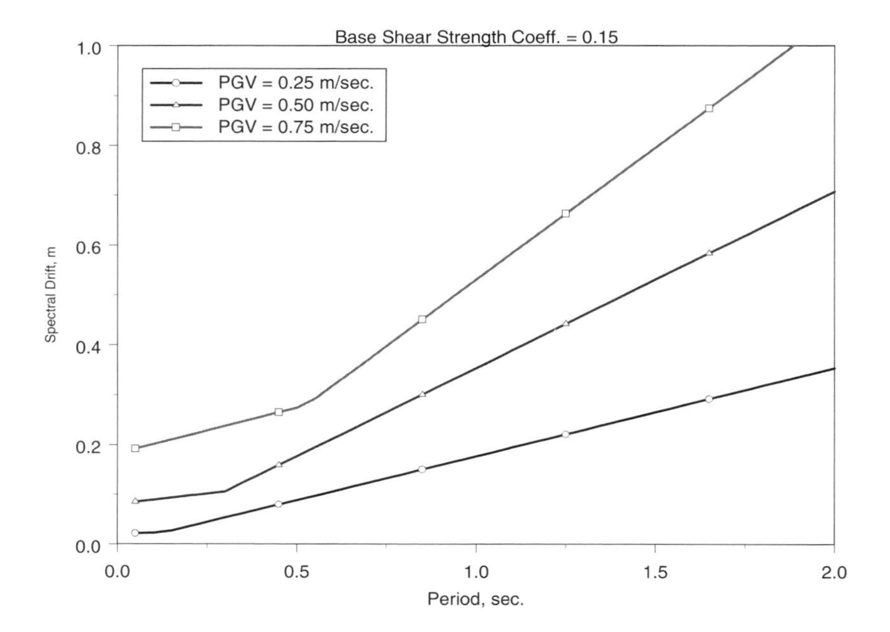

Figure 6

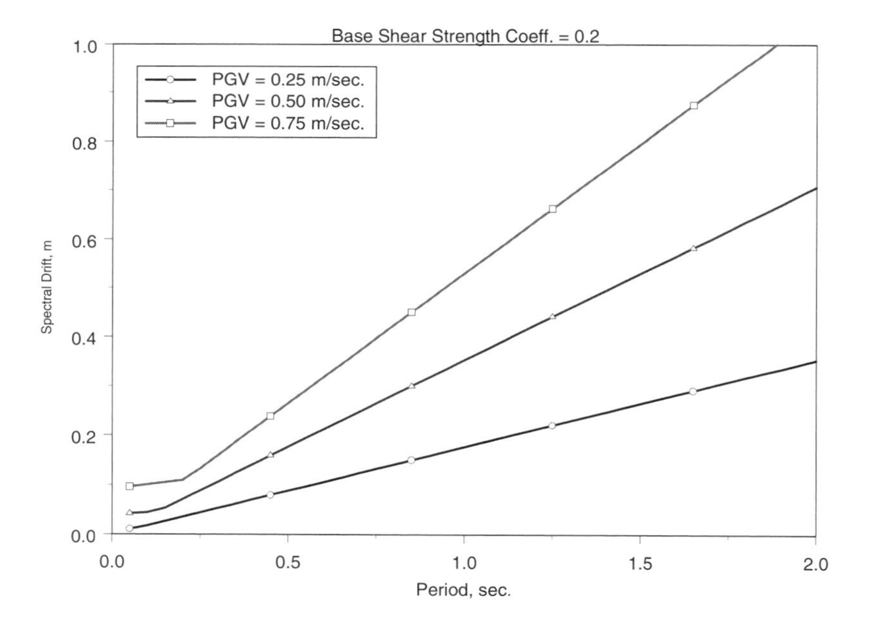

Figure 7

APPENDIX A

REPORT ON DRIFT COMMITTEE ON SPECIAL STRUCTURES STRUCTURAL ENGINEERS ASSOCIATION OF SOUTHERN CALIFORNIA

This committee met with the staff of the Building Division of the Department of Building and Safety on Friday evening, February 27, 1959. Present were Murray Erick, Robert Wilder, Roy Johnston, and S. B. Barnes. R. W. Binder was absent but sent in his written comments. These were read at the meeting and given serious consideration.

The meeting was held in the Hearing Room at the City Hall after dinner at the Redwood House.

The question submitted to the committee was: "What is good engineering practice as related to drift of buildings subjected to lateral forces?"

The first part of the meeting was devoted to a discussion of the extent to which a building department should set limitations of this kind and to the desirability of attempting to legally limit the amount of non-structural damage. The necessity of using "sound engineering judgment", that overworked phrase, was stressed.

Three items were discussed under the category of basic philosophy, namely-protection of health and safety of life, protection of non-structural elements, and protection from motion sickness or discomfort.

It was felt by the committee that the breakage of glass or the breaking up of exterior wall elements which might fall on the public was definitely within the province of regulation. The committee generally felt that the other two items bordered on excessive paternalism. Mr. Hanley Wayne disputed this point of view on the basis that the present law required them to safeguard property.

It was pointed out that some plaster cracks must be expected in case of high wind or severe earthquake even with the most conservative designs. It was noted that isolation of plaster walls, glass or other brittle materials was possible in some degree but that more isolation would result in less damping. It was finally decided that we could not legislate out all plastering cracks but that it would be desirable to avoid breaking up partitions beyond the point of mere plaster repair.

The chairman started the meeting by questioning whether this matter rightfully should come before this committee. He personally feels that this should be a matter for the entire Seismology Committee since it involves general principles. However, the committee proceeded with its discussion and findings, hoping that it was not too far out of line in this respect.

It was further recommended that the advice and opinions of the committee on this matter not be placed in code form but be considered as a portion of a Manual of Good Practice, which could be revised later if found desirable.

Since most experience of record as related to drift in tall buildings has been due to the effect of wind and since such recommendations of experienced engineers are limited to wind, it was decided to make separate recommendations for wind and for earthquake. It was felt by most of the committee that the deflections of buildings subjected to earthquake computed as a statical force are not necessarily a true measure of the actual deflections of these buildings under earthquake shock. It was therefore decided to permit earthquake drift to be twice that for wind. The monetary cost of extreme limitation for earthquake drift was discussed and considered in this recommendation.

Because it is desirable to have uniform agreement between engineers on a state-wide basis, and because the SEAOC is currently proposing a new set of earthquake design criteria to the Uniform Code officials, and since there is at present a committee writing a Manual of Good Practice which will involve drift limitations, it is suggested that this report be sent to the proper committees engaged in this work for their review and comments.

An excerpt of conclusions reached at the meeting, compiled by Tom Brown and Hanley Wayne, is enclosed.

Respectfully submitted,
COMMITTEE ON SPECIAL STRUCTURES,
SEAOSC

S. B. Barnes, Chairman
Murray Erick
Roy Johnston
Robert Wilder
R. W. Binder

Los Angeles, California
March 4, 1959

PRELIMINARY SEISMIC VULNERABILITY ASSESSMENT OF EXISTING REINFORCED CONCRETE BUILDINGS IN TURKEY

Part I: Statistical Model Based on Structural Characteristics

G. Ozcebe, M. S. Yucemen, V. Aydogan and A.Yakut
Structural Mechanics Division, Department of Civil Engineering
Middle East Technical University, Ankara, 06531 Turkey

Abstract: The 1999 earthquakes caused huge damage and economic losses in Turkey. The city of Düzce, hit by the second earthquake of Mw=7.2, suffered widespread damage to many RC buildings. Survey teams conducted post-earthquake evaluations on selected buildings that suffered various degrees of damage. The information collected was analyzed to set up a correlation between the attributes affecting seismic performance and the observed damage. A procedure, developed using a statistical method called discriminant analysis, is presented. The details of the procedure and the content of the database are summarized. The variability of ground motion with respect to the soil properties and the distance to source was incorporated in the improved procedure presented in the companion paper [1].

Key words: Seismic vulnerability, discriminant analysis, reinforced concrete, Düzce earthquake, damage score

1. INTRODUCTION

Up to date procedures on the vulnerability assessment of building structures have primarily focused on the structural system, capacity, layout and response parameters [2-12]. These parameters would provide realistic

S.T. Wasti and G. Ozcebe (eds.), Seismic Assessment and Rehabilitation of Existing Buildings, 29–42.

estimates of the expected performance if the built structural system reflects the prescribed structural and architectural features. In general, the construction practice in Turkey is far beyond reflecting designed structural system, thus violating all assumptions of the usual vulnerability assessment procedures. For this reason, statistical analysis based on the observed damage and significant building attributes would provide more reliable and accurate results for regional assessments. In this context, discriminant analysis technique was used to develop a preliminary evaluation methodology for assessing seismic vulnerability of existing low- to mid-rise reinforced concrete buildings. The main objective is to identify the buildings that are highly vulnerable to damage, that is the seismic performance is inadequate to survive a strong earthquake. Hence, the damage scores obtained from the derived discriminant functions are used to classify existing buildings as "safe", "unsafe" and "intermediate". The discriminant functions are generated based on the basic damage inducing parameters, namely number of stories (n), minimum normalized lateral stiffness index (mnlstfi), minimum normalized lateral strength index (mnlsi), normalized redundancy score (nrs), soft story index (ssi) and overhang ratio (or).

The building damage database used in this study contains 484 buildings, which were evaluated by the survey teams after the 1999 Düzce earthquake. The building inventory was formed entirely by low- to mid-rise reinforced concrete buildings. Figure 1 shows the classification of these buildings according to the number of stories and the observed damage. The observed damage states were determined based on the descriptions given in Table 1.

The description of these parameters and the derivation of discriminant functions are presented in the sections that follow.

This paper serves as the companion paper two the one given in Part II. Therefore, inclusion of the background information on the proposed assessment methodology is given here to provide basis for the improvements that are introduced in the second part.

2. DEFINITION OF THE DAMAGE INDUCING PARAMETERS

In the determination of the estimation variables to be used in the analysis, the basic assumption is that all of the buildings involved in the inventory are exposed to a specific earthquake. In other words, each building stock in itself has faced the same ground motion properties, thus the damage will be evaluated only on the basis of structural responses rather than including the excitation parameters. Considering the characteristics of the damaged structures and the huge size of the existing building stock, the following parameters were chosen as the basic estimation parameters of the proposed method:

- number of stories (n),
- minimum normalized lateral stiffness index (mnlstfi),
- minimum normalized lateral strength index (mnlsi),
- normalized redundancy score (nrs),
- soft story index (ssi),
- overhang ratio (or).

These parameters are briefly defined in the following paragraphs.

Number of stories (**N**): This is the total number of individual floor systems above the ground level.

Minimum normalized lateral stiffness index (**MNLSTFI**): This index is the indication of the lateral rigidity of the ground story, which is usually the most critical story. If the story height, boundary conditions of the individual columns and the properties of the materials used are kept constant, this index would also represent the stiffness of the ground story. This index is calculated by considering the columns and the structural walls at the ground story. While doing this, all vertical reinforced concrete members with "maximum cross-sectional dimension / minimum cross-sectional dimension ratio" less then 7 are considered as columns. All other reinforced concrete structural members are considered as structural walls. The MNLSTFI parameter shall be computed based on the following relationship:

$$\mathrm{MNLSTFI} = \min\left(I_{nx}, I_{ny}\right) \tag{1}$$

I_{nx} and I_{ny} values in Eq.(1) are to be calculated by using Eq.(2).

$$\begin{aligned} I_{nx} &= \frac{\sum (I_{col})_x + \sum (I_{sw})_x}{\sum A_f} \times 1000 \\ I_{ny} &= \frac{\sum (I_{col})_y + \sum (I_{sw})_y}{\sum A_f} \times 1000 \end{aligned} \tag{2}$$

Here:

$\Sigma(I_{col})_x$ and $\Sigma(I_{col})_y$: summation of the moment of inertias of all columns about their centroidal x and y axes, respectively.

$\Sigma(I_{sw})_x$ and $\Sigma(I_{sw})_y$: summation of the moment of inertias of all structural walls about their centroidal x and y axes, respectively.

I_{nx} and I_{ny} : total normalized moment of inertia of all members about x and y axes, respectively.

ΣA_f : total story area above ground level.

Minimum normalized lateral strength index **(MNLSI)**: The minimum normalized lateral strength index is the indication of the base shear capacity of the critical story. In the calculation of this index, in addition to the existing columns and structural walls, the presences of unreinforced masonry filler walls are also considered. While doing this, unreinforced masonry filler walls are assumed to carry 10 percent of the shear force that can be carried by a structural wall having the same cross-sectional area [8, 11, 12]. As in MNLSTFI calculation, the vertical reinforced members with a cross-sectional aspect ratio of 7 or more are classified as structural walls. The MNLSI parameter shall be calculated by using the following equation:

$$\mathrm{MNLSI} = \min\left(A_{nx}, A_{ny}\right) \tag{3}$$

Here:

$$A_{nx} = \frac{\sum (A_{col})_x + \sum (A_{sw})_x + 0.1\sum (A_{mw})_x}{\sum A_f} \times 1000$$
$$A_{ny} = \frac{\sum (A_{col})_y + \sum (A_{sw})_y + 0.1\sum (A_{mw})_y}{\sum A_f} \times 1000 \tag{4}$$

For each column with a cross-sectional area denoted by A_{col}:

$$(A_{col})_x = k_x \cdot A_{col}$$
$$(A_{col})_y = k_y \cdot A_{col} \tag{5}$$

Here [11];
k_x=1/2 for square and circular columns;
k_x=2/3 for rectangular columns with $b_x > b_y$;
k_x=1/3 for rectangular columns with $b_x < b_y$; and
$k_y = 1 - k_x$.
For each shear wall with cross-sectional area denoted by A_{sw}:

$$(A_{sw})_x = k_x \cdot A_{sw}$$
$$(A_{sw})_y = k_y \cdot A_{sw} \tag{6}$$

Here;
k_x=1 for structural walls in the direction of x-axis;

k_x=0 for structural walls in the direction of y-axis; and

$k_y = 1 - k_x$.

For each unreinforced masonry filler wall with no window or door opening and having a cross-sectional area denoted by A_{mw}:

$$\begin{aligned}(A_{mw})_x &= k_x \cdot A_{mw} \\ (A_{mw})_y &= k_y \cdot A_{mw}\end{aligned} \quad (7)$$

Here;

k_x=1.0 for masonry walls in the direction of x-axis;

k_x=0 for masonry walls in the direction of y-axis; and

$k_y = 1 - k_x$.

Normalized redundancy score **(NRS)**: Redundancy is the indication of the degree of the continuity of multiple frame lines to distribute lateral forces throughout the structural system. The normalized redundancy ratio (NRR) of a frame structure is calculated by using the following expression:

$$NRR = \frac{A_{tr}(nf_x - 1)(nf_y - 1)}{A_{gf}} \quad (8)$$

Here;

A_{tr} : the tributary area for a typical column. A_{tr} shall be taken as 25 m^2 if nf_x and nf_y are both greater than and equal to 3. In all other cases, A_{tr} shall be taken as 12.5 m^2.

nf_x, nf_y : number of continuous frame lines in the critical story (usually the ground story) in x and y directions, respectively.

A_{gf} : the area of the ground story, i.e. the footprint area of the building.

Depending on the value of NRR computed from Eq. (8), the following discrete values are assigned to the normalized redundancy score (NRS):

NRS = 1 for $0 < NRR \leq 0.5$

NRS = 2 for $0.5 < NRR \leq 1.0$

NRS = 3 for $1.0 < NRR$

Soft story index **(SSI)**: On the ground story, there are usually fewer partition walls than in the upper stories. This situation is one of the main reasons for soft story formations. Since the effects of masonry walls are included in the calculation of MNLSI, soft story index is defined as the ratio of the height of first story (i.e. the ground story), H_1, to the height of the second story, H_2.

$$SSI = \frac{H_1}{H_2} \tag{9}$$

Overhang ratio (**OR**): In a typical floor plan, the area beyond the outermost frame lines on all sides is defined as the overhang area. The summation of the overhang area of each story, $A_{overhang}$, divided by the area of the ground story, A_{gf}, is defined as the overhang ratio.

$$OR = \frac{A_{overhang}}{A_{gf}} \tag{10}$$

3. STATISTICAL MODEL

The effects of different parameters on seismic damage vary. In order to make a more rational and systematic evaluation of damage inducing parameters in the prediction of seismic vulnerability of structures, a statistical technique, known as discriminant analysis is adopted.

In the most general sense, earthquake damage to buildings is categorized into five levels, namely: none (N), light (L), moderate (M), severe (S) and collapse (C). Because of the nature of available damage data from the 1999 Düzce earthquake, it was necessary to combine the severe damage and collapse states into one group, denoted by (S+C). Furthermore, if none and light damage states are combined into one group, based on the fact that the distinction between these two damage states is not too crucial for vulnerability analysis, then there will be three different damage states, namely: (N+L), (M) and (S+C).

It is possible to evaluate structures at different performance levels according to different objectives. If the main concern is to identify the buildings that are severely damaged or collapsed, the first three damage states (i.e. N, L and M) can be considered as one group and the severely damaged state and collapsed cases as the other group, reducing the distinct damage states into two. Since the main objective is the identification of severely damaged and collapsed buildings for life safety purposes, this classification can be referred as "Life Safety Performance Classification" (LSPC). Similarly, if the main concern is to identify the structures which suffer no damage or light damage during an earthquake, the first two damage states (N and L) can be considered as one group and remaining damage states (M, S and C) as the other group, reducing the distinct damage states into two. This identification is named as "Immediate Occupancy Performance Classification" (IOPC) since the main concern is to identify the buildings that can be occupied immediately after a strong ground motion.

In the discriminant analysis method, first the set of estimation variables that provides the best discrimination among the groups is identified. These variables are known as the "discriminator variables". Then a "discriminant function", which is a linear combination of the discriminator variables, is derived. The values resulting from the discriminant function are known as "discriminant scores". The final objective of discriminant analysis is to classify future observations into one of the specified groups, based on the values of their discriminant scores.

The unstandardized estimate of discriminant function based on six damage inducing parameters is obtained for life safety performance classification by utilizing the SPSS [13] software and the database constituted after 1999 Düzce earthquake. Here, DILS denotes the damage index or the damage score corresponding to the LSPC and the other parameters are as described. The function given in Eq. (11) is referred to as the unstandardized discriminant function, because the unstandardized (raw) data are used for computing this discriminant function

$$DI_{LS}=0.620n-0.246mnlstfi-0.182mnlsi-0.699nrs+3.269ssi+2.728or-4.905 \quad (11)$$

In the case of immediate occupancy performance classification, the unstandardized discriminant function, where DI_{IO} is the damage score corresponding to IOPC, based on these variables is:

$$DI_{IO}=0.808n-0.334mnlstfi-0.107mnlsi-0.687nrs+0.508ssi+3.884or-2.868 \quad (12)$$

A convenient statistical parameter for interpreting the contribution of each variable to the formation of the discriminant function is the loadings or the structure coefficients [14]. The structure coefficient of a discriminator variable is merely the correlation coefficient between the discriminant score and the discriminator variable and the value will lie between +1 and –1. As the absolute value of the structure coefficient of a variable approaches to 1, the communality between the discriminating variable and the discriminant function increases, or vice versa. The structure coefficients that are obtained as an output from the SPSS software are shown in Table 2. Here the number of stories above the ground level (n) has the highest loading (0.738), indicating that it is the best discriminator variable in LSPC. In the case of IOPC, again the number of stories comes out as the best discriminator variable with the loading of 0.789 and the normalized redundancy score is the second best.

4. CLASSIFICATION METHODOLOGY

In the proposed classification methodology, buildings are evaluated according to both performance levels, by using Eqs. (11) and (12), and the final decisions for the damage state of the buildings are achieved by considering the results of the two performance levels simultaneously.

Moreover, the number of stories is the most significant variable in both performance classifications. In order to improve the discriminating contribution of other parameters, new cutoff values are selected depending on the number of stories. For this purpose, a functional relationship is derived between the cutoff values and the number of stories, n, by fitting a least squares curve to the available damage data. In the determination of the cutoff function, two constraints are also imposed at each story level. These constraints are;

the correct classification rate is required to be at least 70 % and,

the maximum classification error related to damage states leading to life loss (i.e. severe damage and collapse) is restricted to be 5 %.

The resulting cutoff functions based on number of stories, corresponding to the two types of classification, are as follows:

$$\begin{aligned} CF(LS) &= -0.090 \cdot n^3 + 1.498 \cdot n^2 - 7.518 \cdot n + 11.885 \\ CF(IO) &= -0.085 \cdot n^3 + 1.416 \cdot n^2 - 6.951 \cdot n + 9.979 \end{aligned} \tag{13}$$

In the proposed classification procedure, firstly the damage scores are obtained by using Eqs.(11) and (12) for the cases of LSPC and IOPC, respectively. Then by comparing these damage scores with the story dependent cutoff values obtained from Eq. (13), the building under evaluation is assigned an indicator variable of "0" or "1". The indicator variable "0" corresponds to none, light or moderate damage in the case of LSPC and none or light damage in the case of IOPC. Similarly, the indicator variable "1" corresponds to severe damage or collapse in the case of LSPC and moderate or severe damage or collapse in the case of IOPC. In the final stage of the classification procedure, the building is rated as "safe" (i.e. "none or light damage") or "unsafe" (i.e. "severe damage or collapse") or "intermediate" depending on the values of the indicator variables obtained from both classification types according to the combinations listed in Table 3.

As observed in Table 3, if the indicator variable is consistently "0" or "1" for both LSPC and IOPC cases, the building is rated as "safe" or "unsafe", respectively. If there is an inconsistency in the classification, in other words if one gives "0" and the other "1" or vice versa, then no final rating is done and the final decision on the seismic safety of the building is left for a more comprehensive detailed seismic evaluation. As the readers may note, in Table 3 all possible ratings are considered, among which the one given in the last row, with an IOPC indicator variable of 0 and LSPC indicator variable of 1, does not have any physical meaning whatsoever. It should be kept in mind that the adopted methodology is a statistical tool and such cases are therefore classified as the cases requiring further study.

Although the decision parameters of the proposed classification method described above are derived from the Düzce damage database, the classification method is applied to the same database in order to check its correct classification efficiency. The resulting output of the proposed classification method is given in Table 4. Out of the 484 buildings forming the seismic damage database, 99 buildings (37+11+51) that correspond to 20.5 % of the entire database, are classified as "intermediate" and left for further detailed evaluation. Among these 99 buildings, only two of them had an IOPC indicator variable of "0" and a LSPC indicator variable of "1". This result actually indicates the success of discriminating ability of the parameters used in the analyses. Out of 122 severely damaged or collapsed buildings, 98 buildings are correctly classified, 13 of them are misclassified and 11 of them are left for further detailed seismic analysis. Thus, the efficiency in identifying the severely damaged or collapsed buildings is increased to 80.3% and among the 484 buildings evaluated only 13 of the severely damaged or collapsed buildings are rated as safe. Thus, the misclassification that may lead to life loss is only 2.7%, i.e. 13/484=0.027.

5. VALIDATION OF THE PROPOSED METHODOLOGY

It is desirable to check the validity of the proposed statistical model by examining the correct classification rates in cases of different databases compiled from different earthquakes. For this purpose, the proposed methodology and the accompanying discriminant functions are applied to damage data assessed from the 1992 Erzincan earthquake and the damage data compiled after 2002 Afyon earthquake.

The classification results according to the proposed classification methodology are presented in Tables 5 and 6 for the Erzincan and Afyon damage databases, respectively.

As it can be observed from these tables, the classification results of the model demonstrate that the correct classification rate for severely damaged and collapsed buildings is quite high. On the other hand, the correct classification rate for none and a light damage state is found to be 96.4 % for the Erzincan database and 75.0 % for the Afyon database. Only 3 buildings forming 9.3 % of the Erzincan database and 22.2 % of the Afyon database cannot be judged. These buildings are identified as "intermediate" and they are the buildings that require further detailed investigations.

Considering the existence of various random factors (such as geotechnical parameters) and sources of uncertainties, these rates are found to be quite satisfactory and support the predictive ability of the proposed statistical model.

6. CONCLUSIONS

A statistical analysis procedure is used to develop a model proposed for the preliminary assessment of the seismic vulnerability of existing reinforced concrete buildings. The procedure uses discriminant analysis technique that yields discriminant functions in terms of the selected estimation parameters. Six estimation parameters, namely number of stories, existence of soft story, normalized redundancy score, degree of overhang, the minimum normalized lateral stiffness and minimum normalized lateral strength indices, are considered for the assessment of seismic vulnerability. Among these parameters the number of stories is found to be the most discriminating parameter for existing low- to mid-rise reinforced concrete buildings.

The proposed classification methodology improves the correct classification rate especially in the cases where life-safety is involved. For the 1999 Düzce earthquake damage database, the correct classification rate in determining the severely damaged and collapsed structures is increased to 80.3 % whereas total misclassification rate that corresponds to the loss in human lives is only 2.7 percent. Besides the increased efficiency and accuracy of the model, a number of buildings are left for further detailed evaluations instead of evaluating them incorrectly.

The validity of the proposed methodology is checked based on the damage data available for the 1992 Erzincan earthquake and for the 2002 Afyon earthquake. Reasonably high correct classification rates are obtained, demonstrating the predictive ability of the proposed seismic vulnerability estimation methodology.

ACKNOWLEDGEMENTS

This study owes its inception to Mete A. Sozen, the Kettelhut Distinguished Professor of Structural Engineering at Purdue University, whose efforts are hereby sincerely acknowledged.

The research work presented in this study is supported in part by the Scientific and Research Council of Turkey (TUBITAK) under grant: YMAU-ICTAG-1574 and by NATO Scientific Affairs Division under grant: NATO SfP977231.

REFERENCES

1. A. Yakut, V. Aydogan, G. Ozcebe and M. S. Yucemen, Preliminary Seismic Vulnerability Assessment of Existing Reinforced Concrete Buildings in Turkey -Part II: Inclusion of Site Characteristics, NATO Workshop, 2003.
2. Yucemen, M. S., Ozcebe, G. and Pay, A. C., Prediction of Potential Damage due to Severe Earthquakes, submitted to Structural Safety for possible publication, in review.
3. Kircher, C., Reitherman, R. K., Whitman, R. V., Arnold, C., Estimation of Earthquake Losses to Buildings, Earthquake Spectra, EERI, vol. 13, no.4, pp.703-720, California, 1997.
4. Hwang, H. H. M., Huo, J. R., Generation of Hazard-Consistent Fragility Curves for Seismic Loss Estimation Studies, State University of New York at Buffalo, Technical Report No. 94-0015, 1994.
5. Wen, Y. K., Hwang H., Shinozuka, M., Development of Reliability-Based Design Criteria for Buildings Under Seismic Load, State University of New York at Buffalo, Technical Report No. 94-0023, 1994.
6. Ozcebe, G., Yucemen, M. S. and Aydogan, V., Assessment of Seismic Vulnerability of Existing Reinforced Concrete Buildings, submitted to Earthquake Engineering and Structural Dynamics for possible publication, in review.
7. Brookshire, D. S., Chang, S. E., Cochrane, H., Olson, R. A., Rose, A., Steenson J., Direct and Indirect Economic Losses from Earthquake Damage, Earthquake Spectra, EERI, vol. 13, no. 4, California, 1997.
8. Sozen, M. A., Hassan, A. F., Seismic Vulnerability Assessment of Low-Rise Buildings in Regions with Infrequent Earthquakes, ACI Structural Journal, vol.94, no.1, pp.31-39, 1997.
9. Gurpinar, A., Yucemen, M. S., An Obligatory Earthquake Insurance Model for Turkey, Proceedings of International Conference on Engineering for Protection from Natural Disasters, pp.895-906, Asian Institute of Technology, Bangkok, Thailand, 1980.
10. Ersoy, U., Ozcebe, G., Lessons from Recent Earthquakes in Turkey and Seismic Rehabilitation of Buildings, S. M. Uzumeri Symposium – Behavior and Design of Concrete Structures for Seismic Performance, SP-197, ACI International, pp. 105-126, 2002,
11. Tankut, T., Ersoy, U., A Proposal for the Seismic Design of Low-Rise Buildings, Turkish Engineering News, Turkish Chamber of Civil Engineers, No. 386, pp. 40-43, November 1996, (in Turkish)

12. Gulkan, P., Sozen, M. A., Procedure for Determining Seismic Vulnerability of Building Structures, ACI Structural Journal, pp. 336-342, May-June 1999
13. SPSS Inc., SPSS Base 11.0 User's Guide, Chicago, Illinois, 2001.
14. Sharma, S., Applied Multivariate Techniques, John Wiley and Sons, 1996.

Table 1. Description of damage states

	STRUCTURAL ELEMENTS	**NON-STRUCTURAL ELEMENTS**
None	No visual sign of damage	No visual sign of damage
Light	Hairline inclined or flexural cracks	Hairline cracks in walls. Flaking of plaster.
Moderate	Concrete spalling	Cracking in walls and joints between panels. Flaking of large pieces of plaster
Severe	Local structural failure	Wide and through cracks in walls
Collapse	Local or total collapse	Crushing of walls or out-of-plane toppling of walls

Table 2. Structure matrix for the cases of LSPC and IOPC

Variables	**Structure Coefficients**	
	LSPC	**IOPC**
N	+0.738	+0.789
nrs	-0.555	-0.594
mnlsi	-0.503	-0.481
ssi	+0.418	+0.092
or	+0.167	+0.284
mnlstfi	-0.076	-0.085

Table 3. Relationships among different classification criteria

Classification	**Indicator Variable**		**Indicator Variable in Classification**
	LSPC	**IOPC**	
SAFE (None or Light Damage)	0	0	0
UNSAFE (Severe Damage or Collapse)	1	1	1
INTERMEDIATE	1	0	2
INTERMEDIATE	0	1	2

Table 4. Classification results for the Düzce damage database

			Predicted Group Membership			**Total**
			0	1	2	
Original Group Membership	**Count**	None or Light Damage	130	44	37	211
		Severe Damage or Collapsed	13	98	11	122
		Moderate Damage	37	63	51	151
	Percent (%)	SAFE (None or Light Damage)	61.6	20.8		100.0
		UNSAFE (Severe Damage or Collapsed)	10.7	80.3		100.0
		INTERMEDIATE			20.5	100.0

Table 5. Classification results for the Erzincan damage database

			Predicted Group Membership			**Total**
			0	1	2	
Original Group Membership	**Count**	None or Light Damage	27	0	1	28
		Severe Damage or Collapsed	0	2	0	2
		Moderate Damage	10	0	3	13
	Percent (%)	SAFE (None or Light Damage)	96.4	0.0		100.0
		UNSAFE (Severe Damage or Collapsed)	0.0	100.0		100.0
		INTERMEDIATE			9.3	100.0

Table 6. Classification results for the Afyon damage database

			Predicted Group Membership			**Total**
			0	1	2	
Original Group Membership	**Count**	None or Light Damage	3	0	1	4
		Severe Damage or Collapsed	1	8	1	10
		Moderate Damage	2	0	2	4
	Percent (%)	SAFE (None or Light Damage)	75.0	0.0		100.0
		UNSAFE (Severe Damage or Collapsed)	10.0	80.0		100.0
		INTERMEDIATE			22.2	100.0

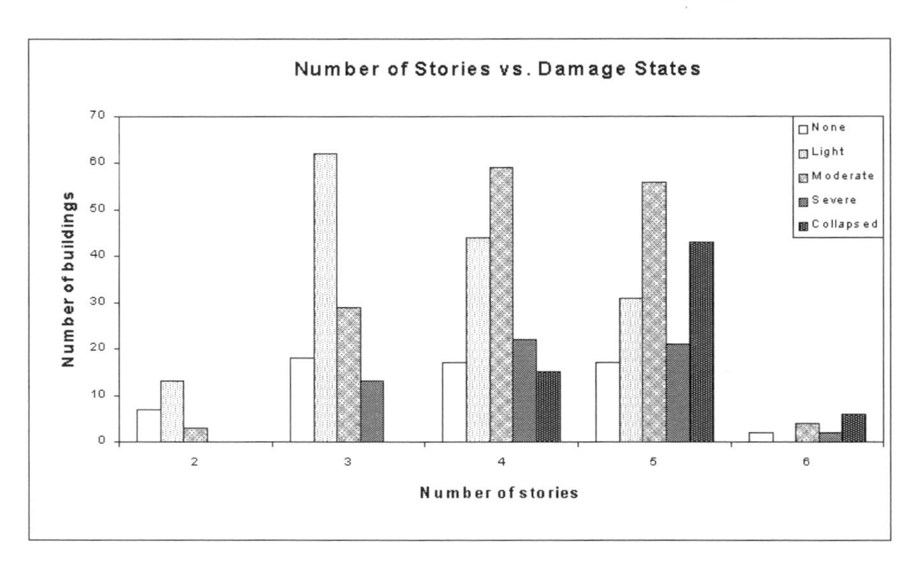

Figure 1. Classification of building data

PRELIMINARY SEISMIC VULNERABILITY ASSESSMENT OF EXISTING REINFORCED CONCRETE BUILDINGS IN TURKEY

Part II: Inclusion of Site Characteristics

A.Yakut, V. Aydogan, G. Ozcebe and M. S. Yucemen
Structural Mechanics Division, Department of Civil Engineering
Middle East Technical University, Ankara, 06531 Turkey

Abstract: The vulnerability assessment method, described in the companion paper [1], relies on a damage score, which is compared with an appropriate cutoff value to identify the buildings as "safe", "unsafe" or "intermediate". The cutoff values are considered to be valid for damaging earthquakes and regions similar to Düzce, where the data was gathered. To generalize the procedure, the variability of ground motion with respect to soil properties and the distance to source needs to be incorporated. This was done by modifying the cutoff values based on the above factors. Sites are classified according to the Turkish Seismic Code's [2] definitions based on the shear wave velocity. Various attenuation relations are used to account for the variation of the ground motion with distance and the soil type.

Key words: Attenuation Model, Building Damage, Site Effect, Vulnerability Assessment

1. INTRODUCTION

This study focuses on modifying the vulnerability assessment procedure developed based on the structural characteristics of buildings located in the city of Düzce. The procedure, which is described in detail in the companion

S.T. Wasti and G. Ozcebe (eds.), Seismic Assessment and Rehabilitation of Existing Buildings, 43–58.

paper [1], relies on the damage cutoff values developed using a statistical analysis approach based on the damage data compiled from Düzce in the wake of 1999 earthquakes. Some selected building attributes are entered into a relation obtained from discriminant analysis to compute a damage score. This damage score is then compared with a cutoff value, which identifies the buildings as "safe", "unsafe" or "intermediate".

The cutoff values recommended are considered to be valid for damaging earthquakes and the regions that have similar distance to source and site conditions to that of Düzce. To apply this procedure to the sites, which have different distance to source and soil properties than Düzce, further modifications must be made to improve the procedure that is presented in the companion paper [1].

2. PROCEDURE

The central point of the study is to capture the relative variation of the ground motion intensity with the distance to source and the soil type. The spectral displacement value was selected as the damage inducing ground motion parameter, as it is a widely used parameter for expressing the vulnerability of buildings. A typical damage curve expressed in terms of the spectral displacement is shown in Figure 1 [3]. It is important to observe that the variation of damage with S_d follows the form of an exponential function. This inference is used to link the change in S_d to the change to be imposed on the cutoff values obtained in [1]. The spectral displacement can be obtained from elastic site spectra computed using available attenuation relations. A number of relations, available in the literature, can be employed to relate inelastic spectral displacement to the elastic one. Although the expressions seem quite different, their influence on the cutoff modifications is shown to be insignificant, especially in the range considered in this study as illustrated in Figure 2 [4,5]. For this reason, equal displacement rule is considered to be adequate.

1. The proposed procedure is developed on the basis of several assumptions, which are listed below:
2. The earthquake magnitude in the region to which the method is applied is similar to the one that affected the reference site, i.e. Düzce.
3. Attenuation relations are believed to represent the variation of the ground motion adequately.
4. Construction practice does not show regional variations.
5. Damage pattern observed in the reference site would be the same for other sites that have same distance to source and soil type.

The steps involved in this procedure can be outlined as follows;

- Step 1: Obtain site-specific response spectra using an appropriate attenuation model.
- Step 2: Calculate spectral displacement at the fundamental periods of interest.
- Step 3: Plot spectral displacement/n as a function of the fundamental period (or n), n representing number of stories considered in the Düzce study.
- Step 4: Convert spectral displacement to a damage index (cutoff value) by assuming an exponential relation.
- Step 5: Normalize all damage indexes at different sites and distances with the damage index obtained for the reference site, i.e. Düzce.
- Step 6: Modify Düzce cutoff values by multiplying them with the cutoff modification coefficients, i.e. normalized values calculated in Step 4.

2.1 Site Classification

Two major parameters used for site classification are the "distance to source (d_S)" and the "soil type (ST)". The sites were characterized by a pair of d_s and ST bins. Five d_s bins were selected in view of the variation in the response spectra with the distance. ST bins were determined based on the shear wave velocity (Vs) of the soil types employed by the Turkish Seismic Code. Twenty different site classes were obtained from the combination of d_s and ST bins, which are illustrated in Table 1. Note that type C2 represents the reference site (Düzce). This way, any region with a certain d_s and ST is assigned a site class according to Table 1, excluding the sites located farther than 50 km from the source. The number of sites can easily be increased by incorporating other distance ranges and soil types (Vs>1000 m/s).

2.2 Attenuation Models

Three attenuation relationships that are suitable for the source mechanism of the North Anatolian Fault were considered. The models developed by Boore et al. [6], Gulkan and Kalkan [7], and Abrahamson and Silva [8] were used to generate site-specific response spectra for all twenty sites included in Table 1. Boore et al., and Gulkan and Kalkan are the most convenient ones because they use the shear wave velocity directly to account for the soil type. For Abrahamson and Silva, however, NEHRP amplification functions were applied on the rock motion to obtain site response spectra. Since the uncertainty in attenuation models can be substantial, using different attenuation models is believed to give a better representation of the actual condition. Among the ones selected, Gulkan and Kalkan's model has been developed based on the local data recorded in Turkey. These models are compared at different distances as shown in Figure 3. Although at short distances Gulkan and Kalkan's model suggest lower estimates as compared to others, at far distances the situation is the other way around.

2.3 Number of Story and Period Relationship

Since the reference cutoff values were obtained as a function of the building height (number of stories), modification factors were also intended for the discrete height levels included in the database. Hence, a relationship between number of stories and the fundamental period was established based on the Turkish Seismic Code formulae. The mean values of the period and the number of stories obtained for the buildings contained in the Düzce seismic damage database are given in Table 2. Although the variation and dispersion of the period with number of stories is large for the buildings in the database, this would not significantly affect the modification factors as will be shown later.

2.4 Calculation of Spectral Displacement

A series of site-specific response spectra computed for a magnitude 7.4 earthquake and a shear wave velocity of 350 m/s is shown in Figure 4. The variations in the spectral ordinates were considered insignificant within the distance bins that were selected. Spectral displacement values were obtained from the calculated spectral accelerations at all periods given in Table 2 for each of the twenty site classes. The spectral displacement normalized with number of stories (corresponding to the building period) is plotted against the number of stories as shown in Figure 5.

This normalization was done to obtain a similar term that would mimic the average drift. The change of S_d with the site class is also evident from these plots. When a linear regression is used to represent data a constant line develops, this is the simplest and the most convenient choice because it leaves out the number of stories. The trend of data implies a nonlinear behavior, so power function was used as an alternative to represent the data as displayed in Figure 6. The modification coefficients were developed for both cases. The influence of the attenuation functions on the calculated response for site C3 is shown in Figure 7. Abrahamson and Silva yields similar results to that of Boore et al., Gulkan and Kalkan, however, provides lower estimates of S_d at all periods.

2.5 Calculation of Modification Factors

Once S_d values for all sites are computed, they are translated into damage terms. In the vulnerability assessment procedure developed for Düzce, there is a reverse relationship between the cutoff value and the damage score of the evaluated building. In other words, as the cutoff value is raised the number of "unsafe" buildings decreases. In view of this relation, the change of the cutoff value (CV) with the normalized spectral displacement was assumed to follow a similar trend observed between damage and S_d/H (Figure 1). Thus, the following function is assumed to reflect the relation between the CV and the normalized spectral displacement (S_d/n);

$$CV = f\left[\frac{1}{1-e^{-S_d/n}}\right] \tag{1}$$

Since the objective is to obtain cutoff modification coefficients (CMC) to be applied on the reference cutoff values (CV_r), the variable of the function in Equation 1 can be used to get CMC values. The CMC values are presented in Tables 3-6 for the three attenuation models employed.

Close inspections of these tables reveal that non-linear and linear formulations of the spectral displacement versus number of story relation provide similar values. The CMC can take values between 0.78-3.90, 0.80-2.14, 0.83-3.03 for Boore et al., Gulkan and Kalkan, and Abrahamson and Silva, respectively. Moreover, among all attenuation models, the one by Gulkan and Kalkan led to narrower range of modification values, meaning that performance differences of the buildings between the sites would be less. The CMC value for reference site class C2 is 1.0 because of the normalization with respect to this site. Obviously, at better site conditions and farther distances cutoff values should be larger. These CMC values were multiplied with the respective reference cutoff values to obtain the cutoff values for other site classes. Modified cutoff values are computed merely from Equation 2, which can handle negative as well as positive values of reference cutoff values.

$$CV = CVR + ABS(CVR)*(CM-1) \tag{2}$$

3. AN APPLICATION OF THE PROPOSED PROCEDURE

As alluded to before, Istanbul is on the verge of being struck by a devastating earthquake, similar to the one that hit Düzce. Assuming that the construction practices in Düzce and in Istanbul are similar, the procedure would provide reasonable results when applied to Istanbul. To see the extent and relativity of the expected damage or the layout of the risk within Istanbul an exercise was undertaken, in which, all buildings in Düzce database were assumed to portray buildings all over Istanbul. In other words, a uniform exposure that is identical to the compiled database for Düzce, is assigned to all districts of Istanbul. The earthquake scenario "Model A" and shear wave velocity estimates of JICA study [9] were employed to model the fault and to classify the sites. The modified cutoff values were applied and all buildings were identified as "safe", "unsafe" or "intermediate" in all districts of Istanbul. It should be pointed out that "safe" buildings represent the structures that would experience none or light damage states, "unsafe" buildings include those that are expected to suffer severe damage or would collapse, and "intermediate" buildings might encompass buildings with all degrees of damage, which can not be clearly identified.

Figures 8-10 display results obtained using Boore et al. [6]. In these figures, results are presented in the form of the ratio of the classified buildings to the total number of buildings. The visual plots indicate some spotty areas, which reflect the local soil profile. The effect of distance to source is clearly observed. The range of safe buildings varies from 38% to 60% depending on the site class. Unsafe buildings constitute 1-40 % and buildings identified as intermediate, which represent buildings that could not be clearly classified as safe or unsafe, have a share of 21-39%. Of the indeterminate buildings, around 50% were moderately damaged, 38% had light or no damage and 10% were severely damaged in Düzce.

The JICA estimates of the heavily damaged building percentages are shown in Figure 11. These results were obtained based on the actual exposures extracted from the data released by the State Statistics Institute of Turkey; the apparent discrepancy is due to this fact.

4. CONCLUSIONS

It has been shown that vulnerability assessment procedures based on observed damage from a particular region can be extrapolated to other sites having similar construction practices and building stock. The variation of ground motion parameters that have known relationship to the damage of buildings are captured using attenuation models that reflect the properties of the sites, i.e. the distance to source and soil type. When the assumptions made are considered to be convincing, which is the case for Istanbul, high-risk areas and vulnerable regions can be identified in a reliable way. This would help determine the rank of regional vulnerability and the mitigation priorities, especially for the mega city of Istanbul for which a large earthquake is due.

This technique is a reasonable theoretical approach that uses available tools to predict the spatial variation of ground motion. Further improvements to the procedure can be made, especially in the intermediate steps, but the end results, which are the modification coefficients, would not be influenced considerably. Besides, the assumptions and approximations already introduced are far beyond the accuracy that would be gained this way.

ACKNOWLEDGEMENTS

The research work presented in this study is supported in part by the Scientific and Research Council of Turkey (TUBITAK) under grant: YMAU-ICTAG-1574 and by NATO Scientific Affairs Division under grant: NATO SfP977231.

REFERENCES

1. Ozcebe, G., M. S. Yucemen, V. Aydogan and A. Yakut, Preliminary Seismic Vulnerability Assessment of Existing Reinforced Concrete Buildings in Turkey -Part I: Statistical Model Based on Structural Characteristics, NATO Workshop, Izmir May 2003.
2. Turkish Ministry of Public Works and Settlement, Turkish Seismic Code, Ankara., 1997.
3. E. Erduran and A. Yakut, Drift Based Damage Functions for Reinforced Concrete Columns, Submitted to Computers & Structures for possible publication, in Review, 2003
4. Craig D. Comartin, A Progress Report on ATC-55: Evaluation and Improvement of Inelastic Seismic Analysis Procedures, Fall, 2002.
5. FEMA 273, NEHRP Guidelines for the Seismic Rehabilitation of Buildings, Federal Emergency Management Agency, October 1997.
6. D. M. Boore, W. B. Joyner and T. E. Fumal, Equations for Estimating Horizontal Response Spectra and Peak Acceleration from Western North American Earthquakes: A Summary of Recent Work, Seismological Research Letters, Volume 68, Number 1, January/February, pp.128-153, 1997.
7. P. Gulkan and E. Kalkan, Attenuation Modeling of Recent Earthquakes in Turkey, Journal of Seismology, 6:397-409. Kluwer Academic Publishers, 2002.
8. N. A. Abrahamson and W. J. Silva, Empirical Response Spectral Attenuation Relations for Shallow Crustal Earthquakes, Seismological Research Letters, Volume 68, Number 1, January/February, pp.94-127, 1997.

9. Japan International Cooperation Agency (JICA) and Istanbul Metropolitan Municipality (IBB), The Study on A Disaster Prevention / Mitigation Basic Plan in Istanbul including Seismic Microzonation in the Republic of Turkey, Final Report, December, 2002.

Table 1. Site classification

Soil Type	Shear Wave Velocity (m/s)	Distance to Source (km)				
		0-4	5-8	9-15	16-25	26-50
A	701-1000	A1	A2	A3	A4	A5
B	401-700	B1	B2	B3	B4	B5
C	201-400	C1	C2	C3	C4	C5
D	<200	D1	D2	D3	D4	D5

Table 2. Period vs. number of stories for Düzce seismic damage database

Number of stories	Period (sec)
2	0.275
3	0.355
4	0.433
5	0.504
6	0.529

Table 3. Cutoff modification coefficients (CMC) for Boore et. al. [6]

		LINEAR					NON-LINEAR				
		Distance (km)					Distance (km)				
N	**Vs (m/s)**	**0-4**	**5-8**	**9-15**	**16-25**	**26+**	**0-4**	**5-8**	**9-15**	**16-25**	**26+**
2-3	**0-200**	0.778	0.824	0.928	1.128	1.538	0.764	0.826	0.959	1.207	1.726
	201-400	0.864	1.000	1.240	1.642	2.414	0.875	1.000	1.239	1.654	2.496
	401-700	0.970	1.180	1.530	2.099	3.177	0.978	1.150	1.468	2.010	3.101
	701+	1.082	1.360	1.810	2.534	3.900	1.075	1.288	1.675	2.329	3.640
4-6	**0-200**	0.778	0.824	0.928	1.128	1.538	0.781	0.825	0.928	1.125	1.535
	201-400	0.864	1.000	1.240	1.642	2.414	0.865	1.000	1.242	1.642	2.424
	401-700	0.970	1.180	1.530	2.099	3.177	0.970	1.182	1.537	2.106	3.204
	701+	1.082	1.360	1.810	2.534	3.900	1.082	1.364	1.824	2.552	3.948

Table 4. Cutoff modification coefficients (CMC) for Gulkan and Kalkan [7]

		LINEAR					NON-LINEAR				
		Distance (km)					Distance (km)				
n	**Vs (m/s)**	**0-4**	**5-8**	**9-15**	**16-25**	**26+**	**0-4**	**5-8**	**9-15**	**16-25**	**26+**
2-3	**0-200**	0.791	0.840	0.931	1.083	1.359	0.748	0.815	0.926	1.099	1.413
	201-400	0.932	1.000	1.126	1.334	1.706	0.892	1.000	1.171	1.431	1.896
	401-700	1.032	1.113	1.263	1.508	1.946	1.006	1.142	1.355	1.678	2.252
	701+	1.115	1.207	1.376	1.652	2.144	1.106	1.265	1.514	1.891	2.558
4-6	**0-200**	0.791	0.840	0.931	1.083	1.359	0.799	0.843	0.932	1.081	1.357
	201-400	0.932	1.000	1.126	1.334	1.706	0.939	1.000	1.121	1.324	1.695
	401-700	1.032	1.113	1.263	1.508	1.946	1.037	1.110	1.253	1.492	1.927
	701+	1.115	1.207	1.376	1.652	2.144	1.120	1.201	1.363	1.630	2.118

Table 5. Cutoff modification coefficients (CMC) for Abrahamson and Silva [8]

		LINEAR					NON-LINEAR				
		Distance (km)					Distance (km)				
n	**Vs (m/s)**	**0-4**	**5-8**	**9-15**	**16-25**	**26+**	**0-4**	**5-8**	**9-15**	**16-25**	**26+**
2-3	**0-200**	0.826	0.917	1.084	1.362	1.887	0.850	0.967	1.185	1.554	2.288
	201-400	0.873	1.000	1.219	1.575	2.236	0.870	1.000	1.240	1.642	2.438
	401-700	0.919	1.077	1.341	1.765	2.542	0.903	1.055	1.329	1.783	2.676
	701+	0.999	1.205	1.539	2.065	3.032	0.947	1.125	1.439	1.957	2.970
4-6	**0-200**	0.826	0.917	1.084	1.362	1.887	0.825	0.917	1.085	1.362	1.894
	201-400	0.873	1.000	1.219	1.575	2.236	0.872	1.000	1.221	1.574	2.241
	401-700	0.919	1.077	1.341	1.765	2.542	0.919	1.078	1.344	1.763	2.550
	701+	0.999	1.205	1.539	2.065	3.032	1.001	1.208	1.545	2.069	3.046

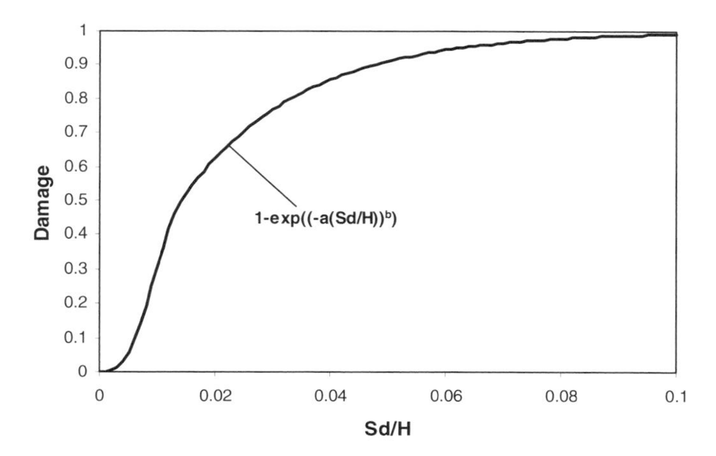

Figure 1. A typical damage curve

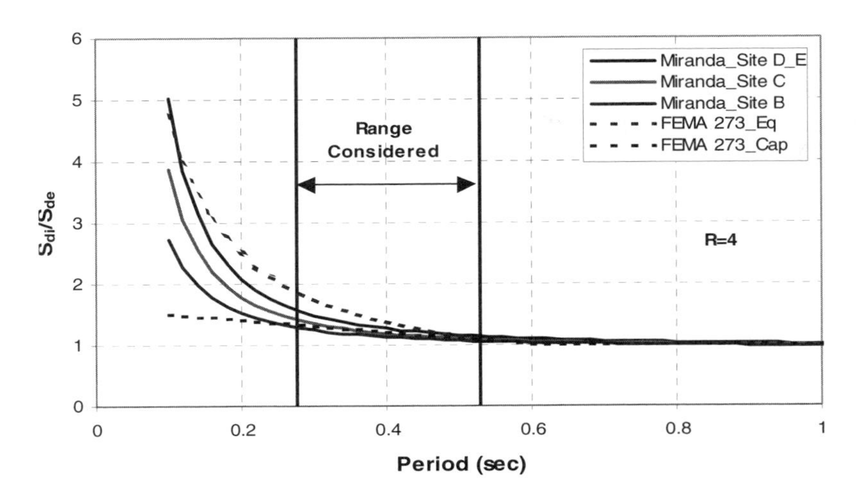

Figure 2. Comparison of S_{di}/S_{de} relations

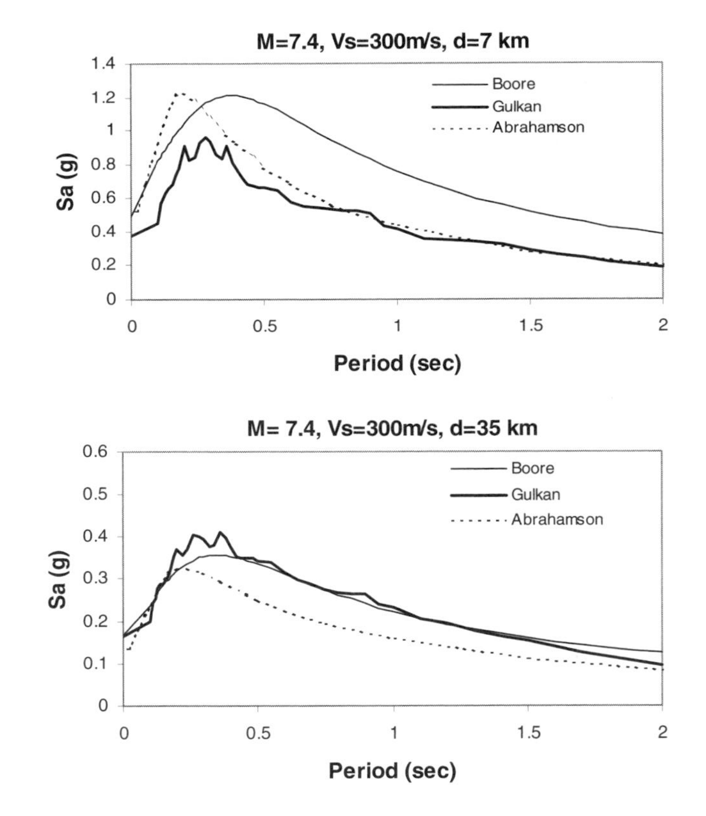

Figure 3. Comparison of the attenuation models

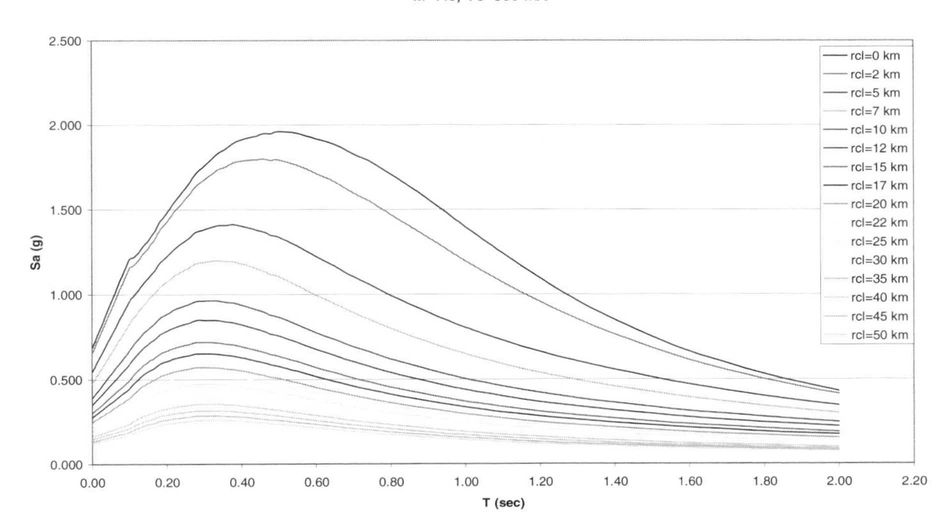

Figure 4. Acceleration response spectra

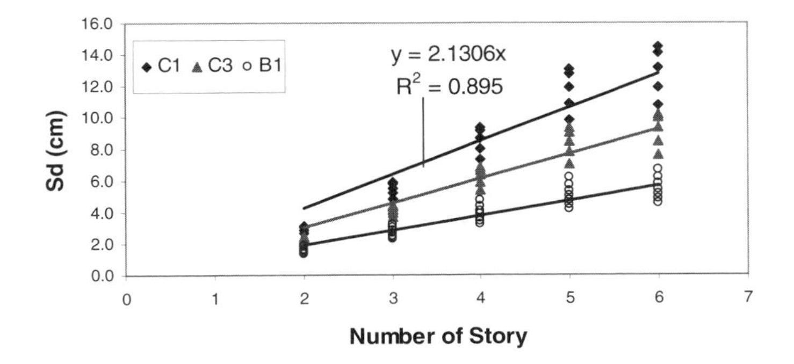

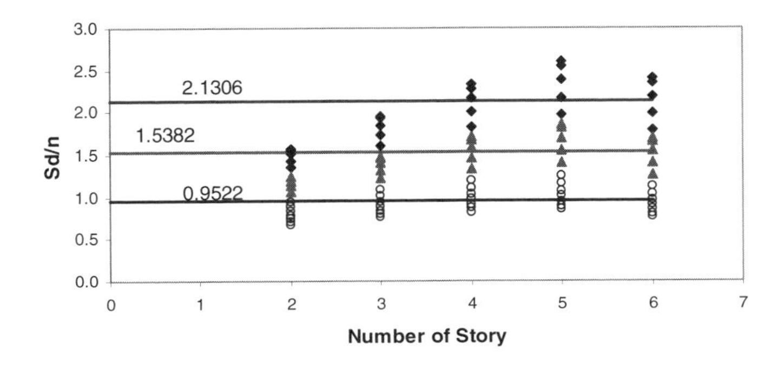

Figure 5. Normalized S_d versus Number of Story (Linear Representation)

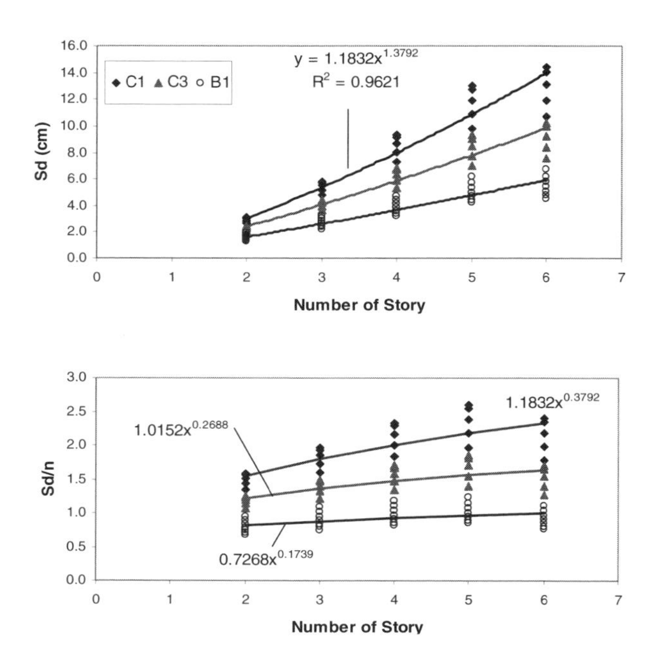

Figure 6. Normalized S_d vs. number of story (non-linear representation)

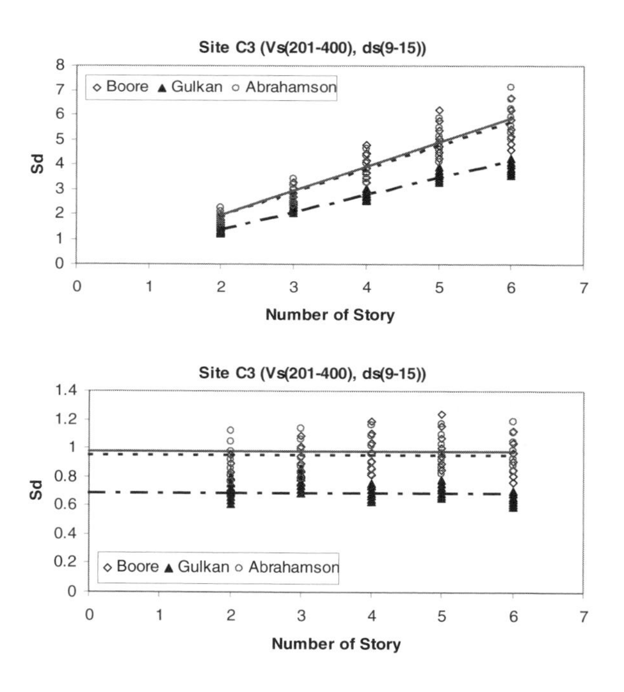

Figure 7. Influence of attenuation relation

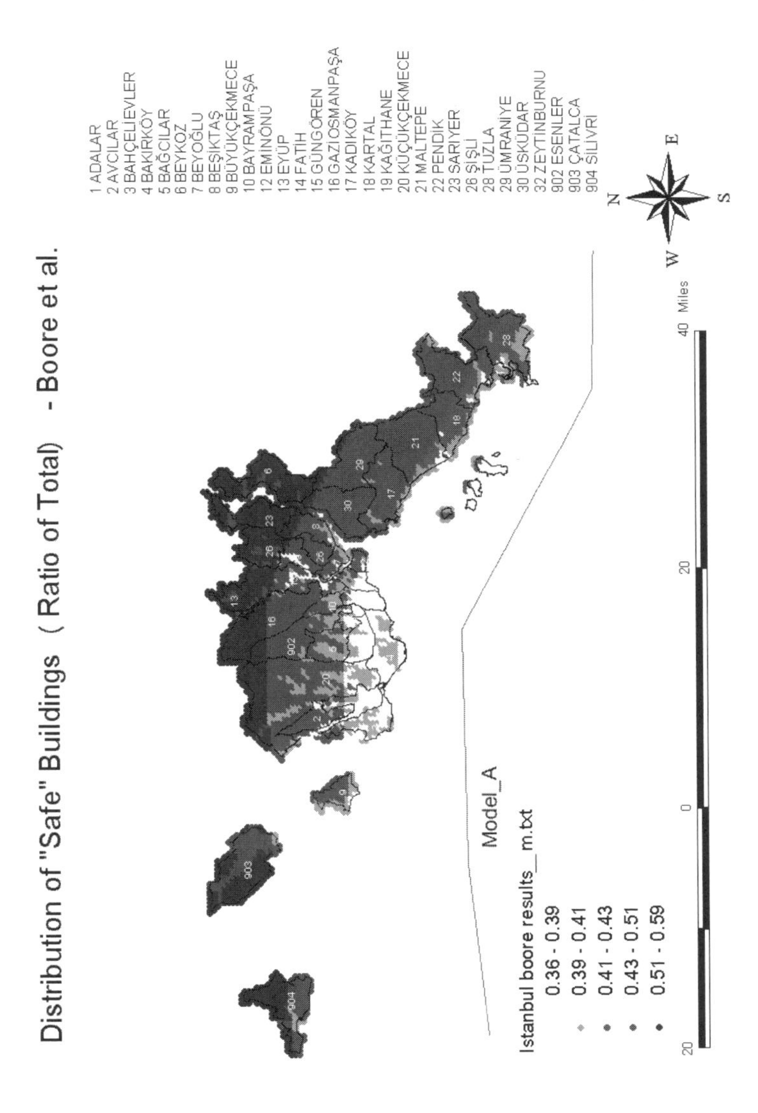

Figure 8. Results using Boore et al. attenuation relationship safe buildings

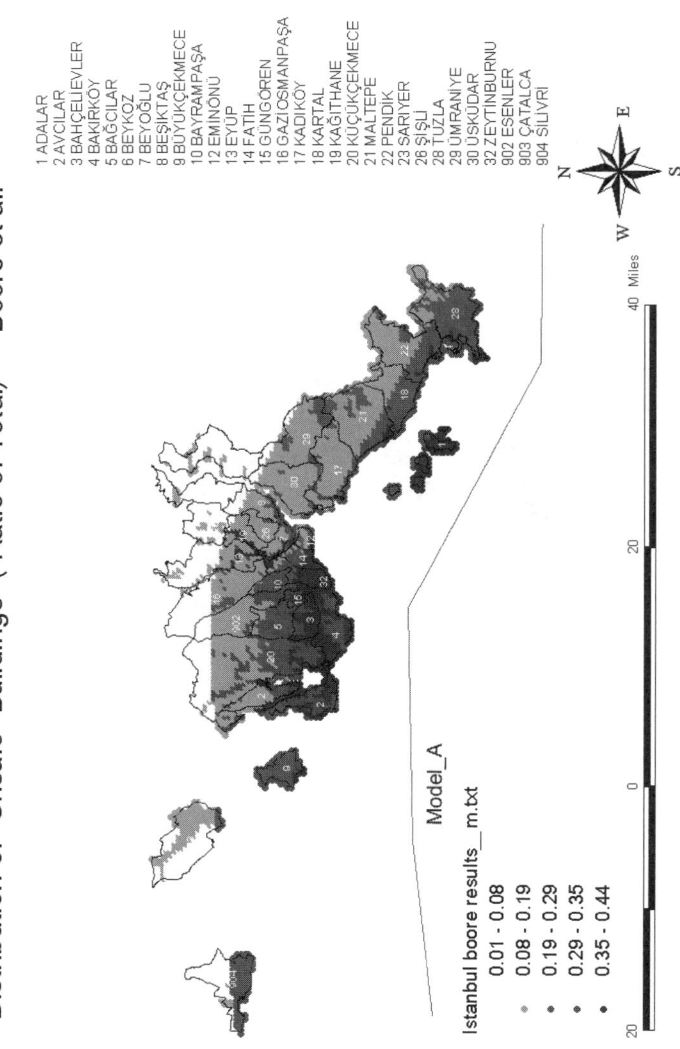

Figure 9. Results using Boore et al. attenuation relationship unsafe buildings

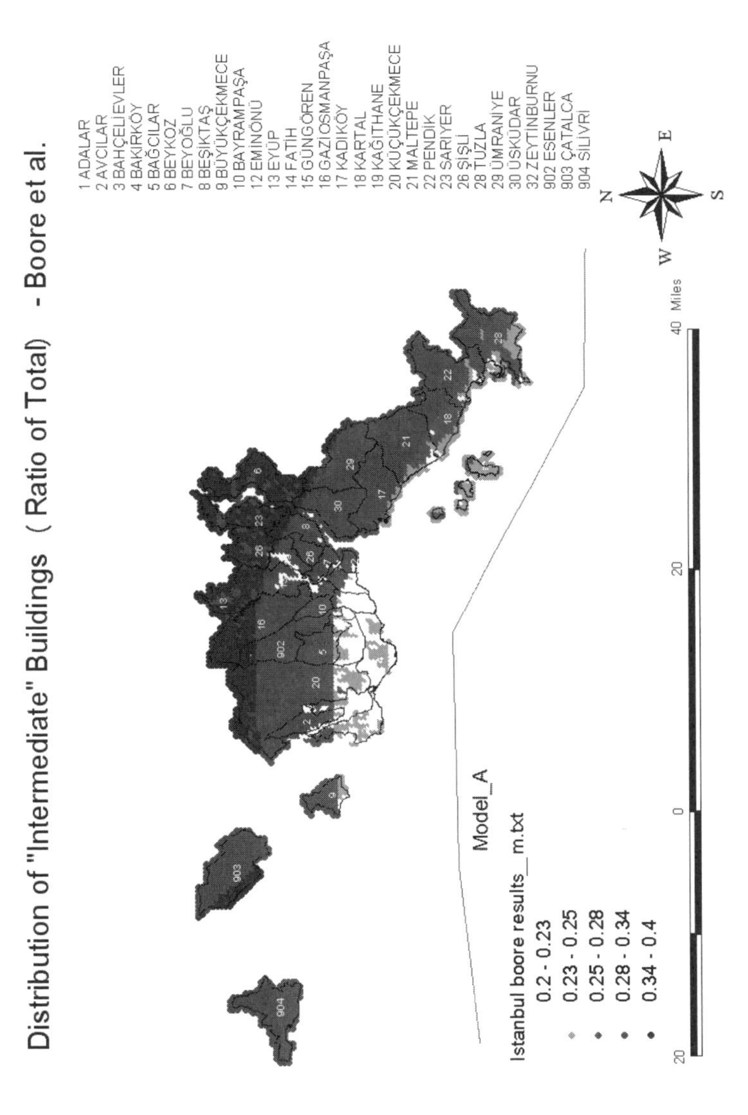

Figure 10. Results using Boore et al. attenuation relationship intermediate buildings

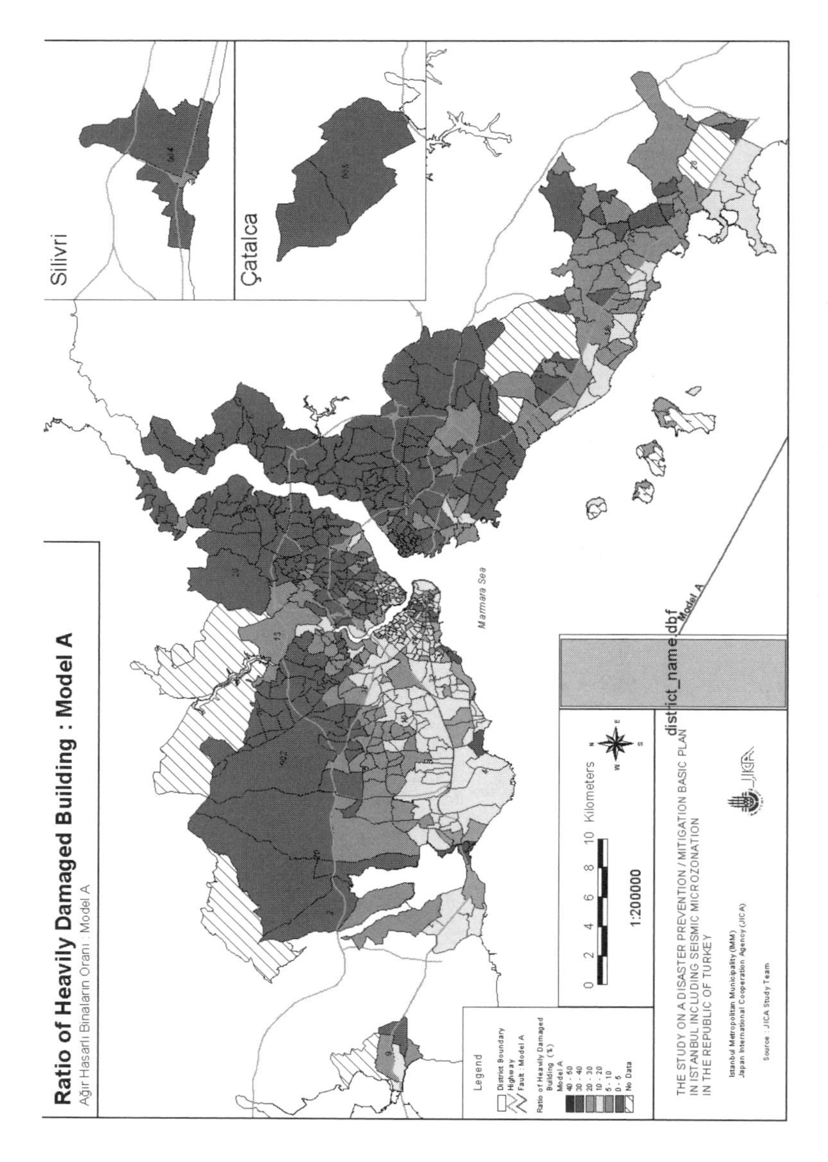

Figure 11. JICA estimates of heavily damaged buildings (from JICA, 2002)

PARAMETERS AFFECTING DAMAGEABILITY OF REINFORCED CONCRETE MEMBERS

Emrah Erduran and Ahmet Yakut
Structural Mechanics Division, Department of Civil Engineering, Middle East Technical University, Ankara, 06531, Turkey

Abstract: Research has been undertaken to develop damage curves for RC members. This paper deals with the first stage of the aimed study, which is the determination of all significant parameters that affect the damageability of RC components. Analytical investigations for columns and beams are presented for a broad range of parameters, emphasizing on their effect on the structural damage. These results will then be used to develop damage curves that are expressed in terms of the inter story drift ratio.

Key words: Reinforced Concrete, Damage Curves, Drift

1. INTRODUCTION

The destructive earthquakes that hit different regions of the world cause loss of huge amounts of economic property and lives. The high economic loss and death toll prompt research dealing with reducing seismic risk in the earthquake prone regions. As a result of the research conducted in the field of earthquake engineering for decades, seismic codes of the countries, which are susceptible to damaging earthquakes are either revised or rewritten to enable the satisfactory seismic performance of the structures. Nevertheless, there are still a lot of structures throughout the world, which are highly

S.T. Wasti and G. Ozcebe (eds.), Seismic Assessment and Rehabilitation of Existing Buildings, 59–76.

vulnerable to seismic action. Identifying structures that have high vulnerability is of critical importance for both reliable loss estimation as a result of an expected earthquake and setting priority criteria for strengthening of structures.

Reinforced concrete frames are amongst the most common construction types in the world. Predicting vulnerability of a whole reinforced concrete frame is not easy to handle due to lack of proper experimental and analytical data. For this reason, the trend has moved towards evaluating the whole structure at the level of its components.

For reinforced concrete frames, the behavior of the beams and columns is an important factor that determines the performance of the whole frame. Thus, predicting the damage level of these components as a result of an earthquake plays a major role in predicting the seismic vulnerability of a reinforced concrete frame structure. In general, the damage state of these components is determined by computing a damage index, which is usually related to ductility, dissipated energy, stiffness degradation and deformation of the members. The deformation such as drift ratio or plastic rotation at the member end is widely used by some guidelines given for seismic evaluation of buildings (ATC-40 [1], FEMA 273 [2]) and in earthquake vulnerability assessment procedures (ACM [3]).

The deformation capacities of both beams and columns are affected by various parameters including concrete strength, yield strength of reinforcement, and cross-sectional dimensions. For decades, researchers have investigated the effects of these parameters on the deformation capacities of the reinforced concrete beams and columns both experimentally and analytically.

Numerous studies carried out on the deformation capacity of the columns dealt with two important terms, the yield displacement and ultimate ductility of the columns. In 1992 Azizinamini et al. [4] tested 12 reinforced concrete columns to investigate the effects of transverse reinforcement on the seismic performance of the columns. At the end of these tests, it was observed that for a constant amount of confinement, flexural capacity of a column increases with axial load, but ductility is reduced substantially. Moreover, for the same amount of transverse reinforcement, reducing the spacing had no significant effect on flexural capacity. However, smaller spacing provided a slightly higher ductility.

In their work, Priestley and Kowalsky [5] aimed to develop dimensionless yield, serviceability and damage control curves for structural walls. During this study they proved that all of the three curvature limit states were largely independent of amount and distribution of longitudinal reinforcement.

In 2002, Paulay [6] stated that the amount of reinforcement used in a section and the gravity induced axial compression do not affect the nominal yield curvature to any significance. The two important terms affecting column yield displacements are the yield strain of the longitudinal reinforcement (ε_y) and the slenderness ratio of the column. The yield displacement is essentially independent of the strength of the section.

Although the conclusions drawn are invaluable for understanding the behavior of the R/C in a better manner, most of these conclusions were qualitative. In this ongoing study, the main aim is to develop damage functions for reinforced concrete components based on the drift ratio. These damage functions will take all the related parameters into account quantitatively. For this purpose, the significant parameters that affect the damageability of these two components, namely beams and columns, were investigated. To develop consistent damage deformation relations, a number of finite element analyses were carried out using the software ANSYS v6.1. In these analyses, the effects of different parameters on the deformation capacities of reinforced concrete columns and beams were investigated by changing only one parameter in each case.

2. ANALYTICAL STUDIES

2.1 Procedure

In the analytical work carried out within the scope of this study, the finite element software ANSYS v6.1 was used. The reinforced concrete members were modeled using 8 node brick elements. The element used is capable of taking cracking and crushing of concrete into account. The longitudinal reinforcement was modeled as smeared throughout the section. According to the study carried out by Barbosa and Ribeiro [7], the difference between modeling the reinforcement as smeared or discrete is not significant in the nonlinear analysis of reinforced concrete members. The modified Kent and Park model [8] was used in modeling the behavior of both confined and unconfined concrete.

In the first part of this study, the parameters affecting the damageability of reinforced concrete columns were investigated. The first step of this part was the verification of the finite element model used. For this purpose a 457 mm by 457 mm column, which was tested by Azizinamini et al. [4] was modeled and analyzed (Figures 1 and 2). Hereafter, this column will be referred to as the *reference column*. The half-height of the column, i.e. the distance between the base of the column and the point of inflection was 1372 mm. The column had N/N_o=0.2 and f_{ck}=39.3 MPa. The load-displacement behavior of the experimental and analytical results for the reference column is shown in Figure 1. The analytical and experimental results match fairly well up to a displacement level of 30 mm. When the displacements exceed 30 mm, the analytical model overestimates the strength of the column; the main reason behind this fact is the difference in the types of loading of the analysis and the test. The analytical load-displacement curve was obtained from a pushover analysis, which is a one-way static procedure. However, the test was carried out under hysteretic loading. The strength degradation due to hysteretic loading could not be taken into account in the static analysis.

Upon verification of the analytical model used, further analyses were carried out to investigate the effects of different parameters on the behavior of the columns. The parameters investigated within the scope of this study are concrete strength, amount and yield strength of longitudinal reinforcement, amount of transverse reinforcement, slenderness and axial load level. Table 1 summarizes the range of the parameters used in the analyses.

The damage criterion used in this study mainly depends on lateral drift levels. The damage levels were defined in terms of the drift corresponding to the maximum load carrying capacity of the column. Although this point is slightly different from the yield point, it will be referred to as the yield drift (δ_y) in this study.

In the second part of the study, the parameters affecting the behavior of reinforced concrete beams were investigated. A portal frame was used for this purpose. The columns of this portal frame were assumed to remain elastic during all stages of the loading. The frame was loaded through the application of lateral displacements to the uppermost nodes of the columns. The damage level of the beam was investigated by monitoring the crack width at a section, which is d/2 units away from the face of the column (d being depth of the beam). The crack width calculated was related to the rotation of the beam. In this study, rotation was defined as the chord rotation between the two nodes, one of which is just on the face of the column and the other one on the section on which the crack width was measured (i.e. d/2 units away from the face of the column) (Figure 4, Equation 1). Analytical modeling of the portal frame is given in Figure 5.

$$\theta = \frac{\delta_2 - \delta_1}{L} \tag{1}$$

There are various relationships recommended for the determination of the crack widths. Almost all of these relationships are based on the tensile strain in the tension steel and the arrangement of longitudinal steel. In this particular study, the crack width formula proposed by Frosch [9] was used. According to Frosch the maximum crack width can be calculated by the relationship given in equation (2).

$$w = 2\varepsilon_s d^* \tag{2}$$

In this equation w is the maximum crack width, ε_s is the strain in tensile steel and d^* is the controlling cover distance given in equation (3).

$$d^* = \sqrt{d_c^{\,2} + (\frac{s}{2})^2} \tag{3}$$

Here, d_c is the clear cover and s is the spacing between two adjacent bars.

In the analyses, a constant value of 100 mm was used for the controlling cover distance in order to eliminate the effect of variations in the reinforcement arrangement.

The parameters investigated within the scope of this study were yield strength of longitudinal reinforcement, concrete strength, amount of tension reinforcement, amount of compression reinforcement, depth of the reinforced concrete beams, and the ratio of beam stiffness to the column stiffness ($(EI)_{beam}/(EI)_{column}$). Table 2 summarizes the range of the parameters used in the analyses.

2.2 Discussion of Results

2.2.1 RC Columns

2.2.1.1 Concrete Strength, f_{ck}

In order to investigate the effect of concrete strength on the deformation capacities of the columns, six pushover analyses for concrete strengths of 10, 14, 16, 20, 25, and 39.3 MPa were carried out. The results of these analyses are shown in Figure 6. The load carrying capacity of the column increases significantly with f_{ck}. Nevertheless, the effect of concrete strength on the yield drift ratio δ_y is insignificant. Moreover, as long as the axial load level and the amount of confinement are constant, the ultimate ductility of the column is also not affected significantly by the variation in f_{ck}. Recalling that the damage criterion used in this study is based on the yield drift ratio and the ultimate ductility level, it can be stated that the concrete strength has no significant effect on the damage level of the reinforced concrete columns provided that the other parameters are constant.

2.2.1.2 Axial Load Level (N/N_o)

Six analyses were carried out to see the effect of the axial load level on the behavior of the reinforced concrete columns (Figure 7). N/N_o ratio varied from 10% to 60% in these analyses. Figure 7 shows that, although the yield drift ratio was almost constant for different axial load levels, the ultimate ductility decreases significantly with increasing axial load. This indicates that columns under large axial loads do not show a ductile behavior and hence may experience high levels of damage beyond yield deformation.

2.2.1.3 Slenderness

The slenderness ratio of the reference column, i.e. the ratio of length of the column (L) to the radius of gyration (i) was 21.123. Seven analyses were carried out to see the effect of slenderness on the deformation characteristics of columns. The capacity curves, the plot of lateral load versus lateral displacement, obtained from these analyses show that the yield drift ratio (δ_y) increases with increasing slenderness ratio (Figure 8). This means that for a given drift ratio, slender columns suffer less damage than standard columns.

2.2.1.4 Amount of Longitudinal Reinforcement (ρ)

Five analyses were carried out for different longitudinal reinforcement ratios to see the effect of ρ on the damage level of the columns. The value of ρ varied between 0.75% and 4%. The results of the analyses indicate that (Figure 9) the amount of longitudinal reinforcement has no significant effect both on the yield drift ratio and ultimate ductility.

2.2.1.5 Yield Strength of Longitudinal Reinforcement (f_{yk})

The results of the analyses carried out for different steel grades show that the yield drift ratio increases with increasing f_y (Figure 10). Thus, the damage level of the reinforced concrete columns is significantly affected by the variation in the yield strength of the longitudinal reinforcement.

2.2.1.6 Amount of Transverse Reinforcement

The effect of amount of transverse reinforcement was investigated by carrying out four analyses for different ρ_s values. The capacity curves reveal that (Figure 11), for a constant axial load, the ultimate ductility increases with increasing ρ_s. The volumetric ratio of transverse reinforcement has no significant effect on δ_y. These results show that the effect of ρ_s is similar to that of N/N_o.

2.2.2 RC Beams

2.2.2.1 Concrete Strength (f_{ck})

Five analyses were carried out for different concrete strengths ranging from 10 MPa to 25 MPa. The crack width - rotation curves given in Figure 12 shows that, once the yield rotation is exceeded, a beam with higher concrete strength suffers more damage than a beam with lower concrete strength for a given rotation level. As the concrete strength increases, the effect of concrete strength becomes less significant.

2.2.2.2 Yield Strength of Longitudinal Reinforcement (f_{yk})

In order to see the effect of yield strength of longitudinal reinforcement on the crack width – rotation curves, five analyses were carried out. In Figure 13, it can be seen that as the yield strength decreases the level of damage increases drastically particularly for low rotation levels.

2.2.2.3 Amount of Tension Reinforcement (ρ)

Six analyses were carried out to see the effect of amount of tension reinforcement on the damageability of reinforced concrete beams. In these analyses, ρ varied from 0.75% to 2%, which is the upper limit according to the Turkish Earthquake Code [TSC 1998]. As a result of these analyses, it was concluded that the amount of longitudinal reinforcement had no significant effect on the crack width - rotation curves, and in turn on the damageability of the beams (Figure 14).

2.2.2.4 Amount of Compression Reinforcement (ρ')

In the analyses carried out in this study, the amount of compression reinforcement varied between 0.3 ρ and 1.0 ρ (ρ is the amount of tension reinforcement). Similar to the amount of longitudinal reinforcement, the amount of compression reinforcement has no significant effect on the damageability of reinforced concrete beams (Figure 15).

2.2.2.5 Depth of Beam (d)

Crack width - rotation curves were examined to see the effect of depth of the beam on the damageability. The results presented in Figure 16 show that, as the depth of the beam increases, the damage suffered increases for a given rotation level.

2.2.2.6 Ratio of Beam Stiffness to Column Stiffness ($(EI)_b/(EI)_c$)

Five finite element analyses were carried out to see the effect of ratio of beam stiffness to column stiffness. The results of the analyses presented in Figure 17 shows that as $(EI)_{beam}/(EI)_{column}$ increases, the damage suffered by the beam increases for a given rotation level.

2.3 Significant Parameters

The parametric study carried out revealed that the most important parameters that affect the deformation limits of reinforced concrete columns are the yield strength of longitudinal reinforcement (f_y), slenderness of the column (L/i), axial load level (N/N_o) and the amount of transverse reinforcement (ρ_s). Of these four parameters, the first two, i.e. f_y and slenderness, affect the yield drift ratio significantly. The axial load level and the volumetric ratio of transverse reinforcement have an effect on the ultimate ductility of the columns. If Figures 6 and 10 are examined carefully, it will be observed that the effects of N/N_o and ρ_s on the deformation capacities of the columns are very similar. Increase (or decrease) in the amount of transverse reinforcement and decrease (or increase) in the axial load level has the same effect on the capacity curves of the reinforced concrete columns. In the light of this discussion a new term, which is defined as the ratio of the amount of transverse reinforcement to the axial load level ($\rho_s/(N/N_o)$) will be introduced. This term is believed to represent the ductility level of the reinforced concrete columns in a good manner. The other two parameters investigated, concrete strength and amount of longitudinal reinforcement have no significant effect on the deformation capacities of the reinforced concrete columns as far as all the other parameters such as axial load level (N/N_o) are kept constant.

In the case of reinforced concrete beams, the amount of tension reinforcement and compression reinforcement do not have significant effect on the crack width – rotation curves of the beams. In contrast to the columns, the deformation capacities of the beams are significantly affected by the concrete strength. Also the ratio of beam stiffness to the column stiffness is another significant parameter that affects the behavior. Nevertheless, the most significant parameters that affect the damage – rotation curves are the yield strength of longitudinal reinforcement (f_{yk}) and depth of the beam. Recalling that, the most important parameters affecting the behavior of columns were f_{yk} and slenderness, it can be concluded that the dimensions of the members and the yield strength of longitudinal reinforcement are important in the deformation capacities of the both reinforced concrete members.

3. CONCLUSION

Predicting the deformation capacities of beams and columns plays a major role in predicting the behavior of reinforced concrete frames. There are several parameters that affect the deformation behavior of these members. Research has been conducted to evaluate the effect of these parameters analytically.

As a result of the finite element analyses carried out, the most significant parameters that affect the deformation capacities of the columns were identified as the axial load level, amount of transverse reinforcement, slenderness of the column, and the yield strength of the longitudinal reinforcement. The first two of these parameters affect the ultimate ductility of the columns and the others affect the yield drift ratio. Also, it was shown that concrete strength and amount of longitudinal reinforcement greatly influence the load carrying capacity, but have no significant effect on the deformation characteristics of the reinforced concrete columns.

Moreover, the damage – rotation curves of the beams were greatly influenced by the variations in concrete strength, yield strength of longitudinal reinforcement, ratio of beam stiffness to column stiffness and the depth of the beam. The effect of amount of tension reinforcement and compression reinforcement on the damage – rotation curves of the beams is proved to be insignificant compared to the other parameters.

Further research will be conducted to propose damage functions for these members based on the drift ratio taking the effect of the aforementioned parameters into account quantitatively.

REFERENCES

1. Applied Technology Council, ATC-40, Seismic Evaluation and Retrofit of Concrete Buildings. California, 1996.
2. Federal Emergency Management Agency (FEMA), 1997, NEHRP Guidelines for the Seismic Rehabilitation of Buildings, FEMA 273
3. Kishi, N., A. Yakut and J. Byeon, "Advanced Component Method (ACM)- An Objective Methodology for the Assessment of Building Vulnerability," Proc. Of Second Asian Symposium on Risk Assessment and Management, Kobe University, Nov. 23-25, pp. 88-97, 2001.
4. Azizinamini et al., Effects of transverse reinforcement on seismic performance of columns, ACI Structural Journal 1992; 89: 442-450
5. Priestley, M. J. N., Kowalsky, M. J., Aspects of drift and ductility capacity of rectangular cantilever structural walls, Bulletin of the New Zealand National Society for Earthquake Engineering 1998, 31(6): 73-85
6. Paulay, T., An estimation of displacement limits for ductile systems, Earthquake Engineering and Structural Dynamics 2002; 31:583-599

7. Barbosa, A. F., Ribeiro, G. O., Analysis reinforced concrete structures using ANSYS nonlinear concrete model, Computational Mechanics: New Trends and Applications, Barcelona, Spain, 1998
8. Kent, D.C., Park, R., Flexural members with confined concrete, ASCE Journal of the Structural Division 1971; 97:ST 7
9. Frosch, R. J., Another look at cracking and crack control in reinforced concrete, ACI Structural Journal 1999; 96:437-442

Table 1. Range of Parameters for Columns

		Longitudinal Rein.		Transverse Rein.		
f_{ck} (MPa)	N/N_o	ρ	f_y (MPa)	ρ_s	f_{ywk} (MPa)	L/i
10.0	0.1	0.0075	220	0.01	454	12.658
14.0	0.2	0.0100	300	0.02		15.873
16.0	0.3	0.0195	375	0.03		21.123
20.0	0.4	0.0300	439	0.04		24.390
25.0	0.5	0.0400	525			28.571
39.3	0.6		600			32.258
						37.037

Table 2. Range of Parameters for Beams

		Longitudinal Reinforcement			
f_{ck} (MPa)	d (mm)	ρ	f_y (MPa)	ρ'/ρ	EI_b / EI_c
10.0	375	0.0075	220	0.30	0.50
14.0	500	0.0100	330	0.50	0.72
16.0	625	0.0125	420	0.70	1.00
20.0	750	0.0150	530	0.85	1.20
25.0		0.0175	650	1.00	1.44
		0.0200			

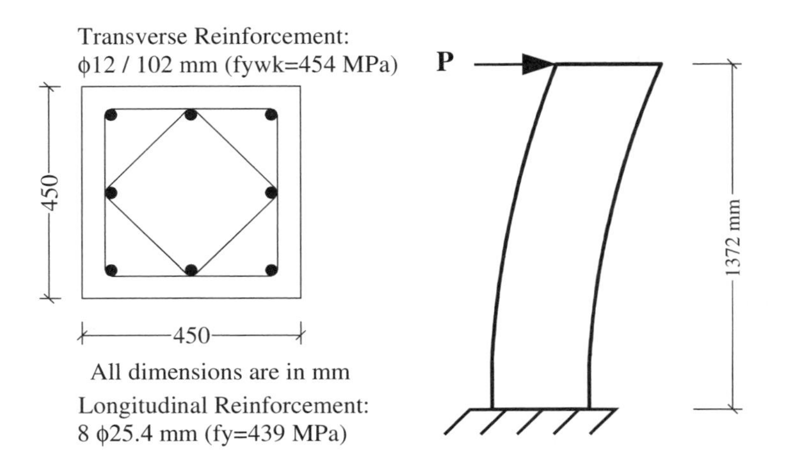

Figure 1. Details of the Reference Specimen

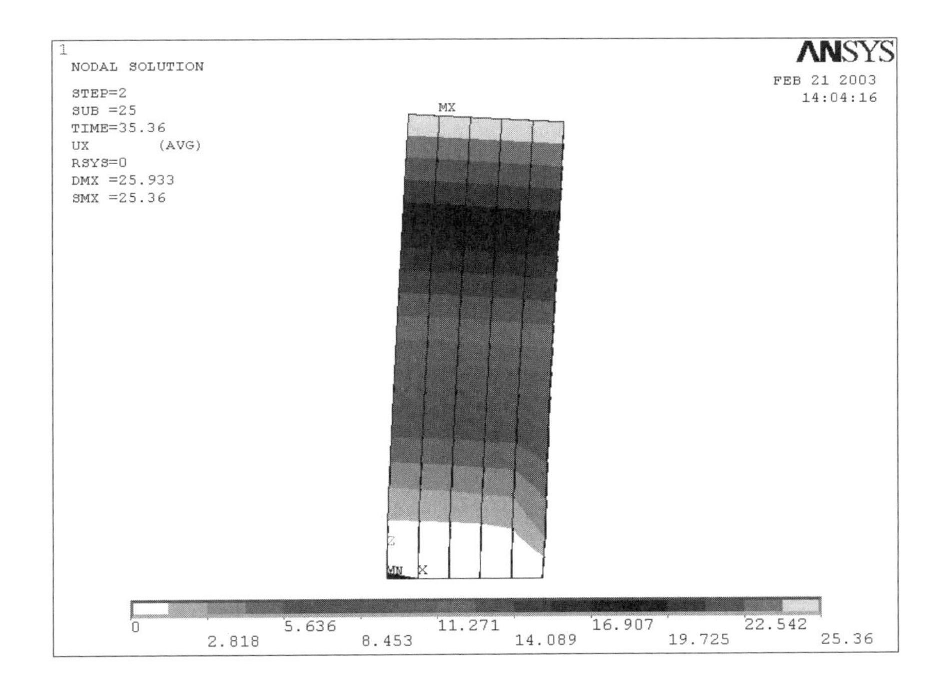

Figure 2. Analytical Modeling of Columns

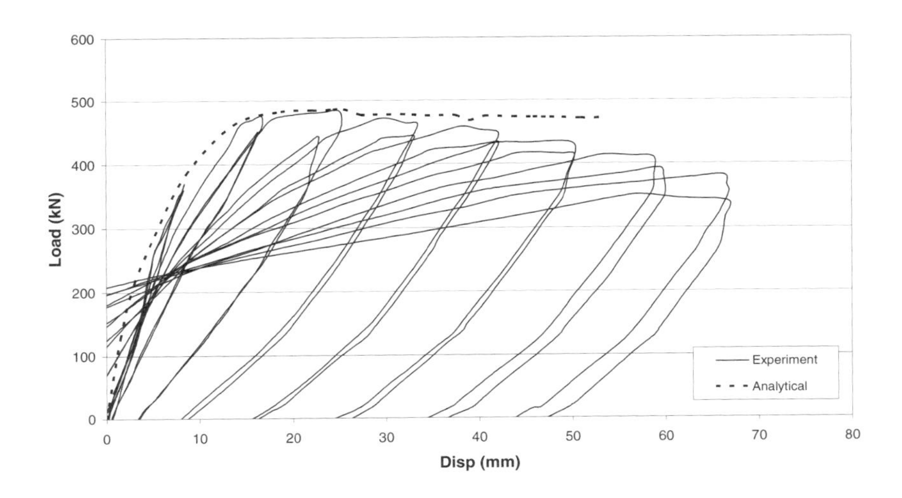

Figure 3. Experimental and Analytical Load-Deflection Behavior of the Reference Column

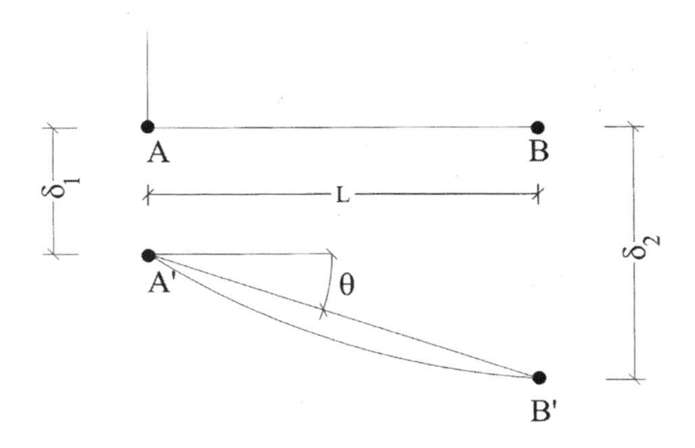

Figure 4. Definition of Chord Rotation

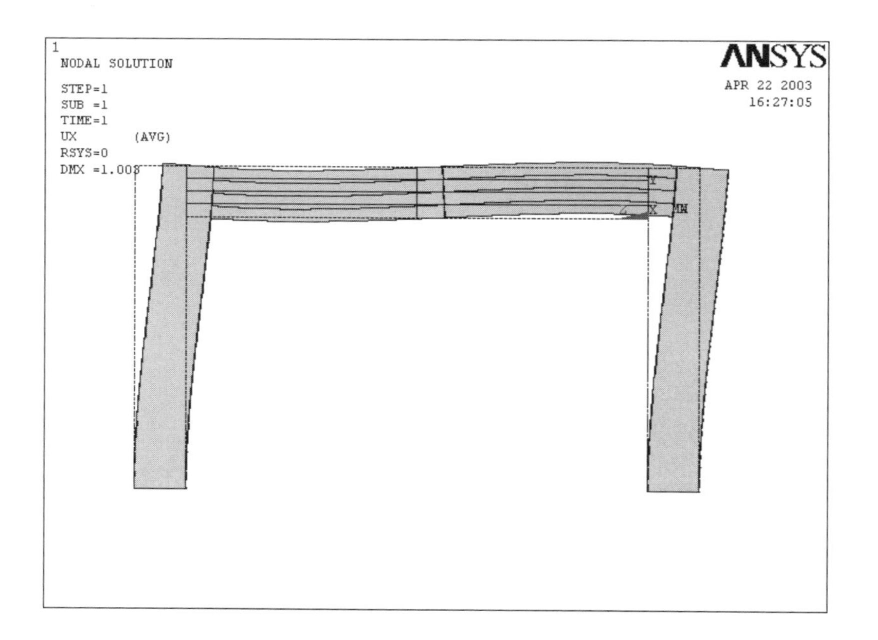

Figure 5. Analytical Modeling of Portal Frame

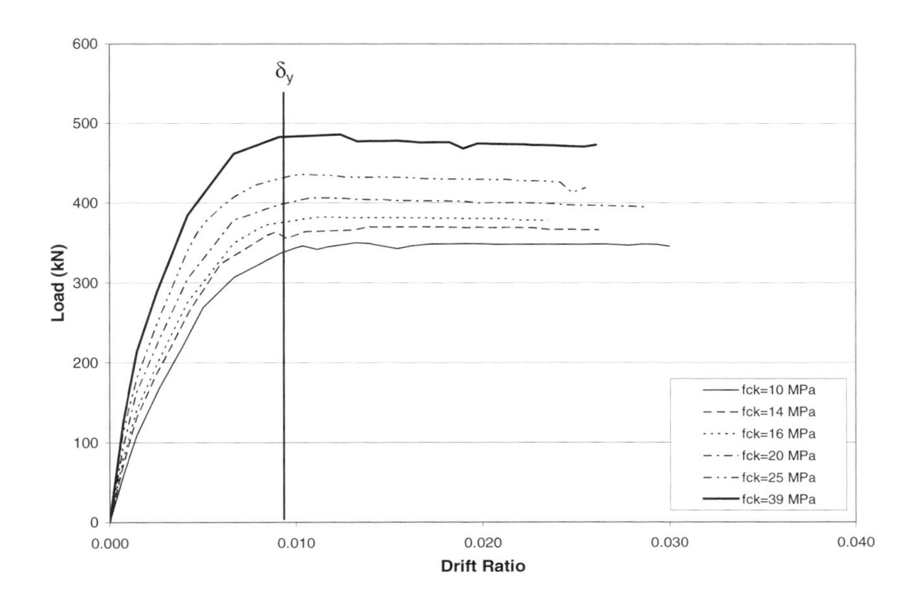

Figure 6. Effect of Concrete Strength on Capacity Curves

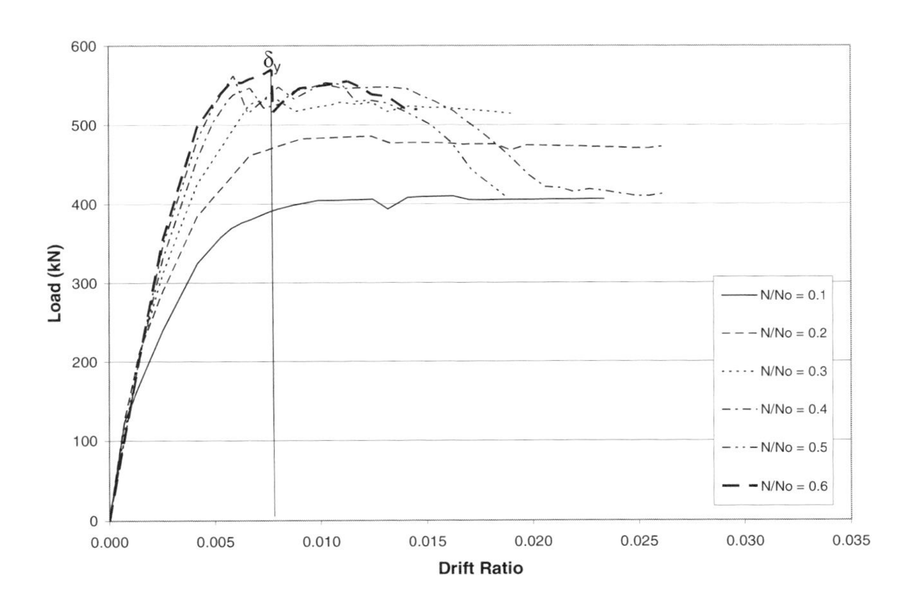

Figure 7. Effect of Axial Load on Capacity Curves

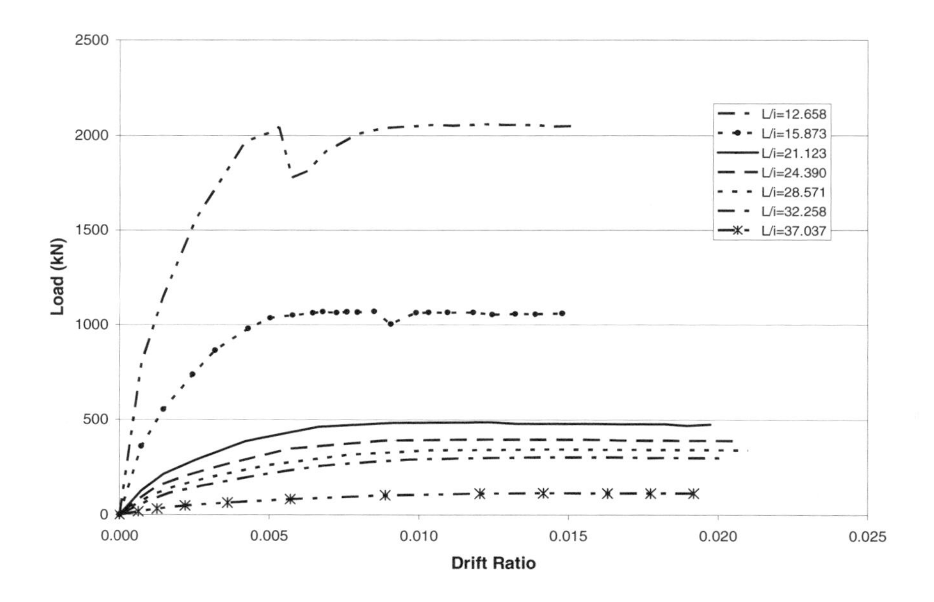

Figure 8. Effect of Slenderness on Capacity Curves

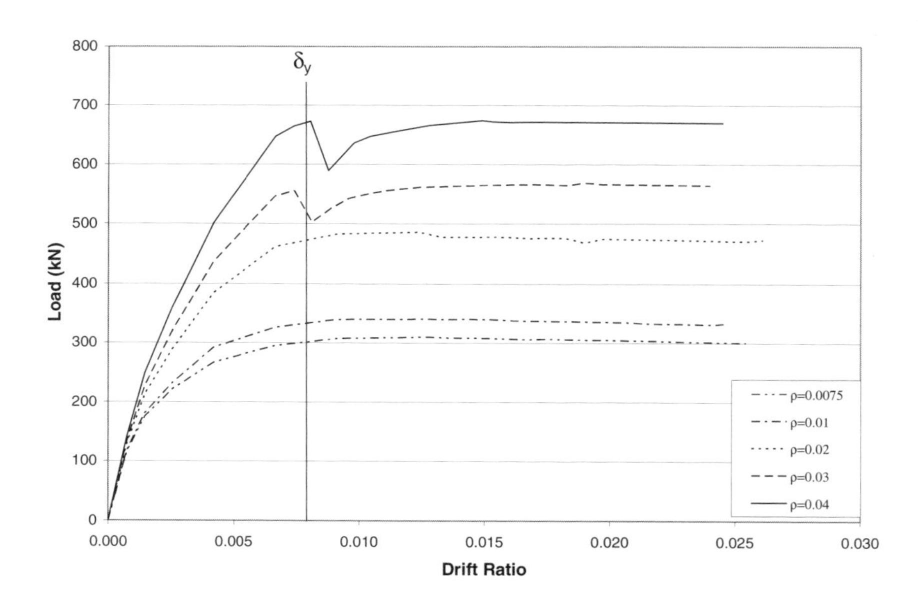

Figure 9. Effect of Amount of Longitudinal Reinforcement on Capacity Curves

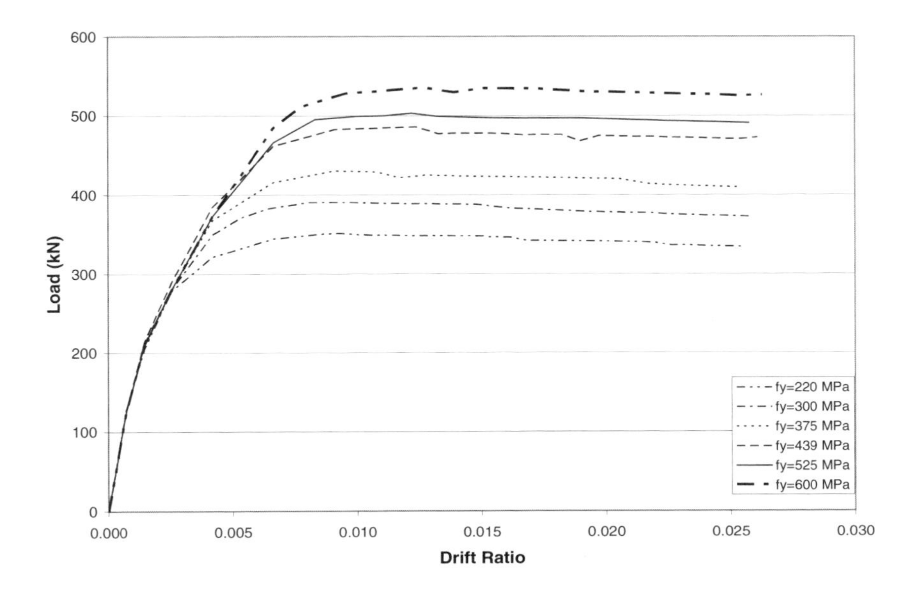

Figure 10. Effect of Yield Strength of Longitudinal Reinforcement on Capacity Curves

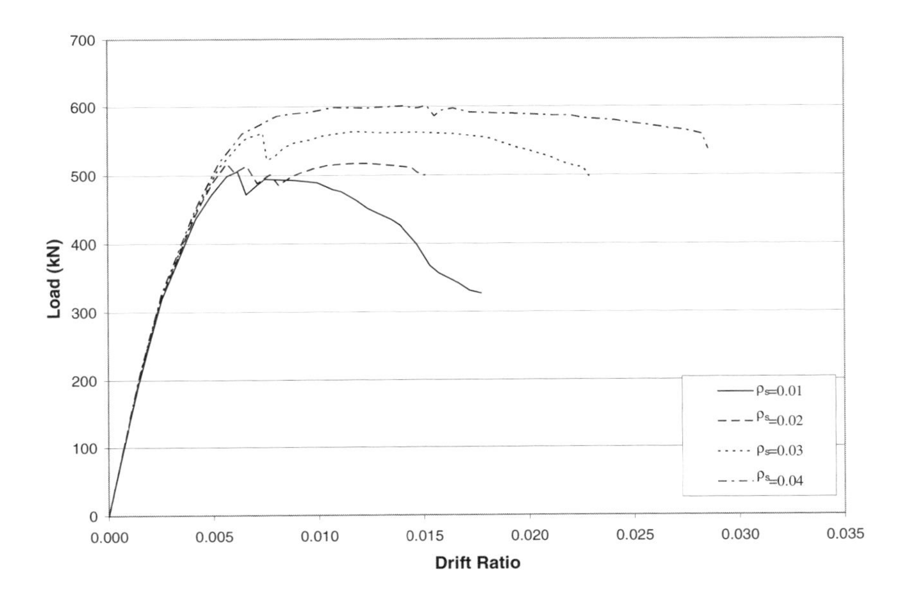

Figure 11. Effect of Amount of Transverse Reinforcement on Capacity Curves

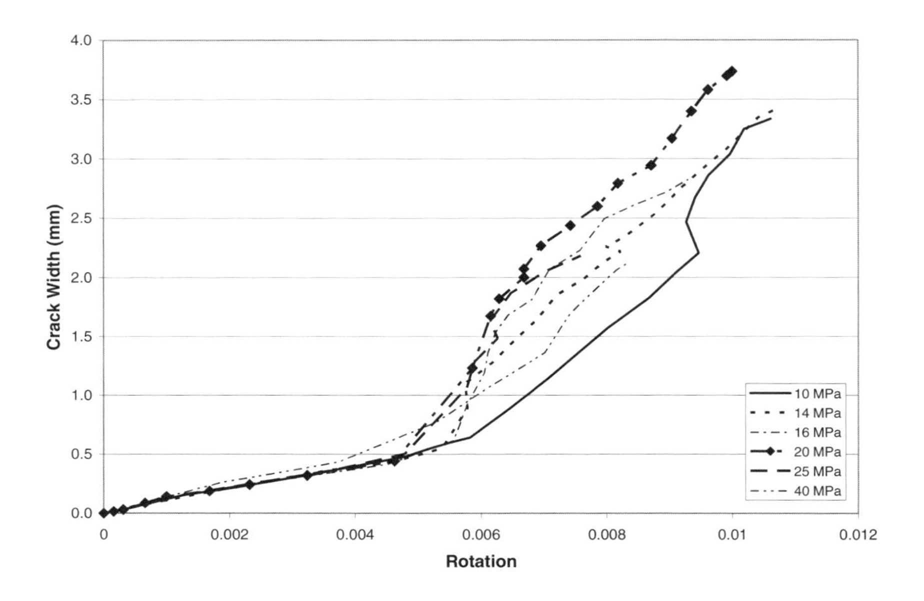

Figure 12. Effect of Concrete Strength on Crack Width - Rotation Curves

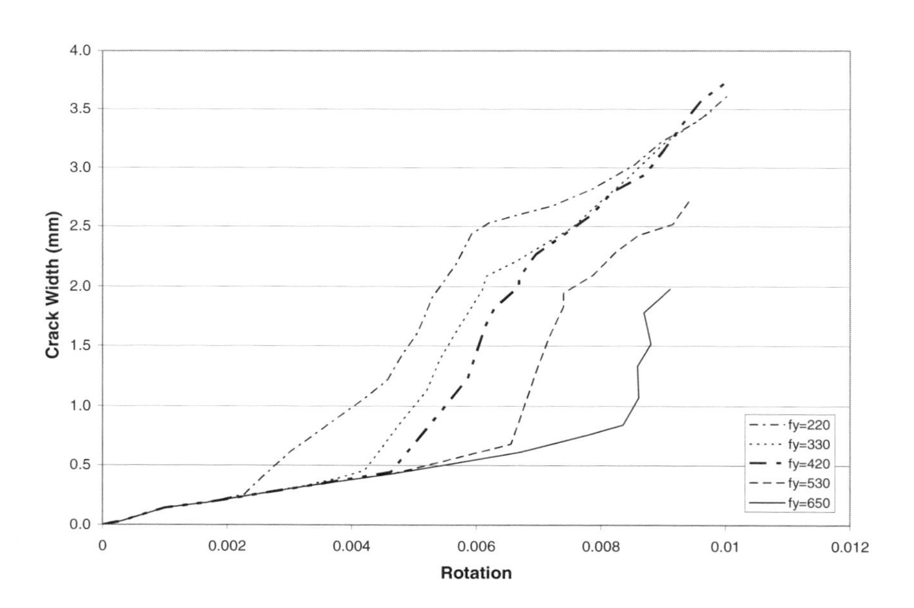

Figure 13. Effect of Yield Strength of Reinforcement on Crack Width - Rotation Curves

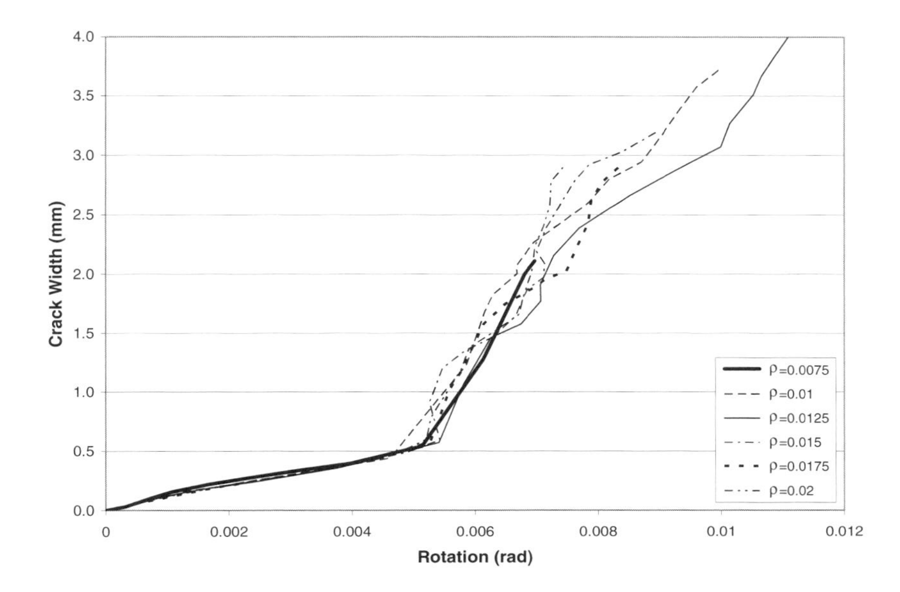

Figure 14. Effect of Amount of Tension Reinforcement on Crack Width- Rotation Curves

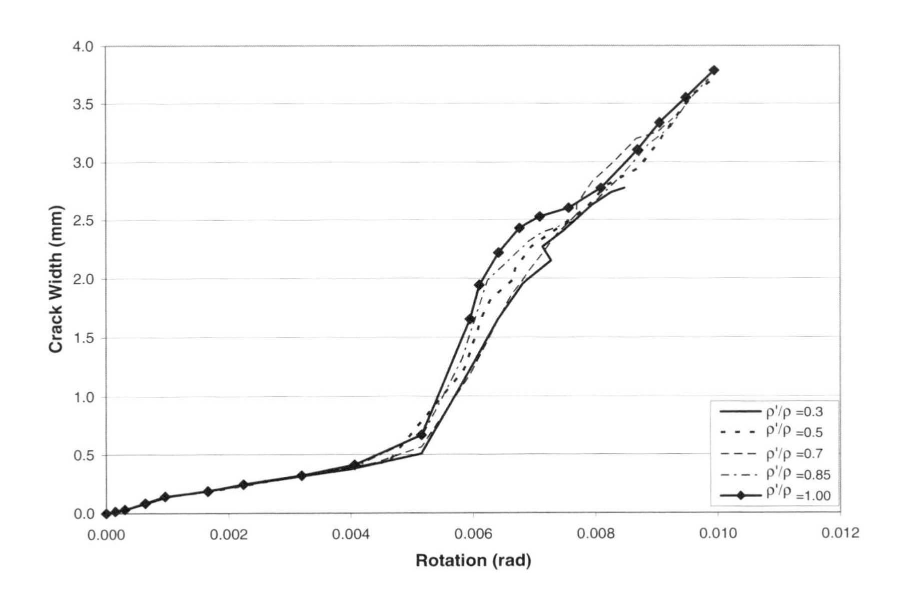

Figure 15. Effect on Compression Reinforcement on Crack Width - Rotation Curves

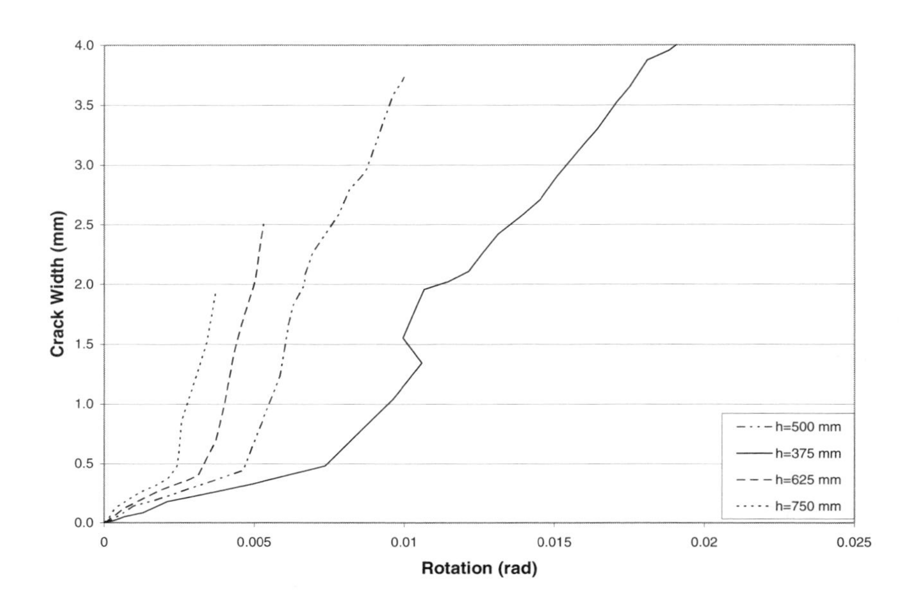

Figure 16. Effect of Amount of Depth of Beam on Crack Width - Rotation Curves

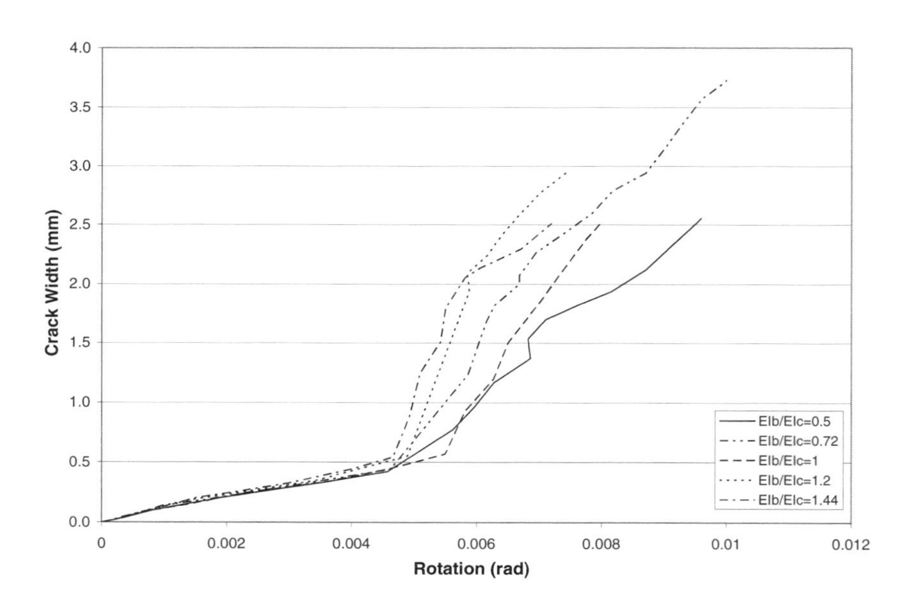

Figure 17. Effect of Ratio of Beam Stiffness to Column Stiffness on Crack Width - Rotation Curves

PEAK GROUND VELOCITY SENSITIVE DEFORMATION DEMANDS AND A RAPID DAMAGE ASSESSMENT APPROACH

Sinan Akkar and Haluk Sucuoglu
Structural Mechanics Division, Department of Civil Engineering, Middle East Technical University, 06531 Ankara, Turkey

Abstract: The effect of peak ground velocity (*PGV*) on the maximum inelastic deformation demand of simple, non-degrading structural systems is studied. Ground motion data sets are assembled for pre-defined ranges of *PGV* and they are used to conduct nonlinear response history analysis of single-degree-of-freedom (*SDOF*) systems. The study is focused on short and intermediate periods of vibration (*T*) and strength reduction factor (*R*) is used to define the lateral capacity of the structure. As part of the study, a simple tool for rapid damage assessment in pre-earthquake evaluation of existing building systems is proposed that combines the ground motion parameter *PGV* with structural properties *R* and *T*.

Key words: Performance-based seismic engineering; Strength reduction factors; Maximum inelastic deformations; Peak ground velocity; Damage assessment

1. INTRODUCTION

In essence, performance-based seismic engineering (PBSE) requires the accurate estimation of maximum inelastic deformation demands that play a key role in design and seismic performance assessment of structural systems.

S.T. Wasti and G. Ozcebe (eds.), Seismic Assessment and Rehabilitation of Existing Buildings, 77–96.

Recent building rehabilitation guidelines [1,2] and individual studies [3, 4, 5, 6, 7] have proposed various methods to estimate maximum inelastic deformation of simple structural systems that are subjected to severe earthquake ground motions. These approximate procedures are based on either the displacement ductility ratio or normalized lateral yield strength ratio (also defined as strength reduction factor) and they are derived as a result of statistical studies by using ground motion data sets that are classified by different criteria. In this respect, statistical studies conducted by Borzi et al. [3], Ruiz-García and Miranda [4] or Aydınoğlu and Kaçmaz [6] use the site class parameter as the primary criterion for categorizing the ground motion data sets. Meanwhile, studies of Nassar and Krawinkler [5] and Vidic et al. [7] make use of scaling factors to bring the chosen ground motion records to a common base. The scaling of ground motions was realized for pre-determined peak ground acceleration (*PGA*) or peak ground velocity (*PGV*) amplitude. In strict terms, scaling of ground motions is not a simple issue and consideration of only one single parameter would alter the most prominent features of the earthquake faulting process. Most attenuation relationships indicate a strong correlation between magnitude, peak ground motion amplitudes and strong motion duration. Thus, more than one ground motion parameter should be considered as a whole during the scaling of earthquake records.

Recent studies have shown clear evidence that the structural damage is well correlated with the *PGV* parameter [8]. Using a large data set from 8 significant California earthquakes, Wald et al. [8] showed that *PGV* is a superior ground motion amplitude parameter for indicating light to severe levels of structural damage (i.e. Modified Mercalli Intensity > VII). These findings are being used efficiently for a rapid evaluation of post-earthquake response and pre-earthquake planning in well-instrumented urban regions [9]. The data from the 1999 Turkey earthquakes also revealed that the horizontal ground motion component of maximum ground velocity has significantly large local and global elastic deformation demands on simple structural systems [10]. In case where the fault-rupture directivity effect is considered, the maximum ground velocity direction turns out to be as critical as the fault-normal component [10]. However, the validity of this finding was not extended to general nonlinear structural behavior. Another important property of *PGV* is its capability of reflecting the characteristics of local site conditions. The research conducted by JICA [11] showed that as the shear wave velocity (v_s) of a site becomes smaller (i.e. when the soil becomes softer), the peak ground velocity attains higher values. Thus, *PGV* can also be used as an indicator of soil type for a given site.

This paper emphasizes the importance of *PGV* on the maximum inelastic deformation demand of short to medium period, non-degrading, single-degree of freedom (SDOF) systems. A relatively large data set that consists of strong ground motions with *PGV* values ranging from 20 cm/s to 130 cm/s is used to conduct nonlinear response history analysis. The structural parameter to define the lateral capacity is chosen as the strength reduction factor, *R*, which implies a constant level of yielding strength for the vibration periods (*T*) of interest. This parameter is believed to be the most practical one to assess the seismic performance of existing structural systems. The resulting maximum inelastic deformations are evaluated in a statistical manner to correlate the deformation demands on structures with the ground motion parameter *PGV*. The mean maximum inelastic to elastic lateral deformation ratios of this study are compared with recently proposed methods to estimate the maximum inelastic deformation demands on simple structural systems. These results are used to derive a tool for a quick structural damage assessment procedure that combines a critical ground motion parameter (*PGV*) with the dynamic structural properties *T* and *R*.

2. STRONG GROUND MOTION DATA SET

A total number of 75 strong ground motion records were used to assemble four data groups with different *PGV* intervals. The first three group contains 20 strong ground motion records in each and the corresponding *PGV* values range from 20 cm/s to 40 cm/s, 40 cm/s to 60 cm/s, and 60 cm/s to 80 cm/s, respectively. These data groups are referred to as Group I, II and III, respectively. The fourth data group (Group IV) includes 15 earthquake ground motion records that have *PGV* values larger than 80 cm/s. The ground motions represent either free-field records or recordings from instruments located at shelters and one-story buildings. The surface wave magnitude (M_s) changes between 5.2 and 7.6 for the overall strong ground motion data set. Distance (*d*) criterion is taken as the closest distance to fault rupture. Almost all of the ground motion records are obtained from stations that are located less than 30 km from the fault rupture. Two exceptional records are from Group I (i.e. 20 cm/s $\leq PGV <$ 40 cm/s bin) that have $d =$ 43.6 km. When the closest distance information was not available, the hypocentral distance was used. The ground motion data set does not contain records with pulse-like signals. It is believed that this constraint reduces the bias in maximum deformation response as the existing dominant pulse period in such excitations alters the structural response significantly [10, 12]. This limitation is the primary reason for the reduced number of ground motion records in the last data group. The worldwide recorded strong ground motion data indicate a clear dominancy of pulse-like signals when the *PGV* value starts to exceed a certain level. Tables 1 to 4 list the most important features of the selected records in this study. The variation of *PGV* as a function of distance (*d*) and surface wave magnitude (M_s) for the overall ground motion data set is shown in the upper and lower frames of Figure 1, respectively.

The strong ground motions listed in Tables 1 to 4 are mostly recordings from dense to firm soil sites. The faulting mechanism of these records is dominantly strike-slip for Groups I and II and reverse faulting for Groups III and IV. Significantly large number of high *PGV* ground motions recorded during the 1994 Northridge and 1999 ChiChi earthquakes causes the dominance of reverse faulting in $PGV \geq 60$ cm/s data groups.

The variation of distance with respect to *PGV* for our ground motion data set displays that closer distances to fault rupture implies an increase in ground velocity. This fact is clearer for Group IV data set in which the large ground velocity records are accumulated within 0 to 10 km distance band. The relationship between the *PGV* and earthquake magnitude, M_s shows that as the earthquake magnitude attains higher values, the strong ground motion records tend to have larger *PGV* values.

3. PEAK GROUND VELOCITY EFFECT ON MAXIMUM INELASTIC DEFORMATION

Nonlinear response history analyses of SDOF systems were performed by using the strong ground data described in the preceding section. The inelastic behavior was simulated by the elastoplastic hysteretic model as it is accepted as one of the benchmark hysteretic models for non-degrading structural systems. At a given period of vibration, the maximum inelastic lateral deformation ($S_{d,i}$) of an SDOF system was computed for a lateral yield strength value that is normalized by the corresponding lateral elastic strength value. This normalized lateral strength parameter is known as strength reduction factor *R* and it is the unique parameter that clearly defines the lateral capacity of existing buildings. The maximum inelastic SDOF deformations computed in this way correspond to inelastic displacement spectra for constant strength.

The nonlinear response history analyses were conducted for periods of vibration ranging from 0.1s to 3.0s. While computing the nonlinear response history analyses, we did not go beyond the periods of vibration greater than 3.0s as the ground motions used in this study are processed by causal filters (http://peer.berkeley.edu/smcat/process.html). The causally filtered data are very sensitive to the filter cut-off frequencies due to significant phase distortions and this causes great variations in the maximum inelastic deformations computed at long periods [13]. A total number of 7 strength reduction factors that have values of R = 1.0, 1.5, 2.0, 3.0, 4.0, 5.0, 6.0 were used in these computations. Explicit from the definition of strength reduction factor, the response history computations for R = 1 represent the elastic case and the corresponding maximum SDOF deformations constitute the elastic spectral displacements, ($S_{d,e}$). The mean maximum lateral deformations and mean maximum inelastic to elastic deformation ratios ($E[S_{d,i}/S_{d,e}]$) were computed for the prescribed data sets at each period of vibration and for every R-value. Figure 2 compares the period (T) and strength (R) dependent variation of mean maximum SDOF deformations for Groups I and IV. The primary observation from these curves is the significant dependence of maximum deformation demand on the *PGV* parameter. The increase in *PGV* is associated with a considerable increase in the mean maximum deformation demands. The mean maximum deformations for Group IV data set are approximately in the order of 5 with respect to the mean maximum deformations computed from Group I ground motion data set. The mean maximum deformation curves of Group IV ground motion data set display very large short period mean inelastic deformation demands for reduced lateral strength values. In general, such short period mean inelastic deformation demands are not common in daily engineering practice. Thus, caution should be exercised while designing or evaluating the seismic performance of structures that are likely to be subjected to high ground velocity ground motions. The comparisons in Figure 2 also reveal that Group I mean maximum inelastic deformation demands are not very sensitive to the changes in lateral strength capacity. This is not the case for mean maximum inelastic deformations of the Group IV ground motion data set, which shows higher sensitivity to the changes in the lateral strength reduction factor. Figure 3 illustrates this observation clearly where the maximum deformation axis of each data set is scaled with respect to the maximum of that group. Group I mean maximum deformation values show a variation within a narrower band while Group IV mean maximum deformations show a larger variation with changes in R-values.

Figure 4 shows the variation of mean inelastic to elastic maximum deformation ratios for Groups I and III. This ratio relates the mean maximum inelastic deformations of SDOF systems to mean maximum deformations of elastic response computed for a pre-defined lateral strength reduction factor R and it is designated by the C_1 factor in the nonlinear static procedure described in [2]. Group I mean inelastic to elastic deformation ratios start to take values equal or less than 1 for periods of vibration approximately longer than 1.5s. The mean inelastic to elastic deformation ratio equal or less than 1 implies the equal displacement rule indicating that the deformation demand of inelastic systems would not be larger than the corresponding deformation demand on elastic systems. This finding is consistent with the results of

Ruiz-García and Miranda [4] and Aydınoğlu and Kaçmaz [6] for strong ground motions recorded on sites similar to the ones considered in this study (i.e. dense to stiff soil sites). However, the mean maximum inelastic to elastic deformation ratios of Group III ground motions show that the mean deformation demand on inelastic systems are larger than the corresponding deformation demands on elastic systems for periods of vibration less than 2.4s. Thus, as the ground velocity takes higher values, the deformation demands on inelastic systems is larger than the elastic systems for very long periods of vibration. This observation does not exist in references [4] and [6] as their ground motion data sets do not consider the variations in *PGV*. Although the site conditions of the ground motions fairly match in all studies, the inelastic deformation amplifications of references [4] and [6] remain lower than the amplifications in this study for high *PGV* data groups. This fact indicates once more how *PGV* could become critical for maximum inelastic deformations on structural systems.

4. A PRACTICAL PEAK GROUND VELOCITY DEPENDENT DAMAGE ASSESSMENT TOOL

The seismic performance of a structure subjected to a severe ground motion can be measured by the observed structural damage. Thus, a damage index can be referred to as a quantitative tool to evaluate the structural performance of existing buildings. A damage index should attain its base value when the structure remains elastic and it should take its maximum value when the structure is in its collapse state.

A damage index is not necessarily used to quantify the damage for post-earthquake seismic performance assessment. Using proper structural parameters, structural damage indices can be established for assessing the damage vulnerability of existing structures for pre-earthquake planning. For example a high index quantity may indicate a more critical structure for a future severe ground motion.

The maximum post-yielding deformation (plastic deformation, Δ_p) experienced by a structure during a severe earthquake ground motion can be accepted as one of the major contributors to structural damage. It can be accepted as a suitable parameter to be used in quantifying the damage, as it is zero when the structure behaves within its elastic limits and it takes larger values as the structure deforms beyond its yielding level. The mean plastic deformations of the ground motion data groups were computed in a similar manner to the mean maximum inelastic deformations as described in the preceding section. Figure 5 presents the variation in mean Δ_p values with respect to period of vibration and strength reduction factor for ground motion data groups I and IV. The plots in Figure 5 indicate the sensitivity of post-yielding deformations to ground velocity. Similarly to the findings of preceding section, Group IV ground motion data that contain the highest PGV values yield significantly larger mean plastic deformations with respect to the corresponding values computed from Group I data set. The curves in Figure 5 also show the changes in mean plastic deformation with respect to the strength reduction factor, *R*. Comparisons of Figure 5 with Figure 2 indicate that mean plastic deformations are more sensitive to the variations in *R*. As the ground motion records get higher *PGV* values (in this case Group IV ground motion data set) the plastic deformation values possess a stronger sensitivity to the reduced lateral strength capacity and they start to follow well-separated paths for different *R*-values.

Figure 6 shows a close up view of mean plastic deformation variation in Group II ground motions for periods of vibration between 0.1s and 1.0s. The mean plastic deformation values follow almost a well-defined, linear trend with respect to the strength reduction factors considered in this study. In order to visualize this linear trend, Figure 6 also shows the first order polynomial fits computed for each *R*-value. Similar to the fits presented in Figure 6, the mean plastic deformation curves of other ground motion groups were represented by linear straight lines for periods of vibration between 0.1s and 1.0s and these fits yielded very high correlation coefficients with respect to the actual data trend. It should be noted that the period interval from 0.1s to 1.0s contains a significantly large percentage of building stock located in earthquake prone zones.

The strong correlation between the PGV parameter and the deformation demands on structural systems together with the linear trend in mean plastic deformations can be combined to derive a simplified approach for structural damage amplification. If high deformation demands are accepted as a direct measure of structural damage, the findings presented here would reveal that an existing structure which has average construction quality within its category will have the least structural damage probability when it is subjected to an excitation of low ground velocity. Thus, taking Group 1 mean plastic deformations as base, one can compute the mean structural damage amplifications (*DA*) in other ground motion groups.

Figure 7 shows such computations for Groups III and IV by using the linear curves fitted on the exact mean plastic deformation data for periods of vibration between 0.1s and 1.0s.These plots show the strength and period dependency of structural damage. According to the general trend of these curves, one can conclude that the higher *PGV* is the stronger is structural damage dependency on *R* and *T*. Besides, the damage amplification of ground motions with smaller *PGV* values (in this case Group III) seems to stabilize rapidly towards longer periods of vibration and attain a constant amplification value regardless of the changes in strength reduction factors. These curves also show that the damage amplification has a rigorous path for periods of vibration less than 0.3s where the increase in structural damage does not follow a decrease in lateral strength capacity. As the period of vibration takes larger values, the damage amplifications follow a more predictable trend (i.e., the damage amplification increases as the structural system lateral strength capacity decreases).

A simple case study is presented to describe the application of computed damage amplification factors in a rapid damage assessment survey for pre-earthquake planning of existing buildings. The case study assumptions listed in Table 5 are believed to be valid for typical frame type building stock in Turkey. The modifications in these assumptions are arbitrary and would not change the general results of this study. The strength reduction factors were defined by considering the variation in the structure story numbers. In common design practice, stiff buildings (short period structures) are designed for high base shear coefficients. The fundamental period estimations were based on the effective period concept defined in [2]. This reference period approximately corresponds to the secant stiffness at 60% of the yielding strength of the structure and it is recommended for seismic performance assessment procedures based on structural deformation. Using a properly defined quantitative damage index for frame type buildings located in a moderate *PGV* seismic zone (in this case 20 cm/s $\leq PGV <$ 40 cm/s), one can prolong the quantitative damage assessment of similar structures of high *PGV* seismic zones by using the amplifications listed in Table 5. These amplification factors are computed from similar curves of Figure 7 by using the assumptions of our case study buildings. Recent seismological studies for well-instrumented seismic zones or deterministic/probabilistic earthquake scenarios have already started to give *PGV* based seismic intensity maps making the proposed damage assessment procedure applicable for engineering based, large scale projects [9, 11].

Figure 8 shows the damage amplification factor plots of our case study buildings. The damage amplification for each bin is plotted at the mid value of corresponding class interval. In other words, for example, the damage amplifications for 40cm/s $\leq PGV <$ 60 cm/s are plotted at 50 cm/s on the *PGV* axis. The curves present the sensitivity of structural damage on story number, *PGV* and lateral strength capacity.

5. SUMMARY AND CONCLUSIONS

The effect of peak ground velocity on maximum inelastic deformations of non-degrading simple structural systems is investigated. The non-degrading behavior is represented by an elastoplastic hysteretic model. Mean maximum inelastic lateral deformations are computed for oscillator periods of 0.1s to 3.0s. The strength reduction factor, *R,* is taken as the system capacity parameter while computing the maximum SDOF deformations. A total number of 7 strength reduction factors, R =1.0, 1.5, 2.0, 3.0, 4.0, 5.0, 6.0 are used in the computations. The ground motion data sets are assembled for different *PGV* groups that contain earthquake ground motion records with PGV intervals of 20 cm/s $\leq PGV <$ 40 cm/s, 40cm/s $\leq PGV <$ 60 cm/s, 60 cm/s $\leq PGV <$ 80 cm/s and $PGV \geq$ 80 cm/s, respectively. In general, the ground motions are recorded on firm to dense soil sites and their distances from fault rupture vary from 0.25 km to 30 km. Owing to their complex nonlinear structural behavior, pulse-like signals are discarded in this study. The ground motion data sets are from earthquakes that have strike-slip or reverse faulting mechanisms and the body wave magnitudes of these earthquakes are between $5.2 \leq M_s \leq 7.6$. It is believed that the data set is a fairly complete ground motion assembly to observe the effects of *PGV* on inelastic deformation demand.

The results of this study revealed that high *PGV* values increase the maximum inelastic deformation demand significantly. Our high *PGV* ground motion data groups yielded considerably larger mean maximum inelastic to elastic deformation ratios than the other studies that used ground motions of similar site classes but lacked a clear classification with respect to *PGV*. This indicates that *PGV* should be a serious concern in estimating the maximum structural deformations.

Mean plastic deformations that can be defined as the post yielding maximum deformations of the same data sets are also computed. Similar to the maximum inelastic deformation, this parameter also shows an increasing trend with respect to increased *PGV* values. Mean plastic deformations are observed to be more sensitive to the changes in strength reduction factors and they follow a linear trend for periods of vibration less than 1.0s for the ground motion data sets of this study. The mean plastic deformations of ground motion data sets with *PGV* values higher than 40 cm/s are normalized with the corresponding values of the 20 cm/s $\leq PGV <$ 40 cm/s data set. This way a damage amplification factor is presented, which assumes that the structures located on low *PGV* seismic zones will be subjected to the least structural damage. The proposed structural damage amplification factor is believed to be versatile as it combines an important ground motion parameter (*PGV*) and key structural properties (*T* and *R*) that are used for assessing the seismic performance of existing structures. This new damage assessment tool is believed to have a wide range of application as the new seismic intensity maps have already started to incorporate *PGV* as one of the main ground motion parameters.

ACKNOWLEDGEMENTS

This study is partially funded by the projects NATO SfP977231 and TUBITAK YMAU-ICTAG1574. The authors express their gratitude for this financial support.

REFERENCES

1. Applied Technology Council (ATC). (1996). "Seismic evaluation and retrofit of concrete buildings." *Report ATC-40*, Applied Technology Council, Redwood City, California.
2. Building Seismic Safety Council (BSSC). (2000). "Prestandard and commentary for the seismic rehabilitation of buildings." *Report FEMA-356*, Washington, D.C.
3. Borzi, B., ElNashai, A.S., Faccioli, E., Calvi, G.M. and Bommer, J.J. (1998). "Inelastic spectra and ductility-damping relationships for displacement-based seismic design," *Report ESEE No. 98-4*, Engineering Seismology and Earthquake Engineering, Civil Engineering Department, Imperial College, London, UK.
4. Ruiz-García, J., and Miranda, E. (2002) "Inelastic displacement ratios for evaluation of existing structures." accepted for publication in *Earthquake Engineering and Structural Dynamics.*
5. Nassar, A.A. and Krawinkler H. (1992). "Seismic demands for sdof and mdof systems," *Report No. 95*, J.A. Blume Earthquake Engineering Center, Stanford University, Stanford, California.
6. Aydınoğlu, N. and Kaçmaz, Ü. (2002). "Strength-based displacement amplification spectra for inelastic seismic performance evaluation," *Report No. 2002-2*, Department of Earthquake Engineering, Kandilli Observatory and earthquake Research Institute, Boğaziçi University, İstanbul, Turkey.
7. Vidic T., Fajfar, P. and Fischinger, M. (1994). "Consistent inelastic design spectra: strength and displacement," *Earthquake Engineering and Structural Dynamics*, 23, 507-521.
8. Wald, D.J., Quitariano, V., Heaton, T.H., Kanamori, H., Scrivner C.W., and Worden, C.B. (1999). "Trinet ShakeMaps: rapid generation of peak ground motion and intensity maps for earthquakes in Southern California," *Earthquake Spectra*, 15, 537-555.
9. Wald, D.J., Worden, B., and Quitariano, V. (2002). "Shakemap: its role in pre-earthquake planning and post-earthquake response and information," *Proceedings SMIP02 Seminar on Utilization of Strong-Motion Data*, 1-20, Los Angeles, California.
10. Akkar, S. and Gülkan P. (2001). "Near-field earthquakes and their implications on seismic design codes," *Report No. METU/EERC 01-01*, Earthquake Engineering Research Center, Middle East Technical University, Ankara, Turkey.
11. Japan International Cooperation Agency (JICA). (2002). "The study on a disaster prevention/mitigation basic plan in Istanbul including seismic microzonation in the Republic of Turkey," *Final Report*, Istanbul, Turkey.
12. Babak, A. and Krawinkler, H., (2001). "Effects of near-fault ground motions on frame structures," *Report No. 138*, J.A. Blume Earthquake Engineering Center, Stanford University, Stanford, California.
13. Boore, D.M. and Akkar, S. (2003). "Effect of causal and acausal filters on elastic and inelastic response spectra," accepted for publication in Earthquake Engineering and Structural Dynamics.

Table 1. Ground motion records for 20 cm/s ≤ PGV < 40 cm/s

GROUND MOTION	M_s	Site [1]	F [2]	d (km)	PGA (cm/s^2)	PGV (cm/s)
PARKFIELD 06/28/66, CHOLAME #5, 085 (CDMG STATION 1014)	6.1	S	SS	5.30	433.21	24.73
LOMA PRIETA 10/18/89, GILROY ARRAY #2, 000 (CDMG STATION 47380)	7.1	S	RO	12.70	360.28	32.92
LOMA PRIETA 10/18/89, GILROY ARRAY #2, 090 (CDMG STATION 47380)	7.1	S	RO	12.70	316.20	39.09
LOMA PRIETA 10/18/89, GILROY ARRAY #3, 000 (CDMG STATION 47381)	7.1	S	RO	14.40	544.48	35.71
COYOTE LAKE 08/06/79, GILROY ARRAY #2, 140 (CDMG STATION 47380)	5.7	S	SS	6.00	332.80	24.89
COYOTE LAKE 08/06/79, GILROY ARRAY #3, 140 (CDMG STATION 47381)	5.7	S	SS	6.00	224.22	28.77
IMPERIAL VALLEY 10/15/79, AERO. MEXICALI, 315 (UNAM/UCSD STATION 6616)	6.9	S	SS	8.50	254.73	24.87
IMPERIAL VALLEY 10/15/79, CHIHUAHUA, 012 (UNAM/UCSD STATION 6621)	6.9	S	SS	28.70	265.30	24.86
IMPERIAL VALLEY 10/15/79, CHIHUAHUA, 282 (UNAM/UCSD STATION 6621)	6.9	S	SS	28.70	249.26	30.14
IMPERIAL VALLEY 10/15/79, EL CENTRO ARRAY 6, 230 (CDMG STATION 942)	5.2	S	SS	13.10 [3]	359.23	20.85
IMPERIAL VALLEY 10/15/79, DELTA, 262 (UNAM/UCSD STATION 6605)	6.9	S	SS	43.60	233.24	26.01
IMPERIAL VALLEY 10/15/79, DELTA, 352 (UNAM/UCSD STATION 6605)	6.9	S	SS	43.60	344.45	33.01
IMPERIAL VALLEY 10/15/79, EL CENTRO ARRAY #2, 140 (USGS STATION 5115)	6.9	S	SS	10.40	309.16	31.49
IMPERIAL VALLEY 10/15/79, EL CENTRO ARRAY #4, 140 (USGS STATION 955)	6.9	S	SS	4.20	476.04	37.42
IMPERIAL VALLEY 10/15/79, EL CENTRO ARRAY #12, 230 (USGS STATION 931)	6.9	S	SS	18.20	114.06	21.81
IMPERIAL VALLEY 10/15/79, EC CO CENTER FF, 002 (CDMG STATION 5154)	6.9	S	SS	7.60	208.64	37.52
NORTHRIDGE 01/17/94, LA - HOLLYWOOD STRG FF, 360 (CDMG STATION 24303)	6.7	S	RN	25.50	351.53	27.51
IMPERIAL VALLEY 10/15/79, CUCAPAH, 085 (UNAM/UCSD STATION 6617)	6.9	S	SS	23.60	302.68	36.31
IMPERIAL VALLEY 10/15/79, SAHOP CASA FLRS, 270 (UNAM/UCSD STATION 6619)	6.9	S	SS	11.10	496.39	31.02
MORGAN HILL 04/24/84, HALLS VALLEY, 240 (CDMG STATION 57191)	6.1	S	SS	3.40	305.73	39.39

[1] S designates soil sites that have shear wave velocities between 180 m/s and 750 m/s

[2] SS, RO, RN designate strike slip, reverse oblique and reverse normal faulting, respectively

[3] Hypocentral distance

Table 2. Ground motion records for 40 cm/s ≤ PGV < 60 cm/s

GROUND MOTION	M	Site [1]	F[2]	d (km)	PGA (cm/s^2)	PGV (cm/s)
SUP. HILLS 11/24/87, EL CENTRO IMP CO CENTER, 090 (CDMG STATION 01335)	6.6	S	SS	13.90	253.42	40.88
SUP. HILLS 11/24/87, EL CENTRO IMP CO CENTER, 000 (CDMG STATION 01335)	6.6	S	SS	13.90	351.06	46.36
CHI-CHI 09/20/99, CHY006, E (CWB)	7.6	S	RN	14.93	357.50	55.41
LOMA PRIETA 10/18/89, CORRALITOS, 000 (CDMG STATION 57007)	7.1	S	RO	5.10	631.51	55.18
SAN FERNANDO 02/09/71, PACOIMA DAM, 254 (CDMG STATION 279)	6.6	S	RN	2.80	1137.4 7	54.30
NORTHRIDGE 01/17/94, PACOIMA KAGEL CANYON, 360 (CDMG STATION 24088)	6.7	S	RN	8.20	424.47	51.49
CHI-CHI 09/20/99, TCU049, W (CWB)	7.6	S	RN	4.48	287.43	47.90
IMPERIAL VALLEY 10/15/79, AERO. MEXICALI, 045 (UNAM/UCSD STATION 6616)	6.9	S	SS	8.50	320.53	42.83
IMPERIAL VALLEY 10/15/79, BONDS CORNER, 140 (USGS STATION 5054)	6.9	S	SS	2.50	577.20	45.22
IMPERIAL VALLEY 10/15/79, BONDS CORNER, 230 (USGS STATION 5054)	6.9	S	SS	2.50	760.05	45.93
IMPERIAL VALLEY 10/15/79, EL CENTRO ARRAY #5, 140 (USGS STATION 952)	6.9	S	SS	1.00	509.27	46.89
IMPERIAL VALLEY 10/15/79, EL CENTRO ARRAY #7, 140 (USGS STATION 5028)	6.9	S	SS	0.60	331.10	47.62
IMPERIAL VALLEY 10/15/79, EL CENTRO ARRAY #8, 140 (CDMG STATION 958)	6.9	S	SS	3.80	590.31	54.26
CHI-CHI 09/20/99, TCU051, W (CWB)	7.6	S	RN	8.27	182.71	49.28
CHI-CHI 09/20/99, TCU082, W (CWB)	7.6	S	RN	5.73	219.14	58.43
IMPERIAL VALLEY 10/15/79, EL CENTRO ARRAY #11, 230 (USGS STATION 5058)	6.9	S	SS	12.60	372.37	42.15
IMPERIAL VALLEY 10/15/79, EL CENTRO DIFF ARRAY, 230 (USGS STATION 5165)	6.9	S	SS	5.30	470.67	40.82
CAPE MENDOCINO 04/25/92, PETROLIA, 000 (CDMG STATION 89156)	7.1	S	RN	9.50	578.43	48.42
WESTMORELAND 04/26/81, FIRE STATION, 090 (CDMG STATION 5169)	5.8	S	SS	13.3 [3]	361.17	48.67
LANDERS 06/28/92, YERMO FIRE STATION, 270 (CDMG STATION 22074)	7.4	S	SS	24.90	240.15	51.49

[1] S designates soil sites that have shear wave velocities between 180 m/s and 750 m/s
[2] SS, RO, RN designate strike slip, reverse oblique and reverse normal faulting, respectively
[3] Hypocentral distance

Table 3. Ground motion records for 60 cm/s □ PGV < 80 cm/s

GROUND MOTION	M_s	Site[1]	F[2]	D (km)	PGA (cm/s^2)	PGV (cm/s)
CHI-CHI 09/20/99, CHY028, W (CWB)	7.6	S	RN	7.30	640.59	72.80
GAZLI 5/17/76, KARAKYR, 000	7.3	S	RN	3.00[4]	596.45	65.40
GAZLI 5/17/76, KARAKYR, 090	7.3	S	RN	3.00[4]	704.36	71.60
NORTHRIDGE 01/17/94, NEWHALL, 090 (CDMG STATION 24279)	6.7	S	RN	7.10	571.92	75.50
NORTHRIDGE, 1/17/94, SYLMAR-CONVERTER STA-EAST, 288 (DWP STATION 75)	6.7	S	RN	6.10	483.67	74.58
NORTHRIDGE EQ 1/17/94, CANOGA PARK - TOPANGA CANYON, 196 (USC STATION 90053)	6.7	S	RN	15.80	412.30	60.79
COALINGA 05/02/83, PLEASANT VALLEY P.P. - YARD, 045 (USGS STATION 1162)	6.5	S	RO	8.50	580.75	60.20
NORTHRIDGE EQ 1/17/94, NORTHRIDGE - SATICOY, 180 (USC STATION 90003)	6.7	S	RN	13.30	467.84	61.48
NORTHRIDGE, 1/17/94, SEPULVEDA VA, 360 (USGS STATION 0637)	6.7	S	RN	8.90	921.16	76.60
NORTHRIDGE 01/17/94, SYLMAR - HOSPITAL, 090 (CDMG STATION 24514)	6.7	S	RN	6.40	592.52	78.20
CHI-CHI 09/20/99, TCU074, W (CWB)	7.6	S	RN	13.70	585.89	73.35
NORTHRIDGE 01/17/94, TARZANA - CEDAR HILL NURSERY A, 360 (CDMG STATION 24436)	6.7	S	RN	17.50	971.19	77.60
KOBE 01/16/95, TAKARAZU, 000 (CUE)	6.9	S[3]	SS	1.20	679.83	68.30
CHI-CHI 09/20/99, TCU076, N (CWB)	7.6	S	RN	2.00	407.93	64.16
CHI-CHI 09/20/99, TCU049, N (CWB)	7.6	S	RN	4.50	246.23	61.20
CHI-CHI 09/20/99, TCU070, N (CWB)	7.6	S	RN	19.10	165.79	62.30
CHI-CHI 09/20/99, WNT, E (CWB)	7.6	S	RN	1.18	939.80	68.80
LOMA PRIETA 10/18/89, SARATOGA W VALLEY COLL, 270 (CDMG STATION 58235)	7.1	S	RO	13.70	325.69	61.50
IMPERIAL VALLEY 10/15/79, EC MELOLAND OVERP FF, 000 (CDMG STATION 5155)	6.9	S	SS	0.50	308.03	71.80
DUZCE 11/12/1999, DUZCE METEOROLOGY STATION	7.3	S	SS	8.20	341.39	60.00

[1] S designates soil sites that have shear wave velocities between 180 m/s and 750 m/s

[2] SS, RO, RN designate strike slip, reverse oblique and reverse normal faulting, respectively

[3] Recorded on soft soil that has average shear wave velocity less than 180 m/s

[4] Hypocentral distance

Table 4. The case study assumptions and corresponding damage amplifications

Assumptions			Damage Amplification (DA)			
Story Number	Period [s]	R	20<PGV<40	40<PGV<60	60<PGV<80	PGV>80
2	0.30	1.5	1	1.37	2.00	2.80
3	0.45	2.0	1	1.47	2.17	3.28
4	0.60	3.0	1	1.36	1.89	3.18
5	0.75	4.0	1	1.35	1.77	3.46
6	0.95	5.0	1	1.35	1.70	3.49

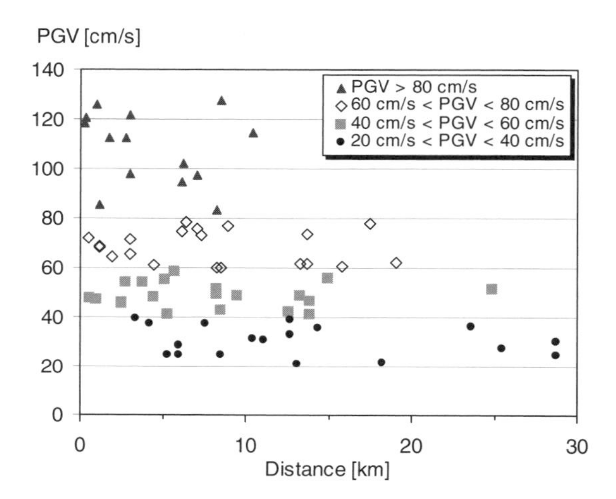

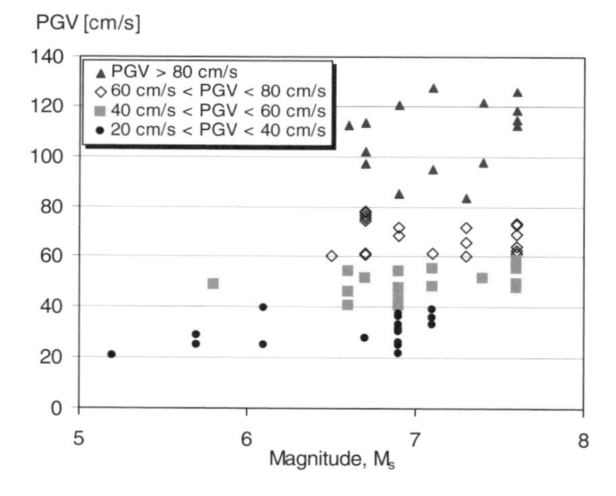

Figure 1. Variation of PGV with respect to distance and magnitude. It should be noted that the two records of Group I that have distances larger than 30 km are not shown in the upper frame

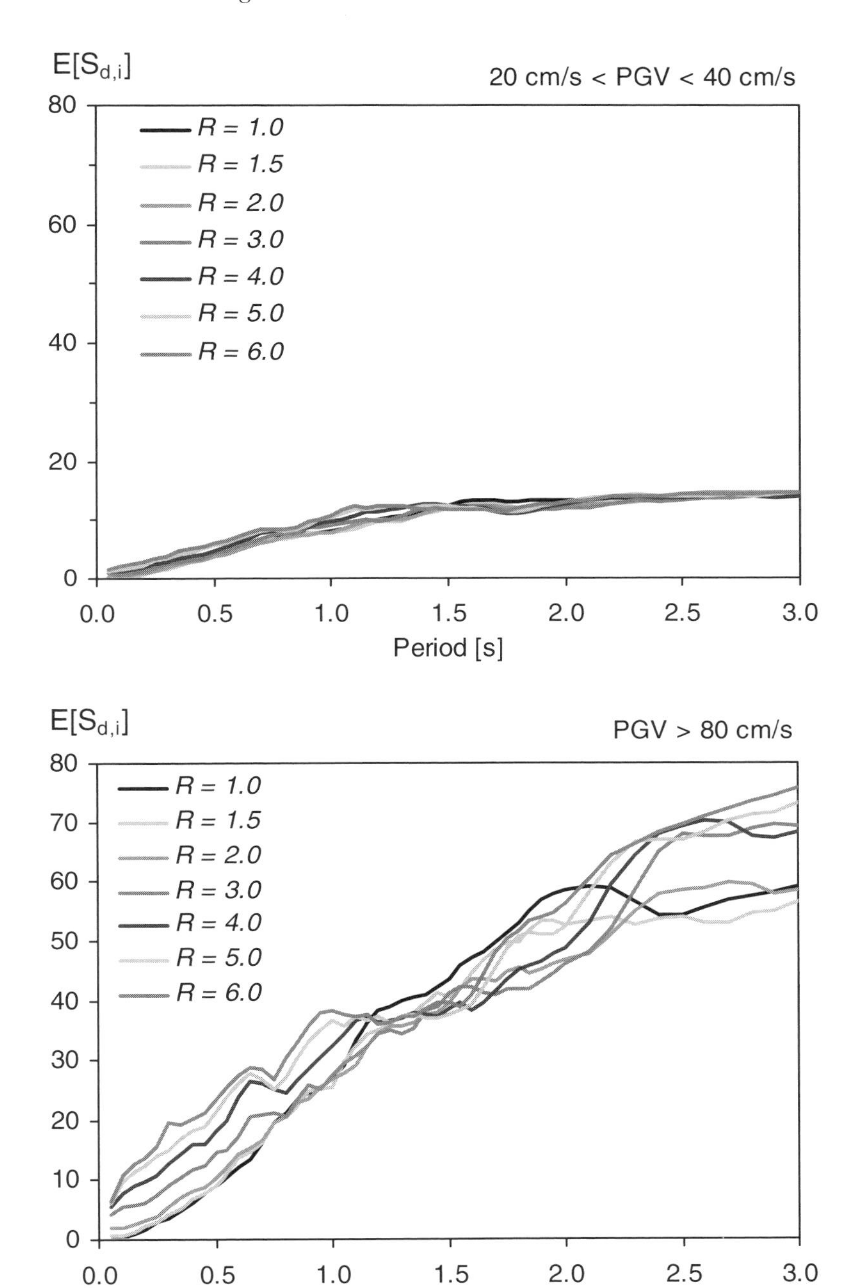

Figure 2. Comparisons of mean maximum deformation demands for Group I and Group IV ground motion records

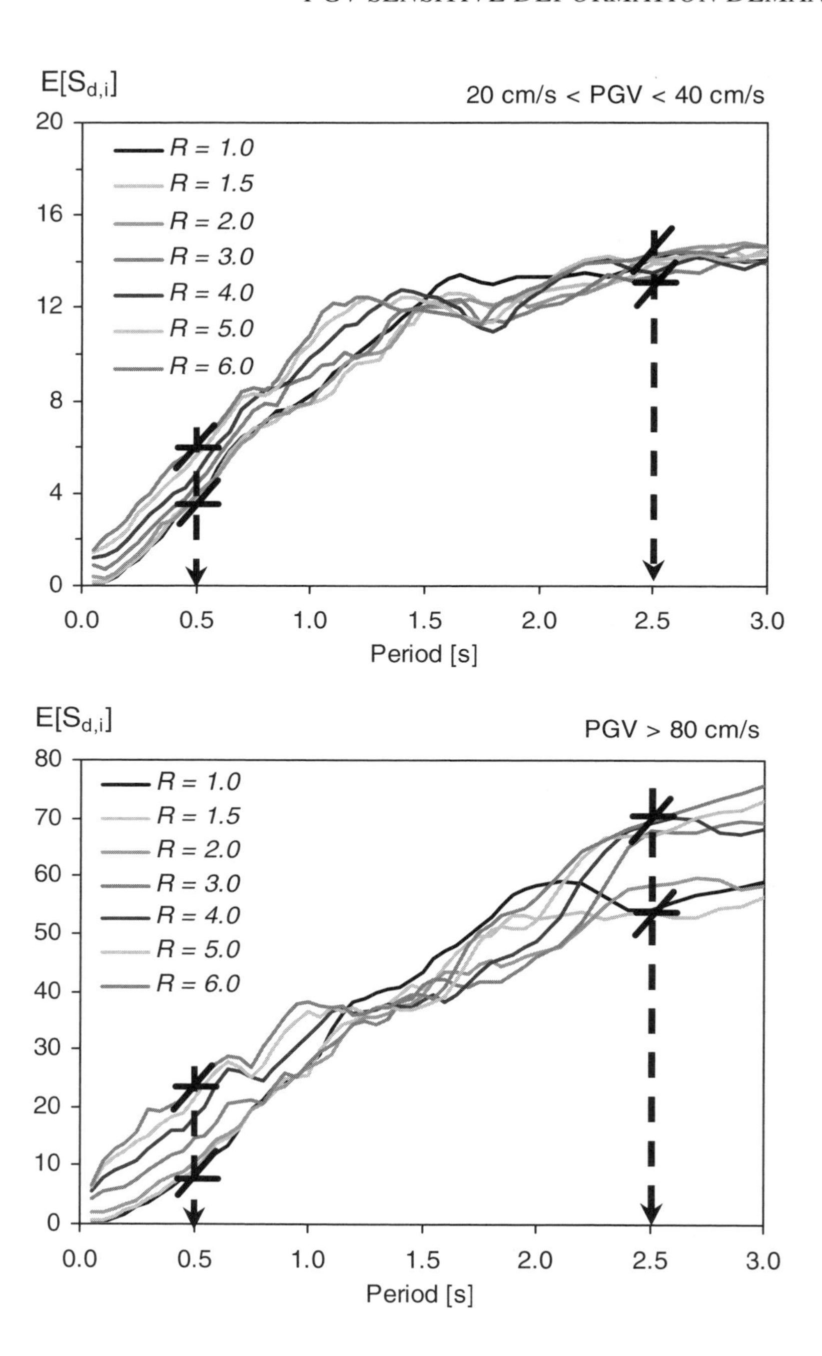

Figure 3. Variation of mean maximum deformations of low- and high-ground velocity data sets with respect to lateral strength reduction. The arrows on the graphs indicate the bandwidth of variation in maximum inelastic deformations at different periods of vibration

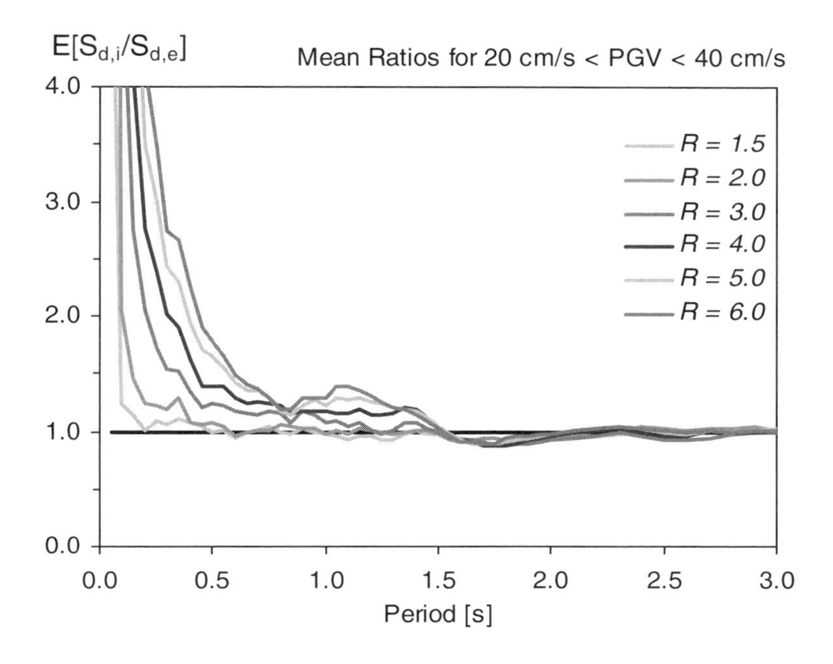

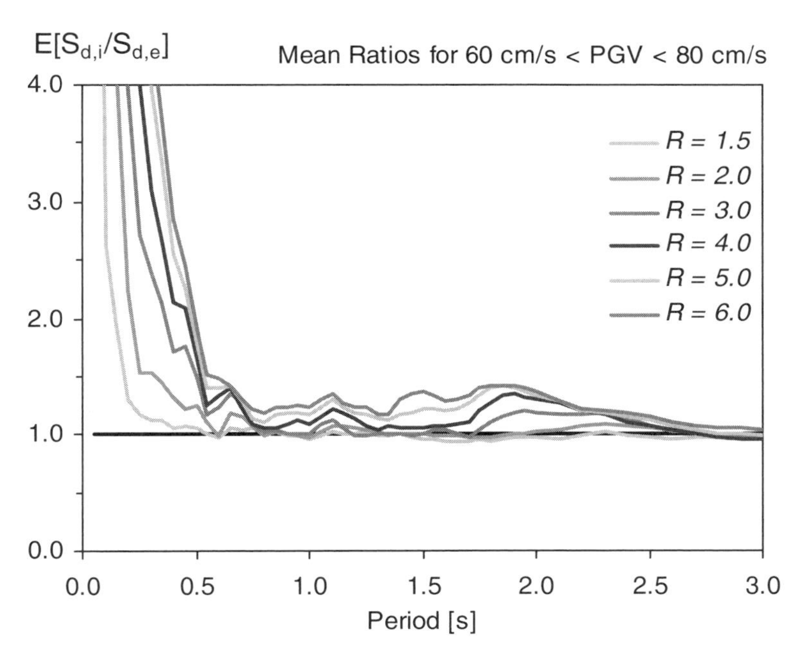

Figure 4. Variation of mean maximum inelastic to elastic deformation ratios for Groups I and III

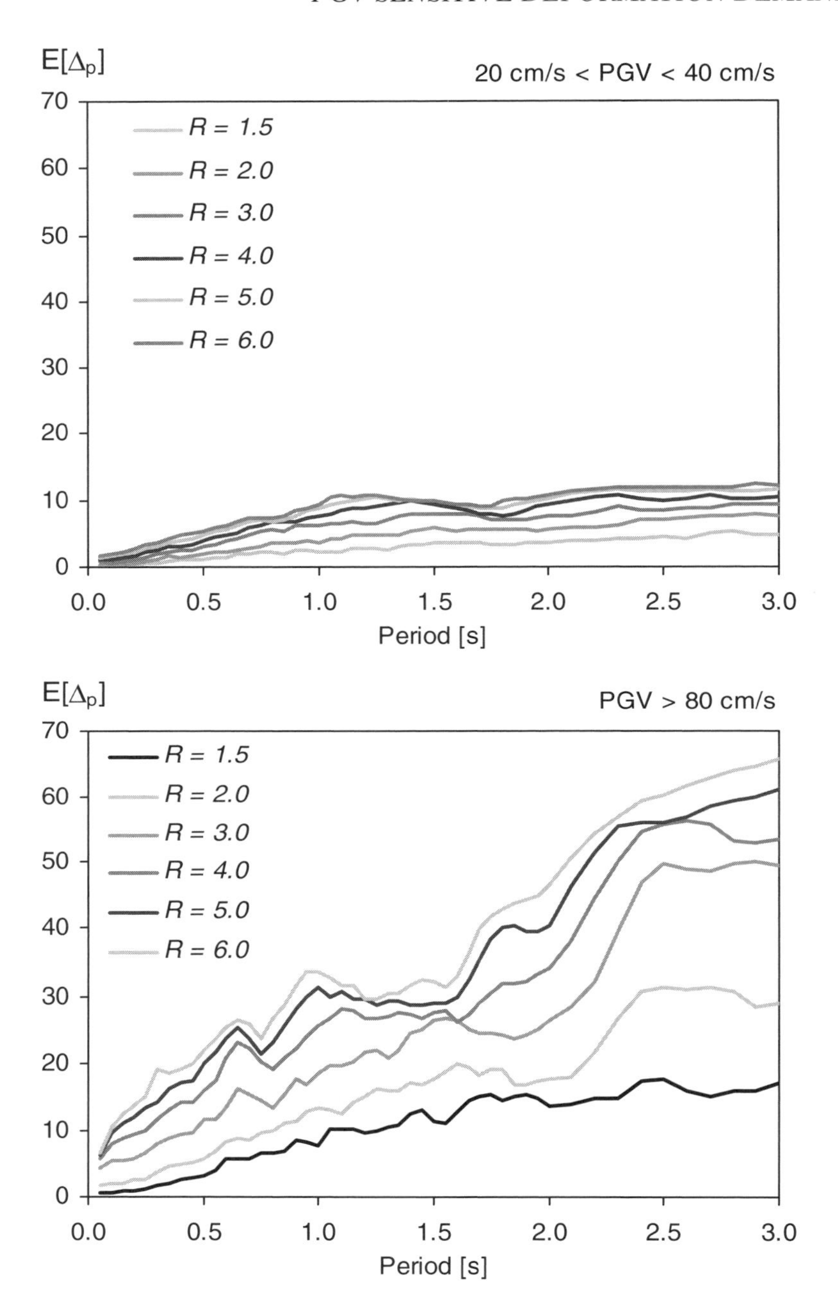

Figure 5. Mean plastic deformation variations of Groups I and IV with respect to T and R

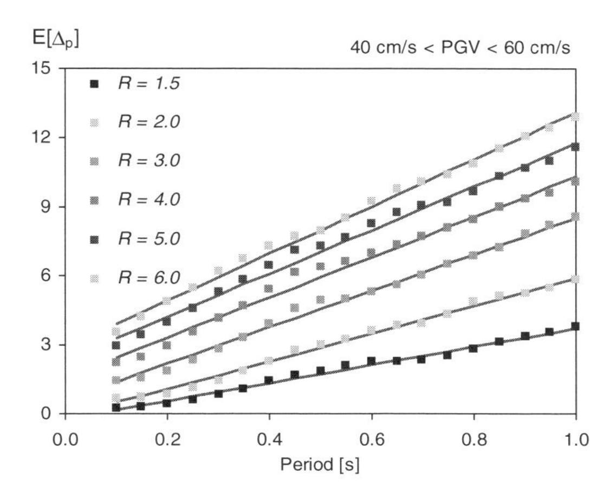

Figure 6. A close up view for the variation in mean plastic deformation of Group II ground motion data set for different R-values. The plots also show the first order polynomial fits computed for periods of vibration less than 1.0s

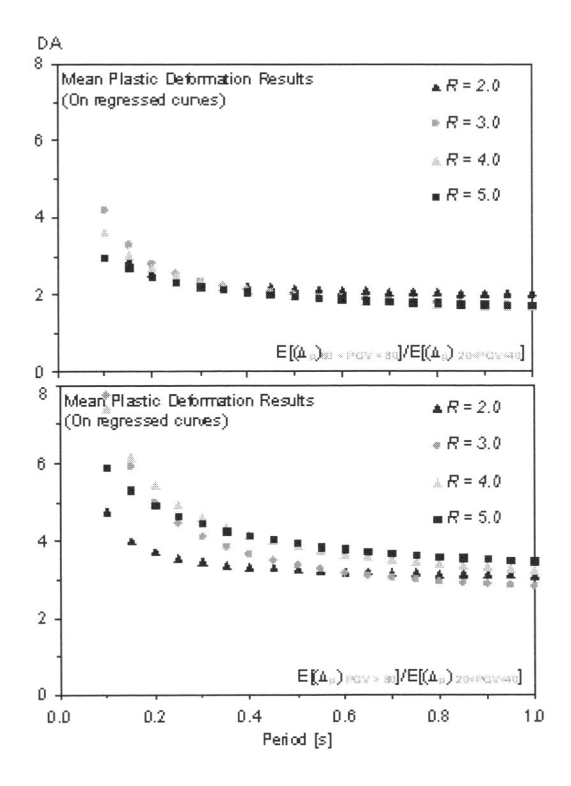

Figure 7. Mean damage amplifications factors of Group III and IV ground motion data sets for various R-values

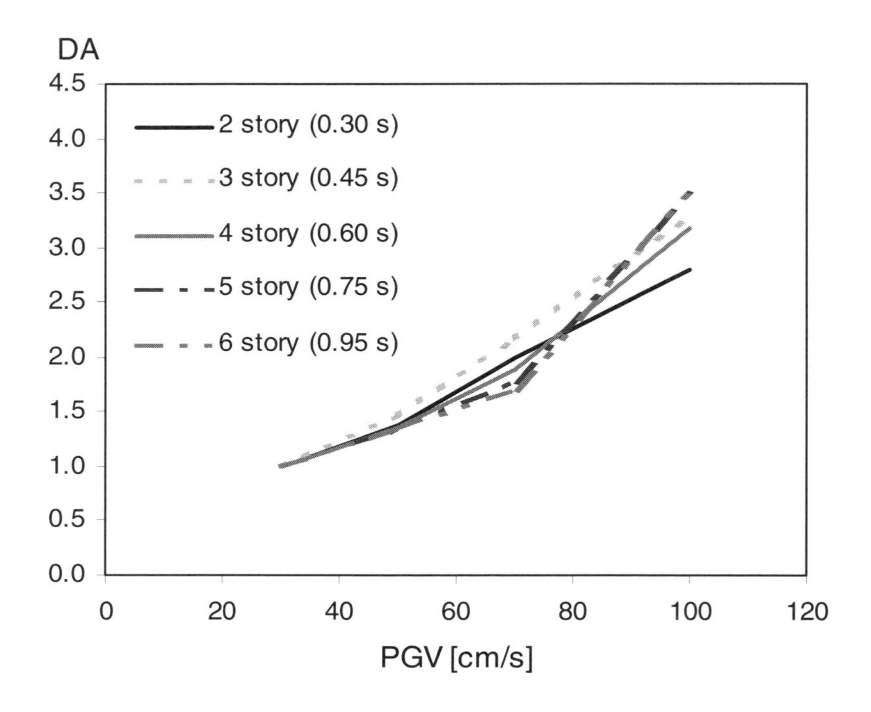

Figure 8. Variation of damage amplification with respect to PGV for the case study frame buildings

SIMPLE SURVEY PROCEDURES FOR SEISMIC RISK ASSESSMENT IN URBAN BUILDING STOCKS

Haluk Sucuoglu and Ufuk Yazgan
Structural Mechanics Division, Department of Civil Engineering
Middle East Technical University, 06531 Ankara, Turkey

Abstract: Cities under significant seismic risk contain a large number of vulnerable buildings. An effective risk assessment measure is to identify the most vulnerable buildings, which may undergo severe damage in a future earthquake. A two-level risk assessment procedure is proposed here. The first level is based on recording building parameters from the street side. In the second level, these are extended by structural parameters measured by entering the ground story. Statistical correlations have been obtained by employing a database of 477 damaged buildings surveyed after the 1999 Düzce earthquake. The results revealed that the parameters observed from the street and measured at the ground story provide strong guidance for identifying those buildings that jeopardize the life safety of their occupants.

Key words: Seismic risk, assessment, buildings, street survey, damage, score

1. INTRODUCTION

The classical engineering approach for providing seismic safety in building structures is to ensure their conformance to the current seismic design codes. This is a valid approach for new buildings. However a

S.T. Wasti and G. Ozcebe (eds.), Seismic Assessment and Rehabilitation of Existing Buildings, 97–118.

majority of the existing buildings in seismic environments do not satisfy modern code requirements. Yet, the ratio of severely damaged or collapsed buildings observed after a severe earthquake is much less than the ratio of substandard buildings. The difference is significant. An optimistic estimation of substandard buildings in Turkey is not less than 90 %, which can be generalized to Istanbul, or other earthquake prone regions in Turkey. On the other hand, the ratio of collapsed or heavily damaged buildings in Düzce after the two consecutive damageable earthquakes in 1999 was 20 % (Sucuoğlu and Yılmaz, 2001). Similar ratios were observed in Gölcük and Adapazarı. A recent loss estimation study for Istanbul (JICA, 2002) revealed that the expected ratio of collapsed buildings under a scenario earthquake of magnitude 7.4 along the Marmara Sea fault is 7 %. Considering these large differences, it may be proposed that a sound risk assessment methodology for effective risk mitigation must be focused on identifying these hazardous buildings in urban environments as the first priority.

A two-level seismic risk assessment procedure is developed in this study for low to medium rise (less than 8 stories) ordinary reinforced concrete buildings. The developed procedure is based on several building parameters that can be easily observed or measured during a systematic survey. The main objective of the procedure is developing a building database, and ranking the buildings in an urban stock with respect to their expected seismic performances under a defined ground excitation.

2. TWO-LEVEL RISK ASSESSMENT PROCEDURE

The first survey level is conducted from the sidewalk by trained observers. In the second survey level, the observers enter the basement and ground stories of the buildings for collecting the simplest structural data. The acquired data is then processed for calculating a risk score for each building.

2.1 Level 1: Observations from the street

A street survey procedure must be based on simple structural and geotechnical parameters that can be observed easily from the sidewalk. The time required for an observer for collecting the data of one building from the sidewalk is expected not to exceed 10 minutes. The parameters that are selected for representing building vulnerability in this study are the following:

1. The number of stories above ground (1 to 7)
2. Presence of a soft story (Yes or No)

3. Presence of heavy overhangs, such as balconies with concrete parapets (Yes or No)
4. Apparent building quality (Good, Moderate or Poor)
5. Presence of short columns (Yes or No)
6. Pounding between adjacent buildings (Yes or No)
7. Local soil conditions (Stiff or Soft)
8. Topographic effects (Yes or No)

Each parameter reflects a negative feature of the building system under earthquake excitations on a variable scale. Evaluating the correlation between observed building damage and parameter variation by using the building data compiled from Düzce assesses the weight of each parameter in expressing a seismic performance score. It is intended to develop a linear combination rule for the selected parameters in order to predict the damage distribution displayed by the collected data as well as possible. Once such a combination rule is developed, it will be possible to rate the seismic performance of reinforced concrete building structures in Turkey by employing a simple walk-down survey procedure. The proposed method bears some similarities with the seismic evaluation procedure developed in FEMA-154 (1988). However it is believed that this method provides a broader description of seismic risk for the multistory reinforced concrete buildings in Turkey, which do not conform to the requirements of modern seismic design and construction codes.

The objective of developing a performance scale for existing buildings is to provide a simple tool, which can be easily implemented by both the building owners and the public administration. If an individual building falls within the lower (high-risk) part of the scale, then a more detailed evaluation will be deemed necessary. The performance scale provides an ordering of the seismic vulnerability of a building stock. The scale can be used to classify low, moderate and high-risk buildings. Low-risk buildings may not require further evaluation, but moderate and high-risk buildings can be subjected to more detailed evaluation procedures before final decisions on retrofitting or removal.

Each vulnerability parameter, to which the damage distribution in the collected building data is found to be sensitive, is evaluated separately in the following paragraphs.

2.1.1 The number of stories

Field observations after the 1999 Kocaeli and Düzce earthquakes revealed that there is a very significant correlation between the number of stories and the severity of building damage. If all buildings were conforming to modern seismic design codes, then such a distribution would not occur, and a uniform distribution of damage would be expected. However if the majority of buildings in the earthquake stricken region lack this basic property, then the increasing number of stories increase seismic forces linearly whereas the seismic resistances do not follow in adequate proportions. Accordingly, damage increases almost linearly with the number of stories. After the two earthquakes in 1999, damage distribution for all 9685 buildings in Düzce is obtained with respect to the number of stories. The results are shown in Figure 1 below, where the number of damaged buildings is normalized with the total number of buildings at a given story number. It can easily be observed from Figure 1 that damage grades shift linearly with the number of stories. As the number of stories increases, the ratio of undamaged and lightly damaged buildings decreases steadily whereas the ratio of moderately and severely damaged buildings increases in an opposite trend. This is a clear indication that the number of stories is a very significant, perhaps the most dominant, parameter in determining the seismic vulnerability of typical multistorey concrete buildings in Turkey.

2.1.2 Presence of a soft story

A soft story usually exists in a building when the ground story has less stiffness and strength compared to the upper stories. This situation mostly arises in buildings located along the side of a main street. The ground stories, which have level access from the street, are employed as a street side store or a commercial space whereas residences occupy the upper stories. These upper stories benefit from the additional stiffness and strength provided by many partition walls, but the commercial space at the bottom is mostly left open between the frame members, for customer circulation. Besides, the ground stories may have taller clearances and a different axis system causing irregularity. The compound effect of all these negative features from the earthquake engineering perspective is identified as a soft story. Many buildings with soft stories were observed to collapse due to a pancaked soft story in past earthquakes all over the world.

2.1.3 Presence of heavy overhangs

Heavy balconies and overhanging floors in multistory reinforced concrete buildings shift the mass center upwards; accordingly increase seismic lateral forces and overturning moments during earthquakes. Buildings having balconies with large overhanging cantilever spans enclosed with heavy concrete parapets sustained heavier damages during the recent earthquakes in Turkey compared to regular buildings in elevation. Since this building feature can easily be observed during a walk-down survey, it is included in the parameter set.

2.1.4 Apparent building quality

The material and workmanship quality, and the care given to its maintenance reflect the apparent quality of a building. A well-trained observer can classify the apparent quality of a building roughly as good, moderate or poor. A close relationship was observed between the apparent quality and the experienced damage during the recent earthquakes in Turkey. A building with poor apparent quality can be expected to possess weak material strengths and inadequate detailing.

2.1.5 Presence of short columns

Semi-infilled frames, band windows at the semi-buried basements or mid-story beams around stairway shafts lead to the formation of short columns in concrete buildings. These captive columns usually sustain heavy damage during strong earthquakes since they are not originally designed to receive the high shear forces relevant to their shortened lengths. Short columns can be identified from outside because they usually form along the exterior axes.

2.1.6 Pounding between adjacent buildings

When there is insufficient clearance between adjacent buildings, they pound each other during an earthquake as a result of different vibration periods and consequent non-synchronized vibration amplitudes. Uneven floor levels aggravate the effect of pounding. Buildings subjected to pounding receive heavier damage in the higher stories.

2.1.7 Local soil conditions

Site amplification is one of the major factors that increase the intensity of ground motions. Although it is difficult to obtain precise data during a street survey, an expert observer can be able to classify the local soils as stiff or soft. In urban environments, geotechnical data provided by local authorities is a reliable source for classifying the local soil conditions.

2.1.8 Topographic effects

Topographic amplification is another factor that may increase the ground motion intensity on top of hills. Besides, buildings located on steep slopes (steeper than 30 degrees) usually have stepped foundations, which are incapable of distributing the ground distortions evenly to structural members above. Therefore these two factors must be taken into account in seismic risk assessment. Both factors can be observed easily during a street survey.

2.2 Level 2: Measurements at the ground story and basement

After the building data is acquired from street surveys and evaluated, buildings falling into the moderate and high risk levels can be identified with respect to their performance scores as explained in the following sections. Observer teams enter into the basements and ground stories of these buildings for collecting more data for further evaluation. Their first task is the confirmation or modification of the previous grading on soft stories, short columns and building quality, through closer observation. The second and more elaborate task is to prepare a sketch of the framing plan at the ground story and measuring the dimensions of columns, concrete and masonry walls. These tasks are expected to consume about two hours of a team consisting of three members. This data is then employed for calculating the following parameters.

2.2.1 Plan irregularity

Irregularity in building plan is a deviation from a rectangular plan having orthogonal axis systems in two directions. Such deviation from plan regularity leads to irregularities in stiffness and strength distributions, which in turn increase the risk of damage localization under strong ground excitations. In earthquake resistant design, regularity in plan is encouraged.

2.2.2 Redundancy

When the number of continuous frames or number of bays in a building system is insufficient, lateral loads may not be distributed evenly to frame members. Especially those frames exhibiting inelastic response during earthquakes suffer from lack of sufficient redundancy, which leads to localized heavy damage. A normalized redundancy ratio is defined by the following expression (Özcebe et al., 2003).

$$NRR = \frac{A_{tr}(n_{fx} - 1)(n_{fy} - 1)}{A_{gf}} \quad (1)$$

Here, A_{tr} is the tributary area for a typical column, A_{gf} is the area of ground floor, n_{fx} and n_{fy} are the number of continuous frames in x and y directions, respectively. Three redundancy scores (NRS) are assigned accordingly.

NRS = 0 when NRR>1 : Redundant
NRS = 1 when 0.5<NRR<1 : Semi-redundant
NRS = 2 when NRR < 0.5 : Weakly redundant

2.2.3 Strength index

The lateral strength of a building is strongly related to the size of its vertical members, among other factors including material strengths, detailing and frame geometry. Since measuring the sizes of vertical members at the ground story of an existing building is possible, a strength ratio SR can be defined as follows (Özcebe et al., 2003).

SR = min (A_{nx}, A_{ny})

$$A_{ni} = \frac{\Sigma(A_{col})_i - \Sigma(A_{sw})_i - 0.1\Sigma(A_{mw})_i}{\Sigma A_f} x100 \quad (2)$$

where

$$(A_{col})_i = k_i \,.\, A_{col}$$

$$(A_{sw})_i = k_i \,.\, A_{sw}$$

$$(A_{mw})_i = k_i \,.\, A_{mw}$$

Here, *i* stands for x or y, k_x is 1/2 for square columns, 1/3 and 2/3 for rectangular columns in weak and strong directions respectively, and 1.0 for concrete and masonry walls in x-direction, k_y=1-k_x. A_{col}, A_{sw} and A_{mw} are the cross section area of each column, shear wall and masonry infilled wall, respectively. A stiffness index SI is described by classifying the strength index SI.

SI = 0 when SI > 0.0025 : strong

$SI = 1$ when $0.0015<SI<0.0025$: moderate
$SI = 2$ when $SI < 0.0025$: weak

3. EVALUATION OF THE DÜZCE DATABASE

A total of 477 buildings were surveyed in Düzce, which survived the 17 August 1999 Kocaeli and 12 November 1999 Düzce earthquakes with some levels of damage. Building damages were classified in four grades, namely none, light, moderate and severe or collapsed. A building with light damage can be occupied with minor repairs after the earthquake whereas a moderately damaged building requires structural repairs. If there is severe damage, then such a building must either be strengthened to upgrade its seismic capacity, or demolished. The damage distribution of the investigated buildings with the number of stories is presented in Table 1.

The variation of damage in 477 buildings with survey parameters is obtained independently for each parameter. The Düzce database did not represent all parameters. Short columns and pounding effects were not surveyed. Moreover, soil conditions were uniform and the topography was flat. Therefore these four parameters are not included in the following evaluation.

3.1 The number of stories

An investigation is conducted on the 477 surveyed buildings in Düzce, to check whether the surveyed building stock represents Düzce building inventory, considering the distribution of damage with the number of stories. The results are shown in Figure 2. The trend in this figure is quite similar to that in Figure 1, which confirms that damage is strongly correlated with the number of stories. Accordingly, it is decided to uncouple this parameter from the others. The data for the other parameters are sorted for each story number separately in order to remove its effect on the other parameters.

3.2 Presence of soft story

Among the 477 surveyed buildings, 234 buildings had soft stories. These buildings are grouped with respect to the damage grades and the number of stories, and then their number is normalized relative to the total number of buildings in each group. The results are presented in Figure 3. For all story numbers, it is evident that the buildings with soft stories exhibit higher severe damage/collapse ratios compared to those with no soft stories.

Notably, almost all severely damaged buildings have soft stories. This is an important observation because if a building with a soft story is vulnerable to seismic damage, it is very likely that this damage will be either moderate or severe, especially when the number of stories exceeds two. It can also be observed that damage distribution among buildings with soft stories does not have a consistent variation with the number of stories. Therefore this parameter can be assessed independently from the number of stories.

3.3 Apparent building quality

The quality classification of 477 surveyed buildings revealed that 63 were good, 391 were moderate and 23 were poor. These buildings are grouped with respect to the damage grades and the number of stories, and then their number is normalized relative to the total number of buildings in each class. The results are presented in Figure 4. The data for 6 story buildings is meaningless. However the data for 3-5 stories reveal that the severely damaged/collapsed buildings have lesser quality than the other damage groups. An increasing effect can also be observed with the number of stories.

3.4 Presence of heavy overhangs

The distribution of damage in buildings with and without heavy overhangs is presented in Figure 5. There were 97 buildings with heavy overhangs among the total of 477. The building ratios are obtained by normalizing the number of buildings in each category with respect to the total number of buildings with or without overhangs for each number of stories. All of the undamaged buildings were free of heavy overhangs. There is a consistently increasing trend in the severely damaged/collapsed building ratios of 2 to 6 story buildings with the story number, with regard to the presence of overhangs. Accordingly, this parameter should be considered in the seismic risk assessment of buildings having more than 3 stories.

3.5 Plan irregularity

The results obtained from the survey data are presented in Figure 6, separately for each number of stories. The number of buildings classified as irregular was 274 among 477. Irregularity in plan does not influence damage distribution in 2 story buildings. In 3 to 6 story buildings, those with irregular plan have a larger share among the severely damaged/collapsed buildings than the ones with regular plan. Therefore plan irregularity should be considered as a parameter in determining the seismic risk of buildings taller than 2 stories.

3.6 Redundancy

The majority of buildings in the Düzce database were classified as weakly redundant (315), whereas 85 were semi-redundant and 77 were redundant. The normalized results are shown in Figure 7. This parameter can only separate the severely damaged and collapsed buildings in the 4 to 6 story groups. Weakly redundant buildings have a share among the severely damaged and collapsed buildings that increases with the number of stories, and becomes notable in 5 and 6 story buildings.

3.7 Strength index

Only 37 buildings among 477 were classified as weak in strength. More than half of the 5 and 6 story weak buildings had collapsed or sustained severe damages according to Figure 8. However the strength index has no influence on the damage distribution of 2-4 story buildings. Therefore this parameter can only be considered for identifying the risk of 5 and 6 story buildings.

4. TWO-LEVEL SEISMIC RISK ASSESSMENT TOOLS FOR ISTANBUL

A practical risk assessment procedure for Istanbul is presented herein, which is based on the data acquired from the two levels of surveys conducted from the street and the ground stories of buildings, respectively. The weight of each building vulnerability parameter is evaluated by statistical procedures, based on the Düzce database. Statistical analysis is conducted by the program package SPSS Version 11, using the "Multivariable Stepwise Linear Regression Analysis" procedure. The results are then smoothed, and the weights of the parameters for which there was no available data (soft story, pounding, topography) are assigned by using engineering judgement. Local soil conditions and associated ground motion intensity in Düzce were uniform. Different intensity zones are described for Istanbul however (JICA, 2002), based on the distribution of peak ground accelerations (PGA) or velocities (PGV) during the scenario earthquake. The effect of ground motion intensity expected in different zones is considered by applying velocity-based conversion factors as explained below.

4.1 Building performance score

Once the vulnerability parameters of a building are obtained from two-level surveys and its location is determined, the seismic performance scores for survey levels 1 and 2 are calculated by using Tables 2 and 3, respectively. In these tables, an initial score is given first with respect to the number of stories and the intensity zone. Then, the initial score is reduced for every vulnerability parameter that is observed or calculated. A general equation for calculating the seismic performance score (PS) can be formulated as follows.

$$PS = (\text{Initial Score}) - \Sigma\, (\text{Vulnerability parameter}) \times (\text{Vulnerability Score})$$

The vulnerability scores are given in Tables 1 and 2, and the vulnerability parameters are defined under the tables.

4.2 Local soil conditions and ground motion intensity

The intensity of ground motion under a building during an earthquake predominantly depends on the distance of the building to the causative fault, and the local soil conditions. Mapping of seismic hazard at micro scale considers both variables. Seismic hazard, or ground motion intensity is mapped in terms of PGA and PGV in the JICA report. PGV usually reflects the effect of soil conditions very well during a large magnitude earthquake (Wald et al., 1999). The correlation of PGV and shear wave velocities of local soils can easily be observed from the associated maps given in the JICA report. Accordingly, PGV is selected to represent the ground motion intensity in this study.

The PGV map in the JICA report has contour increments of 20 cm/s2. The intensity zones in Istanbul are expressed accordingly, in terms of the associated PGV ranges.

Zone I : $60 < PGV < 80$ cm/s^2

Zone II : $40 < PGV < 60$ cm/s^2

Zone III : $20 < PGV < 40$ cm/s^2

The superiority of PGV over PGA can be best observed in the Princes Islands, which are bedrock outcrops. They are in PGV zone II. However if PGA were employed, they would be in zone I due to their proximity to the Marmara fault. It is well documented that the Princes Islands were not severely affected from the strong historical earthquakes.

The differences in ground motion intensities at three PGV zones are reflected in the initial scores given in Tables 2 and 3, according to a study conducted by Akkar and Sucuoglu (2003).

4.3 Testing of risk assessment tools for the Düzce database

Seismic performances of the 477 buildings surveyed in Düzce have been tested with the tools presented in Tables 2 and 3. A cut-off performance score of 50 has been calculated for both survey levels through an optimization analysis to obtain the best prediction. The results revealed that at the level-1 survey (street surveys), 72 % of the severely damaged and collapsed buildings, and 72% of the remaining buildings with lesser damages are identified successfully by using Table 2. These ratios increased to 75 % when level-2 survey results are evaluated by using Table 3.

REFERENCES

1. Sucuoglu H. and Yılmaz, T. (2001). "Düzce, Turkey: A city hit by two major earthquakes in 1999 within three months", Seismological Research Letters, 72(6), 679-689.
2. Japan International Cooperation Agency and Istanbul Metropolitan Municipality (2002). "A disaster prevention/mitigation basic plan in Istanbul including seismic microzonation", Istanbul.
3. Özcebe, G., Yücemen, M.S. and Aydogan, V. (2003). "Assessment of seismic vulnerability of existing reinforced concrete buildings", Proceedings, NATO Science for Peace Workshop, Izmir, Turkey, May 13-14, 2003.
4. Wald, D.J., Quitariano, V., Heaton, T.H., Kanamori, H., Scrivner C.W. and Worden, C.B. (1999). "Trinet Shake Maps: rapid generation of peak ground motion and intensity maps for earthquakes in Southern California", Earthquake Spectra, 15, 537-555.
5. Akkar, S. and Sucuoglu, H., (2003) "Peak velocity sensitive deformation demands and a rapid damage assessment approach", Proceedings, NATO Science for Peace Workshop, Izmir, Turkey, May 13-14, 2003.

Table 1. Damage Distribution of the Investigated Buildings in Düzce

	Damage Observed				
Number of stories	None	Light	Moderate	Severe, Collapsed	Total
2	7	13	3	0	23
3	18	62	29	15	124
4	17	43	60	27	147
5	17	30	56	65	168
6	1	0	4	10	15
Total	60	148	152	117	477

Table -2. Initial and Vulnerability Scores for Level-1 Survey of Concrete Buildings

Story #	Zone I 60<PGV<80	Zone II 40<PGV<60	Zone III 20<PGV<40	Soft Story	Heavy Overhang	Apparent Quality	Short Column
1, 2	90	125	160	-5	-5	-5	-5
3	90	125	160	-10	-10	-10	-5
4	80	100	130	-10	-10	-10	-5
5	80	90	115	-15	-15	-15	-5
6,7	70	80	95	-15	-15	-15	-5

Vulnerability Parameters
Soft story: No (0); Yes (1)
Heavy overhangs : No (0); Yes (1)
Apparent quality : Good (0); Moderate (1); Poor (2)
Short columns : No (0); Yes (1)
Pounding effect : No (0); Yes (1)
Topography effect : No (0); Yes (1)

Table -3. Initial and Vulnerability Scores for Level-2 Survey of Concrete Buildings

Story #	Zone I 60<PGV<80	Zone II 40<PGV<60	Zone III 20<PGV<40	Soft Story	Heavy Overhang	Apparent Quality	Short Column	Pound.	Topog. Effects	Plan Irreg.	Redundancy	Strength Index
1, 2	95	130	170	0	-5	-5	-5	0	0	0	0	-5
3	90	125	160	-10	-5	-10	-5	-2	0	-2	0	-5
4	90	115	145	-15	-10	-10	-5	-3	-2	-2	-5	-5
5	90	105	130	-15	-15	-15	-5	-3	-2	-5	-10	-10
6, 7	80	90	105	-20	-15	-15	-5	-3	-2	-5	-10	-10

<u>Vulnerability Parameters</u>
Soft story: No (0); Yes (1)
Heavy overhangs : No (0); Yes (1)
Apparent quality : Good (0); Moderate (1); Poor (2)
Short columns : No (0); Yes (1)
Pounding effect : No (0); Yes (1)
Topography effect : No (0); Yes (1)
Plan irregularity: No (0); Yes (1)
Redundancy : Redundant (0), Semi-redundant (1), Weakly redundant (2)
Strength Index : Strong (0), Moderate (1), Weak (2)

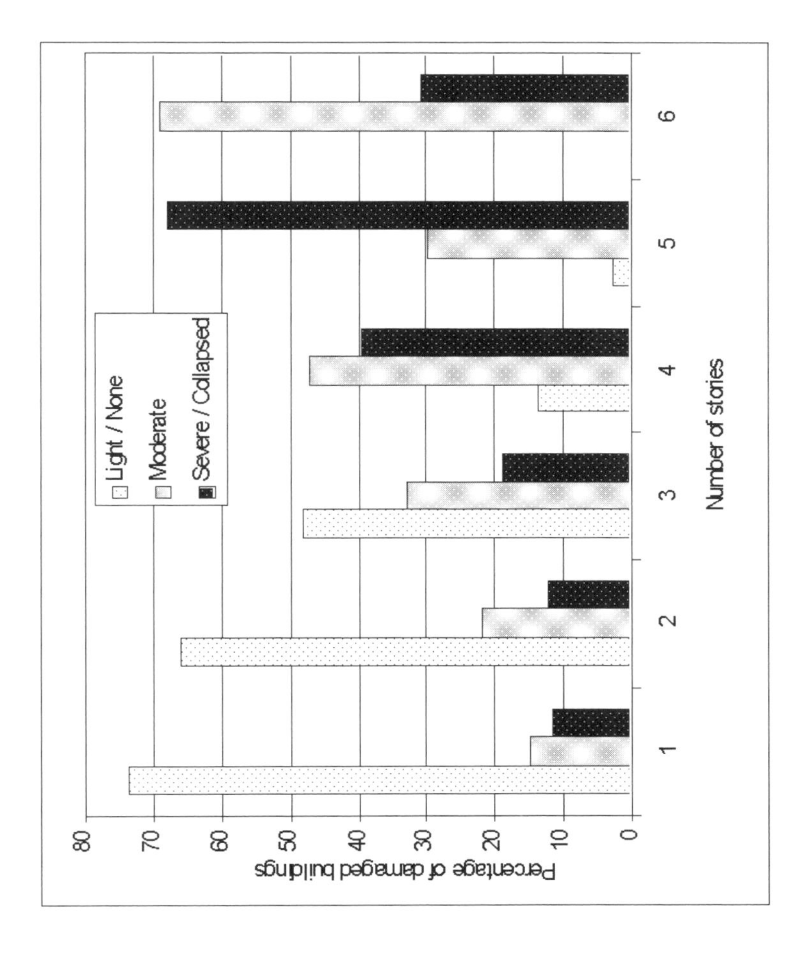

Figure 1. Damage distribution in Düzce after the 1999 earthquakes, with respect to the number of stories

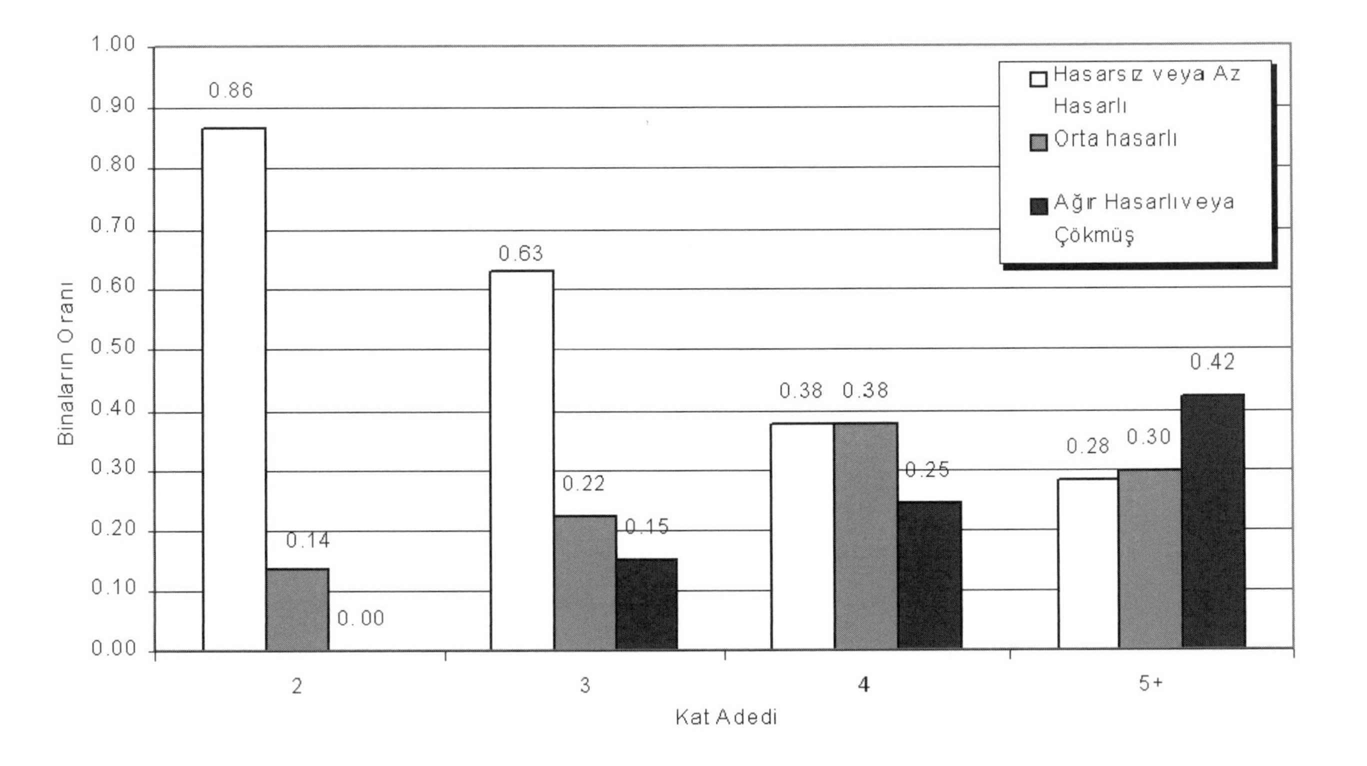

Figure 2. The distribution of damage with the number of stories in 477 buildings

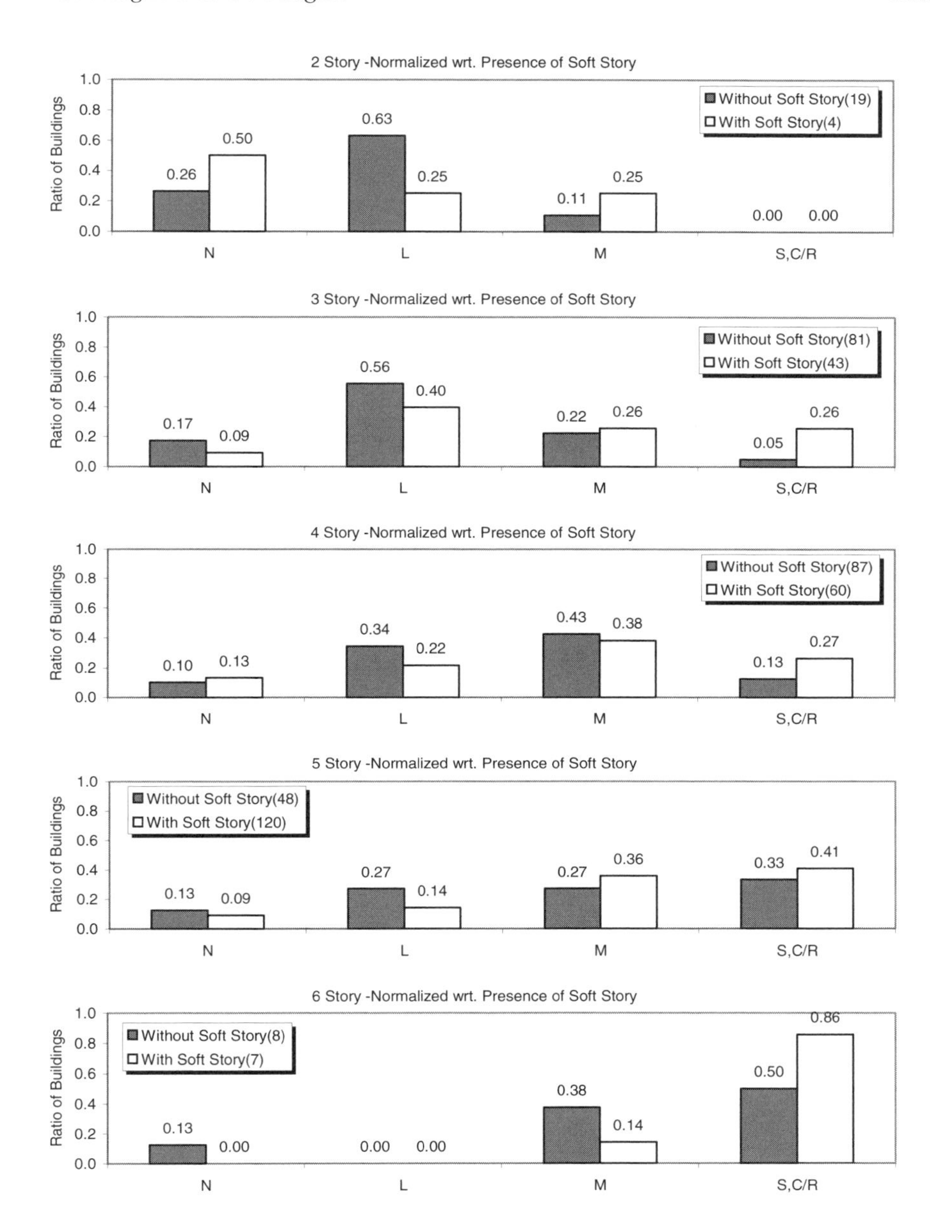

Figure 3. Correlation of damage with the presence of soft story

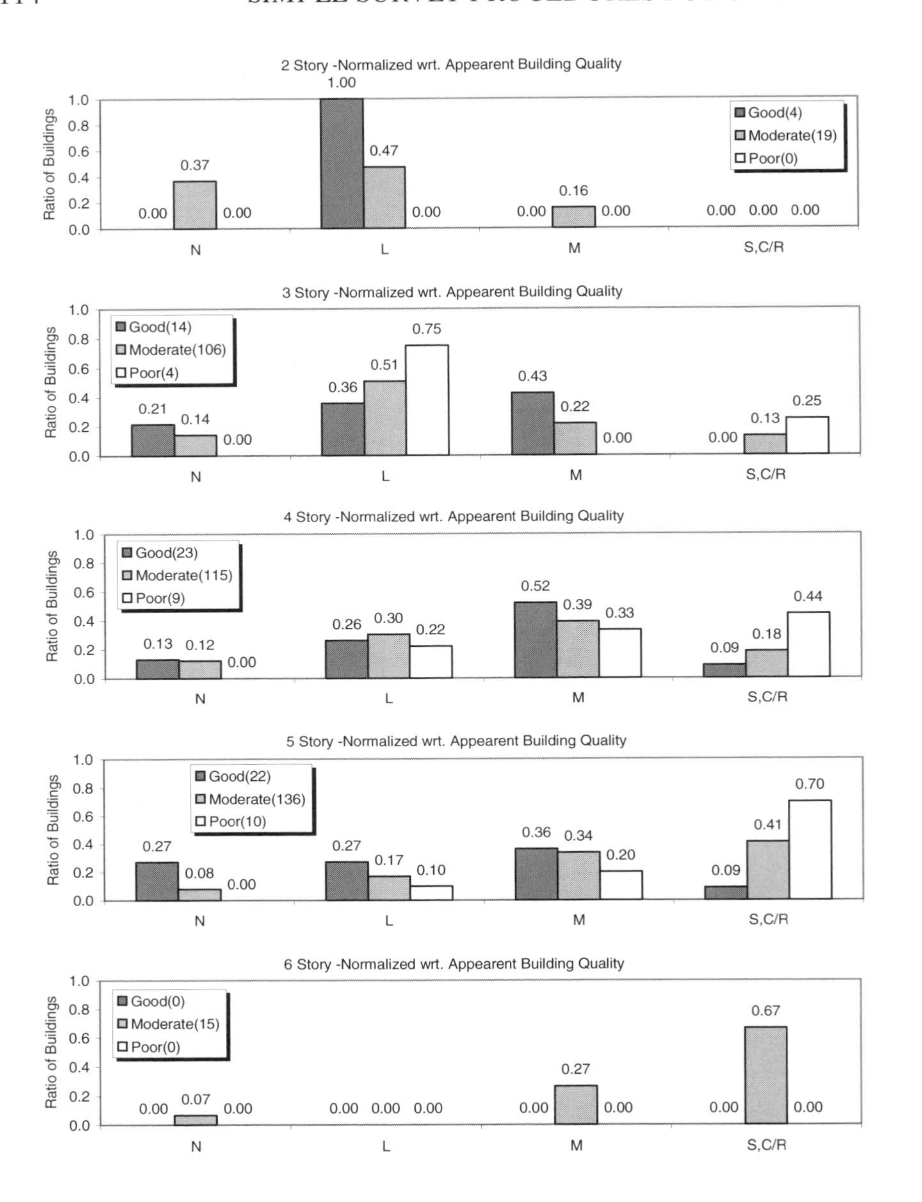

Figure 4. Correlation of damage with the apparent building quality

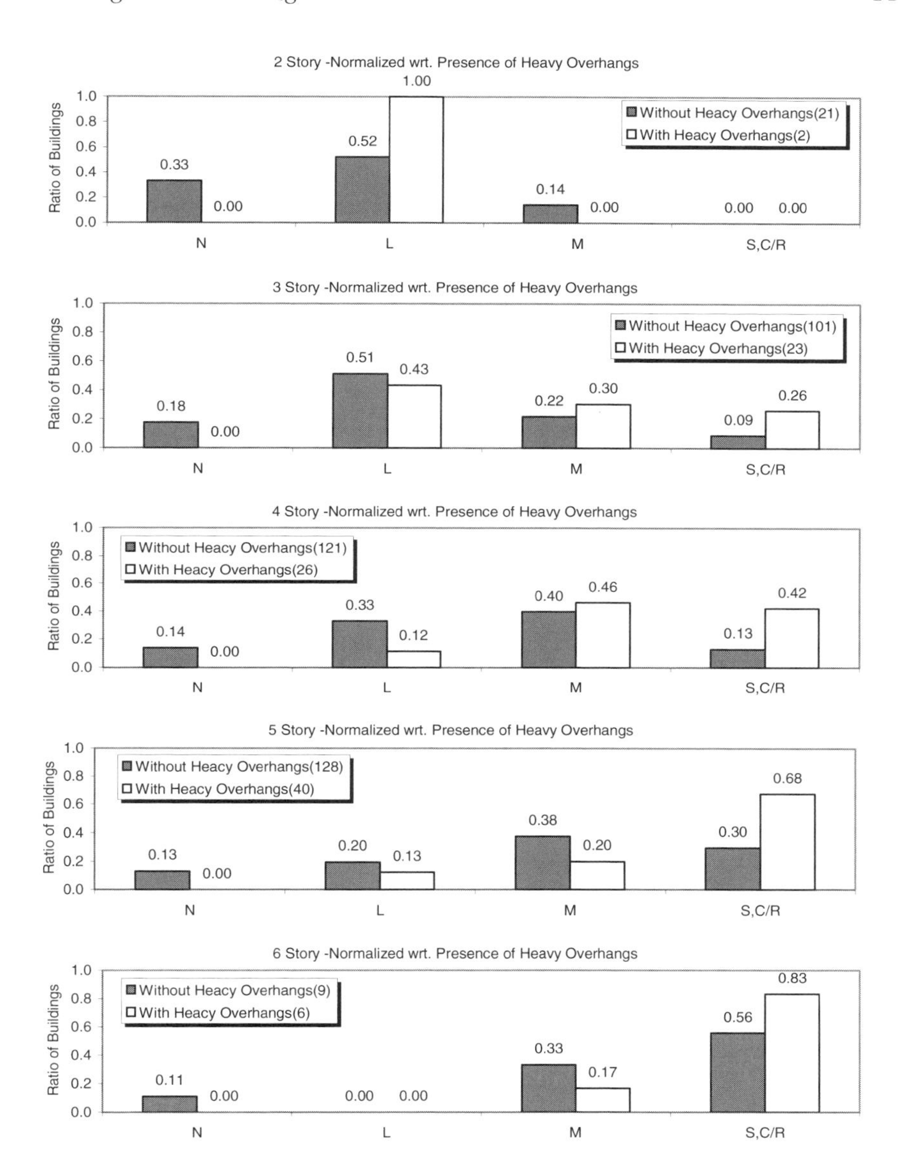

Figure 5. Correlation of damage with heavy overhangs

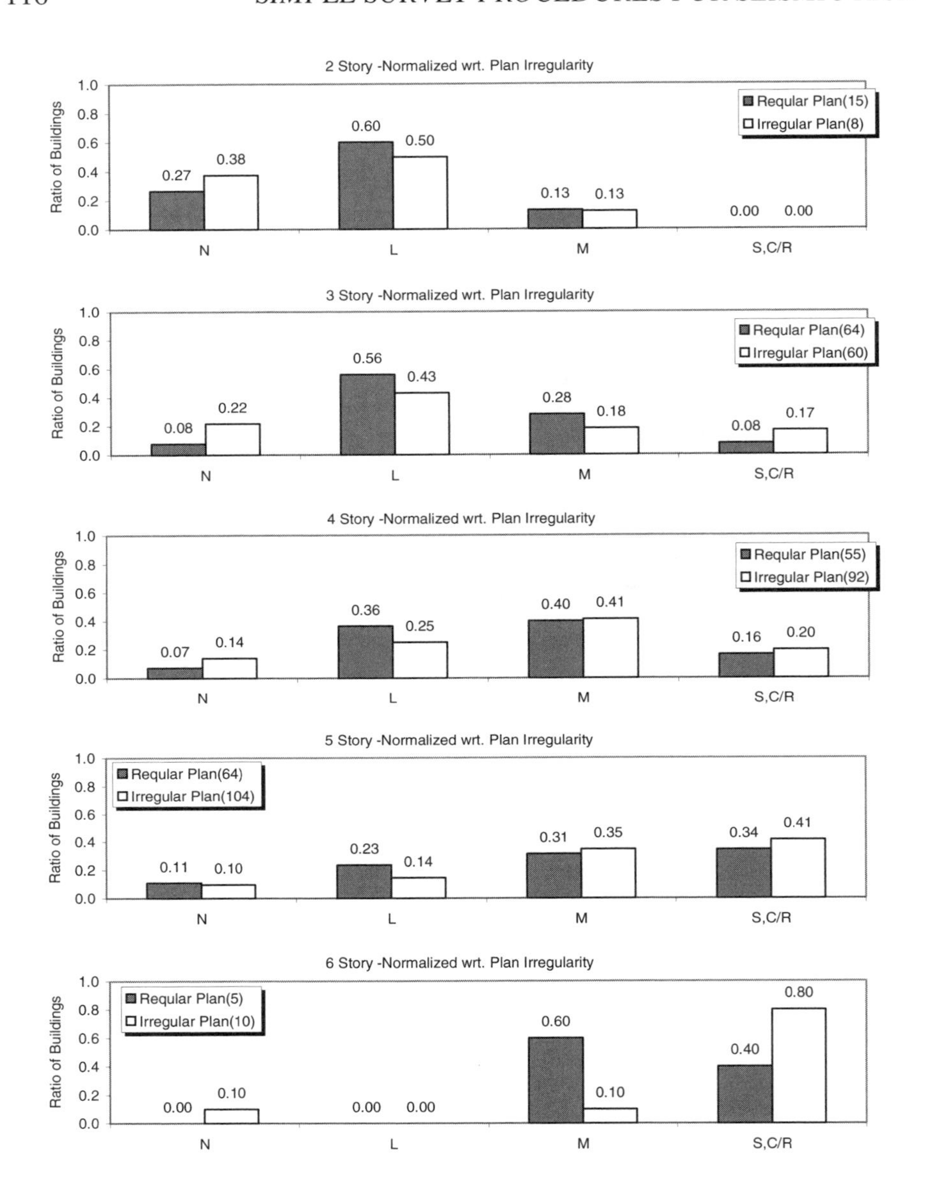

Figure 6. Correlation of damage with plan irregularity

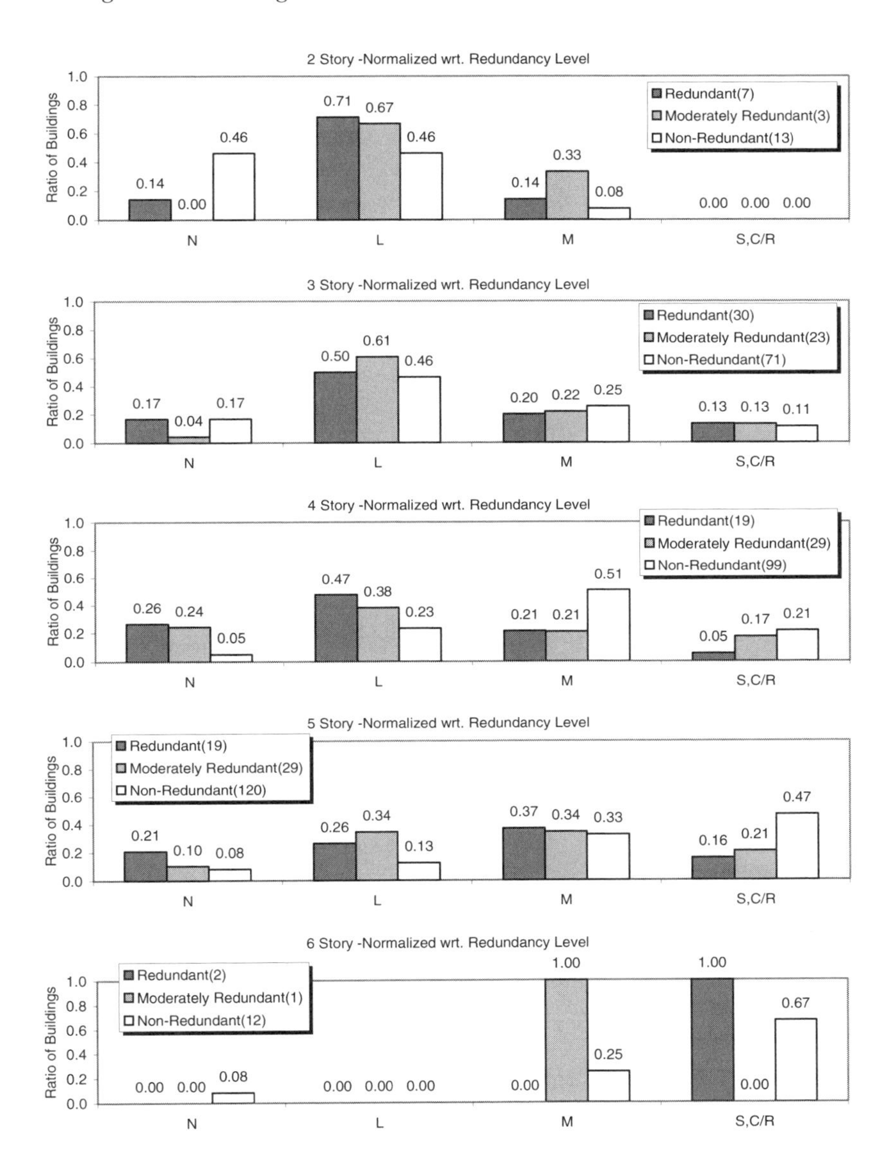

Figure 7. Correlation of damage with redundancy

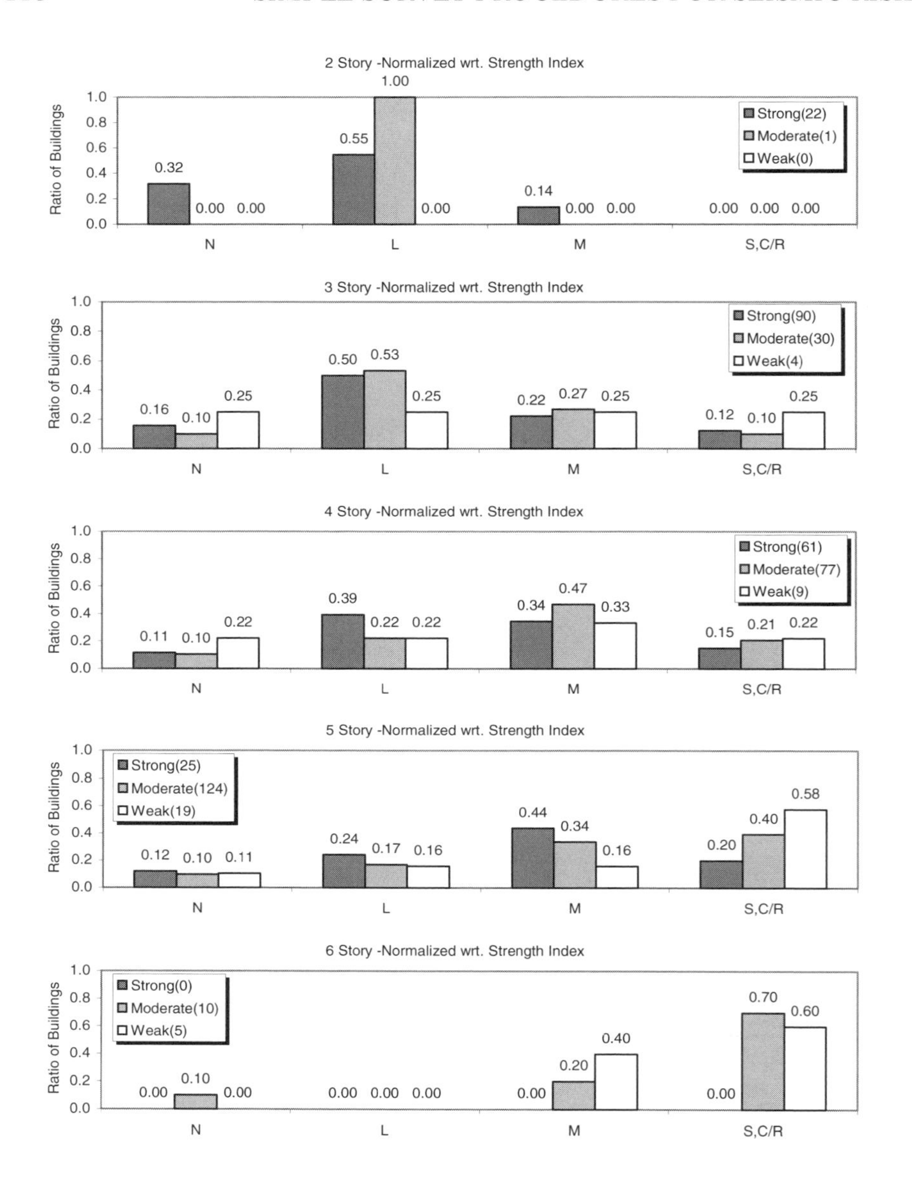

Figure 8. Correlation of damage with the strength index

THE USE OF MICROTREMORS FOR SOIL AND SITE CHARACTERISATION AND MICROZONATION APPLICATIONS

Kyriazis Pitilakis
Aristotle University of Thessaloniki, Greece

Abstract: The Microtremor Exploration Method (MEM) is an innovative, alternative method for the determination of Vs profiles and subsoil geometry for site and soil characterization. Using the Spatial Autocorrelation Method (SPAC) a circular array of few broadband instruments allows the accurate determination of Vs profiles. A Rayleigh wave inversion scheme was applied following the determination of phase velocity dispersion curves for different frequency range windows. The Vs profiles combined with other geotechnical data and simultaneous HVSR measurements allowed the construction of 3D geotechnical maps, describing the Vs velocities of the stratigraphy of the soil formations from the surface till the seismic bedrock.

Key words: microtremors, soil and site characterization, shear wave velocity, stationary random functions, spatial autocorrelation method, site effects, microzonation

1. INTRODUCTION

The geometry of the subsoil structure, the soil types and the variation of their properties with depth, the lateral discontinuities and the surface topography are at the origin of large amplification of ground motion and hence of intensive damage during destructive earthquakes. For this reason the accurate knowledge of the geometry and the shear-wave velocities (Vs)

S.T. Wasti and G. Ozcebe (eds.), Seismic Assessment and Rehabilitation of Existing Buildings, 119–148.

of the alluvial – diluvial deposits and the basement are the key parameters controlling the amplification of seismic motion.

Vs velocities are usually determined in the field using conventional seismic prospecting (e.g. reflection, refraction, borehole seismics) and in the laboratory through dynamic/cyclic tests (RC, CTX), on intact and remolded soil samples. The use of conventional exploration seismic methods for Vs presents many difficulties when deep sedimentary structures need to be investigated. Reflection and refraction tests using artificial sources such as explosives or strong vibrators are very difficult in urban areas. Furthermore, the need of long arrays to reach the desired penetration depth is difficult to be deployed in densely populated areas. Moreover the cost of large scale deep geophysical prospective is very high and therefore in most cases for site effect prospecting studies the depth of the seismic basement is limited to a layer with Vs larger than 1000m/sec and not the real very deep substructure reflector of the incident waves. Additionally, the cost for implementing deep borehole seismics is important, while the reliability of these seismic methods, such as cross-hole and down-hole, at large depths is often questionable, due to practical limitations (e.g. energetic sources, equipment management etc).

In the present paper it is proposed to overcome the aforementioned difficulties by applying the Microtremor Exploration Method (MEM) for the determination of the Vs velocity and the geometry of subsoil conditions. The MEM method has mostly been applied in Japan. Based on the limited published literature, it has the advantage of reaching large exploration depths (Okada et al. 1990; Aki and Chouet 1991; Vernon et al., 1991; Hough et al., 1992; Ling and Okada, 1993; Goldstein and Chouet, 1994; Metaxian et al., 1997; Chouet et al., 1998; Kanno et al., 2000, Apostolidis 2002). The particular method exploits the microtremor ambient noise recordings obtained at stations deployed in a circular array. The measurements are taken simultaneously at all stations, which are operating for a short period of time. The analysis of the microtremor recordings is performed through the Spatial Autocorrelation Coefficient Method (SPAC) introduced by Aki (1957). SPAC is based on the theory of Stationary Random Functions, according to which, microtremor is considered as a stationary stochastic process both temporally and spatially. The SPAC method is an innovative method, which has so far had only limited application in Europe. In the present paper it is proved that it is a very interesting and low cost technique for soil and site characterization, for site effect analyses and microzonation studies.

In the frame of the present study, microtremor measurements were performed at many sites, representative from the geological point of view, both in the EUROSEISTEST experimental site (http://euroseis.civil.auth.gr) and in the city of Thessaloniki, where a complete microzonation study is in evolution, under the responsibility of the Laboratory of Soil Mechanics and Foundations – Research Unit of Soil Dynamics and Geotechnical Earthquake Engineering of the Civil Engineering Department of Aristotle University of Thessaloniki. The city of Thessaloniki with a population of one million inhabitants has a long history of seismic catastrophes; it suffered severe damage during the recent destructive earthquake in 1978 (Mw=6.5). During the last two decades, a considerable amount of research work has been done (a) to investigate the role of site effects on the damage distribution of the 1978 earthquake and to assess future potential damages scenarios and (b) to construct an extensive data base of geotechnical, geophysical and geological data comprising borehole data, SPT, CPT, CH, and DH measurements and adequate laboratory tests to estimate the physical, mechanical and dynamic properties (RC, CTX, TX etc) of all soil formations covering the area (Anastasiadis et al., 2001; Raptakis et al., 1994; 1998). However the detailed sedimentary structure and the full geometry of all soil deposits is not completely defined yet, mainly due to their large thickness which surpass 200m at the coastal zone and the difficulty to conduct expensive deep geophysical prospecting.

The aim of this study is to investigate whether it is feasible to estimate the Vs structure of soil formations within the city, especially for depths larger than 50 m, reaching in many cases the depth of the bedrock. The importance of the investigation is multifold, since the Vs profiles can be used not only for conventional 1D ground response analysis but also in order to construct reliable 2D cross-sections and 3D ground models for theoretical modeling of the wave propagation and site effects studies in a complex soil structure. The efficiency and the accuracy of the method have been validated by comparing the resulting Vs profiles cross-hole Vs measurements at specific sites either in Euroseistest or in Thessaloniki.

2. THEORETICAL BASIS OF THE SPAC METHOD

The fundamental assumption is that microtremors consist mainly of dispersive surface waves, and that we are able to estimate the dispersion curve of phase velocity as a function of frequency. This curve is then used with an inversion scheme in order to get the distribution of shear wave velocity with the depth at the area of measurements.

Two methods are currently used to extract surface waves from microtremors by array records, the frequency (f) – wave-number (k) power spectral density method (e.g. Capon, 1969; Lacoss et al., 1969; Horike, 1985; Matsushima and Okada, 1990; Satoh et al., 2001), and the spatial autocorrelation method (SPAC method). The SPAC method is probably more convenient compared to the f-k method, because it gives equally good results with the f-k method by using less recording stations and arrays of shorter dimensions (Okada, 1987). Therefore, in this study it is decided to apply the SPAC method, which is certainly more attractive within urban areas where limited free spaces are available.

Aki (1957) gave a theoretical basis of the SPAC coefficient defined for ambient noise and developed a method to estimate the phase velocity dispersion of surface waves contained in microtremors using a specially designed circular array. Henstridge (1979) also introduced a licit expression of the relationship between the spatial autocorrelation coefficient and the phase velocity of fundamental mode of Rayleigh waves. Okada et al. (1990) and Okada (1997) extended it to an exploration method that is currently called the SPAC method. For a complete presentation of the method see also Okada (1999).

Recently, Kudo et al. (2002) applied the SPAC method in the heavily damaged area of the 1999 Kocaeli - Turkey earthquake, and confirmed the usefulness of the method for site characterization, even in urban areas. The present work is a pioneering work in Europe (Pitilakis et al.,1999; http://euroseis.civil.auth.gr). (Apostolodis, 2002, Kudo et al., 2002, Raptakis et al., 2002).

The main assumption of SPAC method is that microtremor is an ensemble of the dispersive surface waves, which can be considered as a stationary stochastic process in the time-space domain. This means that microtremor represents the sum of horizontally propagating waves having the same phase velocity for a given frequency and that waves propagating in different directions are statistically independent. When these assumptions are valid, the phase velocity of the surface waves included in microtremors can be related to an azimuthal average of the spatial correlation.

Aki (1965) showed that for a given circular array of stations with one more station at the centre, a space correlation function $S(r, \theta)$ can be defined as:

$$S(r, \theta) = E\,[u\,(x, y, t)*u(x+r\cos\theta, y+r\sin\theta, t)] \tag{1}$$

where $u\,(x,y,t)$ is the velocity observed at point (x,y) at time t, r is the station separation, θ is the azimuth and [*] denotes the ensemble average. An average of this function over all azimuths is given by:

$$S\,(r) \equiv \frac{1}{2\pi}\int_0^{2\pi} S(r,\theta)d\theta \tag{2}$$

The function S(r) can be expressed by the power spectrum, $h_o(\omega)$, of the vertical component via Bessel function, J_o, of the first kind of zero order with the variable $r\omega$:

$$S(r) = \int_{-\infty}^{\infty} h(\omega) J_0 \left(\frac{\omega}{c(\omega)} r\right) d\omega \tag{3}$$

where ω is the angular frequency and $c(\omega)$ is the frequency-dependent phase velocity.

When noise recordings from the array are band-pass filtered over a narrow frequency band centred on a central frequency ω_o and S(r) is computed for the filtered series, then the normalized spatial correlation function is given by an equation of the form:

$$S(r) = J_0 \left(\frac{\omega_o r}{c(\omega_o)}\right) \tag{4}$$

Therefore, the spatial autocorrelation function at the central frequency ω_o is related to the phase velocity c (ω_o) via the Bessel function, Jo.

As is inferred from the above short theoretical description, the phase velocity at a certain frequency ω_o, can be calculated by the spatial autocorrelation coefficient between any pair of stations, S(r), which are part of a circular array and have an azimuthal distribution. The spatial autocorrelation function defined above is a unique quantity of the array location and it reflects the subsurface structure directly below the array.

3. SETTINGS OF NOISE MEASUREMENTS

Array measurements of microtremors were carried out in Eusroseistest (Figure1) and at 16 well – distributed sites within the city of Thessaloniki (Figure 2). The selection of these sites is based on their distinctive features of the geological conditions (IGME 1995), the existing geotechnical data from previous studies, (Anastasiadis et al. 2001), and the location of the sites where borehole seismic prospecting (cross-hole tests) were already performed (Pitilakis et al., 1992). The latter, both in Euroseistest and in Thessaloniki were selected for validation purposes. We tried to cover the whole central city area and to respect as much as possible the necessary reliability conditions for MEM such as flat stratigraphy and less anthropogenic noisy sites.

3.1 Spatial distribution of measurements

Of primary concern was to provide adequate coverage of the urban area, which would allow the design of 2D cross-sections and 3D mapping of the soil structure and the seismic bedrock surface. Therefore, the selected investigation sites were distributed at almost equal distances in the centre of the city, as well as in the east and west parts of the city. Almost all recording stations were installed on top of the Neogene formation (pl. f-c). According to the geological map (Figure 2), the dominant surface formation, within the city, is the Neogene, consisting of mixed phases. The top 10-30 m is composed mainly of sandy clays, loose conglomerates and breccia, while the deeper formation consists of red clayey marl with high stiffness and attenuation (Raptakis et al., 1998). Regarding the neotectonics of the city of Thessaloniki, the area is considered to be tectonically active (Hatzidimitriou et al., 1991); although it has been shown that there are probably active faults in the area, their traces are difficult to be mapped.

Regarding other soil formations met at the surface within the city of Thessaloniki, their thickness is relatively small (3-10m of debris at the historical centre), and they always appear on top of the Neogene formations. Ambient noise arrays are deployed on top of each soil formation in order to have a spherical view of the source information at the examined area. Also one site with surface appearance of the bedrock (outcrop) was selected, in order to assess the Vs velocity and the degree of weathering under outcrop conditions. The same formation of schist underlay the sediments in the whole city area being the rock basement.

Furthermore, the choice of the sites was based on the availability of results from previously conducted geophysical cross-hole and down-hole experiments in order to check the reliability of the method and the accuracy of the estimated soil profiles from the microtremor measurements in a densely urbanized area.

3.2 Arrays geometry and recordings status

All measurements at all sites were taken using circular arrays, consisting of three recording stations in the periphery and one common station, in case when different data sets were obtained at the same place, at the centre of the array (Figure 3). The radii of the circular arrays were individually assigned for each site, based on the expected thickness of the soil formations according to the previous knowledge or in situ considerations. At sites where the depth of the bedrock was expected to be large, we deployed both small arrays, with radius ranging from 10 m to 15 m, and large arrays, whose radius depended on the available space at the selected recording site (usually between 30 m and 50 m). The small array was deployed in order to obtain better resolution at the uppermost layers, while through the large array it is aimed to determine the depth of the bedrock or at least the depth of the lowermost hard soil formation (Vs>750m/s).

The recording instruments were three-component broadband seismometers CMG-40T with standard frequency band from 0.033Hz to 50Hz and Reftek recorders of type 72A-07, in order to get period bandwidths adequate to explore Vs velocity at layers as deep as possible. The sampling interval of the microtremor recordings was 0.008 sec and the recording duration was 30 minutes at each array. The obtained waveforms were band-pass filtered from 0.1 to 25 Hz and corrected for the baseline and the instrument response.

3.3 Accuracy of noise measurements

The interference of anthropogenic noise from small or large vehicles in microtremor measurements could act destructively at the corresponding frequencies carried by this noise, because all the employed stations would not record synchronously the signal generated at very small distances from the array. To obtain the most accurate noise measurements, free of noise due to human activities at small distances from the measuring sites, the selected sites were located as far as possible from high-traffic streets, factories, transportation stations etc and all measurements were carried out in the early morning hours (5-6 am) on Saturday or Sunday.

The analysis leads to the definition of the dispersion curve of surface waves, which is subsequently inverted in order to estimate the subsurface Vs velocity profile. The determination of the Vs velocity at large depths requires the definition of the dispersion curve at long periods, corresponding to large wavelengths adequate to explore sedimentary deposits of considerable thickness. Usually ambient noise recordings are expected to be rich in waves of long periods during winter, as the intense weather phenomena combined with large sea waves have a strong impact during this period of the year. Therefore all measurements were carried out mostly during the winter period (January to April).

4. ANALYSIS OF MICROTREMORS AND DETERMINATION OF VS PROFILES

Only the vertical component data were used to estimate the SPAC coefficients, as we were interested in the Rayleigh-wave part of the ambient noise recordings (Figure 3). The original signal trains were divided into multiple time windows, whose length varied depending on the desired wavelength and hence the penetration depth that would be expected. For sites where the bedrock depth was expected to be large, the duration of the time windows ranged from 40-60 sec in order to determine the phase velocities of the dispersive Rayleigh waves in long periods, while in the opposite case we selected time windows of 10-20 sec duration in order to enhance the resolution of the method at small depths.

The procedure followed to estimate the SPAC coefficients and to determine the Vs profiles is illustrated in Figures 4 and 5 for a single site in the city of Thessaloniki called KYV (Figure 2). It is located at the south part of the city in a small peninsula composed of deep neogenic stiff clay deposits. Three circular arrays with radii of 10m, 20m and 40m were deployed, keeping the same central station for all measurements.

4.1 Definition of experimental dispersion curves

In the first stage of processing, multiple tests were performed to check the validity of the assumption regarding the stationary nature of the microtremor recordings. If the microtremors could be considered as stationary, then they can be described through stationary random functions and the autocorrelation function between two stations, located one at the center and the other on the periphery of the circular array, correspondingly, can be computed as a function of the phase velocity of the contained surface waves in the microtremor signal.

A random function can be considered as stationary when its power spectrum does not vary significantly with time and space. In the case of the present measurements, the validity of the assumption in time was investigated through comparisons of the power spectra estimated for successive time windows of 2.5 min. length and for each station separately. The validation in space was performed through comparisons of the power spectra calculated for different stations and for the same time window. These comparisons revealed the frequency range for which the microtremor signal appears to be stationary and therefore can be used to estimate the phase velocities. The usable frequency ranges of interest were cross-checked through a comparison of the frequency coherency functions estimated for different stations.

At the specific site KYV and for the large array, the inter-distances of the station pairs were R=40m and D=70m respectively. The selected time windows of the recordings were 64 sec. The same shaped power spectra calculated for one time window of the four stations (Figure 4a) and their coherency functions (Figure 4b) imply that the recorded ambient noise is stable in the frequency range from 0.6 to 3 Hz, which in turn is the frequency range where the SPAC method can be used.

In the second stage, which is the main part of the analysis, the autocorrelation functions were calculated for each pair of stations of equal inter-stations distances, R and D, included in the circular array. The selected frequency range for each array was subsequently used to estimate the autocorrelation functions for the corresponding radius R of the array, and the distances D between pairs of stations on the periphery of the array. The autocorrelation functions were finally used to derive the SPAC coefficients for each frequency, which are related to the phase velocity.

In Figures 4c and 4d we present the autocorrelation functions that were calculated for each pair of equally distant stations and the average autocorrelation function for the corresponding distance, which was used to define the SPAC coefficients at site KYV. In Figure 4e the obtained SPAC coefficients are presented at the finally selected frequencies between 1.4 and 1.8 Hz. Open circles correspond to the observations while solid lines are the obtained first kind zero order Bessel functions fitted to the observed SPAC coefficients, using the least squares method. The incapability to estimate the phase velocity at frequencies from 0.6 to 1.4 Hz is due to the small dimensions of the array which could not provide the large wave-lengths necessary to describe the particular frequency band. For the frequency interval between 2 and 3 Hz, we used the intermediate radius array in order to obtain better resolution in the definition of the upper layers.

The inter-distances of the station pairs used in the analysis of the intermediate and small arrays at site KYV, were R=20m and D=35m and R=10m and D=17m respectively. The duration of the time windows was 16 sec. and 32 sec. for the two arrays. The smaller radii in both arrays were reflected in the frequency interval in which the microtremor signal appears to be stationary. In the case of the intermediate array the exploitable frequencies vary from 2.0 to 4.0 Hz, while in the case of the small array the corresponding frequencies range from 4.2 to 7.0 Hz.

The experimental phase velocity dispersion curve of Rayleigh waves resulting from the combination of the results from the data of the three arrays, at site KYV, is presented in Figure 5. Taking into account the fact that phase velocities correspond to the fundamental mode, it is expected that the dispersion curves from all arrays should be parts of the whole dispersion curve without any gap of the velocities. This is true and the whole dispersion curve in Figure 5 nicely illustrates the success of the experiment. The same procedure has been followed for all other sites.

4.2 Inversion procedure and inverted Vs profiles

The experimental dispersion curves defined above were inverted iteratively using the inversion scheme introduced by Herrmann (1987) to reveal the Vs velocity structure below the circular arrays. The employed software has been developed for the surface-wave recorded during P-wave refraction measurements, which is being applied recently in earthquake engineering. This inversion procedure has been widely applied and has given reliable results in many studies conducted in several areas of the world (Mokhtar et al., 1988; Jongmans, 1991) as well as in Greece (Raptakis et al., 1994, 1998, 2000).

For the determination of the Vs models a complete initial soil model with different Vp and Vs velocities, and density for each layer to the half-space is needed. When, after using an iterative procedure, the theoretical dispersion curve is found to fit well with the experimental one, then it is considered that the artificial model well describes the soil stratigraphy and the velocity profile. Figure 6a shows the experimental and theoretical curves at site KYV proving the nice fitting of the two curves and consequently the accuracy of the defined soil profile and velocity.

The reliability of the inverted Vs profile is confirmed by the shape in the form of a delta function, of the resolving kernels, that correspond to different layers of the initial soil models. The determination of the Vs value for each layer is accurate when the peak of the delta function at the particular depth (horizontal axis) corresponds to the depth (vertical axis) of the VS profile. Figure 6b presents the Vs velocity structure for site KYV with the corresponding resolving kernels. Taking into account the relatively small arrays, the penetration depth of about 320 m is found to be very satisfactory.

The inverted Vs soil profiles for the 16 sites in Thessaloniki are given in Figure 7. The penetration depths are reasonably high and almost in all cases the estimating Vs for the deepest layers overpass the velocity of 750 m/sec, which is usually considered as "seismic bedrock" for site effects studies. Thus, these profiles may be combined with the existing geotechnical and geological data to construct reliable 2D cross sections and 3D maps describing the soil stratigraphy in terms of Vs velocity for the whole city.

4.3 General remarks

The analyses of the microtremors recordings at all sites led to some important conclusions concerning the stationariness and the efficiency of the SPAC method. Comparing the power spectra at all different sites, it is concluded that within a large city for inter-station distances up to 50 m and recording time longer than 30 min, the microtremors can be considered as random stationary function. This is an important fact, which allows using ambient noise measurements (microtremors) to define the Vs profile. It is also important to note that although in some cases the instruments were installed close to areas of intense human activity, the stationary character of the recorded noise was not really affected. This means that distances of 40m to 50m, which can be easily found even in densely urbanized areas, are not inadequate to provide array ambient noise measurements of good quality.

The final frequency range providing reliable information on the phase velocity dispersions depends on the radius of the array and on whether this radius is adequate to include the stationary part of the microtremor in the obtained recordings. This frequency range is reflected in the extent of the output phase velocity dispersion curve of Rayleigh waves and was obtained using Bessel functions of the first kind of zero order. The microtremor analysis and the definition of the autocorrelation functions at all sites showed that the SPAC consists of function of frequency and phase velocity in cases of waves that undergo dispersion. The dispersion curve defined by the SPAC method corresponds to one propagation mode of the Rayleigh waves. This was confirmed in all cases where we deployed multiple arrays of different radius at the same site and the whole estimated dispersion curve did not have any gap of the velocities. The large frequency band covered by the dispersion curve is indicative of the dominance of the fundamental mode.

The frequency range of each dispersion curve reflects the range of penetration depths, i.e. curves extended towards low frequencies correspond to sites where the achieved exploration depth was large. The achieved exploration depth H, is related to the distance R between the employed stations included in the circular array through the relation, $4R \leq H \leq 6R$. This empirical relation does not hold only for three sites (AGO, LAZ and MET), where the geological contact between sediments and rock basement is near the surface. In these sites where lateral variation of soil deposits may be dominant, the MEM by SPAC method is probably less accurate, since the main condition of horizontal and flat stratigraphy is not fully satisfied. As a result, at these sites the exploration depths were relatively small. Table 1 presents the final reliable exploration depth at every site together with the corresponding radius of deployed array. Generally, the exploration depths can be considered as significantly large, considering the rather limited free spaces available in the central part of the city of Thessaloniki.

5. VALIDATION OF THE METHOD

To validate the method and the results we performed multiple comparisons mainly with cross-hole data, which a priori, are considered to give the most reliable estimation of the Vs velocity profiles. We selected two representative sites. One was at the Euroseistest experimental site (http://euroseis.civil.auth.gr) and the second in the city of Thessaloniki (location LEU, see Figure 2).

In Euroseistest we performed a very detailed geotechnical and geophysical survey covering the whole valley. In the centre of the valley where the down-hole accelerometric array is located, the bedrock is found at -200m. The surface soils are loose silty sands and silty clays to soft clays. The site is ideal to compare the MEM technique with Vs profiles from other geophysical surveys in a very well known place. Figure 8 presents the final comparisons between MEM, Cross-Hole, SWI (SASW) and laboratory tests (RC). We used a large diameter array (86m) in order to reach the bedrock. Considering this fact and having in mind that the target is to estimate Vs profiles for ground response and site effect analyses, the comparison is very good. An even better "accuracy" could be achieved for the surface soft-loose soils with Vs<200m/s, using a smaller array.

The second validation example is in the city of Thessaloniki where close to the site LEU, near the "White Tower", the landmark of the city, a cross-hole test 30m deep had been conducted in the past (Pitilakis et al., 1992; Raptakis et al., 1994). Figure 9 compares the Vs profiles calculated by the two methods. Vs profiles are in good agreement, describing well the stratigraphy and the Vs velocity of the main soil formations; some local discrepancies are due to the a-priori more detailed cross-hole measurements.

Based on the above comparisons it is proved that the proposed MEM method is reliable and sufficiently accurate in estimating soil stratigraphy and Vs profiles for 1D soil and site characterization purposes in the framework of ground response and site effect analyses.

6. 3D SOIL AND SITE MAPPING

Combining the estimated Vs profiles with existing geotechnical data of the soil formations met with in the city of Thessaloniki (Figure 2), (Anastasiadis et al. 2001), it is possible to construct reliable 2D cross sections and 3D maps of the geotechnical structure of the city of Thessaloniki and to classify the soils in five categories (A to E) according to their physical properties and the range of Vs values.

Formation A1 is a sub-category of the basic formation A, composed by the softest soils, having Vs values of the order of 130 m/sec and thickness of about 10 m. Geotechnically it is characterized as sandy silt (SM) and soft clay (CL). It is found along the shoreline and mainly at the western part of the city. Formation A consists of a the superficial part of the Neogene formations and presents S-wave velocity values ranging from 180 m/sec to 300 m/sec and thickness from 5 m to 35 m. It is mostly found along the shore line of the city, where it presents its lowest Vs values (230 m/sec); its largest thickness is met also at the western part of the city at the port area. It is characterized as clay with low to medium plasticity (CH-CL). In the eastern part of Thessaloniki, the thickness of formation A varies from 5m to 10m and Vs from 250 m/sec to 300 m/sec. It is composed of clay with gravels (CL-CG).

Formation B is underlying formation A; its Vs values are ranging from 300 m/sec to 450 m/sec and the thickness is varying from 20 m to 130 m. Formation C has Vs values varying from 500 m/sec to 800 m/sec and thickness ranging from 50 m to 200 m. Geotechnically, formations B and C are characterized as stiff clays and stiff clays with sands (CL, CL-SC). Formation D is the deepest soil formation, overlying the real rock basement (formation E), or even being the "seismic" rock basement in certain parts of the city, having Vs values varying from 750 m/sec to 1100 m/sec. It can be considered as the "seismic bedrock", for 1D/2D site effect analysis not only because of its large Vs velocity values but also due to its high depth. The exact thickness of this formation cannot be specified everywhere because the depth of the rock basement could not be determined at all sites. However using hydrological deep borehole logs and other available geological and geotechnical information, it is possible to construct a quite accurate model of the deep basin stratigraphy. This has been done in Thessaloniki. The resulting image of the seismic basement is geologically reliable.

The final morphology and the depth of the basement has been also cross-checked and validated using back calculated 1D Vs profiles estimated from the fundamental period deduced from H/V spectral ratios calculated from the same set of ambient noise measurements using the same broad band stations (Apostolidis, 2002). Figure 10 presents the calculated fundamental frequencies in different locations with different thickness of sediments in the city of Thessaloniki. The comparisons with the existing geological, geotechnical data and the results of the MEM profiles are very good, proving once more the reliability of the method described herein.

Based on the combined data we have constructed many 2D cross-sections and 3D maps of the subsoil stratigraphy in the city of Thessaloniki. Both 2D cross sections and 3D maps are proposed using the mean Vs velocities and proper geological – geotechnical judgment. Figures 11 and 12 present two typical cross-sections (see Figure 2). It is worth noticing the rather complex morphology of the alluvial deposits and bedrock. Using many 2D cross-sections it was feasible to produce a set of 3D GIS maps of the 5 basic soil formations. Figure 13 shows a map of the total thickness of the sediments overlying the stiff formation D, while Figure 14 presents the spatial distribution and the iso-depth curves of the basic formation B (stiff clay).

The fact that the geometry of the basin, including weathered bedrock, is complex means that complex site effects are to be expected, since it is well known that the velocity contrast of the sediments and the bedrock, together with their geometry are the most important parameters that control the amplification characteristics of the induced ground motion. 2D or 3D phenomena will certainly appear, producing additional amplification to simple 1D site response (Chavez-Garcia et al., 2000; Raptakis et al., 2000; Makra et al., 2001).

The results presented herein were obtained mainly from array ambient noise (microtremor) measurements analyzed using the Spatial Autocorrelation Coefficient method. MEM is certainly proved to be a powerful tool for soil and site description, especially in a densely populated area, where it is difficult to use conventional seismic prospecting.

7. CONCLUSIONS

Microtremor recordings measured within large densely urbanized cities, such as Thessaloniki, are proved to be stationary random functions, and consequently the SPAC method can be used with spatial extent of 100m and temporal extent of 30 min, without significant perturbation due to biased sources. This stagnancy is a very sensitive parameter and therefore the radii of the circular arrays should be increased, remaining always quite small (<50m), in order to succeed the largest possible exploration depths without violating the assumption of stagnancy. The microtremor measurements in Thessaloniki proved that the unavoidable human activities close to the arrays (e.g. cars, machines etc.) were mainly observed at high frequencies not affecting the recordings for the Vs estimations. Regarding the wave and frequency content of mictrotremors, the determination of dispersion curves at all the examined sites indicates that surface waves basically dominate the recorded signal. In these surface waves one mode is dominant, which is probably the fundamental one, as concluded from the large width of frequency range determination of the dispersion curves at all sites.

The only significant limitation with respect to the array aperture of the method, which was expected theoretically anyway, is the assumption of horizontal layering of soil deposits. The analysis showed that at sites where this assumption is probably violated (based on geological evidences or other information i.e. buried archeological members and discontinuities), it is only possible to determine the Vs values of the surface soil formations, while it is difficult to extract accurate information on the total thickness of the soils deposits and consequently the depth of the bedrock and its Vs velocity. In other words, at sites where the geological contact between soil and rock appears, the proposed MEM using the SPAC method may not be able to calculate the phase velocities for the wavelengths that correspond to the total soil thickness. These geological conditions are usually met at sites where the rock basement is close to the surface. As a result, at sites presenting the previously mentioned conditions, the MEM can be used to investigate only the near surface soil formations and obviously a complementary conventional geophysical survey will be necessary, especially for the determination of the depth/slope and the velocity of the underlining rock basement.

The most significant advantage of the MEM, which stems from the present study, is that it allows the reliable determination of Vs velocity profiles down to large depths (100 m - 300 m) with relatively small arrays apertures (radius<50m). This is an extremely fortunate outcome for ground response studies within large cities, where open spaces, suitable for the deployment of large arrays, are difficult to be found and high energy sources cannot be easily accepted. Compared to other geophysical exploration methods, such as reflection surveys or cross-hole testing, the microtremor array measurements does not require boreholes or any kind of artificial source and therefore has the advantage of exploring large depths with relatively low cost, and in a more convenient way, regarding the necessary field works. Therefore, the MEM can be considered as a valuable low-cost complementary or alternative method/tool for determining Vs velocity and soil stratigraphy, for soil and site characterization, necessary for site response analyses, microzonation studies, design of infrastructures and urban planning.

ACKNOWLEDGEMENTS

The work presented herein is part of the PhD Thesis of my student Dr P. Apostolidis. His efforts and the contribution of Dr D. Raptakis are very much appreciated.

REFERENCES

1. Aki K., 1957. Space and Time Spectra of Stationary Stochastic Waves, with Special Reference to Microtremors. Bull. Earthq. Res. Inst. Tokyo Univ. 25, pp. 415-457.
2. Aki K., 1965. A note on the use of microseismic in determining the shallow structure of the earth's crust, Geophysics 30, pp. 665-666.

3. Aki, K. and B. Chouet, 1991. Characteristics of seismic waves composing Hawaiian volcanic tremor and gas-piston events observed by a near-source array. Journal of Geophysical Research, Vol. 96, No. B4, pp. 6199-6209.
4. Anastasiadis, A., D. Raptakis, and K. Pitilakis, 2001. Thessaloniki's Detailed Microzoning: Subsurface Structure as Basis of Site Response Analysis, PAGEOPH Special Issue on Microzoning, Vol. 158, N.11, pp. 2497-2533.
5. Apostolidis, P., 2002. Determination of the soil structure using microtremors. Application to the estimation of the dynamic properties and the geometry of the soil formations at Thessaloniki city. PhD Thesis, Aristotle University of Thessaloniki, Greece.
6. Capon, J., 1969. High-resolution frequency-wave number spectrum analysis, Proc. IEEE, 57, pp. 1408-1418.
7. Chavez-Garcia, F.J., D. Raptakis, K. Makra and Pitilakis K., 2000. Site effects at Euroseistest-II. Results from 2D numerical modelling and comparison with observations. J. of Soil Dyn. and Earthq. Engnr, 19(1), pp. 23-39.
8. Chouet, B. G. Luca, G. Milana, P. Dawson, M. Martini and R. Scarpa, 1998. Shallow Velocity Structure of Stromboli Volcano, Italy, Derived from Small-Aperture Array Measurements of Strombolian Tremor, Bull. Seism. Soc. Am. Vol. 88, No 3, pp. 653-666.
9. Goldstein, P. and B. Chouet, 1994. Array measurements and modelling of sources of shallow volcanic tremor at Kilauea Volcano, Hawaii. Journal of Geophysical Research, Vol. 99, No. B2, pp. 2637-2652.
10. Hatzidimitriou, P.M., D. Hatzfeld, E.M. Scordilis, E.E. Papadimitriou and Christodoulou A.A., 1991. Seismotectonic evidence of an active normal fault beneath Thessaloniki (Greece), TERRAMOTAE, pp. 648 – 654.
11. Henstridge, J. D., 1979. A signal processing method for circular arrays, Geophysics, 44, pp. 179-184.
12. Herrmann, R., 1985. Computer programs in seismology, vol. III., Saint Louis University.
13. Horike, M., 1985. Inversion of phase velocity of long-period microtremors to the S-wave-velocity structure down to the basement in urbanized areas, J. Phys. Earth, 33, pp. 59-96.
14. Hough, S., L. Seeber, A. Rovelli, L. Malagnini, A. DeCesare, G. Selveggi and A. Lerner-Lam, 1992. Ambient noise and weak motion excitation of sediments resonances: results from Tibel valley, Italy, Bull. Seis. Soc. Am. Vol. 82, No 3, pp. 1186-1205.
15. IGME, 1995. Geological – Geotechnical map of the city of Thessaloniki, scale 1:5000.
16. Kanno, T., Kudo K, M. Takahashi, T. Sasatani, S. Ling and H. Okada, 2000. Spatial evaluation of site effects in Ashigara valley based on S-wave velocity structure determined by array observations of microtremors. Proc. of 12WCEE 2000, pp. 572-580.
17. Kudo K., T. Kanno, H. Okada, T. Sasatani, N. Morikawa, P. Apostolidis, K. Pitilakis, D. Raptakis, M. Takahasi, S. Ling, H. Nagumo, K. Irikura, S. Higashi and K. Yoshida. S-Wave Velocity Structure of EUROSEISTE, Volvi, Greece, Determined by the Spatial Auto-Correlation Method applied for Array Records of Microtremors. Proceedings of Earthquake Engineering Symposium, November, 2002, Japan pp. 15- 28.
18. Kudo, K., T. Kanno, H. Okada, O. Ozel, M. Erdik, T. Sasatani, S. Higashi, M. Takahashi and K. Yoshida, 2002. Site specific issues for strong ground motions during the Kocaeli, Turkey Earthquake of August 17, 1999, as inferred from array observations of microtremors and aftershocks, Bull. Seis. Soc. Am., Vol. 92-1.
19. Lacoss, R. T., E. J. Kelly and M. N. Toksoz, 1969. Estimation of seismic noise structure using arrays, Geophysics, 34, pp. 21-38.
20. Ling, S., and H. Okada, 1993. An extended use of the spatial autocorrelation method for the estimation of the geological structures using microtremors (in Japanese), Proceedings of the 89th SEGJ Conference, pp. 44-48.

21. Jongmans, D., 1991. L' influence des structures geologiques sur l'amplification des ondes sismiques, These de Doctorat, Universite de Liege, Belgium.
22. Makra, K., D. Raptakis, F. J. Chavez-Garcia and K. Pitilakis, 2001. Site effects and Design Provisions: The case of Euroseistest. J. Pure and Applied Geophysics. 158(2001), pp. 2349-2367.
23. Matstushima, T., and H. Okada, 1990. Determination of deep geological structure under urban areas using long-period mictrotremor, BUTSURI TANSA, 43-1, pp. 21-33.
24. Metaxian, J.-P., P. Lesage and J. Dorel, 1997. Permanent tremor at Mesaya Volcano, Nicaragua: wave field analysis and source location, Journal Geophysics Research 102, pp. 22529-22545.
25. Mokhtar, T.A., R. B. Herrmann, D.B. Russell, 1988. Seismic velocity and Q model for the shallow structure of the Arabian shield from short-period Rayleigh waves, Geophysics, 53, No 11, pp. 1379-1387.
26. Okada H., 1997. A new method of underground structure estimation Using Microtremors. Division of Earth Planetary Sciences, Graduate School of Science, Hokkaido University, Japan, Lecture notes.
27. Okada, H., T. Matsushima, T. Moriya and T. Sasasatani, T., 1990. An exploration technique using long-period microtremors for determination of deep geological structures under urbanized areas (in Japanese), BUTSURI-TANSA, 43, pp. 402-417.
28. Okada H., 1999. A New Passive Geophysical Exploration Method Using Microtremors. Division of Earth Planetary Sciences, Graduate School of Science, Hokkaido University, Japan, Lecture notes.
29. Euroseismod, Final Scientific Report, 1999. Development and Experimental Validation of Advanced Modelling Techniques in Engineering Seismology and Earthquake Engineering (Project Co-ordinator K.D. Pitilakis).
30. Pitilakis, K.D., Anastasiadis, A.I., Raptakis, D.G., 1992. Field and Laboratory Determination of Dynamic Properties of Natural Soil Deposits. Proc. 10th World Conference on Earthquake Engineering, Madrid, Vol. 3, pp. 1275-1280.
31. Pitilakis, K., D. Raptakis, K. Lontzetidis, Th. Tika-Vassilikou & D. Jongmans, 1999. Geotechnical & geophysical description of EURO-SEISTEST, using field, laboratory tests and moderate strong motion recordings. J. of Earthq. Eng., 3(3), pp. 381-409.
32. Raptakis, D.G., A.J. Anastasiadis, K.D. Pitilakis and K.S. Lontzetidis, 1994. Shear wave velocities and damping of Greek natural soils. Proc. 10th European Conf. on Earthq. Engng, Vienna, Austria, pp. 477-482.
33. Raptakis, D., Lontzetidis, K., Pitilakis, K., 1996. Surface Waves Inversion Method: A Reliable Method for the In Situ Measurements of Shear Wave Velocity. Proc. 4eme Colloque Nationale de Genie Parasismique et Aspects Vibratoires dans le Genie Civil, Vol. I, pp. 160-169, A.F.P.S, 10-12 Avril 1996, Paris, France.
34. Raptakis, D.G, Anastasiadis, A.J., and K.D. Pitilakis, 1998. Preliminary Instrumental and Theoretical Approach of Site Effects in Thessaloniki. Proc. 11th European Conference on Earthquake Engineering (CDROM), Paris, France.
35. Raptakis, D.G., Chavez-Garcia F., Makra, K.A. and K.D. Pitilakis, 2000. Site Effects at Euroseistest-I. 2D Determination of the Valley Structure and Confrontation of the Observations with 1D Analysis. Soil Dynamics & Earthquake Engineering, 19 (1), pp. 1-22.
36. Raptakis, D., 2000. Soil dynamic testing and geophysical survey for the design of the metro of Thessaloniki. Technical report, Bouygues T. P., (project coordinator K. Pitilakis), pp. 88-103.

37. Raptakis, D., Apostolidis, P., Kudo, K. and K. Pitilakis, 2002. Vs soil structure definition using the inversion of microtremors and seismic prospecting dispersion data (submitted for publication).
38. Satoh T., K. Hiroshi and Shin'ichi Matsushima, 2001. Differences between site characteristics obtained from microtremors S-waves, P-waves, and Codas, Bull. Seis. Soc. Am., Vol. 91, 2, pp. 313-324.
39. Vernon, L., J Fletcher, C. Linda, C. Alan and E. Sembera, 1991. Coherence of Seismic Waves from Events as Measured by a Small-Aperture Array, Journal of Geophysical Research, Vol. 96. No B7, pp. 11981-11996.

Table 1. Day time of measurements and the radii of the deployed circular arrays at all sixteen sites together with the final obtained maximal exploration depths, for which the Vs profiles were determined

CODE	DAY	TIME	ARRAYS	RADII (m)	EXPLOR. DEPTH
KAL	Tuesday	10.0-16.0	2	15, 35	280
KYV	Thursday	10.0-18.0	3	10, 20, 40	320
KRI	Sunday	4.30-11.0	2	10, 35	240
MPO	Sunday	5.0-12.0	2	20, 30	180
KON	Sunday	10-16.0	2	15, 30	250
IPO	Sunday	5.0-9.0	1	10	110
TOU	Saturday	10.0-16.0	2	15, 30	180
LEU	Saturday	4.30-11.0	2	10, 20	130
MET	Wednesday	10.0-16.0	1	10	35
AGO	Sunday	5.0-8.0	1	15	60
LIM	Saturday	7.0-13.0	2	10, 50	180
TEL	Sunday	5.0-8.0	1	15	160
LAZ	Saturday	10.0-16.0	2	15, 30	140
STA	Saturday	5.0-11.0	1	15	80
EUO	Wednesday	10.0-16.0	2	15, 35	220
SST	Monday	12.0-15.0	1	10	80

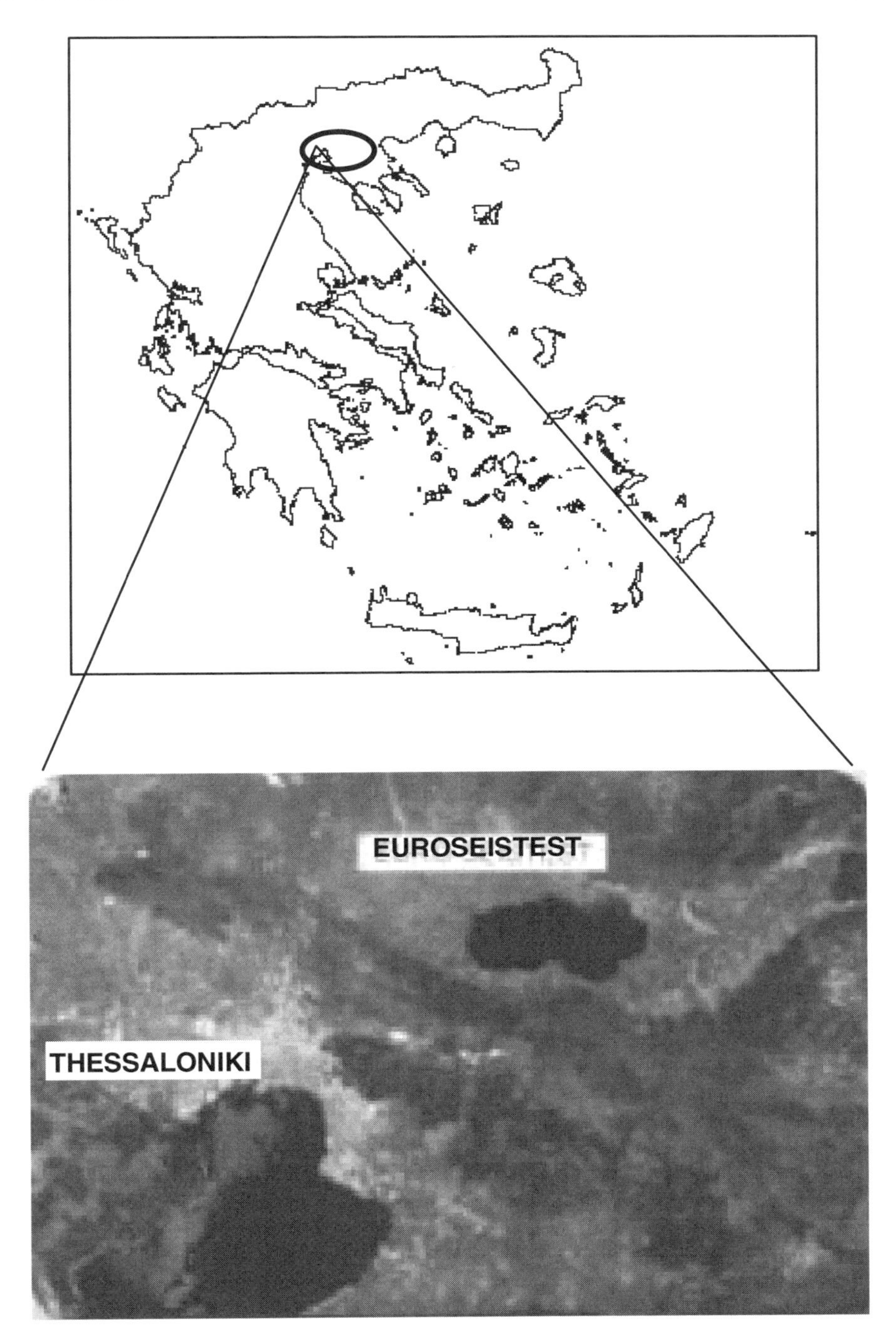

Figure 1. Map of Greece-Thessaloniki and the location of Euroseistest experimental site (http://euroseis.civil.auth.gr)

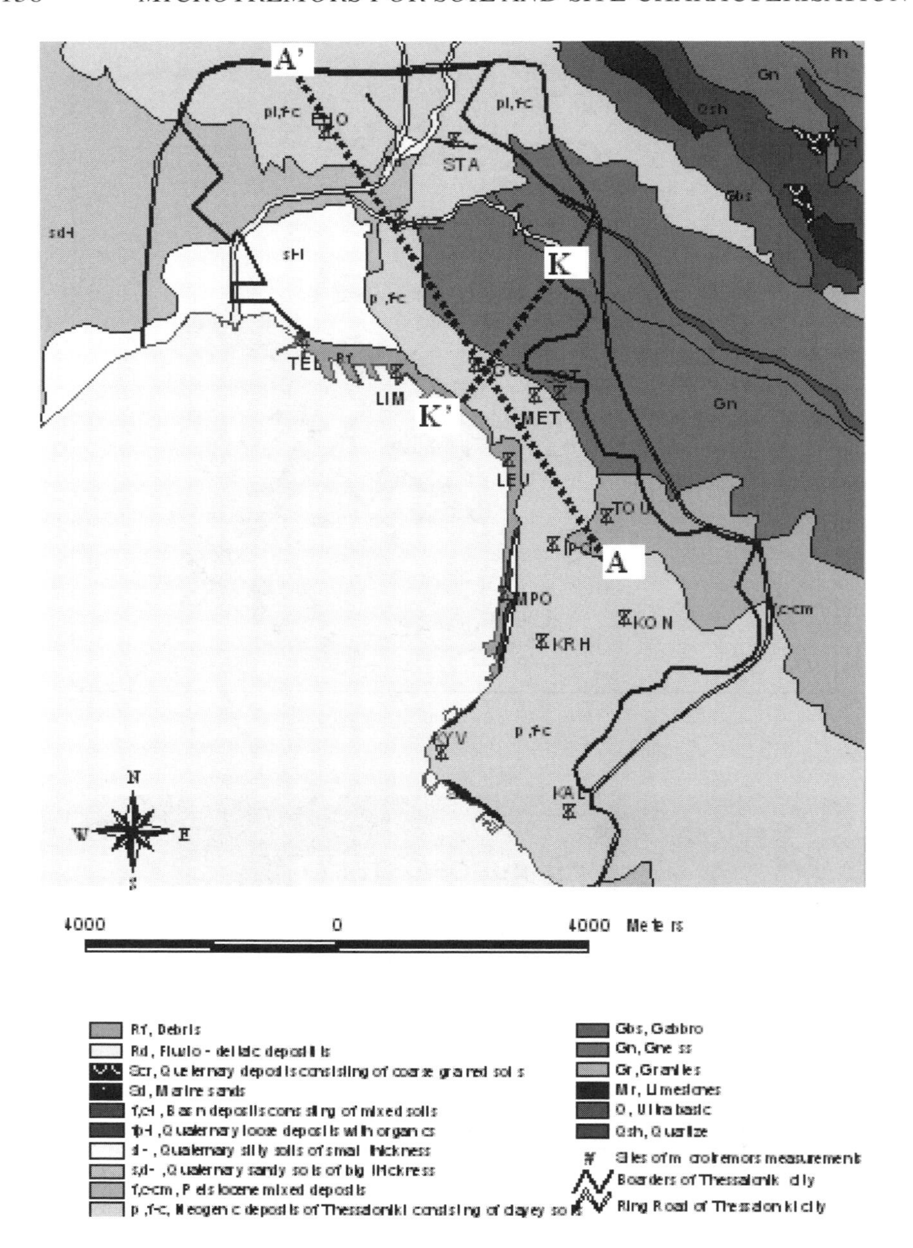

Figure 2. Geological – Geotechnical map of Thessaloniki. Locations of the 16 sites of the microtremor measurements

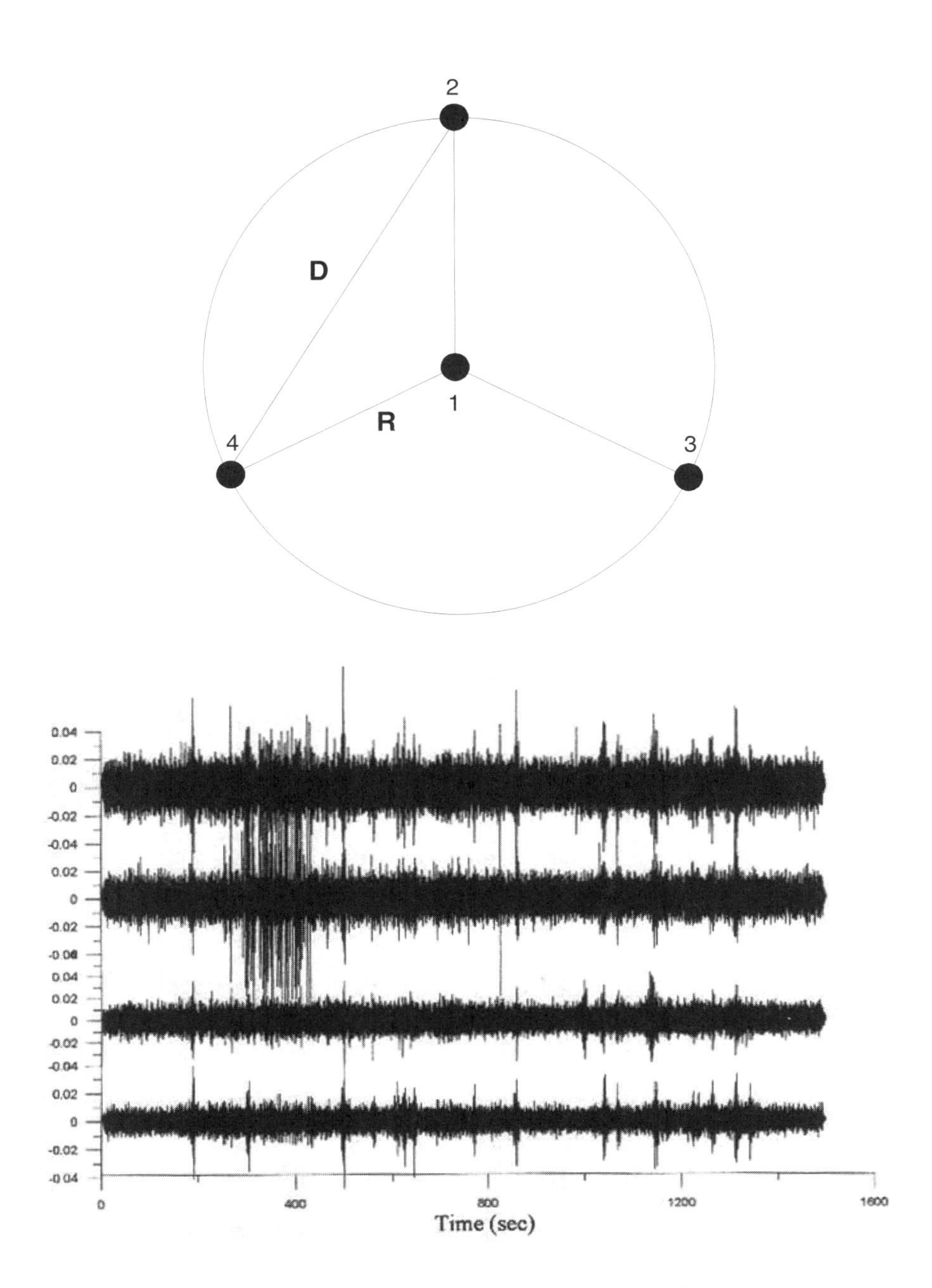

Figure 3. Typical configuration of the circular array of the broad band stations for microtremors detections and microtremor recordings at site KYV

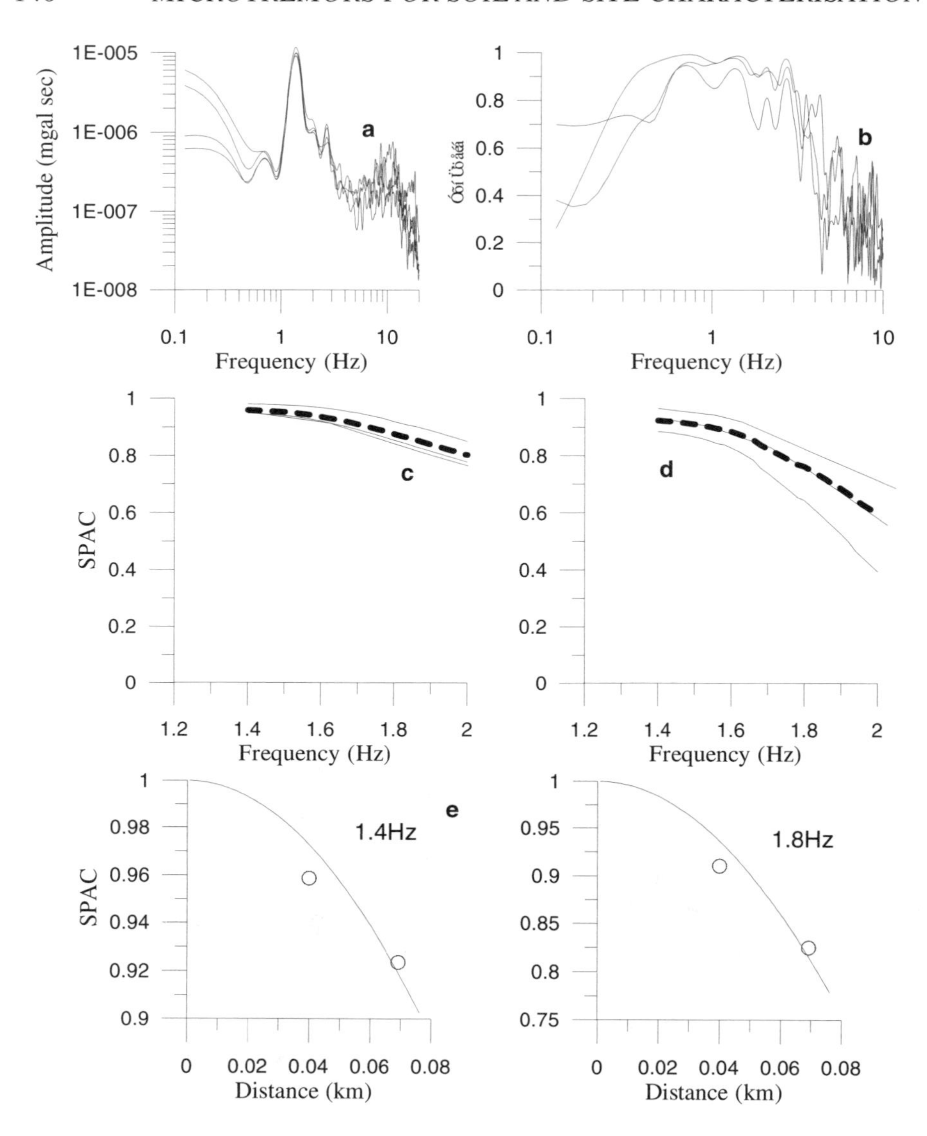

Figure 4. Typical analysis procedure for the estimation of the SPAC coefficients for the large array at site KYV a) Fourier power spectra of the microtremor signal at the four recording stations of the array, b) frequency coherency functions for pairs of stations for R = 40m, c) average autocorrelation functions (dotted lines) and autocorrelation functions (solid lines) of every pair of stations, d) same as in case (c) for the distance D of 70m, e) observed SPAC coefficients (circles) and fitted Bessel functions (continuous line) for representative frequencies 1.4 and 1.8 Hz, respectively

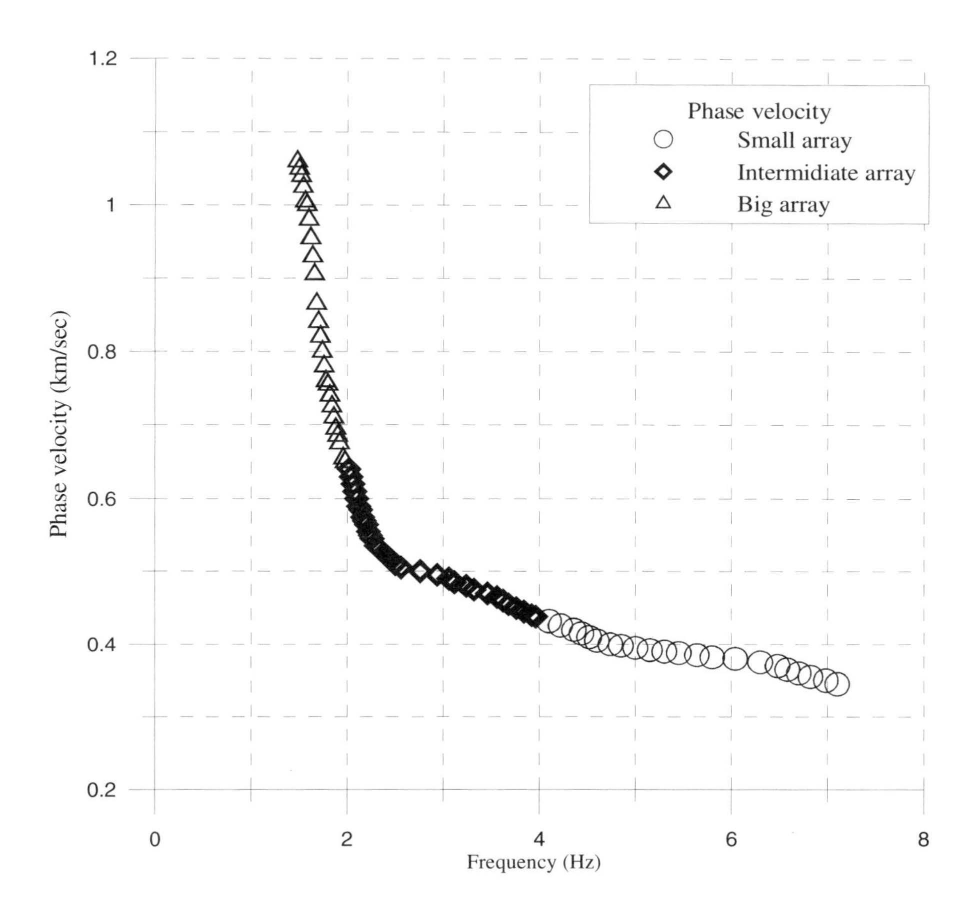

Figure 5. Calculated phase velocity dispersion of Rayleigh waves at site KYV for small, intermediate and large array

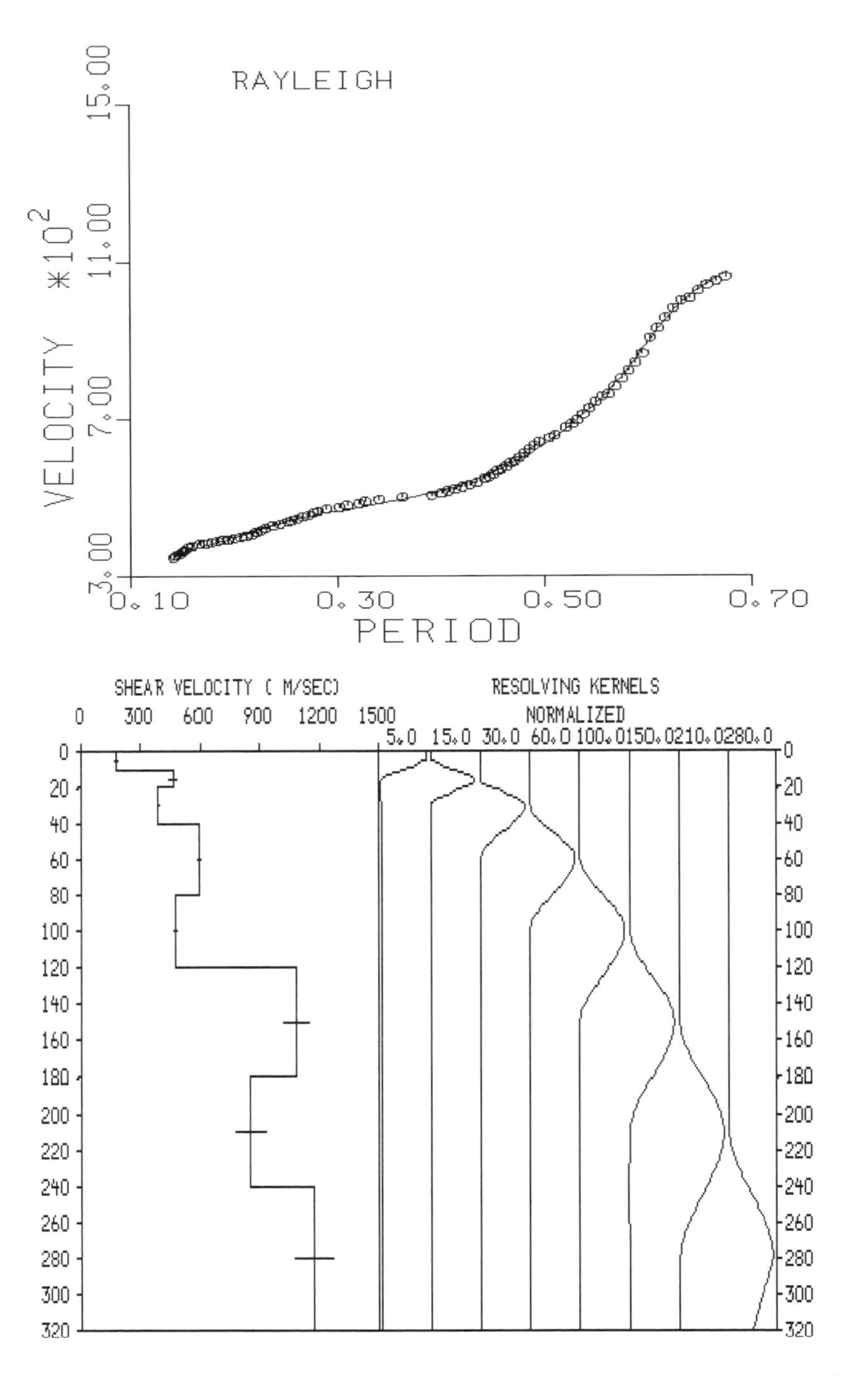

Figure 6. Experimental and theoretical dispersion curves of phase velocities at KYV and final inverted Vs profile with the resolving kernels

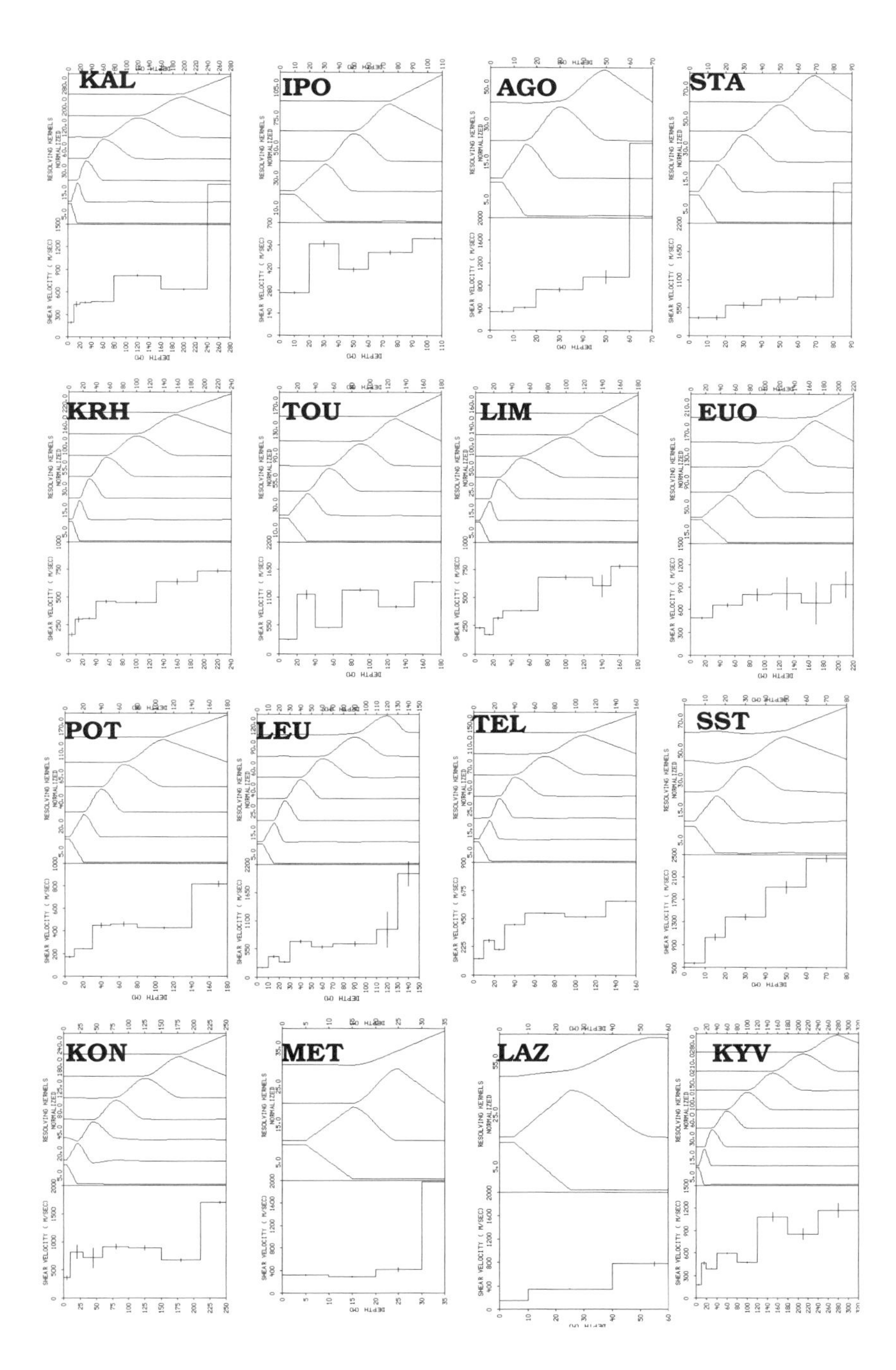

Figure 7. Calculated Vs profiles using microtremor array measurements, at different (16) locations in Thessaloniki

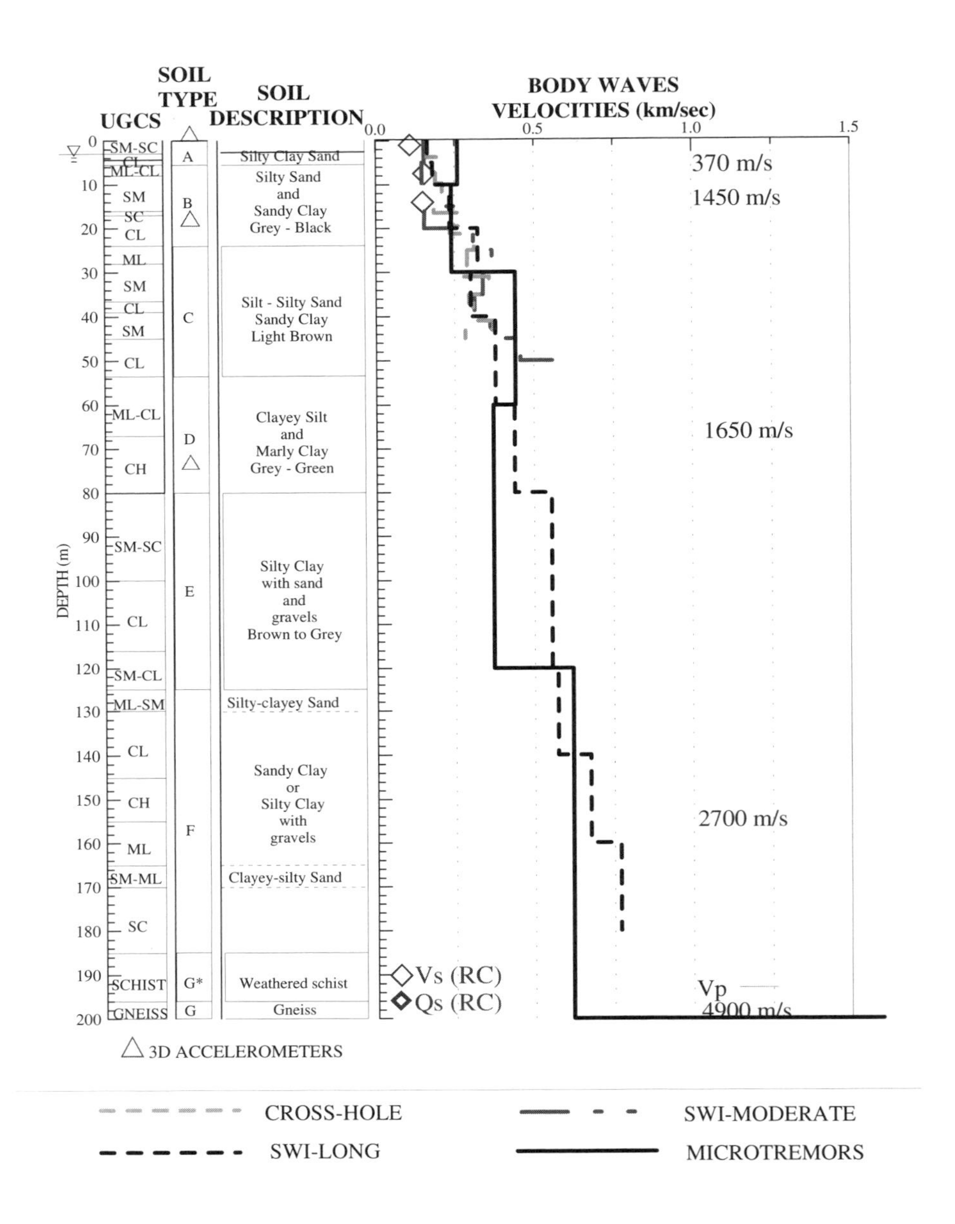

Figure 8. Comparisons of Vs profiles applying different techniques at Eusroseistest experimental site. Microtremor technique with bold line

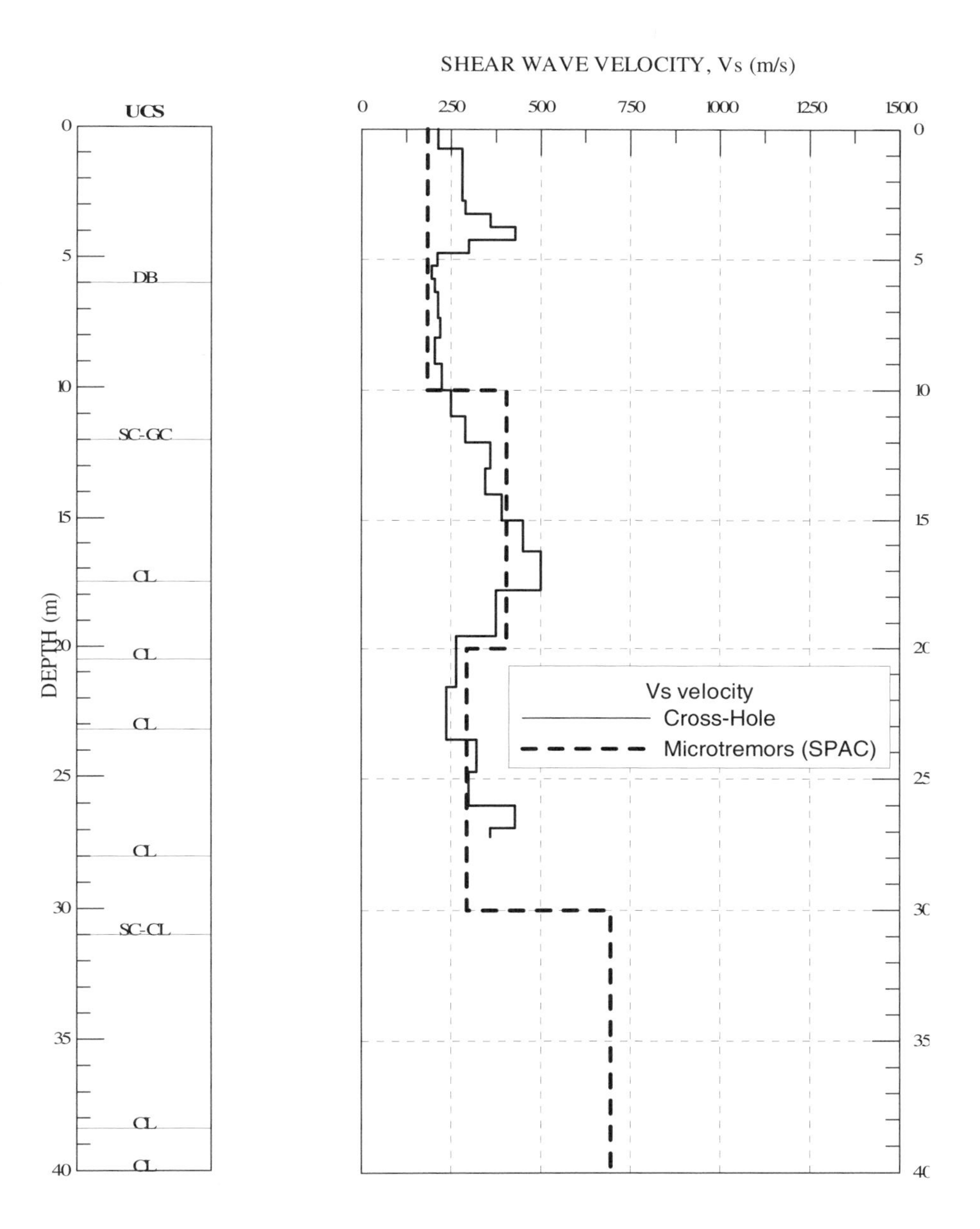

Figure 9. Comparisons of cross-hole Vs measurements and MEM at the location LEU (see Figure 2) in Thessaloniki

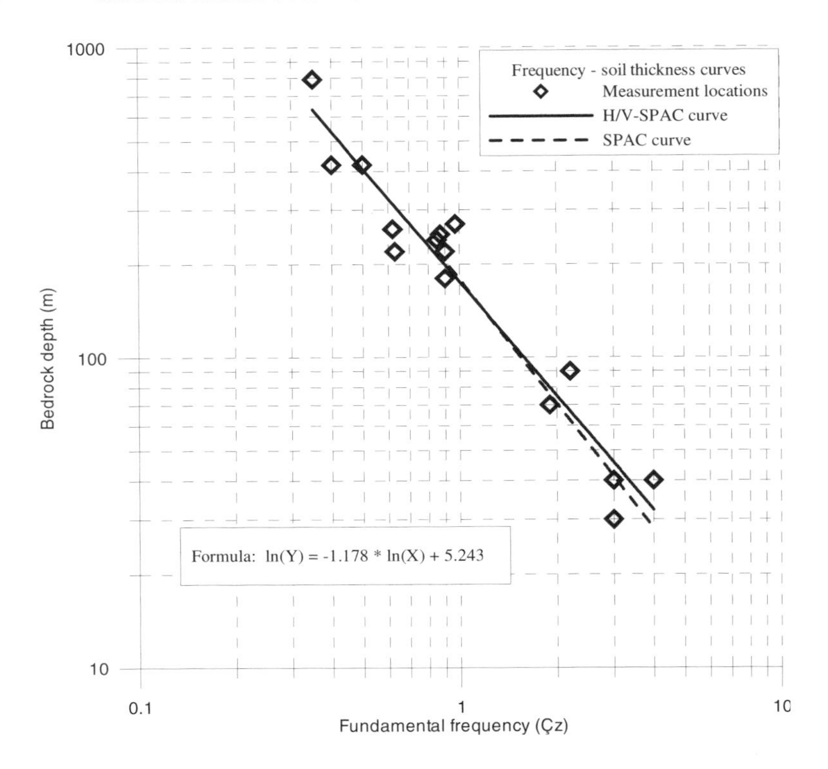

Figure 10. Fundamental periods of soil profiles at the 16 MEM locations in Thessaloniki using ambient noise measurements and applying the Horizontal to Vertical Spectral Ratio (HVSR) technique

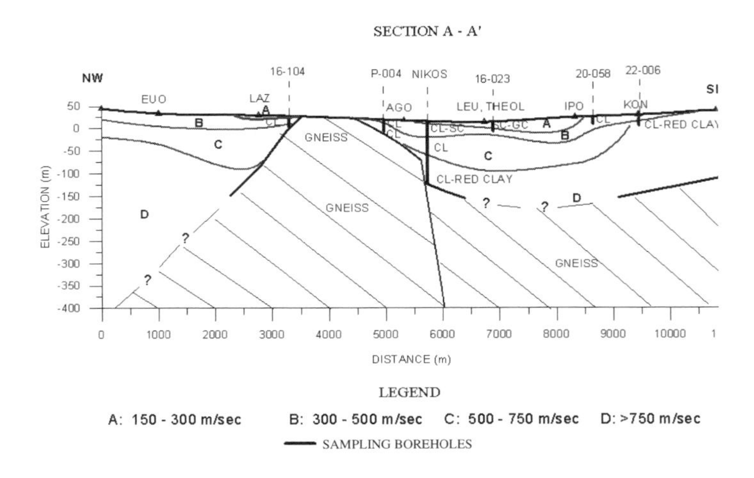

Figure 11. Characteristic 2D cross-section (NW-SE) crossing all soil formations constructed using the present Vs profiles and existing geological and geotechnical data (see Figure 2)

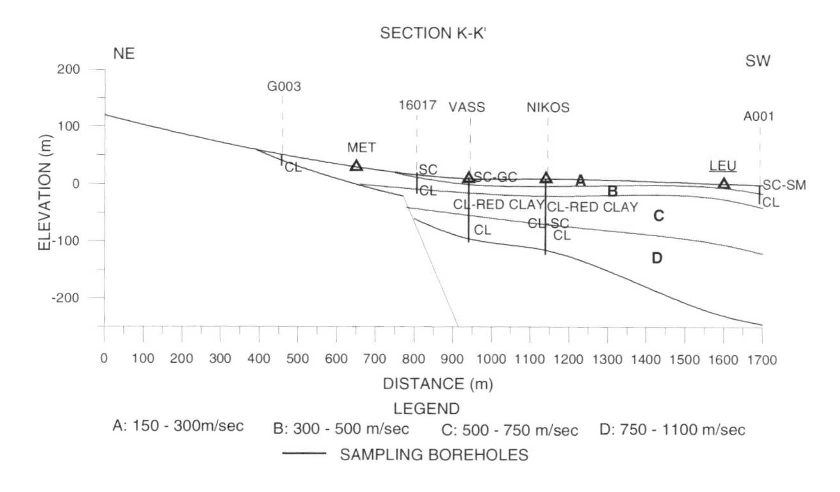

Figure 12. Characteristic 2D cross-section (NE-SW) (see Figure 2)

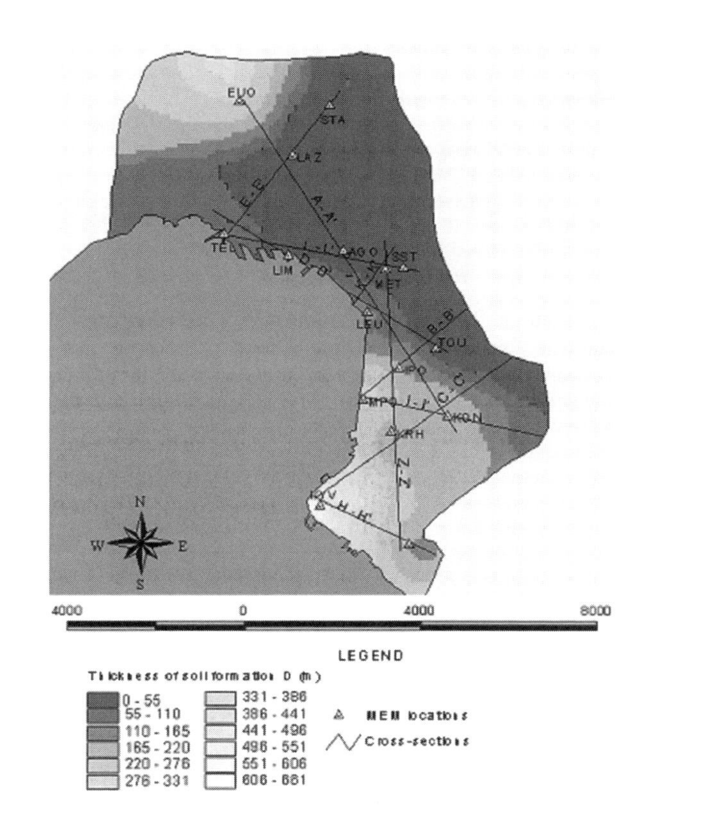

Figure 13. 3D GIS map of the thickness of soil formation B (stiff clays, SC, CL-CH, Vs = 300m/s-500m/s)

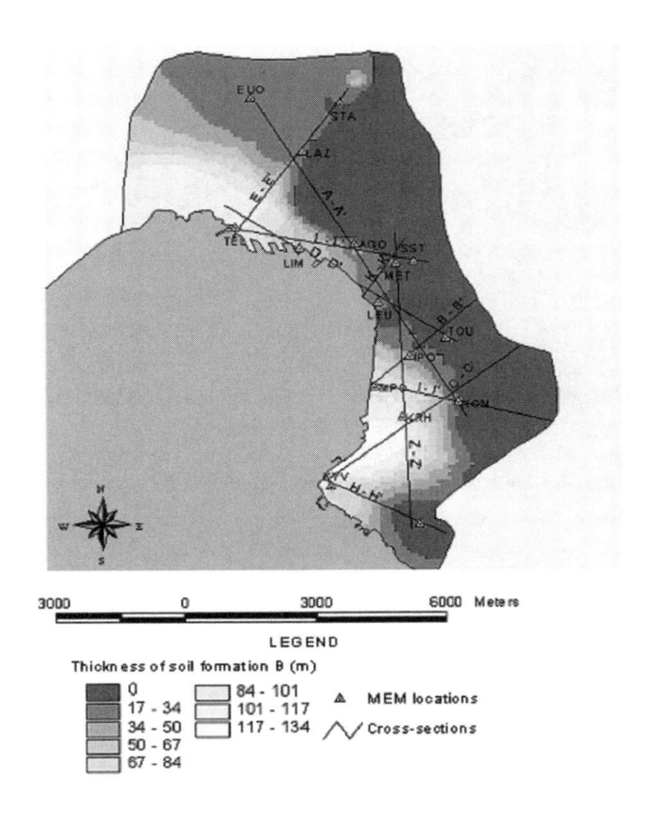

Figure 14. 3D GIS map of the total thickness of the underlain sediments to the upper surface of the formation D, considered as 'seismic bedrock', (Vs larger than 750m/sec)

ESTIMATION OF EARTHQUAKE DAMAGE PROBABILITIES FOR REINFORCED CONCRETE BUILDINGS

M.S. Yucemen[1] and A. Askan[2]
[1]*Structural Mechanics Division, Department of Civil Engineering, Middle East Technical University, Ankara, 06531, Turkey*
[2]*Computational Mechanics Laboratory, Department of Civil and Environmental Engineering, Carnegie Mellon University, Pittsburgh, 15213, PA, USA*

Abstract: Due to uncertainties involved both in the occurrence of earthquakes and in structural response, earthquake damage prediction has to be treated in a probabilistic manner. In this study two statistical methods are presented for the prediction of potential seismic damage to low and mid-rise reinforced concrete buildings in Turkey. These methods are based on the utilization of damage probability matrices and reliability theory. The damage data compiled during recent earthquakes that occurred in Turkey are used to compare the predictions of these two methods.

Key words: Earthquake damage estimation, damage probability matrix, fragility curves, seismic resistance index, reliability theory, Düzce earthquake

1. INTRODUCTION

In recent years, various researchers have been investigating earthquake scenario studies, aiming at the prediction of the extent of possible damage during a future earthquake. In the assessment of earthquake insurance premiums and in earthquake scenario simulations, besides the seismic

S.T. Wasti and G. Ozcebe (eds.), Seismic Assessment and Rehabilitation of Existing Buildings, 149–164.

hazard, it is also important to gather information on the degree of damage that different structures will experience when subjected to earthquakes of different intensities. Even among the same type of buildings subjected to the same earthquake intensity, the degree of damage may differ significantly, depending on the differences in workmanship, material quality, the degree of violation of the earthquake resistance design provisions, soil conditions, etc. Accordingly, the damage to buildings should be investigated based on a probabilistic approach and should be described by a probability distribution. In this study two different probabilistic procedures, which are based on the utilization of damage probability matrices and reliability theory, are presented for the prediction of potential damage to reinforced concrete buildings. The damage data compiled during recent earthquakes that occurred in Turkey are used to compare the predictions of these two methods.

A damage probability matrix (DPM) expresses what will happen to buildings, designed according to some particular set of requirements, during earthquakes of various intensities. In this study, as the first approach, "best estimate" damage probability matrices for each seismic zone are developed by combining expert opinion and the damage statistics compiled from the recent earthquakes occurred in Turkey.

The second approach involves a reliability-based model, which treats the earthquake force and seismic resistance as random variables. This model expresses potential seismic damage in the form of a damage rate distribution, which is a function of modified Mercalli intensity or peak ground acceleration.

2. DAMAGE PROBABILITY MATRICES

2.1 Definition of a Damage Probability Matrix

A damage probability matrix (DPM) expresses what will happen to buildings, designed according to some particular set of requirements, during earthquakes of various intensities [1]. The general form of a DPM is shown in Table 1. An element of this matrix, P_k (DS, I), gives the probability that a particular damage state (DS) occurs when a kth-type structure is subjected to an earthquake of intensity, I. The identification of damage states is achieved in two steps:

a) The qualitative description of the degree of structural and non-structural damage by words. In the damage evaluation forms used by the General Directorate of Disaster Affairs prior to 1994, five levels of damage states were specified. These are: no damage (N), light damage (L), moderate damage (M), heavy damage (H), and collapse (C) states. The above categorization of damage states is also used in this study.

b) The quantification of the damage described by words in terms of the damage ratio (DR), which is defined as the ratio of the cost of repairing the earthquake damage to the replacement cost of the building. For mathematical simplicity it is convenient to use a single DR for each DS. This single DR is called the central damage ratio (CDR). Based on interviews with experts in charge of damage evaluation and similar studies, damage ratios corresponding to the five damage states are estimated by Gurpinar et al. [2] and are shown in Table 1.

2.2 Assessment of Damage Probability Matrices

Depending on the type of structures, different DPM's exist. In the present study only DPM's for conventional reinforced concrete frame buildings are considered. Damage probability matrices can be obtained in the most reliable way from past earthquake data and also by using subjective judgment of experts. Techniques based on theoretical analyses for developing DPM's are also available [1]. In this study, DPM's are obtained by using both empirical results and subjective judgment of experts.

Using the post-event observational data of past earthquakes, P_k (DS, I) values can be calculated from:

$$P_k(DS, I) = \frac{N(DS, I)}{N(I)} \tag{1}$$

In equation 1, N(I) = number of kth-type of buildings in the region subjected to an earthquake of intensity I and N(DS, I) = number of buildings which are in damage state DS, among the N(I) buildings.

In order to obtain the DPM's that are applicable for the four seismic zones of Turkey, previous studies on the assessment of DPM's for reinforced concrete buildings located in different seismic zones of Turkey are examined and revised. The development of DPM's for Turkey was first considered by Gurpinar, et al. [2] and Gurpinar and Yucemen [3]. In these studies, based on the available damage evaluation records from the Denizli (1976) and Bingol (1971) earthquakes, it was only possible to estimate the damage state probabilities in the MMI=VI and MMI=VIII columns of the DPM. Since the empirical data were inadequate for establishing a DPM completely, it was decided to estimate damage state probabilities by making use of the subjective judgment of experts. For this purpose a questionnaire was prepared and sent to thirty engineers experienced in the field of earthquake engineering. Only ten engineers responded and the responses of these ten engineers are averaged to obtain the subjective DPM's. These DPM's give two sets of subjective damage probabilities for reinforced concrete frame buildings constructed in the four different seismic zones of Turkey. The first set corresponds to buildings that are designed and constructed in conformance with the specifications designated in the Code (AC), and in the second set it was assumed that the earthquake resistant design provisions are violated (NAC). In this study the "Code" refers to "Specifications for Structures to be Built in Disaster Areas" which was prepared and put into regulation in 1975 [4]. Because of space limitation, only the "subjective" DPM corresponding to seismic zone I, where the seismic hazard is the highest, is presented here (see Table 2).

Later, Yucemen and Bulak [5] have obtained empirical DPM's by using the post-earthquake damage assessment reports compiled by the General Directorate of Disaster Affairs for the earthquakes of Bingol (1971), Denizli (1976), Erzincan (1983, 1992) and Malatya (1986). Since, generally these reports lack information on the date of construction, it was not possible to classify the buildings as being constructed before or after the 1975 Code. These DPM's are revised and updated by Yucemen [6] based on additional information [7] assessed on these earthquakes. Also the damage assessment reports prepared by various institutions [8-10] concerning the recent earthquakes, namely Dinar (1995), Adapazari (1999) and Düzce (1999) are also utilized, especially for complementing the empirical DPM at higher intensity levels. The damage state probabilities are computed by using equation 1. The resulting empirical DPM is shown in Table 3. In the last row of this table, the number of buildings for which damage assessments were made is given. The empirical values given in Table 3 are assumed to be valid for reinforced concrete structures that are constructed not in accordance with the Code, except for the case of Dinar, where AC and NAC conditions are differentiated.

The values given in Table 3 are used to obtain the empirical damage state probabilities valid for different seismic zones by relating the cities with the seismic zones. Table 4 gives the damage state statistics for zone I with MMI=VI, VIII and IX and for zone II with MMI=VII and VIII. For cases where more than one earthquake data were available for the same zone, weighted average damage state probabilities, based on the number of buildings, are computed. The empirical damage state probabilities given in this table correspond to the NAC case, as explained above.

The DPM's will show differences from zone to zone. Therefore for each zone a DPM is needed. Besides, whether a building has been constructed according to the requirements of the Code [4] or not should be taken into consideration. In selecting these DPM's it is desirable to utilize all of the relevant information in a systematic way. For this purpose the empirical DPM (Table 4) is combined with the subjective DPM's (e.g. Table 2 for seismic zone I) by computing a set of weighted average DPM's. A subjective weight of 0.75 is assigned to empirical values whenever they are available and a weight of 0.25 is given to the subjective DPM's that are reflecting expert opinion.

The DPM's formed in this way for the four seismic zones are called the "best estimate" DPM's and the one corresponding to the first zone, which is the most active seismic zone of Turkey, is shown in Table 5. It is to be noted that under each intensity level there are two columns, one for buildings constructed according to the Code, denoted by AC and the other for buildings violating the requirements of the Code, denoted by NAC. In these tables the term MDR denotes the "mean damage ratio" and is computed from the following relationship:

$$\mathrm{MDR}_k = \sum_{\mathrm{DS}} \mathrm{P}_k(\mathrm{DS}, \mathrm{I}) \times \mathrm{CDS}_{\mathrm{DS}} \qquad (2)$$

Here, $\mathrm{CDR}_{\mathrm{DS}}$ = central damage ratio corresponding to the damage state DS and $\mathrm{P}_k(\mathrm{DS,I})$ is as defined by equation 1. The variation of MDR with MMI for different seismic zones and depending on the degree of conformance with the Code is shown in Figure 1. As observed in this figure, buildings not conforming to the requirements of the Code [4] yield significantly higher mean damage ratios, especially at lower intensity levels. As a matter of fact, the curve for the NAC case forms an upper bound envelope for the MDR's applicable to the different seismic zones.

In Figure 1, it is also observed that regardless of the seismic zone where the structure is located, for all NAC buildings the mean damage ratio versus MMI curves coincide and a single curve is valid for all NAC buildings. This is expected since buildings that are not constructed according to the requirements of the code will experience the same mean damage under the same earthquake intensity, even if they are located in different seismic zones. However, for AC buildings, the mean damage ratios are lower in highly seismic zones compared to those in the less active seismic zones. This may be related to the fact that, since the expected earthquake forces in seismically active regions are quite high, the design resistances of the AC buildings in these zones are accordingly high. However, in less active seismic zones the expected earthquake loads are relatively lower and the buildings are designed accordingly, with less seismic resistance. Hence, in less seismic zones the design seismic capacity is lower and in the case of large-intensity earthquakes, they are expected to experience higher rates of damage compared to those in more active zones, where the design seismic capacity is higher.

3. RELIABILITY-BASED MODEL

3.1 Description of the Model

Earthquake damage prediction involves several uncertainties associated with the earthquake force and the response of the structures. Accordingly the basic concepts of the classical reliability theory form a convenient framework for the development of a reliability-based model for the prediction of damage rate. The model considers building resistance characteristics and ground motion properties peculiar to Turkey by defining an earthquake force index and a seismic resistance index. The damage rate, which is expressed by the probability of occurrence of a certain level of damage, is determined by comparing the earthquake force and seismic resistance indices within the framework of classical reliability theory. The basic assumption in this method is that both the seismic resistance and force indices are random variables. Moreover, statistical analyses of relevant data indicated that both indices are log-normally distributed [11].

The probability of damage, within the framework of classical reliability theory, is defined as follows:

$$\text{Pr(damage)} = \text{Pr(Resistance Quantity} \leq \text{Force Quantity)} \tag{3}$$

Here, the resistance and force quantities will be kept as simple as possible and will be expressed in terms of seismic resistance index and earthquake force index as will be described in the following sections.

3.2 Definition of the Seismic Resistance Index

For the development of a seismic resistance index, the methodology proposed by Shiga [12] is adopted. Shiga utilized three parameters in order to represent the seismic vulnerability of reinforced concrete structures. These are wall-area index (WI), column-area index (CI) and average shear stress (τ_{ave}) in walls and columns. According to Shiga, these parameters were found to be good indicators for the degree of damage to reinforced concrete structures exposed to the 1968 Tokachioki earthquake in Japan. These parameters are defined as follows:

$$WI = \frac{A_W}{\Sigma A_f} \quad (4)$$

$$CI = \frac{A_C}{\Sigma A_f} \quad (5)$$

$$\tau_{ave} = \frac{W}{(A_C + A_W)} \quad (6)$$

Where the following notation is valid:
WI: wall-area index
CI: column-area index
τ_{ave}: average shear stress in walls and columns, in kg/cm^2
A_W: total area of reinforced concrete walls in one direction in the first floor, in cm^2
A_C: total area of columns in the first floor, in cm^2
W: weight above ground level (in kg) and it is taken as $1300.\Sigma A_f$
ΣA_f: total floor area, in m^2

Shiga [12] plotted the average shear stress versus wall and column area indices in order to show the distribution of the damage patterns and states. By using the resulting plots he identified the two critical values of the average shear stress separating the undamaged and damaged buildings; one value shows the shear stress on the walls and the other on the columns. With these two values, he calculated the nominal lateral strength of the structure as "$12.A_C+33.A_W$", in kg.

On the other hand, the nominal lateral force is calculated by multiplying the base shear coefficient with the weight of the structure. The average unit floor weight of the actual buildings in the region was found to be approximately 1300 kg/m^2 and the weight is calculated to be $1300.\Sigma A_f$. Shiga [12] introduced the seismic resistance index C_R through the following equation, where the nominal lateral earthquake force is equated to the nominal strength:

$$C_R \cdot (1300 \cdot \Sigma A_f) = 12 \cdot A_C - 33 \cdot A_W \quad (7)$$

Finally, dividing both sides of equation 7 by the weight of the structure, the following expression for C_R is obtained:

$$C_R = \frac{12 \cdot A_C - 33 \cdot A_W}{1300 \cdot \Sigma A_f} \quad (8)$$

Following the same procedure and using the damage database belonging to the Düzce (Nov. 12, 1999) earthquake, the seismic resistance index is derived in terms of the base shear coefficient and is given as follows [11]:

$$C_R = \frac{18 \cdot A_C + 25 \cdot A_w}{(1300 \cdot \Sigma A_f)} \tag{9}$$

Since the number of buildings in Düzce earthquake damage database is the highest, this database is used as the basis for the derivation of equation 9, and then the coefficients are checked with respect to the Dinar and Erzincan seismic damage databases. In both cases a reasonable fit is observed to that obtained from Düzce [11].

3.3 Definition of the Earthquake Force Index

The earthquake force index is defined as follows [13]:

$$C_S = S(T) \cdot \gamma \cdot \frac{A_{max}}{g} \tag{10}$$

Here:
S(T): characteristic response spectrum values given in the 1998 Turkish earthquake resistant design code [14] as a function of local soil conditions and fundamental period of the building
A_{max}: maximum ground acceleration
g : 9.81 m/s^2
γ : is a reduction coefficient, which is a function of the damping factor [13] and is expressed as follows:

$$\gamma = \left[\frac{1.5}{1+10.h}\right] \tag{11}$$

In equation 11, h is the damping factor; a damping factor of 10% is assumed for reinforced concrete buildings. It should be noted that among the variables in equation 10, Amax is the dominant random variable since it has the highest uncertainty.

As seen in equation 10, the earthquake force index used in this model includes dynamic properties of the buildings such as period and damping effects, soil condition effect, ground motion parameters such as peak ground acceleration, indirectly the return period and intensity of the earthquake.

3.4 Evaluation of Damage State Probabilities

According to the reliability-based approach, the rate of seismic vulnerability, in other words the probability of damage exceeding a certain level, is evaluated by comparing the seismic resistance index and earthquake force index, which are defined in the previous sections. Based on equation 3, the damage probability (also the cumulative damage rate) is expressed as follows:

$$\Pr(\text{damage}) = \Pr(C_R \leq \alpha \cdot C_S) \tag{12}$$

Here: C_R is the seismic resistance index given by equation 9, C_S is the earthquake force index given by equation 10 and α is the damage state factor defining the level of damage. It is derived by using the energy conservation rule and is given as:

$$\alpha = \frac{1}{\sqrt{(2 \cdot d - 1)}} \tag{13}$$

In equation 13, d is the ductility ratio corresponding to the level of damage considered.

Under the assumption of lognormal distribution for the seismic resistance and earthquake force indices, the probability of damage is expressed as follows:

$$\Pr(\text{damage}) = 1 - \Phi(\beta) \tag{14}$$

In equation 14, β is called the reliability index and is to be computed from the following equation:

$$\beta = \frac{\ln(\frac{\mu_R}{\mu_S}) - \ln\alpha - 0.5 \cdot \ln\left(\frac{1+\upsilon_R^2}{1+\upsilon_S^2}\right)}{\sqrt{(\ln(1+\upsilon_R^2) \cdot (1+\upsilon_S^2))}} \tag{15}$$

In equations 14 and 15, Φ: is the standard normal distribution function, μ_R, μ_S are mean values of seismic resistance and earthquake force indices, respectively. υ_R and υ_S, stand for coefficients of variation of seismic resistance and earthquake force indices, respectively.

Values of υ_R vary between 0.38 and 0.62 for the three damage databases under consideration, whereas the value of υ_S is assumed to be equal to 0.4 taking into account all possible uncertainties in the earthquake force index [13].

By substituting these mean values and the coefficients of variation of resistance and force indices computed for a specific earthquake into equation 15, the β value, and by using this β value in equation 14, the probability of damage can be calculated. The resulting damage probabilities are actually cumulative damage rates and they form the basis for the plotting of the fragility curves for seismic damage databases associated with Erzincan 1992, Dinar 1995 and Düzce 1999 earthquakes. Using these fragility curves, DPM's for the three damage databases are derived. In this derivation, for a given intensity level, the differences between the cumulative damage rates corresponding to each damage state give the damage state probabilities at that intensity level. To better view the model, the fragility curve and the DPM for the Düzce 1999 earthquake are given in Figure 2 and Table 6, respectively. Because of space limitation the fragility curves and damage probability matrices for the Erzincan 1992 and Dinar 1995 earthquakes are not given here. However, using the MDR values obtained from the resulting DPM's for the three damage databases, the variation of MDR with MMI is plotted for each damage database and is shown in Figure 3. It is observed from this figure that the reliability-based model yields almost the same degree of damage prediction for Düzce and Erzincan, whereas the damage estimates for Dinar are higher. This is due to the lower amount of reinforced concrete wall areas in the buildings involved in the Dinar seismic damage database than in the other damage databases. It is to be noted that, in this model a mean conversion relationship is used between the peak ground acceleration A_{max} and the MMI scale [15].

4. COMPARISON OF RESULTS

The mean damage ratios (MDR's) determined from each one of the two statistical methods are summarized in Table 7 for the purpose of comparison.

For relatively low intensities, like VI and VII, it is observed that MDR's obtained from the DPM corresponding to AC buildings are quite close to those obtained from the reliability-based model and the value of MDR's are all below 0.3% and 4% for MMI values of VI and VII, respectively. As expected the MDR's obtained from the DPM for NAC buildings give relatively higher results.

For the intensity level of VIII, the reliability-based model yielded lower MDR's. The statistical method based on the use of DPM's can be classified as an "empirical" model of damage assessment, since the only source of information is the observed damage records of past seismic activity. On the other hand, the reliability-based model can be considered as a "semi-analytical" model in the sense that it utilizes a code-based earthquake force index and a seismic resistance index that is partially based on observed damage data. These indices are compared within a reliability framework in deriving the fragility curves and in estimating damage rates.

Another reason for the low MDR's in reliability-based model is the fact that the damage databases used in the analyses belong to earthquakes with considerably high peak ground accelerations. But in implementing the reliability-based model, a specific peak ground acceleration-intensity correlation proposed by Murphy and O'Brien [15] is used. This relationship is a mean conversion curve and in case of an earthquake with certain intensity but with unexpectedly high peak ground acceleration, the corresponding peak ground acceleration obtained from this curve might underestimate the damage ratios. To justify this view, the actual peak ground accelerations of the corresponding earthquakes are utilized in the model and it is observed that the MDR's obtained from the model with the actual peak ground acceleration values are quite close to the MDR values obtained from the DPM for NAC buildings [11].

When the MDR distribution at the intensity level IX is investigated, it is seen that both methods yielded MDR's around 17%-25%, except did the DPM's for NAC buildings. The slight differences between the results of the two methods arise from the points discussed above, for the MMI value of VIII.

5. CONCLUSIONS

Based on the results obtained from the implementation of the two statistical methods to the seismic damage data compiled from the recent earthquakes in Turkey, the following points can be stated:

a) At any certain intensity level, the reliability-based model underestimated the mean damage ratio at high intensities, where the peak ground accelerations obtained from the mean MMI-peak ground acceleration conversion relationship yielded lower A_{max} values than those actually observed in the relevant earthquakes.

b) When the actual (observed) peak ground accelerations are utilized in the reliability-based model, the computed MDR's are found to be very close to the MDR's obtained for NAC buildings given in the best estimate DPM for seismic zone I. This point revealed the fact that majority of the buildings involved in the seismic damage databases are not designed according to the Code [4], as also confirmed by field observations.

c) When the damage rates assessed for the three seismic damage databases are examined, it is easily seen that under a specific intensity level, buildings with a smaller amount of vertical elements, especially structural walls, experience noticeably higher damage than do the buildings with greater amount of walls. This is why the Dinar seismic damage database, in which the buildings have the lowest amount of wall area, contains the highest degree of damage.

d) Considering the various uncertainties and random effects, seismic damage prediction should be carried out within a probabilistic and statistical framework. The two alternative stochastic models provided in this study serve well for this purpose. The seismic damage estimates obtained from the two statistical methods are consistent among themselves and both approaches appear to be the appropriate methods for seismic damage prediction.

REFERENCES

1. Whitman, R V: Damage Probability Matrices for Prototype Buildings, Dept. of Civil Engineering, MIT, R73-57, Cambridge, 1973.
2. Gurpinar, A, Abali, M, Yucemen, M S and Yesilcay, Y: Feasibility of Obligatory Earthquake Insurance in Turkey, Report No.78-05, EERI, METU, Ankara, 1978 (in Turkish).
3. Gurpinar, A and Yucemen, M S: An Obligatory Earthquake Insurance Model for Turkey, Proceedings, International Conference on Engineering for Protection from Natural Disasters, Asian Institute of Technology, Bangkok, Thailand, 1980, pp. 895-906.
4. Earthquake Research Institute, Turkish Government Ministry of Reconstruction and Resettlement: Specifications for Structures to be Built in Disaster Areas, Ankara, 1975.
5. Yucemen, M S and Bulak, S: Assessment of Earthquake Insurance Premiums Based on Statistical Methods, Fourth National Earthquake Engineering Conference, Ankara, September 1997, pp. 699-707 (in Turkish).
6. Yucemen, M S: Prediction of Potential Seismic Damage to Reinforced Concrete Buildings Based on Damage Probability Matrices, Proceedings, 6th International Conference on Concrete Technology for Developing Countries, Vol. 3, Amman, pp. 951-960, October, 2002.
7. Sucuoglu, H and Tokyay, M: Engineering Report of 13 March 1992 Erzincan Earthquake, Turkish Society of Civil Engineers, Ankara, June 1992 (in Turkish).

8. Ozmen, B and Bagci, G: 12 November 1999 Düzce Earthquake Report, Ministry of Public Works and Resettlement, General Directorate of Disaster Affairs, Earthquake Research Department, Ankara, November, 2000 (in Turkish).
9. Wasti, S T and Sucuoglu, H: Rehabilitation of Moderately Damaged Reinforced Concrete Buildings After the 1 October 1995 Dinar Earthquake, METU, EERC, Report No: METU/EERC 99-01, Ankara, April, 1999 (in Turkish).
10. Joint Team of Middle East Technical University, Purdue University, Notre Dame University, University of Texas at Austin: Bolu, Düzce and Kaynasli Building Damage Inventory Report, Ankara, June, 2000.
11. Askan, A : Stochastic Methods for the Estimation of Potential Seismic Damage, M.S. Thesis, Middle East Technical University, Ankara, 2002.
12. Shiga, T: Earthquake Damage and the Amount of Walls in Reinforced Concrete Buildings, Proceedings of 6th World Conference on Earthquake Engineering, India , pp. 2467-2472, 1977.
13. Shibata, A: Prediction of the Probability of Earthquake Damage to Reinforced Concrete Building Groups in a City, Proceedings of 7th World Conference on Earthquake Engineering, Istanbul, pp. 395-402, 1980.
14. General Directorate of Disaster Affairs, Turkish Government Ministry of Public Works and Settlement: Specifications for Structures to be Built in Disaster Areas, Ankara, 1998.
15. Murphy, J R and O'Brien, L J: The Correlation of Ground Acceleration Amplitude with Seismic Intensity and Other Physical Parameters, Bull. Seism. Soc. Am, vol. 67, no.3, pp. 877-915, 1977.

Table 1. Damage Probability Matrix

Damage State (DS)	Damage Ratio Range (%)	CDR (%)	Modified Mercalli Intensity (MMI)				
			V	VI	VII	VIII	IX
None	0-1	0					
Light	1-10	5			Damage State		
Moderate	10-50	30			Probabilities		
Heavy	50-90	70			Pr(DS,I)		
Collapse	90-100	100					

Table 2. "Subjective" Damage Probability Matrix for Seismic Zone I (after [2])

Damage State (DS)	CDR (%)	MMI									
		V		VI		VII		VIII		IX	
		AC	NAC	AC	NAC	AC	NAC	AC	NAC	AC	NAC
None	0	1.0	0.95	0.95	0.70	0.7	0.50	0.50	0.20	0.30	0.05
Light	5	0	0.05	0.05	0.15	0.20	0.20	0.20	0.20	0.30	0.20
Moderate	30	0	0	0	0.10	0.10	0.15	0.20	0.40	0.20	0.40
Heavy	70	0	0	0	0.05	0	0.10	0.10	0.10	0.20	0.20
Collapse	100	0	0	0	0	0	0.05	0	0.10	0	0.15
MDR (%)		0	0.25	0.25	7.25	4	17.5	14	30	21.5	42

AC: According to the Code; NAC: Not According to the Code

Table 3. Empirical Damage State Probabilities for Reinforced Concrete Buildings

Damage State (DS)	CDR (%)	19.8.1976 Denizli MMI=VI	18.11.1983 Erzincan MMI=VI	6.6.1986 Malatya MMI=VII	22.05.1971 Bingol MMI=VIII
None	0	0.49	0.74	0.450	0.040
Light	5	0.37	0.23	0.390	0.430
Moderate	30	0.13	0.03	0.125	0.260
Heavy	70	0.01	0.00	0.035	0.135
Collapse	100	0.00	0.00	0.00	0.135
Number of Buildings		378	112	89	46
MDR(%)		6.45	2.05	8.15	32.9

Damage State (DS)	13.3.1992 Erzincan MMI=VIII	1.10.1995 Dinar MMI=VIII		17.08.199 Adapazari MMI=IX	12.11.1999 Düzce MMI=IX
		AC	NAC		
None	0.31	0.23	0.24	0.040	0.17
Light	0.48	0.31	0.24	0.340	0.16
Moderate	0.09	0.38	0.41	0.270	0.28
Heavy	0.07	0.04	0.05	0.175	0.19
Collapse	0.05	0.04	0.06	0.175	0.20
Number of Buildings	415	39		13240	5420
MDR(%)	15	19.75	23	39.6	42.5

AC: According to the Code; NAC: Not According to the Code

Table 4. Empirical Damage State Probabilities for Reinforced Concrete Buildings Classified According to the Seismic Zones

Damage State (DS)	CDR (%)	ZONE I MMI=VI (NAC)	ZONE II MMI=VII (NAC)	ZONE II MMI=VIII (NAC)	ZONE I MMI=VIII (NAC)	ZONE I MMI=IX (NAC)
None	0	0.54	0.45	0.04	0.30	0.08
Light	5	0.34	0.39	0.43	0.45	0.29
Moderate	30	0.11	0.125	0.26	0.13	0.27
Heavy	70	0.01	0.035	0.135	0.07	0.18
Collapse	100	0.00	0.00	0.135	0.05	0.18
MDR(%)		5.7	8.15	32.9	16.1	40.15

NAC: Not According to the Code

Table 5. "Best Estimate" Damage Probability Matrix Proposed for Seismic Zone I

Damage State (DS)	MMI=V AC	MMI=V NAC	MMI=VI AC	MMI=VI NAC	MMI=VII AC	MMI=VII NAC	MMI=VIII AC	MMI=VIII NAC	MMI=IX AC	MMI=IX NAC
None	1.00	0.95	0.95	0.58	0.70	0.46	0.50	0.28	0.30	0.07
Light	0	0.05	0.05	0.29	0.20	0.34	0.20	0.39	0.30	0.27
Moderate	0	0	0	0.11	0.10	0.14	0.20	0.20	0.20	0.30
Heavy	0	0	0	0.02	0	0.05	0.10	0.07	0.20	0.19
Collapse	0	0	0	0	0	0.01	0	0.06	0	0.17
MDR (%)	0	0.25	0.25	6.2	4	10.4	14	18.9	21.5	40.7

*AC: According to the Code; NAC: Not Ac*cording to the Code

Table 6. DPM for Düzce Damage Database as Generated by the Reliability-Based Model

Damage State	CDR (%)	VI	VII	VIII	IX	X
None	0	0.99	0.88	0.60	0.23	0.05
Light	5	0.01	0.10	0.32	0.44	0.25
Moderate	30	-	0.01	0.06	0.24	0.35
Severe	85	-	0.01	0.02	0.09	0.35
MDR(%)		0.05	1.65	5.1	17.1	41.5

Table 7. Comparison of MDR's Determined from the Two Statistical Methods

Statistical Method	MMI			
	VI	VII	VIII	IX
DPM / AC Buildings (1st Seismic Zone)	0.3	4	14	21.5
DPM / NAC Buildings (1st Seismic Zone)	6.2	10.4	18.9	40.7
Reliability-Based (1992 Erzincan)	0.2	1.7	5.6	17.6
Reliability-Based (1995 Dinar)	0.2	2	7.8	25.3
Reliability-Based (1999 Düzce)	0.1	1.7	5.1	17.1

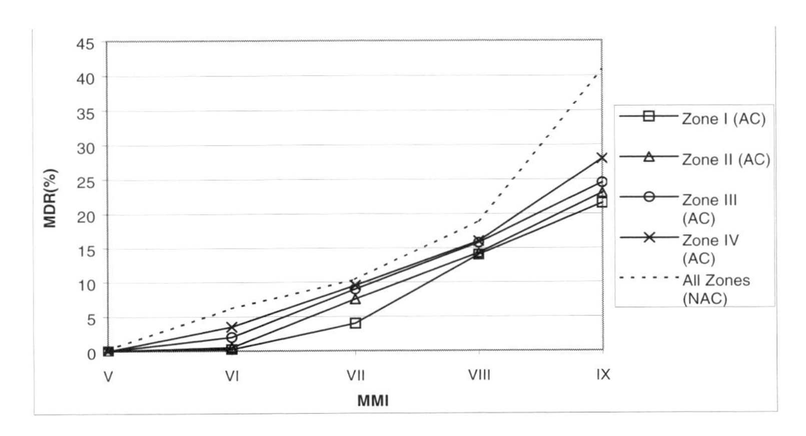

Figure 1. Variation of MDR with MMI for Different Seismic Zones and Design Strategies (AC: According to the Code; NAC: Not According to the Code)

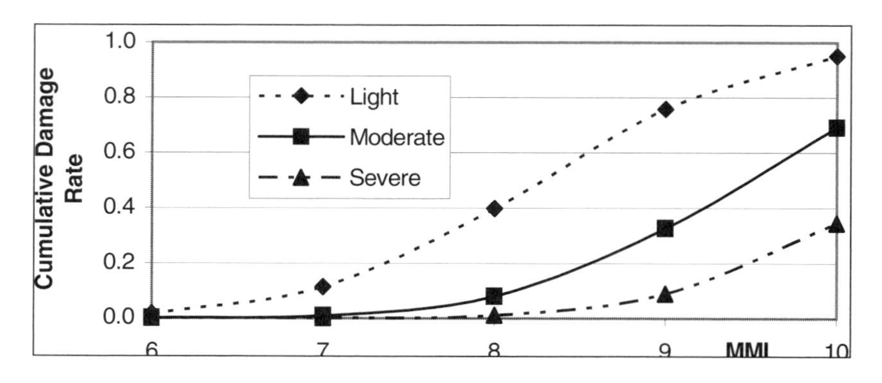

Figure 2. Fragility Curves for the Düzce Damage Database

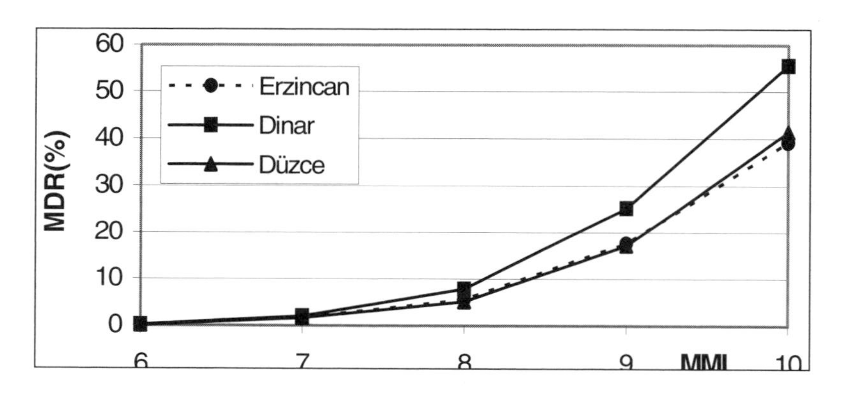

Figure 3. Variation of MDR with MMI for the Three Seismic Damage Databases

JUST HOW PRESCIENT ARE OUR BUILDING DAMAGE PREDICTIONS?

P. Gulkan, B. S. Bakir, A. Yakut and M.T. Yilmaz
Middle East Technical University Ankara 06531, Turkey

Abstract: Adapazari was the scene of spectacular structural damage as well as widespread liquefaction after the earthquake of August 17, 1999. Damage patterns observed are re-examined to investigate whether they are indicative of a consistent trend explicable in terms of the building attributes and/or site conditions. 301 buildings that had collapsed fully have been re-evaluated from their design blueprints. An examination based solely on structural attributes, including data from other sets of building assessment projects, leads us to believe that building collapse is perhaps just too involved to reduce to a few simple guilt pointers. Site effects might have played a major role in the observed damage, because only a conflicting trend between structural attributes and the actual damage can be established.

Keywords: Damage assessment, Site effects, Seismic evaluation

1. INTRODUCTION

The 1999 Izmit earthquake caused significant damage in many urban districts, including the city of Adapazari. The widespread liquefaction and peculiar site effects are considered as other major factors in addition to structural parameters that caused significant damage to many buildings.

S.T. Wasti and G. Ozcebe (eds.), Seismic Assessment and Rehabilitation of Existing Buildings, 165–192.

After the earthquake, the Adapazari Municipality conducted a comprehensive damage assessment survey in the city. The results of this survey are provided at district level within the central municipality area. The total number of buildings investigated in 26 districts covering an area of about 20 km^2, was 23,914. The damaged buildings were classified into two damage grades according to the criterion whether the building can be feasibly repaired (light or moderate damage) or has to be demolished and removed (collapse or heavy damage). Corresponding damage statistics are presented in Table 1.

A damage survey was conducted during the summer of 2002, nearly three years after the actual event. By that time, extensive repairs had been done in surviving buildings, and many others had been demolished and removed. Homeowners were reluctant to permit another round of examinations to be conducted in their property, so the decision was made to include those buildings that had collapsed during the earthquake, and had led to loss of human life. When such is the case, the law requires expert witnesses to prepare an affidavit that will establish culprits. Because of their proximity to the site the Department of Civil Engineering, Sakarya University (SU), had served in this capacity for many buildings on the basis of design blueprints and structural calculations, so the decision was reached to utilize that information, extracted through a damage assessment form, assuming that the drawings were an accurate replication of the as-built structure. The data template used for this purpose was modified from a post-earthquake damage assessment form developed for engineered buildings in Turkey (1). The coordinates of the buildings were determined by a hand-held GPS device, accurate to about 10 m. This was judged to be an acceptable error.

None of the buildings in this data set existed at the time of our studies, so there remain a good number of points that must be known if accurate projections for the cause of the damage were to be identified.

2. BUILDING STOCK INFORMATION

The buildings reported by SU were situated in the locations in Figure 1. This figure is deliberately superposed on the analog map of the city because even with its poor legibility it allows urban features such as major roads and the Sakarya River banks on the east. The same information is repeated in Figure 2 where the districts in the city have been numbered, and the number of stories of these collapsed buildings coded to allow a rapid assessment of the building heights. Building heights are as much a function of the economics of the corresponding usage as it is of zoning laws enacted by the city government. In central urban areas in Turkey there is commonly a mixed form of occupation, where small businesses (small grocery shops, barber shops, professional offices, etc.) are dispersed among the residences under the same roof. The ground story is usually made as free of obstructions to human traffic as possible, and this leads to weak or soft stories at the level where the seismic demand is largest. Figure 3 shows that half of the buildings examined in this study were five stories in height. The next largest group is for six story buildings comprising 32 percent of the total, and this is typical for much of recent construction.

The information contained in Figures 1-3 is useful to establish the overall architectural characteristics of the housing stock in Adapazari. We believe that extrapolations can safely be made from these figures for reaching general conclusions. There is also reason to believe that most of the buildings in the sample group dated from the post-1975 period, so that their earthquake designs were done according to the provisions of the 1975 edition of the code.

The total number of building in the data set was 301. Interpretation as absolute numbers of normalized graphs in the following will be facilitated when those figures are multiplied with this sum. All buildings in the data set were reinforced concrete frames.

The current Turkish earthquake regulations make a clear definition of structural irregularities that lead to added earthquake vulnerability. In general, plan irregularities are classified as type A, with sub-group designated as A1, A2, etc., and irregularities in elevation are of type B. These are each described, and follow universal procedures for quantification. The questions listed under "Irregularities" in the form used have been designed to clarify the prevalence of the distribution of these among the sample of buildings examined by SU.

It is instructive to examine first the distribution of soft stories. A soft story occurs when the lateral rigidity of the horizontal load resisting members at any level is such that the average lateral drift under the design load at that level is more than 50 percent larger than the same quantity calculated for the next story. This property needs a set of calculations to be made. With taller columns and no infill or structural walls, this situation usually is encountered at the ground level in Turkey. In Figure 4 we have shown the distribution of this situation in normalized fashion among the buildings in the sample set.

It can be assumed that taller buildings had more recent dates of construction, and as building height limitations had been eased in recent times owners took advantage of the full height permitted in the zonation decisions. It is quickly seen that while (older) 3-story buildings were evenly distributed with respect to existence of soft stories, taller buildings were about 70 percent that way, and the seven story subset was entirely rated as having the property. In general, the situation for identifying a weak story where the lateral strength at a given story differs by more than 20 percent that at the neighboring story did not arise because designers accounted for that by varying the reinforcement details.

Torsional irregularity is defined in the Turkish code as the situation when the average lateral translation under the design forces at a given story differs by more than 20 percent the largest lateral situation at the same story. This usually occurs at the extreme ends of the floor slabs when L-, T- or similar shapes in plan exist. Land ownership patterns sometimes force architects to design such buildings when they wish to take advantage of utilizing the entire buildable footprint area of the land. Figure 5 describes the distribution of this property among our sample set.

Plan irregularity refers to the situation when long projections in plan exist for the building. In addition to the increased likelihood of causing torsional irregularity this situation creates re-entrant corners where stresses become concentrated, and may cause unexpected distress. The data in Figure 6 describes this situation. Again, taller buildings seem to be more likely to possess this characteristic.

From the structural viewpoint, mezzanines can prove to be trouble spots because the lateral strength of short and tall columns are different by large amounts, and the extra inertia effects transmitted to columns at their mid-height are difficult to account for. As Figure 7 shows, this situation appears not to have been a widespread practice.

While the post-collapse damage form used has entries for material quality, workmanship and similar features it was not possible to conduct any tests on actual material coupons or cylinders to answer this with clarity. We have relied on the intended material qualities on the calculation sheets, which were generally specified as 20 MPa. It is known that the actual strengths may differ by large margins from their intended values. We have also no factual data on whether plain or deformed bars had been used.

3. ASSESSMENT PROCEDURE

Available procedures to assess the expected vulnerability of reinforced concrete buildings can be employed to evaluate the collapsed buildings in order to examine the reasons behind their unsatisfactory performance. These evaluations, in fact, would be based on the information gathered through filling the assessment forms using the structural and architectural drawings rather than as-built features. Several procedures developed specifically for reinforced concrete buildings in Turkey are explained here briefly because they are applied to the buildings in the database for re-evaluating their performance.

3.1 Column and Wall Indexes

Our experience with the seismic vulnerability of the housing stock in Turkey has taught us that the normalized column and wall areas a given building possesses can serve as a useful index [2,3]. The central idea for expressing vulnerability is expressed as the weight of the structure (which is proportional to its total floor area) divided into the sum of the cross-sectional areas of the columns and walls at the base. This format is attractive because of the ease with which the data can be acquired. Lengthy calculations are not necessary. The outcome is a crisp numerical index that can be matched against a derived yardstick so that a judgment about the vulnerability can be stated. The paper by Shiga et al. [2] is difficult to generalize because it was derived exclusively in relation to a particular group of buildings constructed in accordance with the Japanese practice of the 1960s. Recalibrating their format can best be accomplished by testing its predictive success against observed damage in a collection of buildings with dimensional and material properties based on random choices made during the design and construction stages. This way, the question of whether a rational explanation of the empirical evidence is contained in the theory can be answered with confidence.

Conventional design procedures focus on arriving at proper component dimensions that will accommodate prescribed forces, but a more important issue is to keep structural displacements within prescribed limits. This idea has gained popularity in recent years as the central principle of performance based earthquake design. The concept of limiting ground level drift to prescribed limits was tested against empirical data collected in Erzincan following a major earthquake there in 1992. The rationalization of the concept of linking column and wall areas to drift, and its comparison with actual data are contained in [3]. Basically, the column index is 0.5 times the total cross sectional area of columns at the base plus 1.0 times the area of structural walls divided by the total floor area of the building. Filler wall index is defined as 0.1 times the area of filler walls at the base divided by the total floor area. Because orientation of planar members such as walls is direction dependent, it is possible to express this ratio in the two principal directions of the building. The result of this exercise for the set of buildings in our sample has produced the diagrams in Figures 8 and 9. We stress that the directions have been defined arbitrarily.

Conventional wisdom suggests that, when the column index falls below 0.0025-0.003 the structure is rated as being vulnerable under conditions in

Turkey. Both figures indicate that this was the median figure for the buildings in our population, so a clear indication of their collapse is difficult to rationalize on the basis of that information alone.

In support of this observation Figures 10-12 show the results of a survey on 162 buildings that served as service facilities for a major bank in Turkey. The buildings were first entered on a form similar to that used for Adapazari buildings, and then each was analyzed in the linear as well as the nonlinear mode. A good many other considerations entered the eventual decision regarding the adequacy of the load-carrying system but the figures encapsulate the decision in terms of the two indexes.

We must stress the essential difference between information contained in Figures 8-9 and 10-12. The first set is for buildings that collapsed in Adapazari in 1999. The second set is for buildings situated in many cities in western Turkey that were analyzed with state-of-the-art techniques for their vulnerability. The scatter in the first set suggests that while a column index of 0.004 would appear as the demarcation line for the onset of vulnerability, this is not a confident threshold, and outliers outnumber members of the population. Similarly, many buildings with similar indexes in Adapazari survived the same ground motion, so parameters not contained in Figures 8-9 must have played roles that do not come through the index number alone. It is possible to combine undesirable structural features with the index information in an attempt to gain increased wisdom for foreknowledge of whether a given building will perform adequately, but all buildings reflected in Figures 8 and 9 collapsed. Without a reference to an adequate number of buildings from the rest of the building stock in Adapazari, this will also not help in a meaningful way.

3.2 Capacity Index

With the same premise as other index methods of building evaluation, a new approach was developed based on the results of analytical investigations and field observations of building performance from major earthquakes. The method relies on determining the base shear capacity of reinforced concrete buildings from the dimensions and layout of vertical structural components such as columns and shear walls as well as in-fill walls. It can be considered as an improvement that takes into account the strength of concrete and evaluates the structure based on the expected shear capacity. FEMA 310 Tier 1 [4] evaluation and many other preliminary assessment procedures require that the shear stress on concrete be checked against certain allowable values. The same principle is adopted here for the whole structure, checking total base shear capacity of the building against a prescribed empirical value calibrated to the observed and computed building performance data. The central point in this method is that the building would experience damage when the yield base shear capacity of an idealized bilinear system is exceeded. The general construction practice in Turkey leads to buildings that do not have adequate ductility and consequently have low deformation capacities. This reinforces the thought that a strength-based criterion would be appropriate for evaluating the performance of existing buildings. The objective is to perform a quick and simple evaluation to arrive at a decision about the expected seismic performance of the buildings for which detailed evaluation procedures could not be applied due to lack of data.

A number of buildings typical of mid-rise reinforced concrete construction in Turkey were subjected to pushover analysis to obtain their pushover curves. These buildings were extracted from two databases; The first database was compiled from a population of nearly five-hundred buildings located in the city of Düzce that were investigated after the 1999 earthquakes. The other database is comprised of service and housing buildings that belong to a state agency that is responsible for the distribution of natural gas in Turkey, which were analyzed to evaluate their expected performance if a major earthquake were to occur. A typical idealized pushover curve is shown in Figure 13, where the code base shear value (V_{code}), yield base shear capacity (V_y) and ultimate base shear (V_u) are displayed.

It is well known that the contribution of the in-fill walls to overall seismic resistance of reinforced concrete buildings is vital, especially for buildings that have poor deformation capacities, as is the case in Turkey. This enhancement of the strength is generally proportional to the area of the filler walls in a given floor. To take this effect into account, the buildings were modeled both with and without inclusion of the in-fill walls in an attempt to derive a relationship that can be used to determine the increase in capacity due to presence of the in-fill walls. For this reason, the in-fill wall area normalized with the floor area is plotted against the yield base shear capacity normalized with corresponding capacity of bare frame system as shown in Figure 14. The increase in the yield shear capacity can be estimated approximately using the following equation based on the normalized in-fill wall area:

$$\frac{Vy_w}{Vy} = 14\frac{A_w}{A_f} + 1.0 \qquad (1)$$

Here, Vy_w and Vy are the yield base shear capacities with and without in-fill walls, respectively, A_w is the total area of in-fill walls on the ground floor, and A_f is the ground floor area.

In the light of this information, an alternative evaluation criterion may be based on an index we call "Capacity Index," defined as the ratio of the yield base shear capacity of the building to the code base shear value, i.e. V_y/V_{code}. The yield base shear capacity can be obtained either from the pushover curve or using the gross shear capacity (V_{cap}) of all lateral load resisting members computed from the concrete areas effective in each direction. There is reason to believe that there exists a relationship between the gross shear capacity and the yield base shear capacity of the buildings. This relationship, given in Equation (2), was obtained from the analyses of 23 selected mid-rise buildings from the two databases. Figure 15 presents the relationship between V_{cap} and V_y.

The buildings studied revealed that, when the capacity index falls below 1.65, the building may be deemed unsafe, or would perform unsatisfactorily. The capacity indexes computed for the buildings analyzed are shown in Figure 16, along with the respective damage state of each building contained in the Düzce database. Only buildings with moderate or a lower damage state were analyzed and a line was drawn from the minimum of the mean "capacity index" computed for the three damage states included, namely None, Light and Moderate. This evaluation essentially intends to incorporate life safety performance criterion using a strength-based index.

$$V_y = \frac{V_{cap}}{1.4e^{0.065n}} \qquad (2)$$

In Equation (2) n stands for the number of stories.

The capacity index method applied to the buildings in Adapazari database is shown in Figure 17. Approximately 31 percent of the buildings were classified as unsafe. The remaining 69 percent could not be captured by this method. This observation is in line with what the column index method suggests.

The method was also applied to the 162-building set discussed previously, to test its efficiency. Figures 18-20 show that with a good success rate those buildings that are deemed unsafe were captured. In Figure 18, 27 percent of the buildings that were judged safe from analytical assessment were classified as unsafe when capacity index was applied. Among the buildings that were identified as inadequate and need to be strengthened based on the analytical performance evaluation, 78 percent were decided to be inadequate by this method. The success rate with respect to analytical evaluation was 86 percent for buildings that were classified unsafe.

Among many reasons that might have played important role in collapse of the buildings in Adapazari, one obvious outcome appears to be that the structural system of the building could have been responsible for only a portion of the poor performance and closer attention should be devoted to the influence of site effects, which is discussed in the following section.

4. INFLUENCE OF LOCAL SITE EFFECTS

4.1 Background

Adapazari is located at the edge of a sedimentary basin, which is a former lake bed, underlain by the thick sediments of clay (Figure 21). A Quaternary alluvium layer, up to a depth of about 15 m and primarily consisting of silt and fine sand deposited by the Sakarya River and its tributaries overlay the lake sediments. Bedrock formation descends sharply through the north beneath alluvia and reaches depths in excess of 200 m within the city limits. Variation of the bedrock depth beneath the city is depicted in Figure 22. Most of the city is situated over the Quaternary alluvial sediments of the basin, where damage was highly concentrated during the 17 August earthquake.

Soil liquefaction related phenomena are known to have had substantial influence on the seismic response and modes of damage of buildings during the 17 August earthquake [Bakir et al., 2002]. Accordingly, since the main goal of the present work is to ascertain the effects of local ground conditions on the trends of building damage distribution for a post-fact case, the impact of such phenomena has to be incorporated in the evaluation process. In this study, the spectral accelerations are assessed disregarding the impact of soil liquefaction on surface response, as recommended by NEHRP. However, such potential effects are evaluated subsequently, through contrasting the building damage distribution to the distribution of those areas having relatively higher liquefaction potential in the city.

4.2 Idealized Soil Profile and Properties

The idealized soil column developed by Bakir et al. (2002) for Adapazari was utilized. The model combines the highly consistent stratifications observed in the deep borehole logs and the generally loose silty, sandy character of the surface soils.

Available deep borehole logs, idealized soil profile and corresponding variation of the shear wave velocity of the model are presented in Figure 23. Depending on depth to bedrock, the idealized profile was truncated or extended at the base in 1-dimensional site response analyses. The shear wave velocities of surficial deposit and gravel layer intersecting the deep clay deposits were determined by employing empirical relations. The model was capable of capturing the site response appropriately in the interval of fundamental vibration period range of most buildings in Adapazari.

4.3 Development of Idealized Response Spectra

Based on the preliminary information obtained from the analytical model studies and using the idealized soil profile, numerous analyses were conducted with SHAKE for a series of combinations of surficial soil characteristics and alluvium depth. The east-west component of 17 August earthquake Adapazari record, which was the only available lateral component from that station, was assigned as rock motion in the analyses. The surficial soil thickness was presumed to be 10 m, which is the depth of a considerable portion of the boreholes. The results were compiled to develop an idealized codification of site-specific spectra that can be utilized to assess the spectral accelerations associated with the 17 August earthquake throughout the city. Averaging and regression techniques were utilized in the idealization process.

The codification is classified for two main categories of site conditions identified depending on the presence of soft surficial deposits. Such deposits, which impose a significant de-amplification, particularly over the short period range of the response spectra, are identified as follows:

1. Sands and silty sands with uncorrected SPT-$N_{45} \leq 30$
2. Silt- Clay mixtures with SPT-$N_{45} \leq 10$

Sites having such surficial soil deposits to the extent of at least 50% of upper 10 m of borings (i.e., 5 m of total thickness) are referred to here as "soft sites". Such deposits can develop relatively large damping capacities during strong shaking, and cause significant reductions in seismic demand on common building structures in Adapazari.

Sites that do not comply with the definitions of "soft site" given above, are referred to here as "stiff site". Idealized site-specific spectra for such sites, for which the spectral response is essentially dependent on the alluvium thickness, are defined by the following equations:

$$
\begin{aligned}
0 < T \le T_A \quad &: \quad SA(T) = 0.4 + \frac{0.5}{T_A} T \quad (g) \\
T_A < T \le T_B : &\quad SA(T) = 0.9 \quad (g) \\
T_B < T \quad &: \quad SA(T) = 0.9 \left(\frac{T_B}{T} \right)^n \quad (g)
\end{aligned}
\tag{4}
$$

Here, T_A, T_B and n are provided in graphical form in Figure 24 as a function of depth to bedrock at the site locality. A set of sample spectra generated utilizing the developed codification is shown in Figure 25.

For the soft sites, an idealized spectrum cannot be reliably developed utilizing equivalent linear site response approach, and true nonlinear analyses are required. On the other hand, a relatively conservative spectrum for such sites can be defined for the period range up to 0.5 s, regardless of depth to bedrock, as the soft surficial deposit dominates the response. Such a spectrum is representative of site conditions corresponding to the boundary state between soft and stiff site definitions and forms an upper bound envelope spectral response for soft sites. The spectral response is observed to be very sensitive to further reductions in surficial soil stiffness, decreasing drastically at short period range, as surficial deposits get softer. Spectra representative of soft sites are defined in Figure 25.

4.4 Overview of Building Stock and Damage Distribution

The distribution of the ratio of collapsed and heavily damaged buildings to the total number of buildings is plotted in Figure 26 on the basis of districts numbered in Figure 22. It can be observed that only few buildings collapsed or experienced heavy damage in the southernmost districts of the city, underlain by stiff and shallow soil deposits. Through the north, over the deep alluvium, damage concentration increases significantly in the central districts where the buildings are generally taller compared to the rest of the city. The total number of collapsed buildings exceeds in this case the subset that had become the subject of litigation, and could be examined more closely.

The heavily damaged central part, which constitutes the business district as well as the bulk of residential and state buildings, had been developed rapidly over the past two decades. A great majority of the buildings constructed in this period are 4-5 story, rarely 6-7 story apartment buildings, typically with a high entrance floor. Almost without an exception they have a reinforced concrete framing system with hollow brick infill walls. This category of buildings experienced the greatest impact from the earthquake, a considerable portion of which either totally collapsed or damaged heavily. Some of these buildings were classified as heavily damaged due to foundation displacements beyond tolerable limits, despite minor or no structural damage. Relatively older buildings, mostly 1- or 2-story, are either stone or brick masonry, or traditionally built with timber frame and brick infill. Such buildings, which constitute the majority in the outskirting districts of the city, are relatively fewer in the central section. Collapses in this category of buildings were notably less. With few exceptions, all of the buildings in Adapazari are built over shallow foundations, while a majority of the buildings with 3 or more stories has mat foundations. Evidently, the consistently high ground water level has been the primary controlling parameter over the foundation depth, which ranges between 1-1.5 m throughout the city, irrespective of building characteristics.

Two strongly contrasting modes of building damage were observed in the city depending on the stiffness of surficial soils. As a general trend, collapsed or structurally damaged buildings had no or comparatively less foundation displacements, while those buildings subjected to various forms of foundation displacements sustained relatively less or no structural damage at all. This observation suggests that sites which were soft or softened due to soil liquefaction, provided means of natural base isolation that reduced seismically induced forces transmitted to the superstructure during the earthquake. As a result, the number of collapse cases, and hence loss of life, must have been reduced to a certain extent over such sites.

4.5 Assessment of Local Site Effects on Structural Damage

To overview the potential effects of local sites on the observed structural damage, the distribution of collapsed and heavily damaged buildings is contrasted to the relevant soil stiffness data in the top 10 m of borehole logs in Figure 27. The percentages of liquefaction susceptible soils (sand and silty sand) in boreholes are indicated by gray scale circles and deposits, either sand or clay, that comply with the soft site definition given in Section 4.3 are enclosed in squares.

From Figure 26, redone as a palimpsest, the heaviest concentrations of damage in the city are observed to be localized on deep alluvial deposits and generally coincide with surface soils that are either stiff or least sensitive to liquefaction (districts 7, 9, 21 and 22). Within the area enclosed by the dashed line in the figure, foundation displacements of various forms and levels were commonly encountered. Within this area, where the sites at borehole locations are generally either highly susceptible to local soil liquefaction or already soft, the building damage is relatively reduced, despite the concentration of buildings with higher story numbers with potentially greater vulnerability to the excessive foundation displacements or seismically induced forces. On the other hand, the general reduction trend observed in damage concentration through outskirting districts located over the deep alluvial soils can be largely attributed to the decreasing story numbers and thus to the generally reduced building vulnerability.

Utilizing the idealized spectra presented in Figure 25 and depth to bedrock contours shown in Figure 22, distributions of spectral accelerations are computed for 0.2 s and 0.5 s periods, respectively representative of the fundamental period ranges of varieties of masonry (1–3 story) and reinforced concrete building (3-7 story) categories encountered in the city. Variations of spectral accelerations, smoothed and grouped by an Inverse Distance Weighting algorithm, are contoured in Figure 27 for 0.2 s and in Figure 28 for 0.5 s. The areas identified as soft sites according to the criteria set, as well as the locations of collapsed buildings provided by Sakarya University are superimposed. The soft sites were developed by the same algorithm, however with a greater resolution and by accepting 150 m as the representative radius for a boring.

As can be observed in Figure 27, two spectral acceleration zones were identified corresponding to the 0.2 s period in the city. The distribution is almost uniform with 0.9g in the southern section on relatively shallower deposits of alluvium and 0.8g in northern section where depth of alluvium reaches 200 m. It is to be noted that no buildings represented by the 0.2 s period (the masonry category) were included in the set of collapse cases investigated by Sakarya University. In fact, collapses in this category were markedly rare and more or less uniformly distributed throughout the city. This occurrence can be attributed to the smaller seismically induced forces in this category due to smaller building masses.

Spectral accelerations corresponding to the 0.5 s period, range between 0.6g and 0.9g, increasing through north with increasing alluvium depth (Figure 28). Nearly all of the collapsed buildings are located within the highest spectral acceleration zone, clearly indicating the influence of alluvium depth on seismically induced force levels imposed over the buildings. Note that, due to the inherently greater building masses in this category, a greater impact in terms of seismically induced forces would result from spectral variations, compared to the category represented by 0.2 s. The other significant observation from Figure 28 is that the locations of collapsed buildings rarely coincide with the patches representing soft sites, which is consistent with the post-earthquake observations of significantly reduced levels of structural damage over soft surficial deposits.

We reach the following conclusions from Figures 26-28:

- Ground response at a specific site in the city was controlled by the two major geotechnical factors: the alluvium thickness, which is highly variable throughout the city, and the presence of soft surficial deposits.
- Distribution of spectral accelerations corresponding to 0.2 s period was almost uniform throughout the city, whereas for 0.5 s period the spectral response was relatively varied with the maximum level overlapping the zone where alluvium thickness is greater.
- Significant de-amplification of spectral response for the period interval up to 0.5 s is predicted at the sites defined as "soft" according to the idealized site-specific spectra.

5. CODA

During post-event surveys most earthquake engineers have experienced the puzzling syndrome of observing seemingly identical adjacent buildings in widely disparate states of damage. Our current wisdom is such that we can usually rationalize our way out of this bewildering state by invoking arcane engineering minutiae with varying degrees of convincement. When, however, one is faced with the task of foretelling which of those identical buildings would experience the heavier damage in some future earthquake our authority suffers a humbling disconcertment: one simply does not know. We may know the damage outcome of an earthquake in the aggregate, but picking individual candidates for abject collapse on the basis of their visible attributes is an inexact science. Site conditions provide illumination, but secrets of structural framing seem to be safely shielded from simplistic prediction.

ACKNOWLEDGMENT

The major part of the work reported in this article was carried out during the course of a multi-party study titled "Microzonation for Earthquake Risk Mitigation," funded by the Government of Switzerland and executed through the World Institute for Disaster Risk Management (DRM) to produce a microzonation manual for local governments in Turkey to follow when making land use decisions with the aim of minimizing seismic risks. The Government of Turkey was represented in this project through the General Directorate of Disaster Affairs. Other data was extracted from consultancy work we have carried out for T. Is Bankasi, a major bank, and BOTAS, the operating entity for the natural gas distribution pipeline network in Turkey.

It is hoped that the presented results and associated discussion will provide a useful overlap with the aims and objectives of NATO Project SfP977231.

REFERENCES

1. Gulkan, P., and A. Yakut, " An Expert System for Reinforced Concrete Structural Damage Quantification," in M.A. Sozen Special Publication, ACI SP-162, Ed. J.K. Wight and M.E. Kreger, pp. 53-71, 1996.
2. Shiga, T., A. Shibata, and T. Takahashi, "Earthquake Damage and Wall Index of Reinforced Concrete Buildings," Proceedings of the Tohuku District Symposium, Architectural Institute of Japan, No. 12, Dec., 1968, pp. 29-32 (in Japanese).
3. Gulkan, P., and M.A. Sozen, "Procedure for Determining Seismic Vulnerability of Building Structures," ACI Structural Journal, 96, 3, May, 1999, pp. 336-342.
4. FEMA 310, " Handbook for the Seismic Evaluation of Buildings – A Prestandard, Federal Emergency Management Agency, Virginia, US, January 1998.
5. Bakir, B.S., H. Sucuoglu, and T. Yilmaz. "An Overview of Local Site Effects and the Associated Building Damage in Adapazari during the 17 August 1999 Izmit Earthquake", Bulletin of the Seismological Society of America, Vol.92, No.1, 2002, pp.509-526

Table 1. Building damage statistics in Adapazari

District	Collapsed and Heavily Damaged Buildings	Moderately and Slightly Damaged Buildings	Undamaged Buildings	Total
1-Maltepe	0 (0.0%)	38 (2.4%)	1514 (97.6%)	1552
2-Hizirtepe	7 (0.6%)	40 (3.3%)	1177 (96.2%)	1224
3-Sirinevler	15 (3.7%)	2 (0.5%)	391 (95.8%)	408
4-Gulluk	14 (3.3%)	3 (0.7%)	403 (96.0%)	420
5-Mithatpasa	100 (5.3%)	55 (2.9%)	1735 (91.8%)	1890
6-Yenidogan	130 (24.6%)	69 (13.0%)	330 (62.4%)	529
7-Pabuççular	185 (28.8%)	51 (7.9%)	407 (63.3%)	643
8-Akincilar	151 (20.3%)	155 (20.8%)	439 (58.9%)	745
9-Yenicami	83 (25.4%)	36 (11.0%)	208 (63.6%)	327
10-Çukurahmediye	68 (18.6%)	83 (22.7%)	214 (58.6%)	365
11-Semerciler	220 (24.2%)	88 (9.7%)	600 (66.1%)	908
12-Tigcilar	60 (11.4%)	150 (28.5%)	317 (60.2%)	527
13-Yenigun	338 (16.5%)	215 (10.5%)	1490 (72.9%)	2043
14-Tepekum	15 (1.7%)	25 (2.8%)	868 (95.6%)	908
15-Seker	248 (12.5%)	188 (9.5%)	1550 (78.0%)	1986
16-Cumhuriyet	133 (15.4%)	145 (16.8%)	586 (67.8%)	864
17-Orta	114 (13.8%)	84 (10.2%)	626 (76.0%)	824
18-Yahyalar	48 (8.3%)	44 (7.6%)	488 (84.1%)	580
19-Yagcilar	151 (7.1%)	154 (7.2%)	1832 (85.7%)	2137
20-Kurtulus	60 (10.7%)	44 (7.8%)	457 (81.5%)	561
21-Istiklal	205 (40.8%)	67 (13.3%)	230 (45.8%)	502
22-Karaosman	227 (30.0%)	117 (15.5%)	413 (54.6%)	757
23-Ozanlar	95 (8.6%)	84 (7.6%)	924 (83.8%)	1103
24-Sakarya	86 (8.6%)	63 (6.3%)	849 (85.1%)	998
25-Tekeler	56 (8.0%)	70 (9.9%)	578 (82.1%)	704
26-Tuzla	35 (8.6%)	6 (1.5%)	368 (90.0%)	409
Sum	2844 (11.9%)	2076 (8.7%)	18994(79.4%)	23914

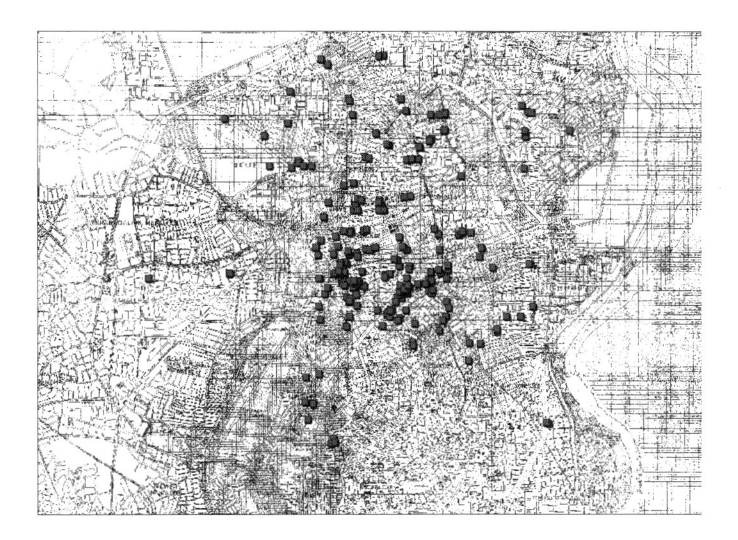

Figure 1. Building Locations

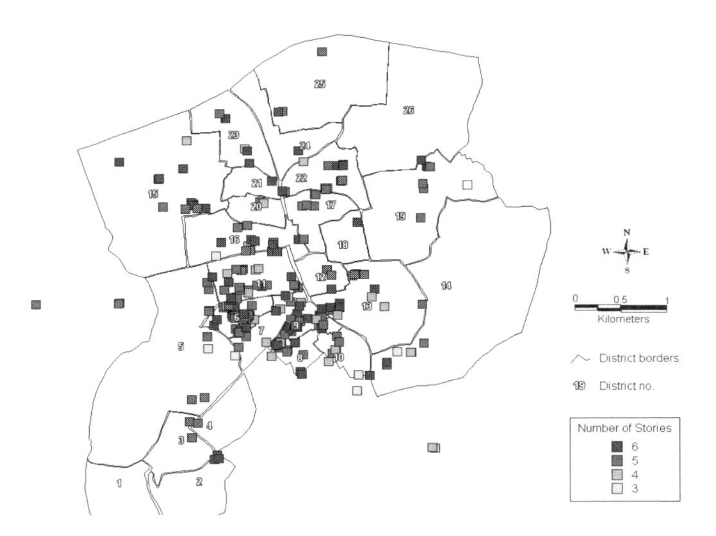

Figure 2. Building Locations Differentiated according to Height

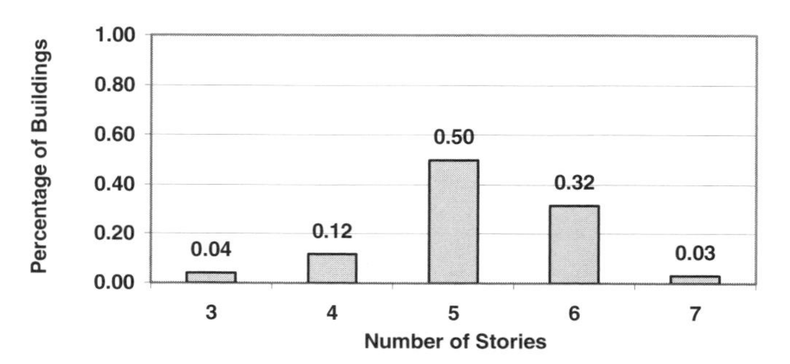

Figure 3. Distribution of Building Height in the Sample

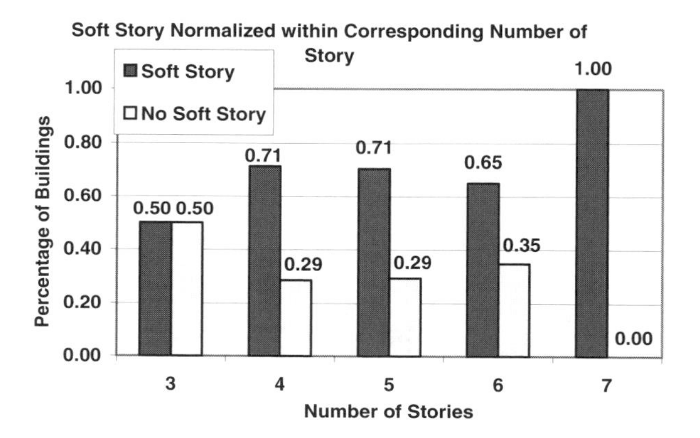

Figure 4. Distribution of Soft Stories

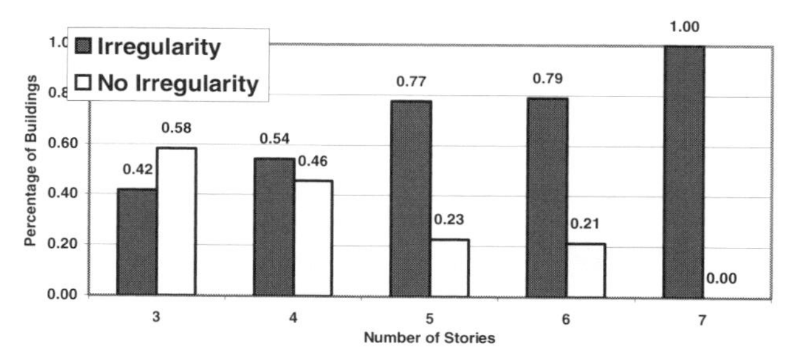

Figure 5. Existence of Torsional Irregularity

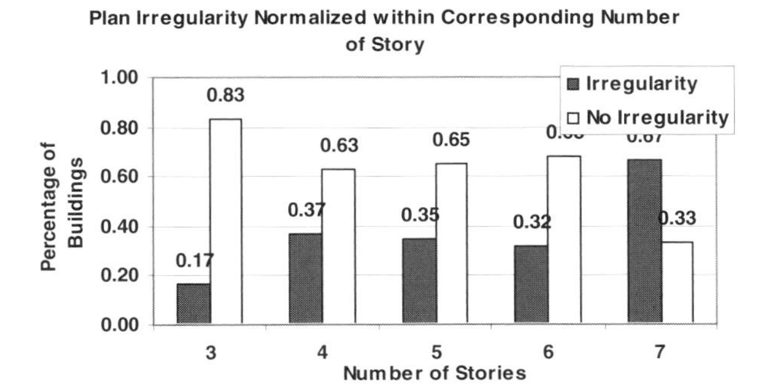

Figure 6. Existence of Plan Irregularity

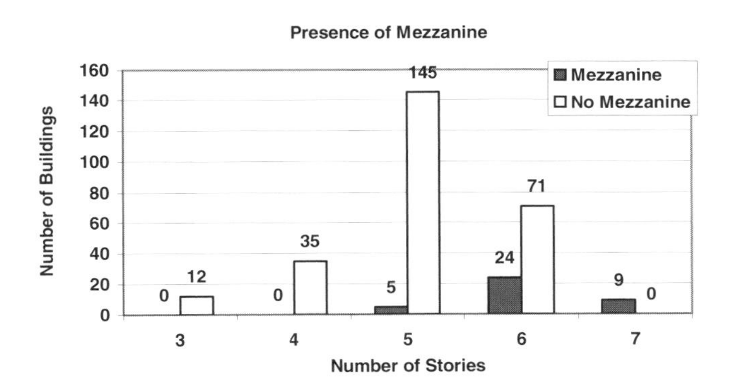

Figure 7. Occurrence of Mezzanine Floors at Ground Level

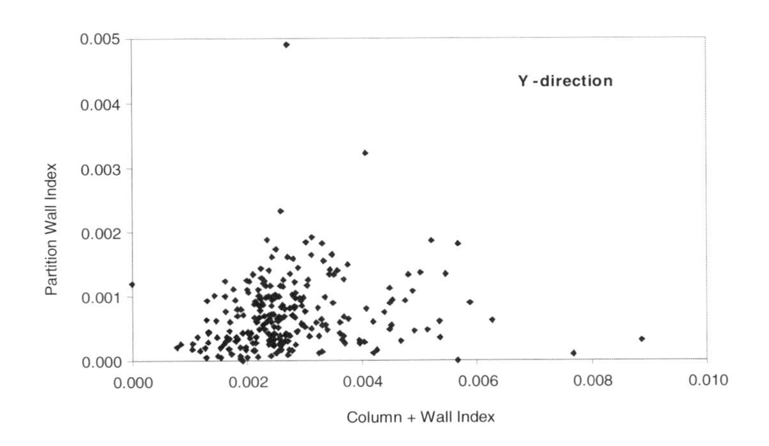

Figure 8. Wall and Column Indexes in y-Direction

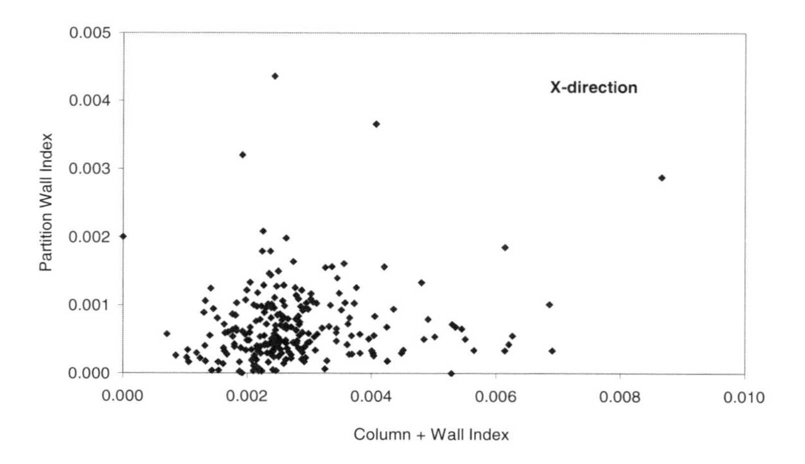

Figure 9. Wall and Column Indexes in x-Direction

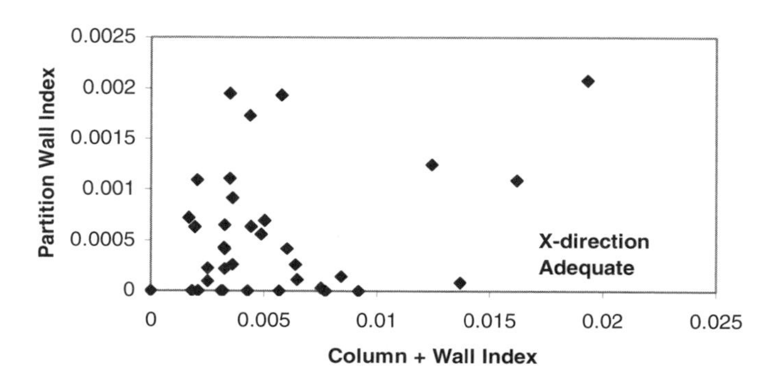

Figure 10. Buildings Rated as Adequate

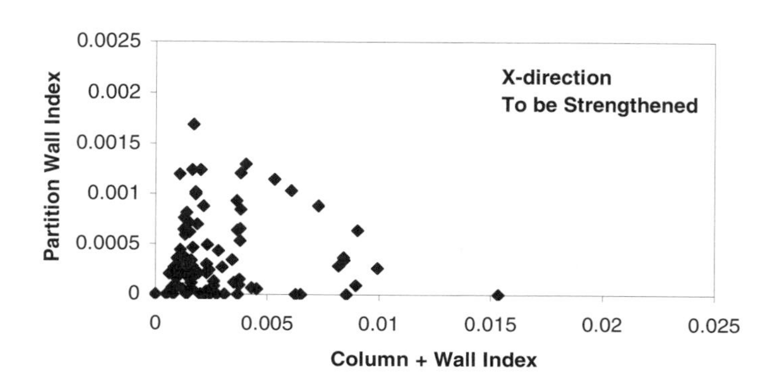

Figure 11. Buildings Rated as Requiring Strengthening

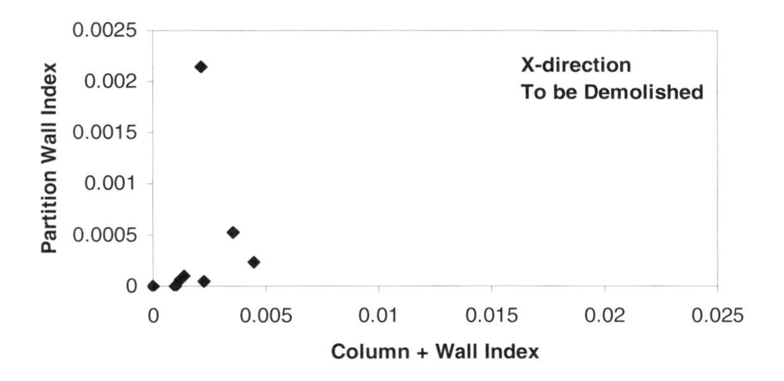

Figure 12. Buildings Rated as Requiring Demolition

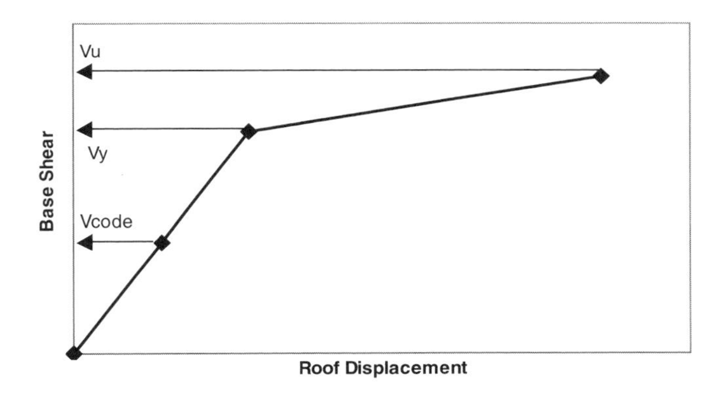

Figure 13. An Idealized Pushover Curve

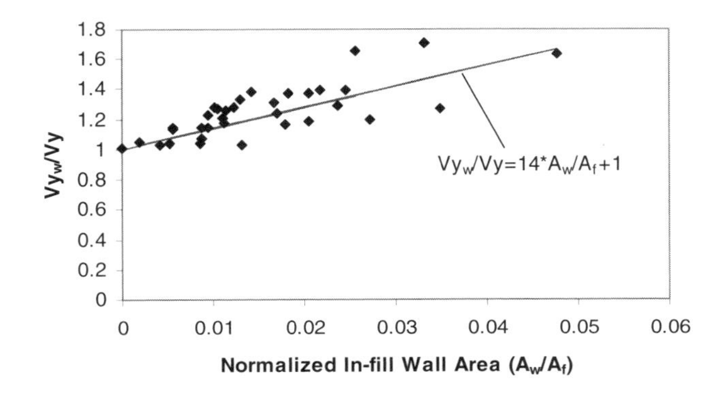

Figure 14. Effect of In-fill Wall on Shear Capacity

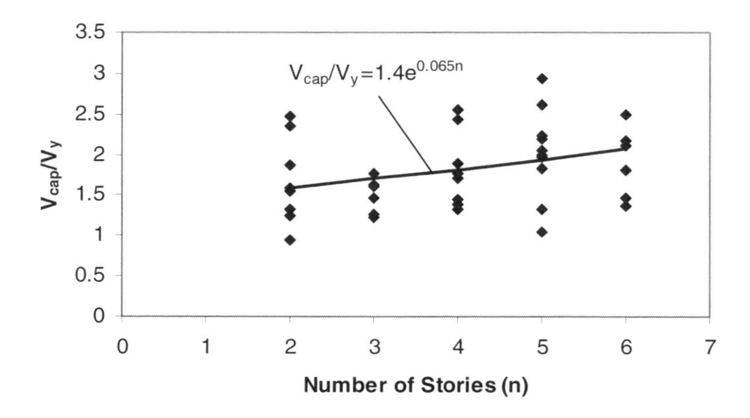

Figure 15. Yield Base Shear Capacity versus Height

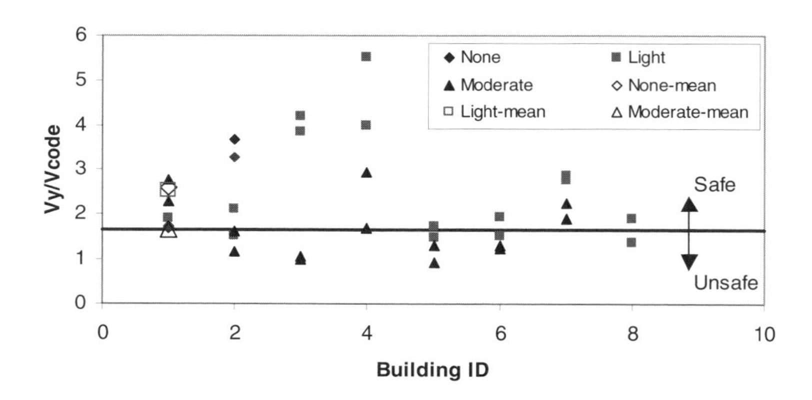

Figure 16. Capacity Index for Selected Buildings

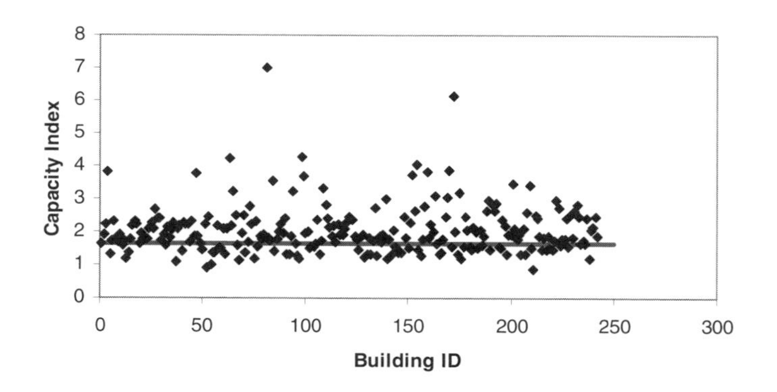

Figure 17. Capacity Index for Adapazari Building Database

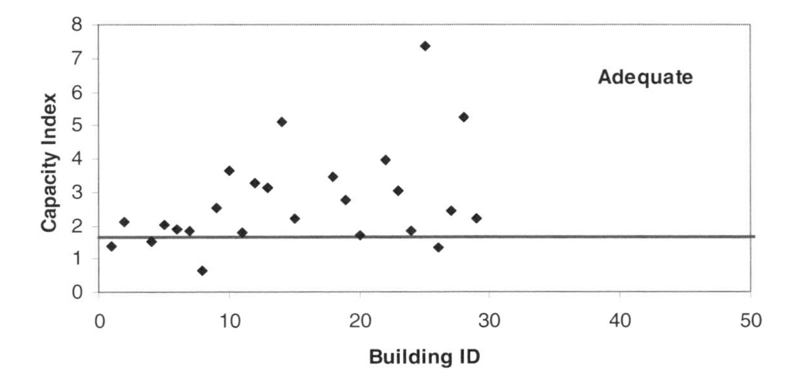

Figure 18. Capacity Index of Buildings Rated as Adequate

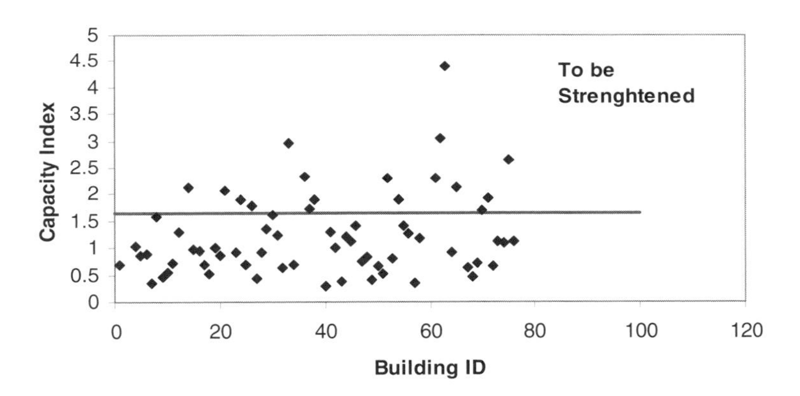

Figure 19. Capacity Index of Buildings Rated as Inadequate

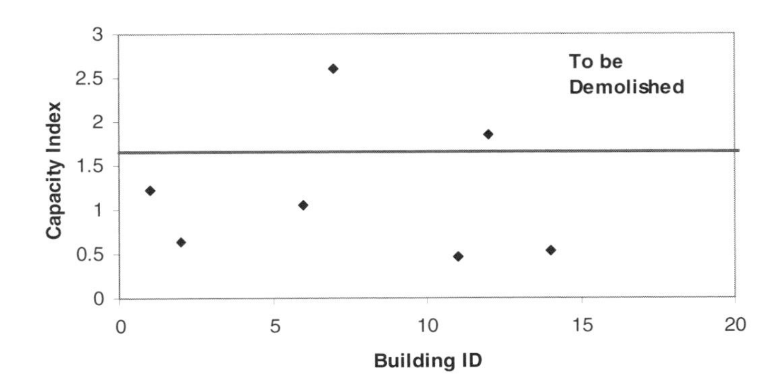

Figure 20. Capacity Index of Buildings decided to be Demolished

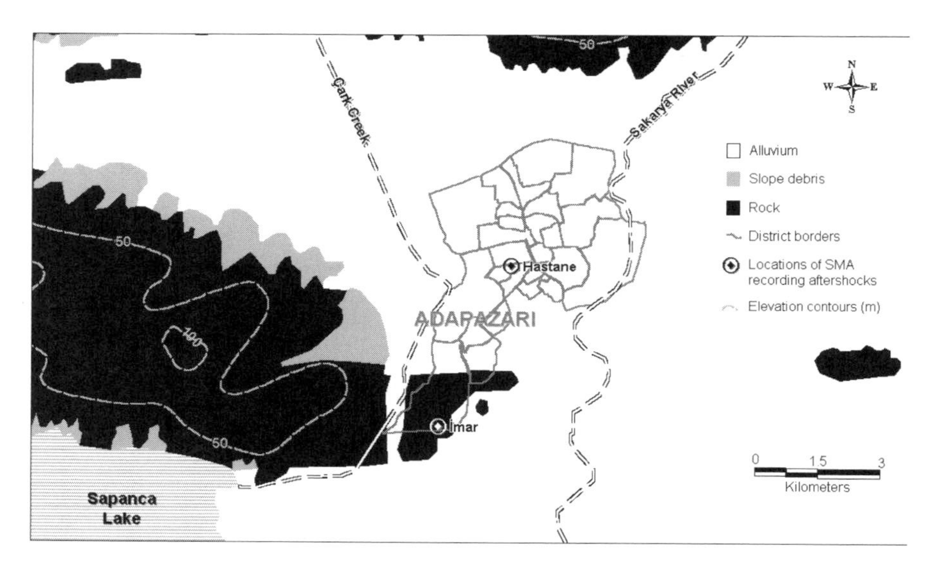

Figure 21. Main Geological Features of Adapazari Area (after Bakir et al., 2002)

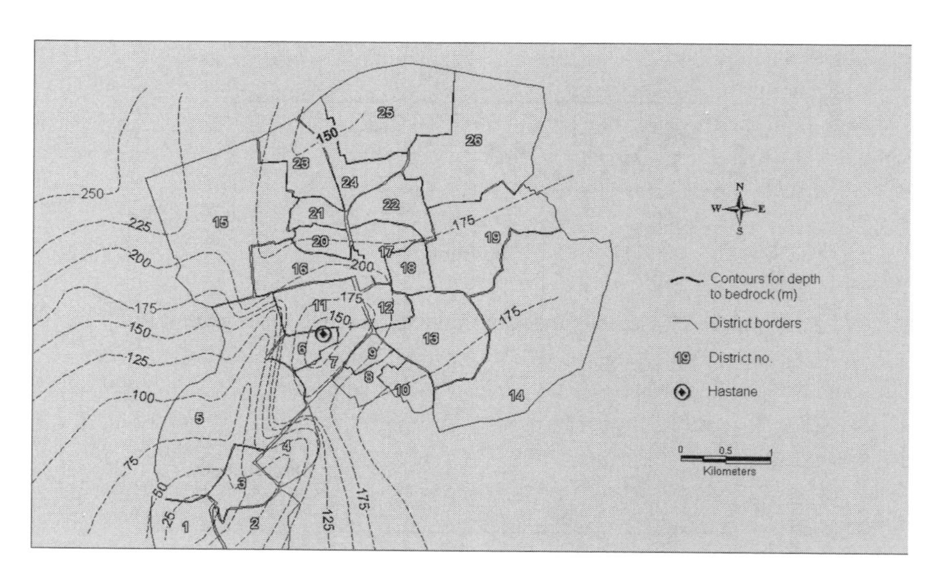

Figure 22. Variation of Bedrock Depth and Central Municipality Districts in Adapazari (after Bakir et al., 2002)

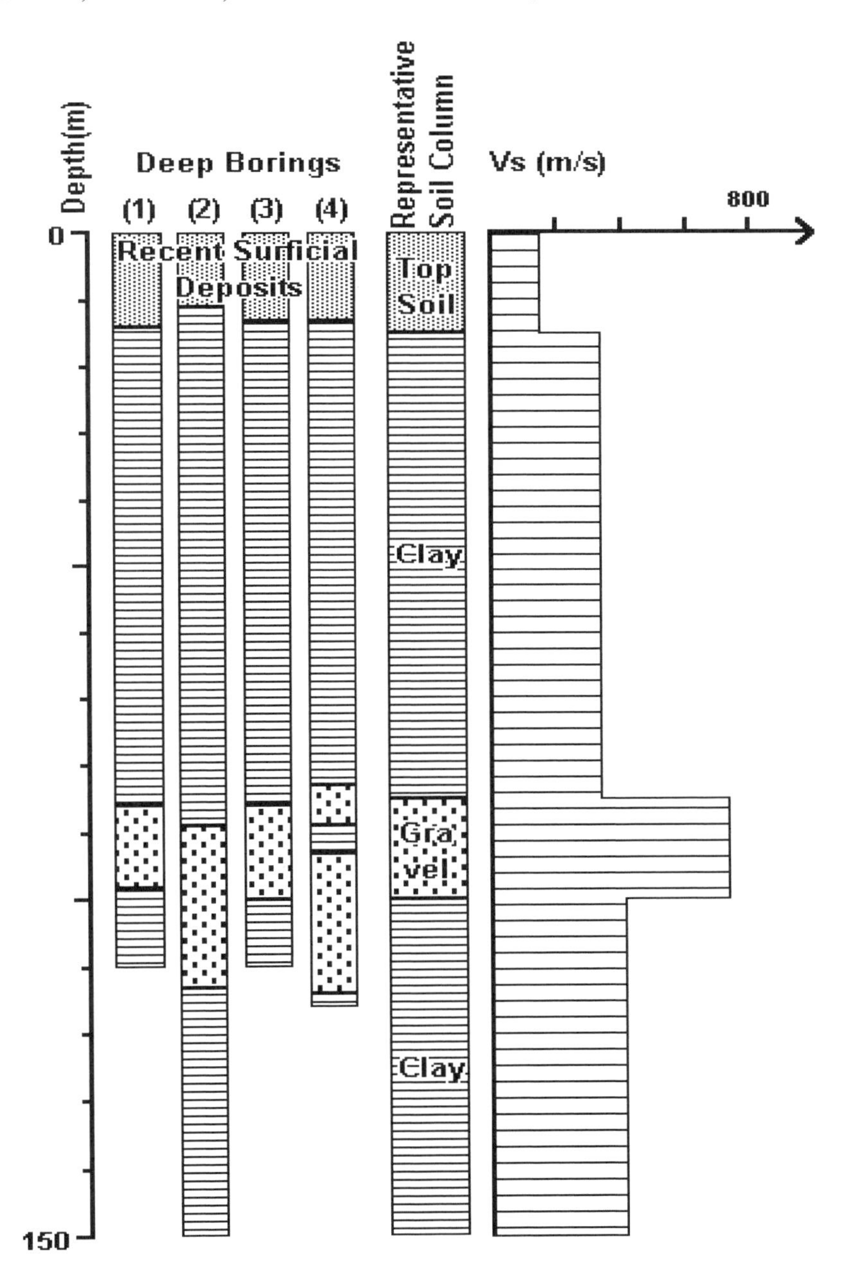

Figure 23. Deep Borehole Logs, Idealized Soil Profile and Variation of Shear Wave Velocity

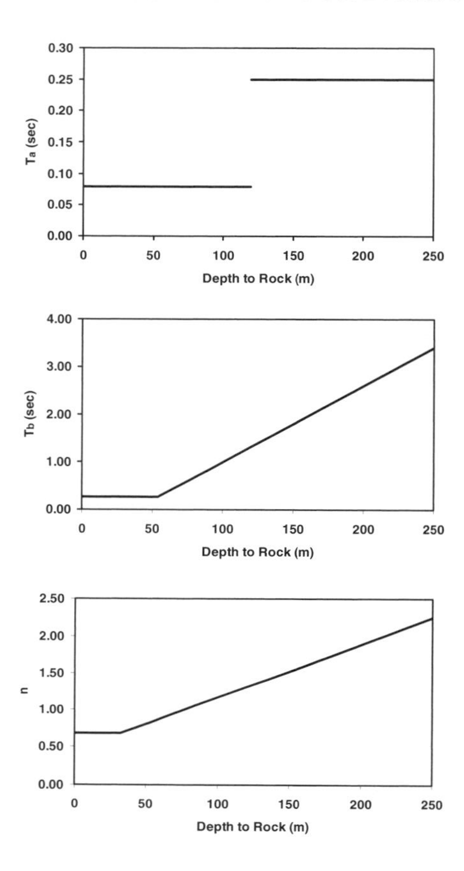

Figure 24. Set of Curves to construct Site-Specific Spectra for Stiff Sites in Adapazari

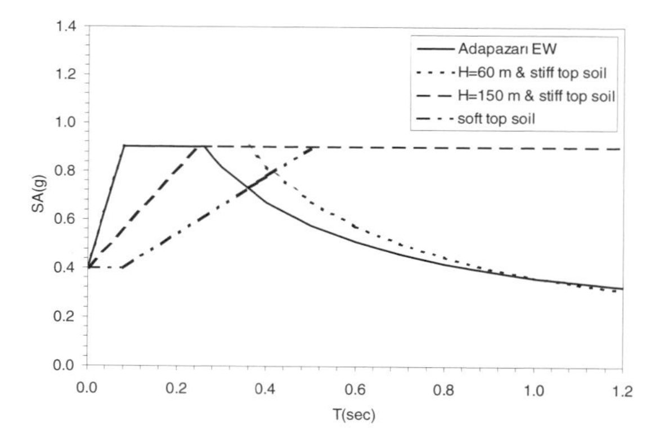

Figure 25. Sample Spectra for Soft and Stiff Sites (Outcrop spectrum is the smoothed response spectrum of Adapazari record. Soft site spectrum is representative of upper bound envelope for spectral response)

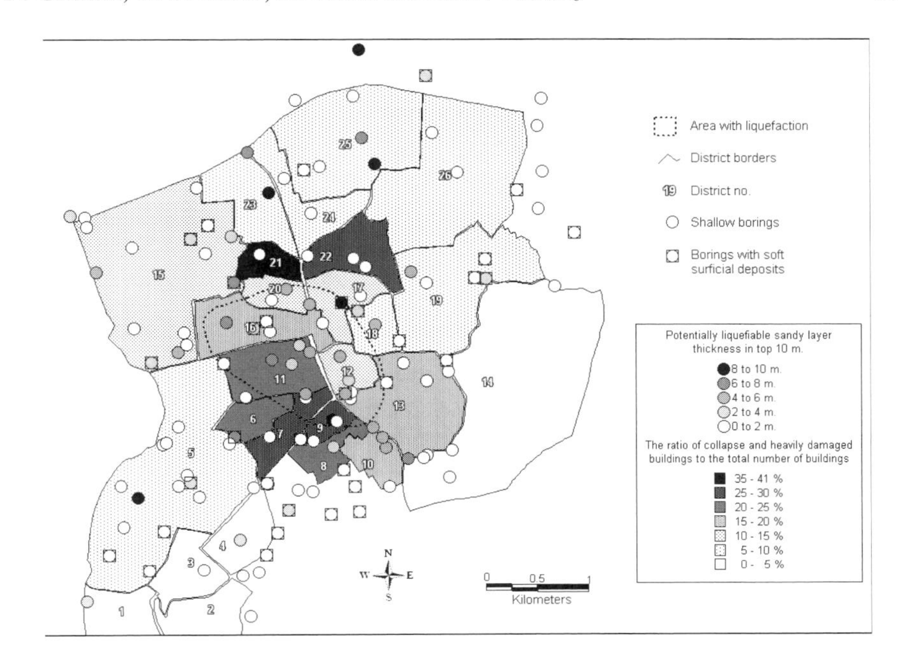

Figure 26. Borehole Locations and Comparison of Collapsed and Heavily Damaged Building Distribution to the Relevant Soil Stiffness Data in the Top 10 m of the Borehole Logs

Figure 27. Variation of Spectral Acceleration for T = 0.2 s (dark gray patches mark the soft sites)

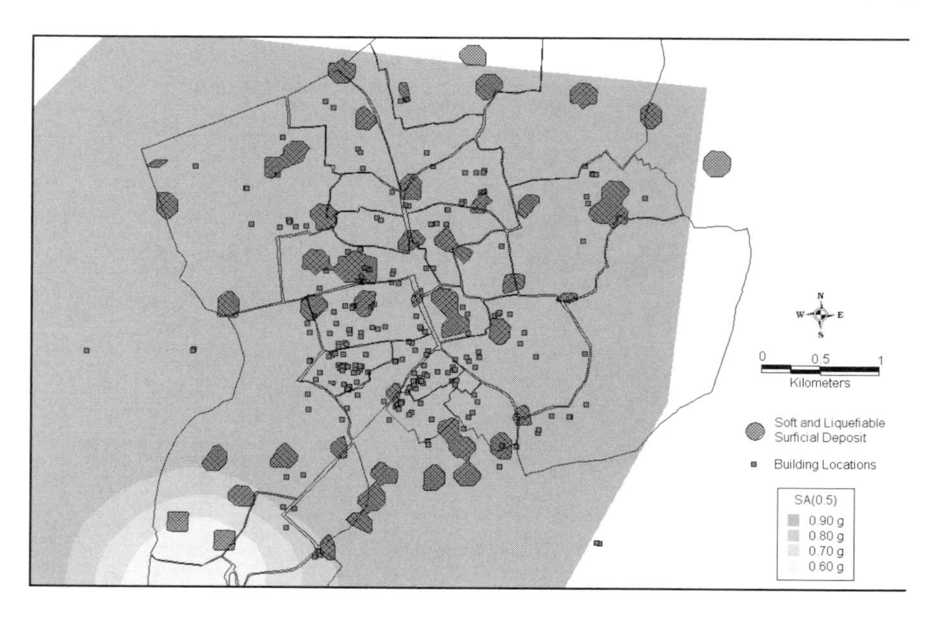

Figure 28. Variation of Spectral Acceleration for T = 0.5 s and Locations of Collapsed Buildings (dark gray patches mark the soft sites)

CONDITION ASSESSMENT TECHNIQUES USED FOR NON-BUILDING STRUCTURES

Emphasis on measurement techniques

Ahmet Turer
Structural Mechanics Division, Civil Engineering Department,
Middle East Technical University, Ankara, Turkey

Abstract: This paper attempts to summarize common condition assessment techniques, focusing on 'bridges' and 'historic monuments' and giving brief examples of condition assessment and measurement techniques from previous projects. Bridges have special importance among other civil engineering structures not only because they are the weakest chain links of an operational transportation system, but also they have demanding loading conditions of truck traffic, cyclic temperature, dynamic loading, and harsh environmental conditions (e.g. freezing and thawing, salting etc.). Condition assessment techniques are spread over a wide range from visual inspection to changing the bridge to an on-site laboratory.

Key words: condition assessment, measurement, historic structures, bridge, evaluation

1. INTRODUCTION

1.1 Factors affecting the existing condition of a structure

The condition of an *existing* structure is usually a lot different than its initial condition when it was first built. Environmental conditions, changes

S.T. Wasti and G. Ozcebe (eds.), Seismic Assessment and Rehabilitation of Existing Buildings, 193–214.

in soil and support conditions (such as settlement or scour), and operational demand causes degradation in structural properties which may also be termed as an "aging process".

The "as-built" condition would have significant differences compared to the structure in the drawings. There are many parameters that may not be controlled during construction, especially in concrete bridges. Material properties (e.g. ultimate strength, modulus of elasticity), geometric dimensions (e.g., length, thickness, slope), and structural properties (e.g., unintended composite action) are the most common parameters, with a large scale of uncertainties. Additional parameters about a bridge (such as cracking, bolt slip, effects of splicing, support conditions, soil characteristics, and many other parameters) that are not well known bring in further uncertainty to the "as-built" condition.

The condition of a bridge would naturally change over time because of external effects. The "aging process" causes changes in material properties (e.g. alkali-silica reaction), structural elements (rusting of rebars, girders, truss members, scouring), and original geometry (support settlement, creep). Maintenance work carried out with certain intervals during the life time of a bridge, which may often be poorly documented, also contributes greatly to structural parameter and condition changes in a bridge.

1.2 Need and relevance of condition assessment

Frequently, the structural performance of a structure is questioned for a variety of reasons: an over load pass permit, seismic performance level determination, maintenance need and precedence, general appraisal determination for inventory, etc. The most common way to determine the structural performance of a structure is to construct a finite element model and evaluate the response under anticipated loading simulations. Other relatively simple methods are also used with a coarse assessment precision.

However, it would be naive to claim that the behavior of an existing bridge/structure would be known, simulated, or closely replicated just by generating a nominal finite element model without considering the "as-is" or "current" condition. The need and necessity to conduct a finite element model updating (calibration) using field measured data becomes evident in order to assess the existing condition of a structure. The existing condition assessment must be handled under a Structural Identification (St-Id) framework.

Although assessment techniques might be applied to most civil engineering structures in general, different types of structures require different approaches. In this study, relatively unconventional structures such as bridges and historic monuments will be addressed.

2. STRUCTURAL CATEGORIZATION

Bridges and historic monuments can be briefly grouped under their most common types, in order to categorize specific assessment tools that may be applied to each type.

2.1 Types and behavior and of bridges

Bridges might be very generally categorized under four headings in accordance with their geometric and load carrying mechanisms. These are:

- Beam bridges
- Arch bridges
- Cable-stayed bridges
- Suspension bridges

Bridges might also be categorized with regard to the material used in their construction:

- Concrete bridges
- Steel bridges
- Stone bridges
- FRP bridges

Combination (such as Concrete/Steel: concrete deck on steel stringers, Steel/FRP, Steel/Stone: stone piers, steel bridge)

2.2 Historical monuments

Historical monuments are traditionally made from stone blocks in a masonry/friction type of building construction. Arches are very commonly used if openings and internal spaces (such as domes) are needed. In newer historic structures, shear key locks and metal strips are used to keep the blocks together and increase shear transfer. An example is given in Figure 1 illustrating the improved shear transfer between arch stones in the Old Mostar Bridge which was constructed in 1566 and located in Bosnia-Herzegovina.

Non-building historic structures have a broad range of types but a common type is in the form of large scale statues and sculptures used for representation of kings or gods. Open air theaters can also be categorized under this heading. The detailed grouping of non-building historical structures is outside the scope of this paper.

3. ASSESSMENT TECHNIQUES

The techniques used for condition assessment of bridges vary from very primitive to the most advanced methods. A list of methods and tools used for condition assessment and health monitoring of bridges is shown in Figure 2. Condition assessment techniques can be categorized under four headings:

- Visual and visual related
- Diagnostic load testing
- Non-destructive evaluation (NDE)
- Long-term continuous monitoring

Assessment techniques used for historical monuments can be categorized under the following headings:

- Visual inspection (usually for stability and deterioration)
- Material sampling for lab tests

Each title will be explained briefly giving an example application.

3.1 Visual inspection

Visual inspection is among the most commonly used assessment methods due to its simplicity and ease. The general view and critical details of a structure reveal important clues about its current condition. Critical inspection criteria such as rusting, crack formation, excessive deflection, and spalling show potential problems about a bridge (Figure 3). However, an objective and quantitative assessment is not possible due to the judgmental nature of the method. Different experience and backgrounds of inspectors cause inconsistencies in evaluation results. Different inspectors can rate the same bridges for different appraisal values. Furthermore, visual inspection is only superficial and possible visual obstacles might block inspection.

An alternative to direct visual inspection is photogrammetry which studies sub-millimeter changes in location using high resolution cameras. Pictures of a structure might be taken in certain intervals (e.g., every month) and changes in deflected shape or crack openings can be investigated. The results of photogrammetric measurements would yield hints about the current condition of a bridge.

3.2 Diagnostic Load Testing (Controlled Tests)

Controlled load tests should normally be included under Non-Destructive Testing (NDT); however, loading tests might also be destructive. Diagnostic load tests usually follow two stages: (1) instrumentation, (2) measurement of responses under controlled loading (such as a crawl speed truck crossing the bridge). The axle weights of the test truck are measured before testing. The truck is usually loaded with sand to provide extra weight which would improve the signal/noise ratio by increasing structural response. Deflections and strains can be measured for steel stringer bridges. For concrete bridges measurement of strains is usually erroneous since crack formation releases strains in tension zone. Strain measurement on concrete surfaces should be practiced only in negative bending regions under a bridge. Concrete strain might also be measured at the upper surface of deck in positive bending regions.

At least two strain gages should be placed in steel girders in order to capture the bending moment (M) and axial force (N) induced strains. Past experiences with concrete deck on steel stringer bridges showed that measurement might be erroneously converted into M and N values especially for "non-compositely designed" bridges. A non-compositely designed bridge has certain amount of chemical bonding between the deck and the girders which might be lost by a certain amount over time. The neutral axis location would shift upwards when full composite action exists (Figure 4(a)). The bridge might experience axial forces due to support or other geometric conditions. A compressive axial force in a positive bending region would cause the "zero strain" location to shift downwards (Figure 4(b)). The strain profile obtained using the top and bottom flange strain measurements can also be duplicated if the composite action is partially lost, as shown in Figure 4(c). Even if the strain profile is successfully obtained for the girder from strain measurements, determination of sectional forces highly depends on assumptions made for concrete material properties, effective deck width, existing composite action percentage, transverse load transfer, etc.

Deformations under controlled load tests are usually measured at maximum deflection points (normally at midspan). Deflection quantities might possibly be obtained from a sparse and accurate strain profile measurement by double integration if the strain profile measurements are obtained with sufficiently spaced intervals. Calculation of deflection from strain measurements can be useful for bridges that are too high or have traffic/water underneath.

3.3 Non Destructive Testing (NDT)

Non-Destructive Testing methods used on bridges are usually preferred for their non-intrusive nature. Most common methods of NDT used on bridges can be listed as follows:

- Micro-sampling, Core-sampling
- Ultrasound
- Modal testing

Micro sampling and Core sampling are usually used for steel and concrete bridges or samples are taken from concrete and steel parts of a single bridge if it includes both materials in its construction. The material properties obtained from samples are normally localized and do not mutually represent the general nature of material used in a bridge. Therefore, the number and location of samples taken from a bridge need to be in such quantity and variation as would constitute a representative set of existing overall properties.

Ultrasound techniques are very commonly used in NDT related activities. The theory is based on the travel of sound waves in continuous media and their reflection from natural boundaries and possible cracks that exists in a member. Critical members and details such as girders, welding, or pins of truss members can be checked using ultrasound techniques.

Modal testing is relatively new compared to other NDT techniques. The theory uses the free dynamic vibration *response* of a structure which is normalized using a known dynamic *input force* vector at a predefined location (Figure 5). Forced vibration, ambient vibration, and impact modal testing are among different versions of the same theory. The basic dynamic equilibrium equation (Equation 1) is converted to Equation 2 after a Laplace transform and system matrix [B(s)] is obtained (Equation 3). The system transfer function [H(s)] is the inverse of system matrix [B(s)]. The matrix of measured frequency response functions (FRFs) is equal to the system transfer function [H(s)] which is also equal to $[B(s)]^{-1}$.

$$[M]\cdot\ddot{x} + [C]\cdot\dot{x} + [K]\cdot x = F(t) \tag{1}$$

$$\{[M]\cdot s^2 + [C]\cdot s + [K]\}\cdot\{X(s)\} = \{F(s)\} \tag{2}$$

$$\{[B(s)]\}\cdot\{X(s)\} = \{F(s)\} \tag{3}$$

The measured FRFs (=H(s)) are represented using various modal parameters and analytical models (Equation 4). The resulting system of modal variables is used to identify structural parameters such as mode shapes, modal frequencies, and modal flexibility. The system transfer function H(s) becomes the flexibility matrix when ω is equal to zero as H(s) in Equation 4 becomes equal to $1/[K] = [K]^{-1}$.

$$[H(s)]_{s=j\omega} = [H(j\omega)] = \frac{1}{\left\{-[M]\cdot\omega^2 + [C]\cdot j\omega + [K]\right\}} \quad (4)$$

Modal flexibility matrices, mode shapes, and modal frequencies are used for condition assessment of bridges. Comparison of measured parameters against analytical models also reveals additional parameters within the context of structural identification (St-Id). Modal testing can be conducted with certain intervals and results can be compared against each other to track any changes in structural behavior.

3.4 Instrumented (long term) health monitoring

Instrumented long term monitoring watches for daily and seasonal changes and tries to capture a general pattern of behavior. Any changes in the "normal" response or "behavior" pattern raises a flag pointing out to potential differences in structural parameters, which is a probable indication of structural damage. A schematic representation of long term bridge health monitoring system is shown in Figure 6. [6]

Instrumentation studies for long term monitoring are similar to short term monitoring, however the gages should be suitable for being stable over long periods of time and under changing environmental conditions (such as temperature swings between summer and winter). The gages should be of a rugged type and have low-energy consumption rates. An example to bridge instrumentation is shown in Figure 7. The power supply for long term bridge monitoring is usually obtained through rechargeable batteries and solar panels which charge the batteries during daylight hours.

3.5 Examples to some assessment applications to historic structures

3.5.1 Curve centroid assessment of historic Old Mostar Bridge

Mostar bridge, a limestone arch bridge, located in the Mostar city of Bosnia and Herzegovina used to span the Neretva river. The bridge construction took 9 years and completed in 1566 for the Ottoman emperor Suleyman the Magnificent. The master builder was a Turkish engineer, Hayruddin, who was a student of the celebrated architect-engineer Koca Sinan. After 427 years of its completion, the bridge was demolished by heavy tank artillery on November 9, 1993 during the Bosnian civil war (1992-1995).

Restoration and rebuilding related work has begun in 1997 as a joint venture by international organizations, which brought up many questions about the geometry and original design of the Mostar Bridge. Previous studies on the bridge and existing pictures helped to extract x-y coordinates of the general bridge geometry (Figure 8) as a subset of visual inspection methods. One of the disputes about the bridge was in determining whether the bridge had two centers or one center for the main arch. Historical facts show that ogival arches are commonly employed by Koca Sinan. The numerical techniques used by Middle East Technical University, one of the participant organizations, to determine the arch centers uses an optimization routine. The objective function for optimization is defined by the sum of errors between each individual radius and *average of all calculated radius values* (R_{bar}) relative to an arbitrary centroid of a generic circle. The R_{bar} value is calculated using Equation 5. The objective function (OF) is then evaluated using Equation 6 for each iteration step of the optimization process. As different centroid coordinates (x_R, y_R) are selected, R_{bar} and OF are computed again. As the sum of errors between each radius and R_{bar} term decreases, the analytical curve defined by centroid (x_R,y_R) and radius R_{bar} becomes closer to the measured arch coordinates.

$$\overline{R} = \frac{\sum_{i=1 \to n} \left| \left(\sqrt{(x_i - x_R)^2 + (y_i - y_R)^2} \right) \right|}{n} \tag{5}$$

$$OF = \sum_{i=1-n} \left| \left(\overline{R} - \sqrt{(x_i - x_R)^2 + (y_i - y_R)^2} \right) \right| \tag{6}$$

The optimization results for curve fitting are shown in Figure 9 for North and South faces of the bridge. The optimum radius values for North and South faces are calculated as 15.01 m and 15.57 m, respectively. The centroid location is calculated 2.97 m and 3.66 m below the edges of arch for North and South faces respectively. The results are also summarized in Table 1. The horizontal distances between centroids (0.44m and 1.23m) are found to be relatively small compared to the arch radius (15.01 and 15.57m).

Closer comparison of measured data against the optimally fitted curve shows that the measured (actual) arches are side-swayed towards the left approximately 14 cm on both North and South faces relative to the best fit curve. The symmetric sway in both directions indicates a "clock-wise" torsional deformation when looked down from above. Consistent torsional deformation on both sides of the arch on both faces may be an indication of the East support shifting downstream (the North face is on the upstream side).

Analytical studies combined with measured visual data might reveal important information about the existing condition of a bridge. In the Mostar Bridge a side sway of about 14 cm is detected and a torsional distortion caused by a possible East support downstream shift may therefore be concluded.

3.5.2 Assessment of Nemrut dagi statues (Commagene)

The artificial tumulus of Nemrud Dagi, a World Heritage Site, is located within the boundaries of Adiyaman province at 47.73.47E and 420.39.24N UTS global coordinates (Figure 11).The Nemrud dagi monuments belong to Commagene civilization which rules the area about 2000 years ago. The region is seismically active in Zone #1 (the most severe zone) according to Turkey seismicity maps (Figure 12). When the available fault maps are examined, the site location is calculated to be as near as 5 to 6 km away from the fault line passing at North-West side of the monuments (Figure 13). No doubts about the seismic activity of the region (over the last 2000 years) remain when the ruins of statues are examined as seen in Figure 15.

The statues are in general composed of 5 to 6 layers of large stone blocks piled on top of each other forming sitting position human figures about 8 meter tall on the average, including heads (see Figure 14). There are 5 statues lined up in a row sitting over a large base rock. There is one eagle and one lion statue (smaller size) located on either side of the 5 human figures. A group of 9 statues exists on East and West terraces of a tumulus which is located at the very top of Nemrut dagi.

Visual inspection of the structural condition of statues reveals that the heads and shoulders of all statues have fallen down on the East side, and most of the blocks have been dislocated on the west side of the tumulus (Figure 15). [7] The base rock under some of the statues has been eroded leaving large spans of the stone blocks in air generating a "beam-like" condition (Figure 16). Stones are brittle in nature, and crack formation in the tension zone might cause the heavy statue to lose stability and roll down the hill after disintegration.

Further evaluation of the stone blocks shows shear cracks which makes the system vulnerable to various instabilities (Figure 17). The stones located on top of each other do not have shear keys in between, and rely on frictional forces only. Many dislocated stone blocks are observed at the site indicating stones movement during past earthquakes, and these might reach an unstable condition in the future if no precautions are immediately taken.

In addition to the main human and animal statues at the historic Nemrut site, there are additional stelae figures located next to the main statues (Figure 18). The snow load accumulating behind the statues in winter causes them to fall forwards since the bases of the stelae are very small and potentially unstable. It is observed that fallen and broken stelae figures have been lifted up into position again and again only to continue falling down next season. Amateurishly conducted epoxy + concrete repairs utilizing short steel rods (smooth on the surface) was not successful keeping the pieces together (Figure 19). Stone deterioration on the surface was serious due to harsh environmental conditions and touching/climbing of these monuments by visiting tourists.

Visual inspection of the remains proved to be successful in locating most of the important problems and emergencies. Material testing in laboratory is necessary to obtain material properties of the stone used for the statues. Structural tests should be carried out on small scale lab models using shaking table to assess structural behavior to various seismic excitations. Field tests to measure the dynamic behavior of statues can also be conducted if required.

4. CONCLUSIONS

In this paper, various condition assessment techniques are studied and a few examples are given for the illustration of selected items. Modern and historic bridges are used for instrumented and visual condition assessment examples.

Visual inspection results about historic Nemrut dagi monuments are given. The Nemrut dagi monuments are located in a seismically active zone few kilometers away from fault zones. The restoration studies should be made professionally considering the seismicity of the region. The stone blocks do not have shear keys and rely on frictional forces for lateral stability. Re-erection of the statues, by placing the fallen pieces on top of each other, should be avoided before any structural dynamic analysis and laboratory experiments for possible earthquakes. Re-erection of the statues would be a crime since they will collapse in the next earthquake as they did before permanently damaging healthy stone blocks. Decision should be made by local authorities, scientists, and engineers whether or not leave the organized pieces on ground or re-erect using applicable strengthening methodologies. "Preservation" should be the primary goal with minimal intrusion.

The experiences gained from numerous projects show that visual techniques are generally superficial and subjective, whereas scientific, measurement based tools are usually more objective and reliable. Assessment should be made to capture the (a) global and (b) local conditions/properties. Any assessment missing one of these two may not be accepted as a successful assessment.

Condition assessment tools are much more extensive and sophisticated in technological background and range of application than those presented in this paper. However, a subset of the most commonly used assessment tools are presented for non-building structures with an emphasis on bridges and historic monuments.

ACKNOWLEDGEMENTS

This paper includes experience gained from previous projects which have been collaborative work with many researchers. The author would like to thank the University of Cincinnati team, Ohio DOT, and Federal Highway Administration for their help and support at various levels. The World Monuments Fund supported trip expenses to Nemrut dagi site for the Turkish team. Author would also acknowledge contributions made and support provided by the rest of the Turkish team (Dr. E. Caner Saltik, Dr. T. Topal, Dr. A. Tavukcuoglu, and Dr. E. Erder)

REFERENCES

1. Rehabilitation of the old bridge of Mostar, design - photogrammetry – calculations, General Engineering s.r.l., 2001
2. Butler County Bridge (BUT-732-1043), 6 Bridge Project – PhD Thesis, Univ. of Cincinnati, A. Turer, 2000
3. Report: "Instrumentation of Concrete Box Girder Bridge Ramp I, Over I-85 Span #4, Between Piers #4 & #5 At Fulton County, Atlanta, Georgia", 1998
4. "Stresses and damage prediction of steel stringer bridges using finite element modeling techniques and UILD method", M.S. Thesis, Univ. of Cincinnati, A. Turer, 1997
5. "Modal Space, Back to Basics", Pete Avitabile, Experimental Techniques by the Society for Experimental Mechanics, 1998.
6. FP6-IST-1 "Bridge Monitoring" proposal, A.Turer, et al., 2003.
7. Nemrut dagi, 2002 Final Field Report, World Monument Fund, 2002.

Table 1. Comparison Table of Calculated Parameters for Old Mostar Bridge

		North Face	South Face
Double Centroid	Rbar=	15.01 m	15.57 m
	vertical distance to centroid from arch edge=	2.98 m	3.66 m
	horizontal distance between double centroid=	0.44 m	1.23 m
Single Centroid	Rbar=	14.80 m	14.73 m
	vertical distance to centroid from arch edge=	2.86 m	2.99 m
	horizontal distance between double centroid=	0 m	0 m

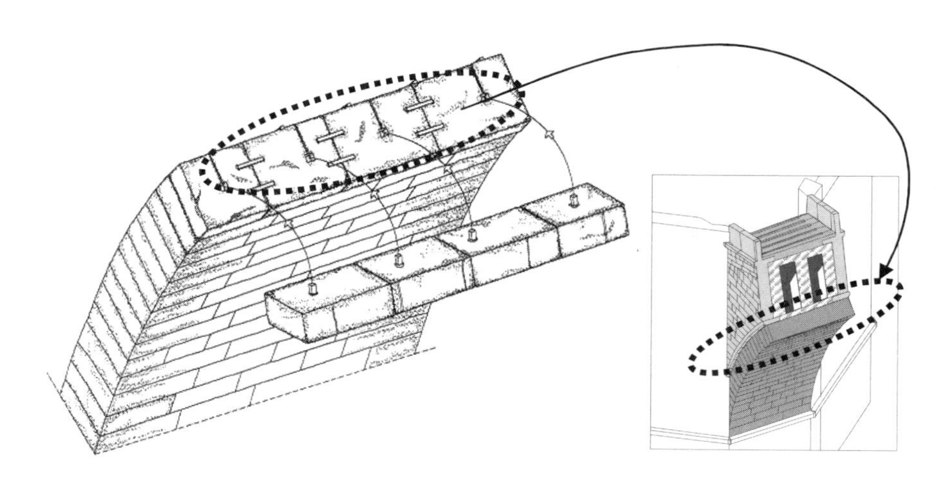

Figure 1. Shear lock keys of main arch, Old Mostar Bridge [1]

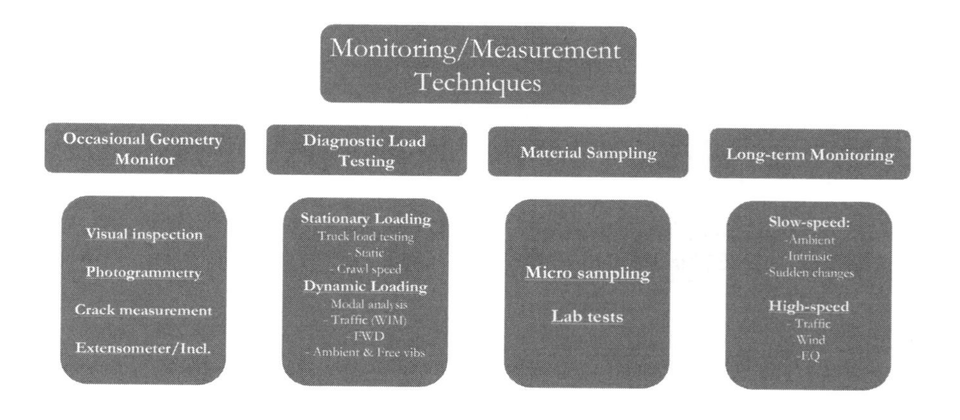

Figure 2. Monitoring Table

Figure 3. Visual inspection of highway bridges [2], [3]

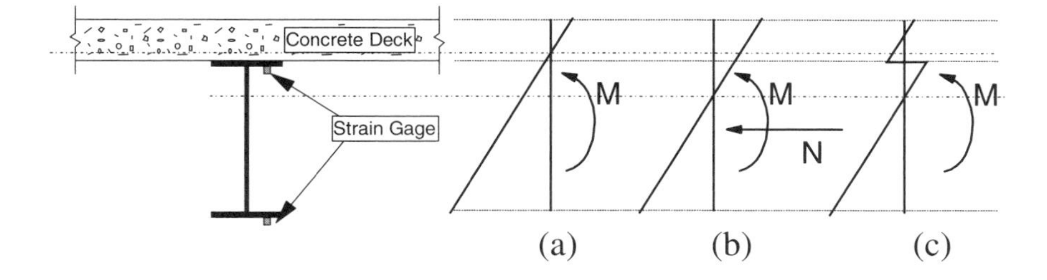

Figure 4. Possible strain diagrams and resulting section forces]

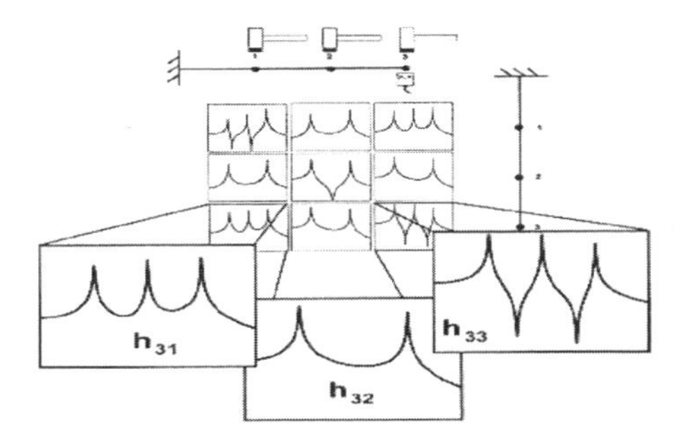

Figure 5. Impact modal testing to obtain FRFs [5]

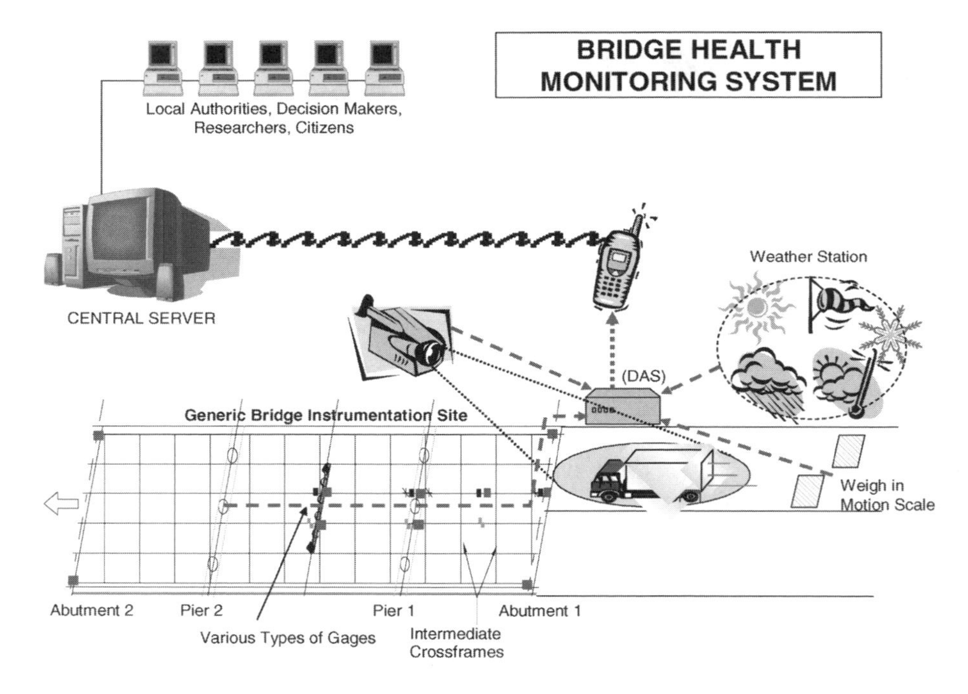

Figure 6. Long term health monitoring of highway bridges

Figure 7. Bridge monitoring studies for short and long term measurement

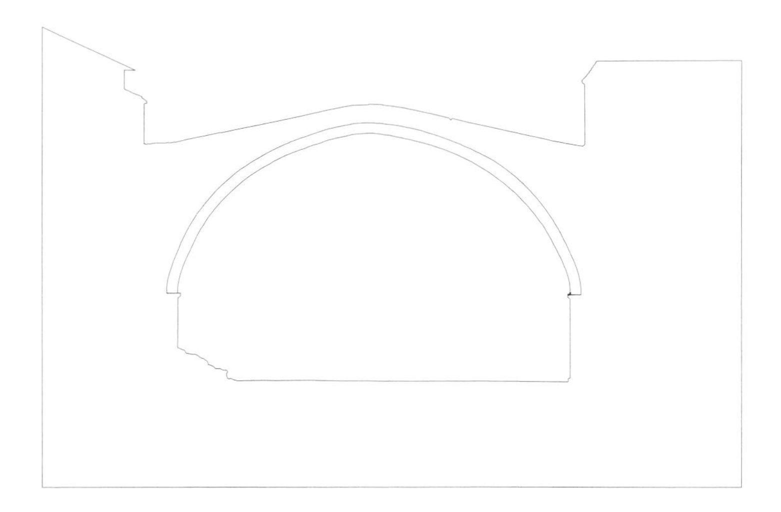

Figure 8. Old Mostar Bridge Geometric View

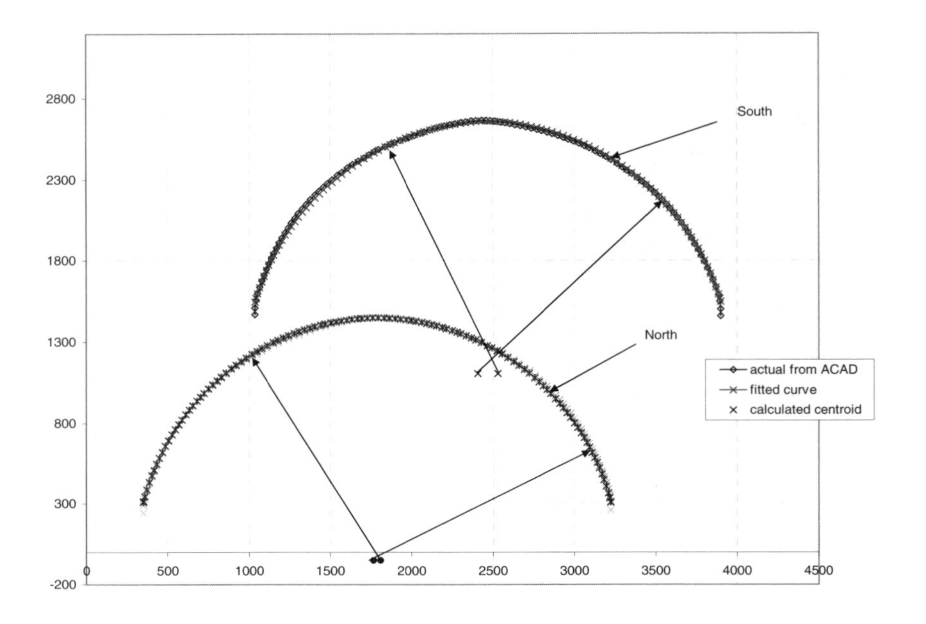

Figure 9. A general view of the x-section of the old Mostar Bridge

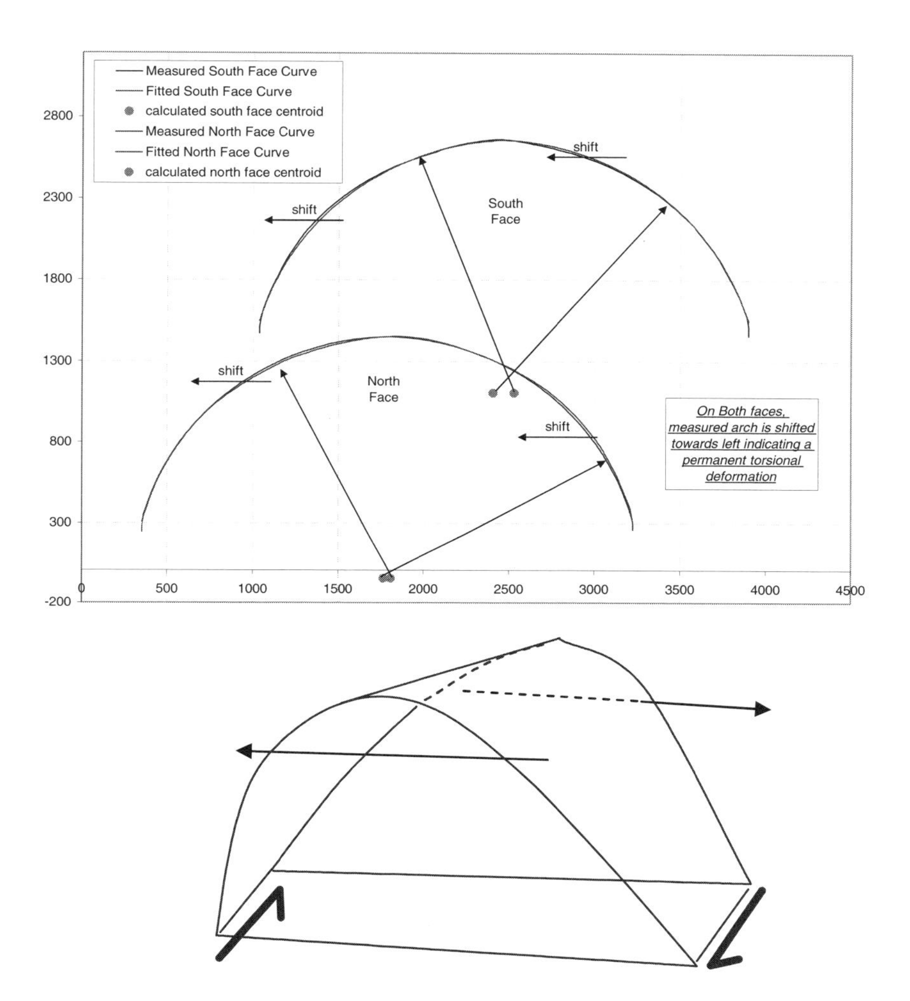

Figure 10. Circular (double radius) curve fit studies to Mostar Bridge arch

Figure 11. Location of Nemrut dagi relative to nearby cities

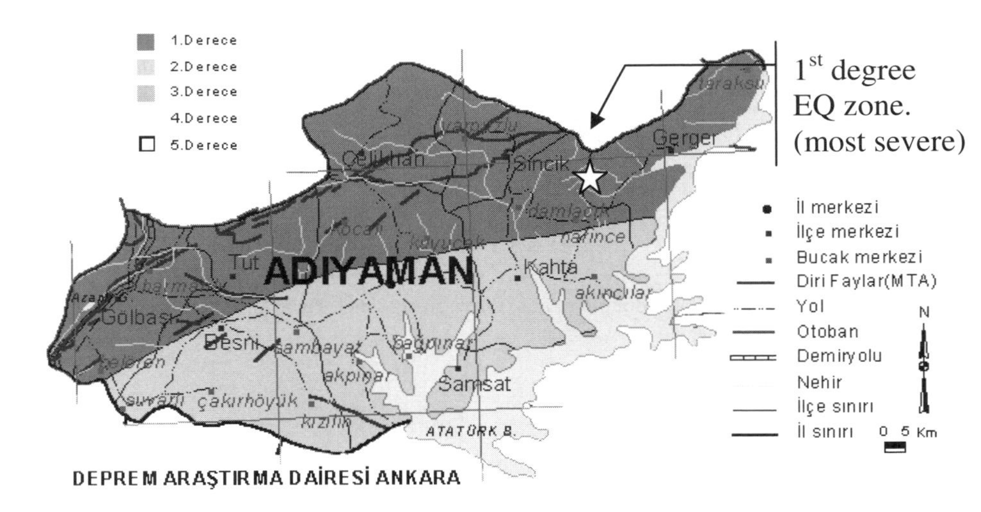

Figure 12. Seismic map of Adiyaman

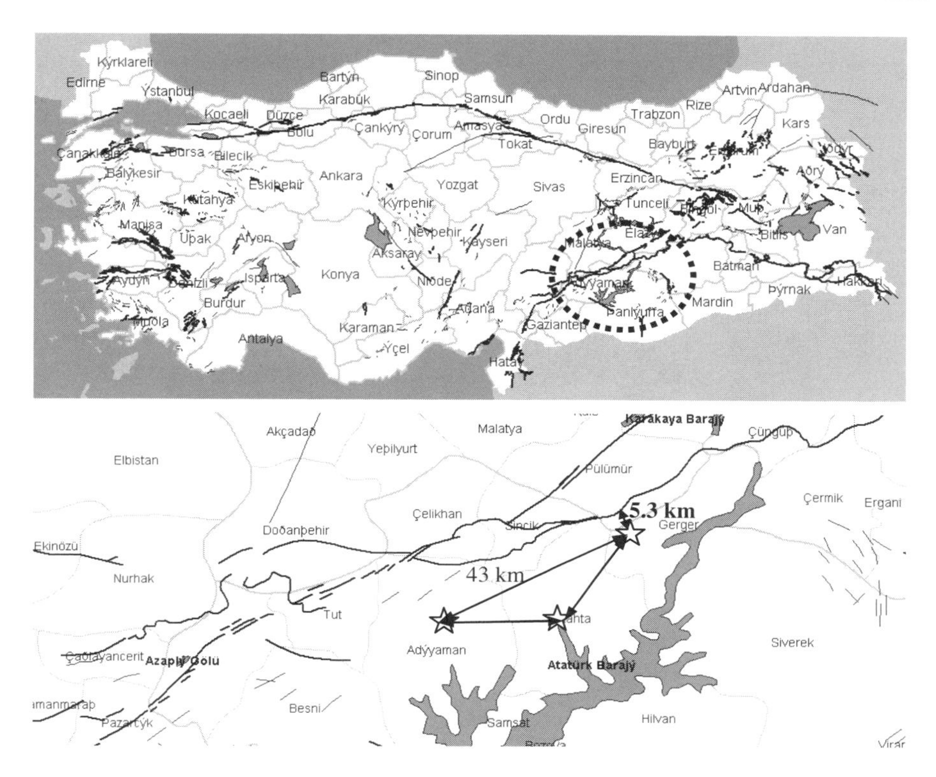

Figure 13. Major active faults of Turkey (South-East)

Figure 14. General view of the East terrace statues

Figure 15. General view of the West terrace statues

Figure 16. Stone over eroded base rock in bending.

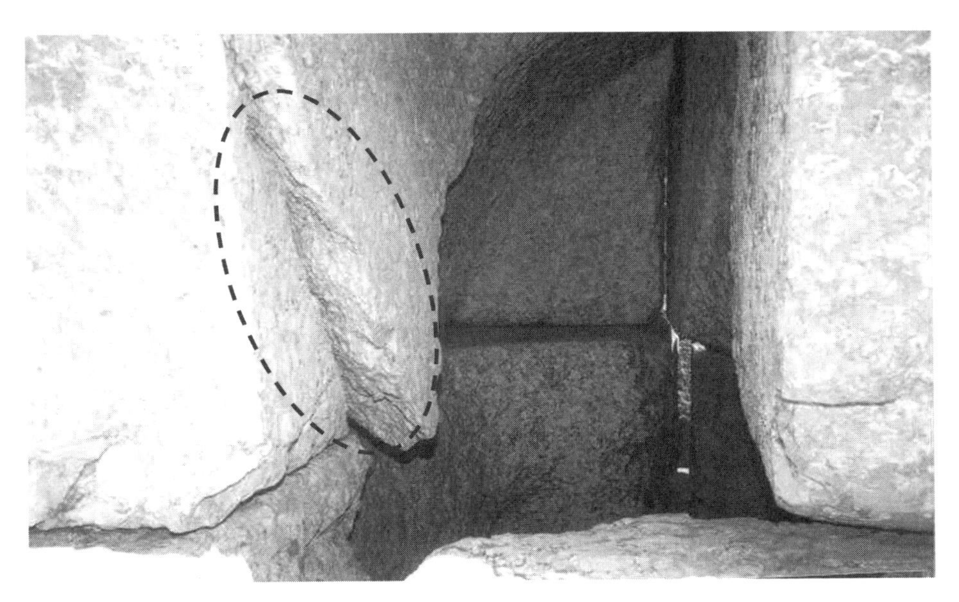

Figure 17. Shear crack and dislocated stone blocks

Figure 18. General view of stelae figures.

Figure 19. Poor strengthening attempt to connect broken pieces of stelae.

SEISMIC ASSESSMENT OF EXISTING RC BUILDINGS

Michael N. Fardis, Telemachos B. Panagiotakos, Dionysis Biskinis and Antonis Kosmopoulos
Structures Laboratory, Department of Civil Engineering, University of Patras, Greece

Abstract: Two procedures are presented for seismic assessment of individual RC buildings on the basis of member seismic chord rotation demands. In the (simpler) "preliminary" evaluation procedure, inelastic chord rotation demands are estimated from linear analysis, while forces for brittle failure modes are computed in a capacity-design fashion. In the "final" or "detailed" evaluation, inelastic chord rotation demands and member shears are determined via nonlinear static (pushover) analysis. This procedure also requires more information on as-built materials and reinforcement. Verification criteria at three performance levels are proposed, tuned to be on the conservative side for "preliminary" evaluation.

Keywords: concrete buildings, deformation capacity, seismic assessment, seismic evaluation, ultimate deformation

1. INTRODUCTION

In the light of current knowledge and modern codes, the majority of the building stock and of other types of structures in most of the world are substandard. This applies especially to seismic regions, as even there seismic design of structures is relatively recent. In those regions the major part of the seismic threat to human life and property comes from old buildings.

S.T. Wasti and G. Ozcebe (eds.), Seismic Assessment and Rehabilitation of Existing Buildings, 215–244.

Due to the increasing decay of the infrastructure, frequently combined with the need for structural upgrading to meet more demanding design codes (especially against seismic loads), structural retrofitting is becoming more and more important and receives today considerable emphasis throughout the world. In recognition of the importance of the seismic threat arising from existing substandard buildings, the first standards for structural upgrading to be promoted by the international engineering community and by regulatory authorities alike are for seismic rehabilitation of buildings. This is the case, for example, of Part 3: "Strengthening and Repair of Buildings" of Eurocode 8 (i.e. of the draft European Norm for earthquake-resistant design, under finalisation in late 2003, derived from the 1996 pre-Standard), which is the only one among the set of 58 draft-Eurocodes that addresses the problem of structural upgrading. It is also the case of the recent (2001) ASCE draft standard on "Seismic evaluation of existing buildings" and of the 1996 "Law for promotion of seismic strengthening of existing reinforced concrete structures" in Japan.

A detailed seismic assessment (or evaluation) of an individual building, not only determines the potential need for seismic retrofitting, but also identifies the particular weaknesses and deficiencies to be corrected through retrofitting. For this reason recent years have seen a worldwide shift from rapid screening and empirical evaluation methods, to fundamental assessment procedures based on a direct or indirect comparison of the inelastic deformation demands to the corresponding deformation capacities.

New approaches have been proposed recently for the seismic assessment of existing RC buildings, which are partly or fully displacement-based. The rationale behind displacement-based approaches is well known: The earthquake does not represent for the structure a set of given lateral forces to be resisted, as considered in forced-based seismic design or assessment, but a demand for accommodation of a given energy input or of given imposed dynamic (ground) displacements. Therefore displacements, rather than forces, represent a much more rational basis for the seismic design or assessment of structures. After all, structures collapse not due to the earthquake lateral loads per se, but due to gravity loads, acting through the lateral displacements caused by the earthquake (P-Δ effects).

This paper presents two procedures, of different sophistication and complexity, for the seismic assessment of individual RC buildings. Both are based on seismic displacements and use chord rotations at member ends as the primary deformation measure for the verification of ductile failure modes. In the simpler procedure, which is appropriate for preliminary evaluation of buildings, inelastic chord rotation demands may be estimated from linear elastic analysis, while forces for the verification of brittle failure modes are computed on the basis of equilibrium in a capacity-design fashion. In the more advanced procedure, appropriate for final or detailed evaluation, inelastic chord rotation demands and member shear forces are determined via nonlinear static (pushover) analysis. The detailed evaluation procedure is also more demanding with respect to the amount and type of information on material properties and quantity of reinforcement needed for its application, as well as in other computational and verification aspects.

Verification criteria at three different performance levels are proposed for each one of the two evaluation procedures. These criteria are tuned to be on the conservative (safe) side for the “preliminary evaluation”. Therefore, if the assumptions made regarding material properties and quantity of reinforcement are also on the conservative side, then for a building which is evaluated as safe or adequate on the basis of the “preliminary’ evaluation, there is no need to proceed to application of the “final” or “detailed” evaluation procedure.

It is emphasized that the “preliminary’ evaluation is not a rapid screening procedure which has at least one order of magnitude less requirements in engineering effort and time than the “final” or “detailed” evaluation: it is less demanding than this latter procedure only regarding the type and amount of information about the as-built structure and the computational tools and expertise necessary for its application. This being so, the “final” or “detailed” evaluation requires about double the amount of engineering effort and time of the “preliminary’ evaluation.

The general approach of the proposed procedures is similar to that of the ASCE (2000) prestandards and of the relevant Eurocode 8 part (CEN, 2003) currently under development. The procedures described in this paper are more specific and focused to RC buildings and are complete with all the detailed tools necessary for their application.

2. PRELIMINARY EVALUATION PROCEDURE

2.1 Introduction

The “preliminary” evaluation procedure of frame or dual RC buildings uses linear-elastic analysis of the building (equivalent static under lateral forces with triangular heightwise distribution, or modal response spectrum analysis) in the two orthogonal horizontal directions, on the basis of code-specified 5%-damped elastic response spectra and secant-to-yield member stiffnesses. The analysis model includes all concrete members as prismatic elements, but masonry infill panels are not explicitly and individually modeled. RC members are evaluated by comparing chord rotation demands from the analysis to estimates of the corresponding (ultimate) deformation capacities, calculated on the basis of default estimates of material properties and amount of reinforcement. The possibility of member shear failure is assessed on the basis of capacity-design upper limits of shear force demands.

Detailed knowledge of material properties and of the amount and arrangement of reinforcement is not required in the "preliminary" evaluation procedure. Material properties may be assumed equal to the nominal values specified in the design documents. If such documents are not available, low default values may be assumed, representative of the prevailing practice at the time of construction. The amount and arrangement of reinforcement may also be taken from design documents, without further confirmation in situ. If such documents are not available, the original design may be emulated on the basis of the applicable codes and the prevailing practice at the time of construction. Conservative assumptions should be employed in such an emulation; this may entail design for only gravity loads, without engineered earthquake resistance.

2.2 Estimation of seismic chord rotation demands at member ends

Estimation of chord rotation demands under the design seismic action is based on the fact that RC structures without masonry infills typically have an effective predominant period of response to strong ground motions in the velocity-controlled part of the response spectrum, where the equal displacement rule between elastic and inelastic Single-Degree-of-Freedom systems applies well. Then member chord rotations may be estimated from a 5%-damped elastic analysis, equivalent static with inverted triangular distribution of lateral forces or of the modal response spectrum type. The criteria for use of an equivalent static instead of modal response spectrum analysis are the ones usually applied in seismic design of new structures: Heightwise regularity of storey mass, stiffness and strength and fundamental period less than twice the corner period between the acceleration- and velocity-controlled regions of the spectrum, or 2 sec., whichever is more critical. Results of such elastic analyses on average well represent the peak inelastic demands, $\theta_{E,m}$.

For this approach to yield good estimates of inelastic chord rotation demands, the structure should be considered in the analysis with member elastic rigidity equal to the secant rigidity of the RC member at yielding of both ends in antisymmetric bending, i.e. as:

$$EI_{ef}=M_yL/6\theta_y \quad (1)$$

In general four values of EI_{ef} are computed, considering positive or negative bending at each end, and averaged into a single EI_{ef} value. The yield moment M_y can be computed from first principles. The chord-rotation θ_y at yielding is computed through Eqs. (2), accounting for flexural and shear deformations (1st and 2nd term) and for anchorage slip of bars (3rd term).

For columns or beams (from 1336 tests, 198 of which without slip of the reinforcement from the anchorage zone):

$$\theta_y = \phi_y \frac{L_s}{3} + 0.00275 + a_{sl} \frac{0.2\varepsilon_y d_b f_y}{(d - d')\sqrt{f_c}} \tag{2a}$$

For walls of rectangular section (65 tests, all with slip of reinforcement):

$$\theta_y = \phi_y \frac{L_s}{3} + 0.002 + a_{sl} \frac{0.2\varepsilon_y d_b f_y}{(d - d')\sqrt{f_c}} \tag{2b}$$

For walls of non-rectangular section (barbelled, T-shaped, etc., from 40 tests, all with slip of the reinforcement):

$$\theta_y = \phi_y \frac{L_s}{3} + 0.00225 + a_{sl} \frac{0.2\varepsilon_y d_b f_y}{(d - d')\sqrt{f_c}} \tag{2c}$$

In Eqs. (2) ϕ_y is the yield curvature, computed from first principles, or empirically as $\phi_y = 1.85 f_y / E_s d$, L_s is the shear span $\approx L/2$, f_y and f_c (in MPa) the strengths of steel and concrete, d_b the diameter of tension reinforcement, d, d' the depth to the tension and compression reinforcement respectively and a_{sl} a zero-one variable, equal to 0 if slip of the longitudinal bars from the anchorage zone is not possible, or to 1 if it is.

Eqs. (2) fit the data with a median of the ratio of the experimental value to the one predicted essentially equal to 1.0 and a coefficient of variation of 38.5% for Eq. (2a) (29.7% for the 198 tests without slip), of 35.8% for Eq. (2b) and of 30.5% for Eq. (2c). The fit to the data is shown in Figure 1.

A simpler alternative, which is actually more appropriate for the preliminary evaluation procedure, is to take the effective member stiffness, EI_{ef}, equal to 25% of the uncracked gross-section stiffness, $E_c I_g$ (Eq. (3) below). Nonetheless, this alternative is twice less accurate than Eqs. (1) and (2) as it fits experimental results with a coefficient of variation of 71%, instead of about 40% for Eqs. (1), (2).

$$EI_{ef} = 0.25 E_c I_g \tag{3}$$

The version of the equal displacement rule adopted here was developed on the basis of about 1500 nonlinear dynamic analyses of several fairly regular bare RC frame or dual structures, from 3 to 12 storeys, designed to EC2 and EC8 (ENV versions of mid-90's) and to different combinations of peak ground acceleration and Ductility Class, as well as to different versions of capacity design. (Panagiotakos and Fardis, 1999). Peak inelastic chord rotations computed at member ends from these analyses were divided by the corresponding elastic values obtained from an equivalent static or modal response spectrum analysis using the 5%-damped elastic spectrum. The ratio of these values seems relatively insensitive to the details of the structural configuration (at least for fairly regular geometries) and to the intensity of the ground motion (from the design earthquake to about twice that level). The inelastic-to-elastic chord rotation ratio depends systematically on the type of element (horizontal or vertical) and its elevation in the structure.

Table 1 presents mean values of this ratio for chord rotations and displacements, top or equivalent (mean). Figure 2 shows the mean value of the ratio of the inelastic chord rotation to the elastic separately for the three types of buildings, a common linear fit corresponding to the values at roof and base level in Table 1, and standard deviations of the individual data points with respect to these linear mean fits. The scatter about the mean reflects the effect of the time-history of the seismic motion – for given 5%-damped response spectrum – as well as model uncertainty. The standard deviation is lower in the more critical lower stories (at the base: 0.15 for the columns, 0.1 for the beams) and increases with height due to the effect of higher modes (at the roof: 0.25 for the columns, 0.3 for the beams).

For the purposes of the preliminary evaluation procedure, a single representative value of the ratio of inelastic chord rotations to the elastic ones is appropriate. Such a single value may be taken equal to 1.0, meaning that the equal displacement rule may be applied to chord rotations. 95%-fractiles of inelastic chord-rotation demands are equal to 1.25-times the elastic demands in columns, or 1.45-times these demands in beams, whilst "mean + one-standard deviation values" are 1.15-times the elastic demands in columns and 1.35-times such demands in beams.

The conclusions above were derived for new RC structures, which typically satisfy the strong-column/weak beam rule of capacity design. Nevertheless, the nonlinear analyses on which these results were based exhibit plastic hinging and inelastic deformations both in columns and beams. For this reason the conclusion above (namely that an elastic analysis using the 5%-damped spectrum and the secant-to-yield stiffness of all members in antisymmetric bending, can be used to estimate member peak inelastic chord demands) is expected to hold in approximation also for existing structures without significant engineered earthquake resistance, under ground motions inducing global ductility demands significantly higher than 1.0. Detailed numerical results may differ from those above, if there is very strong tendency for concentration of inelasticity in a single story. For such cases the more cumbersome "Detailed (final) Evaluation procedure" of 3 below, based on nonlinear static analysis may be used.

Old RC buildings are often infilled in all storeys except the bottom one. Ground storey columns are normally not designed for the concentration of inelastic seismic deformations expected there. Several thousands nonlinear dynamic analyses of partially infilled multi-storey buildings by Panagiotakos and Fardis (1999) led to the conclusion that peak inelastic chord rotations at the ends of ground storey beams and columns may be estimated from the 5%-damped spectrum and an equivalent static analysis of the elastic structure, with the infills in all infilled storeys modelled as rigid diagonal struts, by applying the factors of Table 2.

2.3 Member ultimate chord rotation in flexure

Member chord rotation demands (mean values, 95%-fractiles, or mean-plus-one-standard-deviation values) should be compared to the corresponding capacities, i.e. to the ultimate chord rotation, θ_u of the member under cyclic loading. Eq. (4) below was developed for the mean ultimate chord rotation of RC members cyclically loaded until flexure-controlled failure:

$$\theta_u = \alpha_{st}\left(1-0.4a_{cyc}\right)\left(1+0.45a_{sl}\right)\left(1-0.35a_{wall}\right)\left(0.325^{\nu}\right)\cdot$$

$$\cdot\left[\frac{\max(0.01,\omega')}{\max(0.01,\omega)}f_c\right]^{0.2}\left(\frac{L_s}{h}\right)^{0.425}30^{\left(\alpha\rho_{sx}\frac{f_{yw}}{f_c}\right)}\left(1.4^{100\rho_d}\right) \tag{4}$$

where:

- ast: coefficient for the steel of longitudinal bars, equal to 0.018 for ductile hot-rolled or heat-treated (tempcore) steel, or to 0.01 for brittle cold-worked steel;
- acyc: zero-one variable, equal to 0 for monotonic loading or to 1 for cyclic, with at least one full reversal of loading at the peak amplitude
- asl: zero-one variable, equal to 0 if slip of the longitudinal bars from the anchorage zone is not possible, or to 1 if it is (cf. Eq. (2)).
- awall: coefficient, equal to 1.0 for shear walls or to 0 for beams or columns.

- ν = N/bhcfc: axial load ratio normalized to the width b of the compression zone, to section depth h and to fc;
- ω, ω': mechanical reinforcement ratios, ρfy/fc, of the tension and compression longitudinal reinforcement (not including any diagonal bars); in shear walls the entire vertical web reinforcement is included in the tension steel;
- Ls/h = M/Vh: shear span ratio at the member end;
- fc: uniaxial concrete strength (MPa);
- ρsx = (Asx/bwsh): ratio of transverse steel parallel to the direction (x) of loading (sh=stirrup spacing);
- α: confinement effectiveness factor, equal to:

$$\alpha = \left(1 - \frac{s_h}{2b_c}\right)\left(1 - \frac{s_h}{2h_c}\right)\left(1 - \frac{\sum b_i^2}{6b_c h_c}\right) \qquad (5)$$

(b_c, h_c = dimensions of confined concrete core, b_i= distances of restrained longitudinal bars on the perimeter);

- ρd: steel ratio of diagonal reinforcement (if any) in each diagonal direction.

Eq. (4) was fitted by regression to the results of 1158 monotonic or cyclic tests to failure of flexure-controlled beam-, column- or wall-specimens. Wall specimens (of rectangular, barbelled or T section) were 59 and the rest were beam or column specimens. The ratio of experimental values to the predictions of Eq. (4) has a median of 1.0 and a coefficient of variation of 45.2% (30.3% for the wall specimens alone, or 44.7% for all the rest). The fit to the data is shown in Figure 3, separately for all the data or for the cyclic test data alone.

The ultimate chord rotation given by Eq. (4) is considered as an expected value and denoted by θ_{um}. Due to the large scatter, in the verification of chord rotations the lower characteristic (5%-fractile) value, $\theta_{uk,0.05}$, or the "mean-minus-one-standard deviation" value, $\theta_{u,m-\sigma}$, of the deformation capacity may be used instead of θ_{um}. The 5%-fractile and the "mean minus-one-standard deviation" values are equal to:

$$\theta_{uk,0.05} = 0.425\theta_{um} \qquad (6a)$$

$$\theta_{u,m-\sigma} = 0.55\theta_{um} \qquad (6b)$$

Eq. (6a) is depicted in Figure 3 with a dashed line.

Eq. (4) was developed mainly for RC elements with seismic detailing. The database used for the development of Eqs. (4) and (6) covers a very wide range of the parameters considered of prime importance in member proportioning and detailing for earthquake resistance, such as stirrup spacing and confining reinforcement ratio, axial load ratio, compression-to-tension reinforcement ratio, diagonal reinforcement ratio, shear span ratio. To check whether earthquake-resistant detailing or lack thereof is reflected by Eq. (4), a total of 40 specimens cyclically tested to flexure-controlled failure were identified as not intended for use in earthquake resistant construction (from recent studies of the cyclic deformation capacity of old-type RC members under cyclic loading, or of the cyclic behaviour of such members before seismic retrofitting). These specimens were not included in the database from which Eq. (4) was developed. For specimens without intentional seismic detailing subjected to cyclic loading, the confinement effectiveness factor α of Eq.(5) is set equal to zero in Eq.(4), to reflect the lack of stirrup closing with 135° hooks and the sensitivity of cyclic behaviour to improperly closed stirrups.

The comparison of the predictions of Eq. (4) with the experimental data for the members without intentional seismic detailing shows that, although the fit of Eq. (4) to monotonically tested members without seismic detailing is on average almost as good as that of the overall database, Eq. (4) overpredicts on average by 13.5% the deformation capacity in the 40 cyclic tests of non-seismically detailed members (see Figure 4). So it seems that there are certain features of members not intentionally designed and detailed for earthquake resistance, which may not be fully reflected in the mechanical and geometric parameters of the specimens and their reinforcement considered as control variables of the regression analyses, yet are quite important for their ultimate deformation under cyclic loading. The available data may not be enough for quantification of these effects. At the present time this analysis shows that Eq. (4) may be applied for the prediction of the expected cyclic deformation capacity of flexure-critical members without seismic detailing, with a multiplicative correction factor k_u of 0.865. If the same correction factor is applied to Eq. (6a), the 5% fractile of the deformation capacity of old-type components is indeed obtained.

2.4 Shear capacity and demand in RC members

The database from which Eqs.(4), (6) were derived extends to elements with shear ratios, M/Vh, as low as 1.5 for beams or columns or 1.0 for walls. Nonetheless the database does not contain shear-critical members, in which at a certain point during the test the nominal flexural capacity M_u exceeds L_s times the nominal shear capacity, V_R, as this decreases with the cyclic flexural ductility ratio, $\mu_\theta=\theta/\theta_y$. The possibility of pre-emptive shear failure should be evaluated separately from the check of flexure-controlled failure at member ends, as development of the chord rotation capacity θ_u there presupposes that the member does not fail earlier in shear.

Assessment of RC members in shear is based on the check that maximum shear force during the response, $V_{E,max}$, does not exceed shear capacity, V_R:

$$V_{E,max} \leq V_R \tag{7}$$

With units: MN, m, the expression fitted to V_R as a function of plastic chord rotation ductility demand, $\mu_\theta^{pl}=\theta^{pl}/\theta_y$, ($\theta^{pl}=\theta-\theta_y$) at the member end where shear is checked, is (Biskinis et al., 2003):

$$V_R = \frac{h-x}{2L_s}\min\left(N, 0.55A_c f_c\right) + 0.16\left(1-0.055\min\left(5, \mu_\theta^{pl}\right)\right)\cdot \left[\max(0.5, 100\rho_{tot})\left(1-0.16\min\left(5, \frac{L_s}{h}\right)\right)\sqrt{f_c}A_c + V_w\right] \tag{8}$$

In Eq.(8) N is the (compressive) axial load, x the compression zone depth, A_c is the cross-sectional area, ρ_{tot} the total ratio of longitudinal steel and h is the depth of the cross-section (equal to the diameter D in circular sections).

In rectangular sections the contribution of transverse reinforcement, V_w, is taken equal to:

$$V_w = \rho_w b_w z f_{yw} \tag{9}$$

with b_w denoting the width of the web, ρw the ratio of transverse steel, $z \approx d-d' \approx 0.9d$ the internal lever arm (d: effective depth) and f_{yw} the yield stress of transverse reinforcement.

For circular sections:

$$V_w = \frac{\pi}{2}\frac{A_{sw}}{s} f_{yw}(D-c) \tag{10}$$

with A_{sw} denoting the cross-sectional area of a circular hoop, s its spacing and c its concrete cover.

Eqs. (8)-(10) were fitted to 41 tests on columns with circular section and to 133 tests on columns, beams or walls with square or rectangular section (total of 174 specimens). 24 specimens are representative of members without seismic detailing ("old" or "nonconforming" type of member). Eqs. (8)-(10) fit the data with a mean and median of the ratio of experimental to calculated values equal to 1.0. The median is also equal to 1.0 for the subsets of specimens with circular or rectangular section, as well as for those with new ("conforming" to modern codes) or old ("non-conforming") type of detailing. The coefficient of variation of the ratio of experimental to predicted values is 13.6%.

A linear elastic analysis in the nonlinear regime may overestimate $V_{E,max}$ and N. In this case $V_{E,max}$ at end i may be estimated from equilibrium of the member, considering that plastic hinges develop at its ends 1 and 2, or at the ends of the other members framing into the same joint, if the flexural resistance of the latter is more critical in flexure:

At both ends 1 and 2 of columns, bending moments, $M_{1,d}$, $M_{2,d}$, are considered that correspond to plastic hinging in the beams or the columns – whichever takes place first (Figure 5):

$$M_{1,d} = \gamma_{Rd} M_{Rc,1} \min(1, \frac{\sum M_{Rb}}{\sum M_{Rc}})_1 \quad (11a)$$

$$M_{2,d} = \gamma_{Rd} M_{Rc,2} \min(1, \frac{\sum M_{Rb}}{\sum M_{Rc}})_2 \quad (11b)$$

where:

- MRc,1, MRc,2 : column flexural resistance at end 1 or 2
- ΣMRc, ΣMRb: sum of column or beam moments around joint at end 1 or 2

γ_{Rd} = safety coefficient, with value that depends on the performance level at which shear failure is checked (see 2.5 below).

Then the shear force demand is calculated as (i=1, or 2):

$$V_{E,CD,i} = \min\left[\frac{M_{1,d} + M_{2,d}}{l_{cl}}, V_{E,i}\right] \quad (12)$$

where $V_{E,i}$ is the value of the shear force from the linear elastic analysis and l_{cl} the column clear height.

In beams, bending moments, $M_{1,d}$, $M_{2,d}$, at both ends 1 and 2 are considered that correspond to plastic hinging in the beams or the columns – whichever takes place first (Figure 6):

$$M_{1,d} = \gamma_{Rd} M_{Rb,1} \min(1, \frac{\sum M_{Rc}}{\sum M_{Rb}})_1 \tag{13a}$$

$$M_{2,d} = \gamma_{Rd} M_{Rb,2} \min(1, \frac{\sum M_{Rc}}{\sum M_{Rb}})_2 \tag{13b}$$

where:

- MRc,1, MRc,2 = column flexural resistance at end 1 or 2;
- ΣMRc, ΣMRb, γRd = as above.

Then the shear force demand is calculated as (i=1, or 2):

$$V_{E,CD,i} = \min\left[V_{g+\psi_2 q,oi} + \frac{M_{1,d} + M_{2,d}}{l_{cl}}, V_{E,i} \right] \tag{14}$$

In Eq. (14) $V_{g+\psi 2q,oi}$ is the shear force at end i due to the simultaneously acting transverse loads, $g+\psi_2 q$, considering the member as simply-supported; l_{cl} is the beam clear span. When i =1 the moment $M_{1,d}$ should be determined considering tension at the top and moment $M_{2,d}$ at end 2 tension at the bottom.

In walls, the shear force not only at the end i, but also anywhere along the height, may be calculated as:

$$V_{E,CD} = \frac{\gamma_{Rd} M_{Rw}}{M_{Ew}} V_E \tag{15}$$

where:

- VE = wall shear from the elastic analysis;
- MEw = wall bending moment at the base from the elastic analysis;

M_{Rw} = wall flexural resistance at the base.

In walls and columns the values of the shear resistance V_R in Eq. (8) and those of M_{Rw}, M_{Rc}, in Eqs. (15) (13) and (11) should be based on consistent values of the axial force N. As both the demand, $V_{E,max}$, and the supply, V_R, increase with N, it makes sense to neglect the fluctuation of N during the seismic response and to use at both places its value $N_{g+\psi 2q}$ due to gravity loads $g+\psi_2 q$ alone.

The flexural capacities of beams, columns or walls, M_R

2.5 Member evaluation criteria

Different member evaluation criteria are proposed, depending on the performance level at which assessment takes place: Following the trend in recent US documents for performance-based seismic evaluation of buildings (FEMA 273/274, FEMA 356), three such levels are considered:

"Operational" or "Limited Damage"

"Life Safety" or "Significant Damage"

"Near Collapse"

Of course these three performance levels are checked under increasing intensity of seismic action (seismic hazard). Normally the building will be assessed for a single performance level under the corresponding seismic hazard level, depending on its importance classification.

In all the verifications, expected (mean) values of material properties are used, except in the calculation of the design value of the shear force capacity, V_{Rd}, (calculated on the basis of design values of material properties, f_{cd} and f_{yd}, incorporating material factors γ_m), wherever explicitly noted.

2.5.1 "Operational" or "Limited Damage" performance level

All RC members are required to remain elastic.

Expected values of chord rotation demands at member ends from the elastic analysis, θ_E, are checked against the expected value of chord rotation at yielding from Eq.(2):

$$\theta_{E,m} \leq \theta_y \quad (16)$$

Shear is checked through Eq. (7). To avoid pre-emptive shear failure, the value of shear resistance is calculated from Eqs. (8)-(10) on the basis of design values of material properties, f_{cd} and f_{yd}, equal to the nominal or default values, f_c and f_y, divided by material (partial) factors γ_m. The so-computed design value of shear resistance is denoted as V_{Rd}.

As this performance level may be checked for a lower intensity earthquake (hazard level), implying low value of the elastic shear force demand from the analysis, $V_{E,i}$, in Eqs. (12), (14) term $V_{E,i}$ is ignored.

2.5.2 "Life Safety" or "Significant Damage" performance level

Normally this verification takes place for a hazard level that corresponds to the design seismic action of new structures.

The proposed verification of members against ductile, flexure-controlled failure is:

For ductile primary members:

$$\theta_{Ek,0.95} \leq \theta_{uk,0.05} \qquad (17)$$

where $\theta_{Ek,0.95}$ is the upper characteristic value of the chord rotation demand, determined according to 2.2, and $\theta_{uk,0.05}$ the lower characteristic value of the chord rotation capacity from Eq. (6a), with the reduction factor k_u= 0.865 applied for "old-type" or "non-conforming" members (see Figure 4).

For ductile secondary members:

$$\theta_{Ek,0.95} \leq \theta_{u,m-\sigma} \qquad (18)$$

where $\theta_{u,m-\sigma}$ is the mean-minus-one-standard-deviation value of chord rotation capacity from Eq. (6b), with application of the reduction factor k_u= 0.865 for "old-type" or "non-conforming" members.

Shear is again checked through Eq. (7), with $V_{E,max}$ equal to $V_{E,CD}$ from Eqs. (10), (12), (15) and proposed value of γ_{Rd} =1.2 in Eqs. (11), (13) and (15). The design value of shear resistance, V_{Rd}, calculated on the basis of design values of material properties, f_{cd} and f_{yd}, incorporating material (partial) factors γ_m, is used in this verification. Moreover, brittle primary members are checked with the shear resistance divided by a capacity reduction factor of 1.25, to reflect the difference in structural importance with secondary members.

2.5.3 "Near Collapse" performance level

Normally this verification will take place for a very severe hazard level, corresponding to about 1.5 times the design seismic action of new structures.

The proposed verification of members against ductile, flexure-controlled failure is:

For ductile primary members:

$$\theta_{Ek,0.95} \leq \theta_{u,m-\sigma} \qquad (19)$$

where $\theta_{u,m-\sigma}$ is the mean-minus-one-standard-deviation value of chord rotation capacity from Eq. (6b), with the reduction factor k_u= 0.85 for "old" or "non-conforming" members.

For ductile secondary members:

$$\theta_{Ek,0.95} \leq \theta_{u,m} \tag{20}$$

where $\theta_{u,m}$ is the mean value of chord rotation capacity from Eq. (4), with the reduction factor k_u= 0.85 applied to "old" or "non-conforming" members.

Shear is checked again through Eq. (7), with $V_{E,max}$ equal to $V_{E,CD}$ from Eqs. (10), (12), (15) and proposed value of γ_{Rd} =1.2 for both primary and secondary members. The design value of shear resistance, V_{Rd}, calculated on the basis of design values of material properties, f_{cd} and f_{yd}, incorporating material (partial) factors γ_m, is used in the verification of primary members, whilst the expected value of shear resistance, V_R, calculated on the basis of mean values of the nominal or assumed default values of material properties, f_c and f_y, without material (partial) factors γ_m, is used for secondary members.

3. FINAL (OR DETAILED) EVALUATION PROCEDURE

3.1 Introduction

The procedure proposed for final and detailed evaluation of low- to medium rise frame or dual RC buildings is based on a nonlinear-static (pushover) analysis of the building in the two orthogonal horizontal directions. The nonlinear analysis in each direction is carried out up to a target displacement obtained from the code-specified 5%-damped spectra and the fundamental period estimated through the Rayleigh quotient, using secant-to-yield member stiffnesses. The structural model includes all concrete members as prismatic elements with the possibility of developing point-hinges at the two ends, as well as masonry infills as equivalent elasto-plastic diagonal struts. RC members are evaluated by comparing chord rotation demands from the analysis to estimates of the corresponding (ultimate) deformation capacities, calculated on the basis of material properties and amount of reinforcement estimated from the construction documents and in-situ measurements. The possibility of member shear failure is assessed on the basis of shear force demands from the analysis, taking into account the effect of cyclic flexural deformations on shear strength. The possibility of joint shear failure is also checked.

In the following, only the differences with the preliminary evaluation procedure described in 2 are highlighted.

One of the differences with the preliminary evaluation procedure lies in the required knowledge of material properties and of the amount and arrangement of reinforcement. Material properties should be inferred from the design documents, along with sporadic sampling from the as-built structure and testing, to an extent that depends on the completeness and reliability of such documents. Mean values of material properties may be taken equal to the nominal values, unless the data collected from the sampling and testing are sufficient to support a statistically robust estimation of a mean value. The amount and arrangement of reinforcement should be taken from design documents, checked with sporadic exposure from the as-built structure, the extent of which should also depend on the completeness and reliability of such documents.

3.2 Estimation of seismic chord rotation demands at member ends

Chord rotation demands are estimated from a nonlinear static (pushover) analysis, carried up to a work-equivalent horizontal displacement of the building, defined as $\delta_{eq} = \sum m_i \delta_i^2 / \sum m_i \delta_i$ (with δ_i denoting horizontal displacement of mass m_i), given by the 5%-damped elastic spectrum at the fundamental period T_1 of the structure estimated through the Rayleigh quotient. The target displacement (work-equivalent mean displacement) is obtained by applying on the elastic 5%-damped spectral displacement corresponding to the fundamental period T_1 the modification factors in Table 5. These factors were derived from the approximately 1500 nonlinear dynamic analyses by Panagiotakos and Fardis [1999] referred to in 2.2. Upper characteristic (95%-fractile) values listed also in Table 4 reflect variability due to the details of the ground motion for given elastic response spectrum, as well as model uncertainty.

The structure should be considered in the analysis with member elastic rigidity equal to the member secant rigidity at yielding of both ends in antisymmetric bending, i.e. according to Eq. (1), with chord rotation at yielding from Eq. (2).

Nonlinear static (pushover) analysis can also be used for the estimation of chord rotation demands in fully or partially infilled RC buildings, including the members of an open ground story. Then, the nonlinear static analysis should also include the infills as elasto-plastic diagonal compression struts. The target top displacement can still be estimated from the 5%-damped elastic spectrum, with the multiplicative factors in Table 4.

The nonlinear static (pushover) analysis is much more accurate than linear elastic analysis, if there is strong tendency for concentration of inelasticity in a single storey. It holds also certain advantages as far as estimation of internal forces in the structure is concerned. Internal forces are important for the brittle (or force-controlled) failure modes and elements, as e.g. in shear-critical members and beam-column joints. Moreover, the chord rotation capacity of columns or walls, as well as the values of their yield moment, flexural resistance and effective stiffness, depend on their axial force N.

3.3 Assessment of RC members in shear

When the analysis is nonlinear static (pushover), the acting shear force $V_{E,max}$ in Eq. (8) is the "actual" value from the analysis, taking into account the simultaneously acting transverse loads, $g+\psi_2 q$ (for beams). Then shear can be checked at any step of the analysis, using the current values of V_E on one hand, and of $\mu_\theta^{pl} = (\theta-\theta_y)/\theta_y$ and N in Eq. (8) on the other.

In columns or walls Eq. (8) should be checked for the two extreme values of N during the response.

3.4 Assessment of beam-column joints in shear

Slippage of beam or column bars within joints contributes with a fixed-end rotation to the chord rotation at member ends. In the present procedure this fixed-end rotation is taken into account directly: a) in the last term in the calculation of θ_y through Eqs. (2) (reducing the member effective rigidity, $EI_{ef}=M_yL/6\theta_y$) and b) in an apparent increase of member chord rotation capacity in Eq.(4).

In addition to contributing to these effects, beam-column joints develop very high shear forces (and stresses) in their core, running the risk of pre-emptive shear failure. This is more so in existing RC buildings, which typically have no shear reinforcement in the core of beam-column joints.

The maximum shear force that can develop in the joint is determined from the capacity of the beams or columns framing into it (whichever is weakest) to deliver shear by bond along the extreme beam or column bars passing through. If beams are weaker in flexure than columns, i.e. if: $\Sigma M_{yb} < \Sigma M_{yc}$ (ΣM_{yb}= sum of yield moments of beams framing into the joint; ΣM_{yc}= corresponding sum for columns), then the beams govern the shear input in the joint and the horizontal shear force V_{jh} in the joint is:

$$V_{jh} = (A_{sb1} + A_{sb2})f_y - V_c = (A_{sb1} + A_{sb2})f_y - \frac{\sum M_{yb}}{h_{st}} \frac{L_b}{L_{bn}} =$$
$$= \sum M_{yb} \left(\frac{1}{z_b} - \frac{1}{h_{st}} \frac{L_b}{L_{bn}} \right) \qquad (21)$$

where A_{sb1}, A_{sb2} denote the cross-sectional area of the beam top and bottom reinforcement, V_c the column shear at beam plastic hinging, h_{st} the storey height, L_b and L_{bn} the theoretical and clear span of the beams and z_b=d-d_1≈0.9d the beam internal lever arm. Then the shear stress demand in the joint is:

$$v_j = \gamma_{Rd} \frac{V_{jh}}{b_j h_c} \qquad (22)$$

In Eq. (20) γ_{Rd} is a safety factor that depends on the performance level at which joint shear failure is checked (see 3.5 below), h_c is the column cross-sectional depth in the horizontal direction in which the joint is checked and b_j the width of the joint in the transverse horizontal direction, taken as:

$$b_j = \min(\max(b_c, b_w), 0.5h_c + \min(b_c, b_w)) \qquad (23)$$

with b_c and b_w denoting the width of the column and the beam in the direction normal to h_c.

If $\Sigma M_{yb} > \Sigma M_{yc}$, then the columns control the shear input in the joint. Usually column vertical bars are the same above and below the joint. If the total cross-sectional area of all vertical bars at the two extremes of the column section and of the joint core is denoted by $A_{sc,tot}$, the vertical shear force in the joint core is:

$$V_{jv} = f_y A_{sc,tot} + N_{top} - V_{b,min} \qquad (24)$$

where N_{top}: axial force in the column above and $V_{b,min}$: minimum beam shear force on either side of the joint, approximately equal to:

$$V_{b,min} \approx \min\left(\frac{\sum M_{yc}}{L_b} \frac{h_{st}}{h_{st,n}} - V_{g+\psi_2 q,b} \right) \qquad (25)$$

$V_{b,min}$ may take also negative values. In Eq. (25) h_{st} and $h_{st,n}$ are the theoretical and the clear storey height – average value – and $V_{g+\psi 2q,b}$ the shear force at the beam end due to gravity loads alone.

As:

$$\sum M_{yc} \approx f_y A_{s,tot} z_c + 0.5h_c \left(N_{top}\left(1-\nu_{top}\right) + N_{bot}\left(1-\nu_{bot}\right)\right) \qquad (26)$$

where ν=N/$A_c f_c$ and z_c (internal lever arm of the column) ≈0.9d≈0.8h_c, Eq. (26) yields:

$$V_{jv} \approx \sum M_{yc}\left(\frac{1}{z_c} - \frac{1}{L_b}\frac{h_{st}}{h_{st,n}}\right) + \max V_{g+\psi_2 q,b} \qquad (27)$$

Then the shear stress in the joint core is (with h_b: beam depth):

$$v_j = \gamma_{Rd}\frac{V_{jv}}{b_j h_b} \qquad (28)$$

Diagonal tension cracking of the joint core will take place when the principal tensile stress under the combination of v_j and of the mean vertical compressive stress in the joint, $\nu_{top} f_c$, exceeds the tensile strength of concrete, f_{ct}. This takes place when:

$$v_j \geq v_c = f_{ct}\sqrt{1 + \frac{\nu_{top} f_c}{f_{ct}}} \qquad (29)$$

According to Priestley [1997], in exterior joints with bars bent vertically towards the joint core (instead of outwards into the column above and below) confinement due to the bent bars increases the joint shear stress at diagonal cracking by 50% over the value v_c in the right-hand-side of Eq. (29).

Diagonal cracking of the joint core seldom has catastrophic consequences, especially if beams of significant cross-section frame into the joint from more than two sides (i.e. beam-column joints not found to satisfy Eq. (29)) do not need to be retrofitted. The real threat is crushing of the unreinforced joint core due to diagonal compression. This takes place if v_j exceeds the limit:

$$v_j \geq v_{ju} = nf_c\sqrt{1-\frac{v_{top}}{n}} \tag{30}$$

with: $n=0.7-f_c(MPa)/200$: reduction factor on f_c due to simultaneous transverse tensile strains.

3.5 Member evaluation criteria

Only the differences with the evaluation criteria proposed in 2.5 for the preliminary evaluation procedure are noted here. The evaluation criteria are also listed in Table 3, where they may be compared to those proposed for the preliminary evaluation procedure.

On the demand side (left-hand-side of the verification inequality), given that the nonlinear static analysis method employed for checking performance at the "Life Safety" (or "Significant Damage") and the "Near Collapse" levels is – overall – more accurate than the elastic analysis employed by the preliminary evaluation procedure, a lower safety margin may be warranted. Therefore, the mean chord rotation demand, $\theta_{E,m}$, enters in the criteria (Eqs. (31)-(34) below corresponding to Eqs. (17)-(20)), instead of the 95% characteristic value, $\theta_{Ek,0.95}$. Moreover, shear force demands, $V_{E,max}$, are directly estimated from the nonlinear analysis, without applying any γ_{Rd}-coefficient. Such a coefficient is employed only in the assessment of joints, assuming the values proposed in 2.4 for the member assessment in shear, depending on the performance level at which joint shear failure is checked. In all the analyses, expected (mean) values of material properties are used.

As far as capacity is concerned (right-hand-side of the verification inequality), there is no differentiation with the preliminary evaluation: the 95%-fractile, mean-minus-one-standard-deviation or expected (mean) value of deformation capacity, design or mean value of force capacity of brittle members, etc. are used as in the preliminary evaluation, always calculated on the basis of mean values of material properties, except for the design value of force capacity of brittle members which is calculated on the basis of design values of material properties. The difference in knowledge level between the detailed (final) and the preliminary evaluation affects the value of deformation capacity only to the extent that the mean material strengths to be inferred with the help of in-situ measurements will normally be higher than the assumed default values, or the nominal ones specified in original construction documents, and therefore will give higher capacity estimates.

3.5.1 "Operational" or "Limited Damage" performance level

No difference; Eq.(16), repeated below, applies.

$$\theta_{E,m} \leq \theta_y \tag{16}$$

Shear is checked through Eq. (7), using the inelastic shear force demand from the analysis, $V_{E,i}$. The design value of shear resistance, V_{Rd}, calculated on the basis of design values of material properties, f_{cd} and f_{yd}, incorporating material (partial) factors γ_m, is used.

3.5.2 "Life Safety" or "Significant Damage" performance level

The verification of members against ductile, flexure-controlled failure is:
For ductile primary members:

$$\theta_{E,m} \leq \theta_{uk,0.05} \tag{31}$$

where $\theta_{uk,0.05}$ is the 5% fractile value of chord rotation capacity from Eq. (6a), with the reduction factor k_u= 0.865 applied for "old-type" or "non-conforming" members.
For ductile secondary members:

$$\theta_{E,m} \leq \theta_{u,m-\sigma} \tag{32}$$

with $\theta_{u,m-\sigma}$ denoting the mean-minus-one-standard-deviation value of chord rotation capacity from Eq. (6b), again with application of the reduction factor k_u= 0.865 for "old-type" or "non-conforming" members.

Shear is again checked through Eq. (7), using the inelastic shear force demand from the analysis, $V_{E,i}$. As in the preliminary evaluation, the design value of shear resistance, V_{Rd}, calculated on the basis of design values of material properties, f_{cd} and f_{yd}, incorporating material factors γ_m, is used in this verification. Again, brittle primary members are checked with the shear resistance divided by a capacity reduction factor of 1.25, to reflect the difference in structural importance with secondary members.

3.5.3 "Near Collapse" performance level

The proposed verification of members against ductile, flexure-controlled failure is:
For primary members:

$$\theta_{E,m} \leq \theta_{u,m-\sigma} \tag{33}$$

where $\theta_{u,m-\sigma}$ is the mean-minus-one-standard-deviation value of chord rotation capacity from Eq. (6b), with the reduction factor k_u= 0.85 for "old" or "non-conforming" members.

For secondary members:

$$\theta_{E,m} \leq \theta_{u,m} \tag{34}$$

where $\theta_{u,m}$ is the mean chord rotation capacity from Eq. (4), with the reduction factor k_u= 0.85 for "old" or "non-conforming" members.

Shear is checked through Eq. (7), using the inelastic shear force demand from the analysis, $V_{E,i}$. As in the preliminary evaluation, the design value of shear resistance, V_{Rd}, calculated on the basis of design values of material properties, f_{cd} and f_{yd}, incorporating material (partial) factors γ_m, is used in the verification of brittle primary members, whilst the expected value of shear resistance, V_R, calculated on the basis of mean values of the nominal or assumed default values of material properties, f_c and f_y, without material (partial) factors γ_m, is used for secondary ones.

4. CONCLUSIONS

In this paper two procedures of different sophistication and complexity are proposed for the seismic assessment of existing RC buildings. Both are based on seismic displacements and the associated chord rotation demands at member ends. In the simpler procedure, that may be used for preliminary evaluation of buildings, inelastic chord rotation demands may be estimated from linear elastic analysis, according to rules developed on the basis of over one thousand nonlinear dynamic analyses, while forces for the check of brittle failure modes are computed in a capacity-design fashion, from equilibrium of forces at development of flexural plastic hinges. In the more advanced procedure, which may be used for final or detailed evaluation, inelastic chord rotation demands and member shear forces are determined via nonlinear static (pushover) analysis. In this procedure, brittle shear failure of joints is also checked, on the basis of capacity-design estimates of joint shear stresses. In both procedures members are evaluated in shear on the basis of empirical expressions for the shear force capacity of RC members, as affected by the magnitude of cyclic ductility demands, developed on the basis of a large volume of test results. Also in both procedures members are evaluated in flexure by comparing chord rotation demands to the corresponding yield or ultimate capacities, established through empirical expressions derived from over a thousand cyclic test results. Verification criteria at three different performance levels (Operational or Limited Damage, Life Safety or Significant Damage, and Near Collapse) are proposed for each one of the two evaluation procedures. The two procedures differ also in the amount and type of information on material properties and quantity of reinforcement required for their application.

ACKNOWLEDGEMENT

The contribution of the NATO Science for Peace Program to the support of this work is acknowledged.

REFERENCES

1. ASCE (2000). Prestandard for the seismic rehabilitation of buildings. Prepared by American Society of Civil Engineers for the Federal Emergency Management Agency (FEMA Report 356), Reston, VA.
2. ASCE (2001). Seismic evaluation of existing buildings. ASCE draft Standard, 4th Ballot, American Society of Civil Engineers, Reston, VA.
3. Biskinis D., Roupakias G. and Fardis M.N. (2003). Cyclic deformation capacity of shear–critical RC elements. Paper No. 199, *Proceedings, fib 2003 Symposium: Concrete Structures in Seismic Regions*, Athens.
4. CEN (1996). European prestandard ENV 1998-1-4:1996: Eurocode 8: Design provisions for earthquake resistance of structures. Part 1-4: Strengthening and repair of buildings. Comite Europeen de Normalisation, Brussels.
5. CEN (2003). Draft European Standard prEN 1998-3: 200x Eurocode 8: Design of structures for earthquake resistance. Part 3: Strengthening and repair of buildings. 3rd Project Team Draft (Stage 34) Doc. CEN/TC250/SC8/N343. Comite Europeen de Normalisation, Brussels.
6. Japan Building Disaster Prevention Association (1996). Law for promotion of seismic strengthening of existing reinforced concrete structures and related commentary (in Japanese).
7. Panagiotakos, T.B., Fardis, M.N. (1999): Estimation of Inelastic Deformation Demands in Multistorey RC Buildings, *J. of Earthquake Engineering and Structural Dynamics*, Vol. 29, 501-528.
8. Priestley, M.J.N. (1997): Displacement-based Seismic Assessment of Reinforced Concrete Buildings, J. of Earthquake Engineering, *IC Press,* Vol. 1, No. 1, 157-192.

Table 1. Mean value of ratio of inelastic chord rotation to elastic and of inelastic drift ratio to elastic

	Beam chord rotation	Column or wall chord rotation	Drift
Roof	1.25	1.08	0.85
Base	1.15	0.87	-
Mean	1.20	0.95	1.05

Table 2. Inelastic-to-elastic chord rotation ratio in open ground-storey of partially-infilled buildings

Beams	Existing columns (assessment)		Upgraded columns (redesign)	
	Bottom	Top	Bottom	Top
0.7	0.95	1.2	0.7	0.8

Table 3. Summary of member evaluation criteria

Performance level:	"Operational" - "Limited Damage"	"Life Safety" - "Significant Damage"		"Near Collapse"	
Evaluation procedure:	Preliminary or Detailed	Preliminary	Detailed	Preliminary	Detailed
Ductile primary members	$\theta_{E,m}\leq\theta_y$	$\theta_{Ek,0.95}\leq\theta_{uk,0.05}$	$\theta_{E,m}\leq\theta_{uk,0.05}$	$\theta_{Ek,0.95}\leq\theta_{u,m-\sigma}$	$\theta_{E,m}\leq\theta_{u,m-\sigma}$
Ductile secondary members	$\theta_{E,m}\leq\theta_y$	$\theta_{Ek,0.95}\leq\theta_{u,m-\sigma}$	$\theta_{E,m}\leq\theta_{u,m-\sigma}$	$\theta_{Ek,0.95}\leq\theta_{u,m}$	$\theta_{E,m}\leq\theta_{u,m}$
Brittle primary members	$V_{E,max}\leq V_{Rd}$	$V_{E,CD}(\gamma_{Rd}=1.2)\leq V_{Rd}/1.25$	$V_{E,max}\leq V_{Rd}/1.25$	$V_{E,CD}(\gamma_{Rd}=1.2)\leq V_{Rd}$	$V_{E,max}\leq V_{Rd}$
Brittle secondary members	$V_{E,max}\leq V_{Rd}$	$V_{E,CD}(\gamma_{Rd}=1.2)\leq V_{Rd}$	$V_{E,max}\leq V_{Rd}$	$V_{E,CD}(\gamma_{Rd}=1.2)\leq V_R$	$V_{E,max}\leq V_R$

Table 4. Mean and 95%-fractile building-averages of inelastic-to-elastic mean drift ratio

	Mean	95%-fractile
mean (work-equivalent)	1.05	1.35

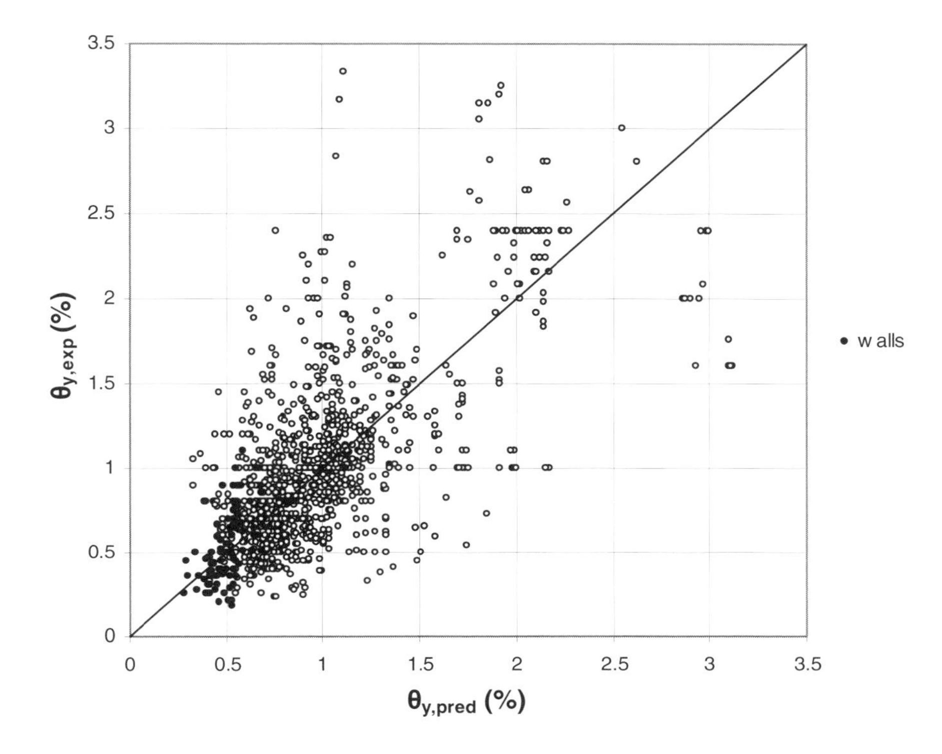

Figure 1. Comparison of predictions of Eqs. (2) with the experimental data from which they were derived

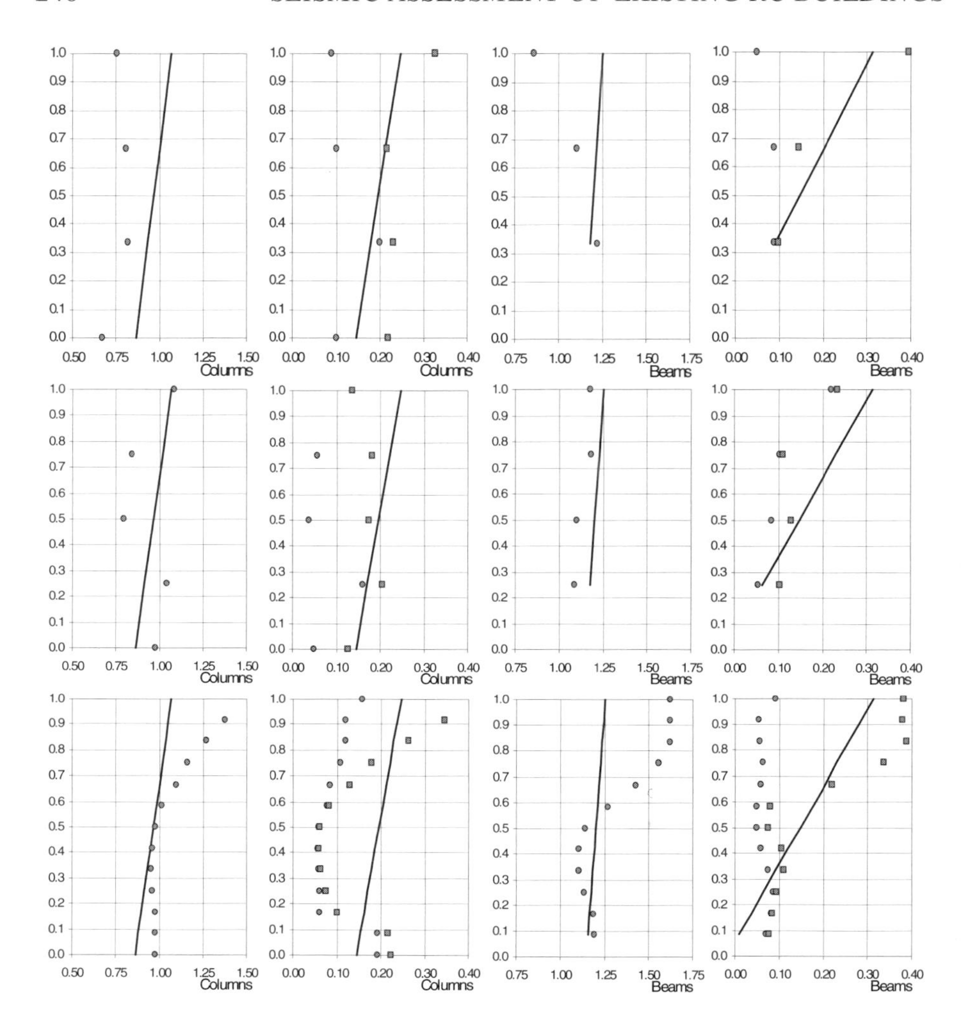

Figure 2. Ratio of inelastic chord rotation from nonlinear dynamic analyses, to elastic value from linear static analysis with lateral forces of triangular heightwise distribution. Top: Vertically irregular 3-storey buildings; middle: regular 4-storey buildings; bottom: regular 12-storey buildings. 1st and 3rd column of figures: Mean value of ratio over large number of cases of same number of storeys and average linear fit for all 3 types of buildings. 2nd and 4th column of figures: circles: standard deviation of ratio over large number of cases of same number of storeys; squares: standard deviation of ratio over large number of cases of same number of storeys with respect to linear fit to the mean values (line in figures of 1st and 3rd columns); line: average linear fit to the squares for the 3 types of buildings

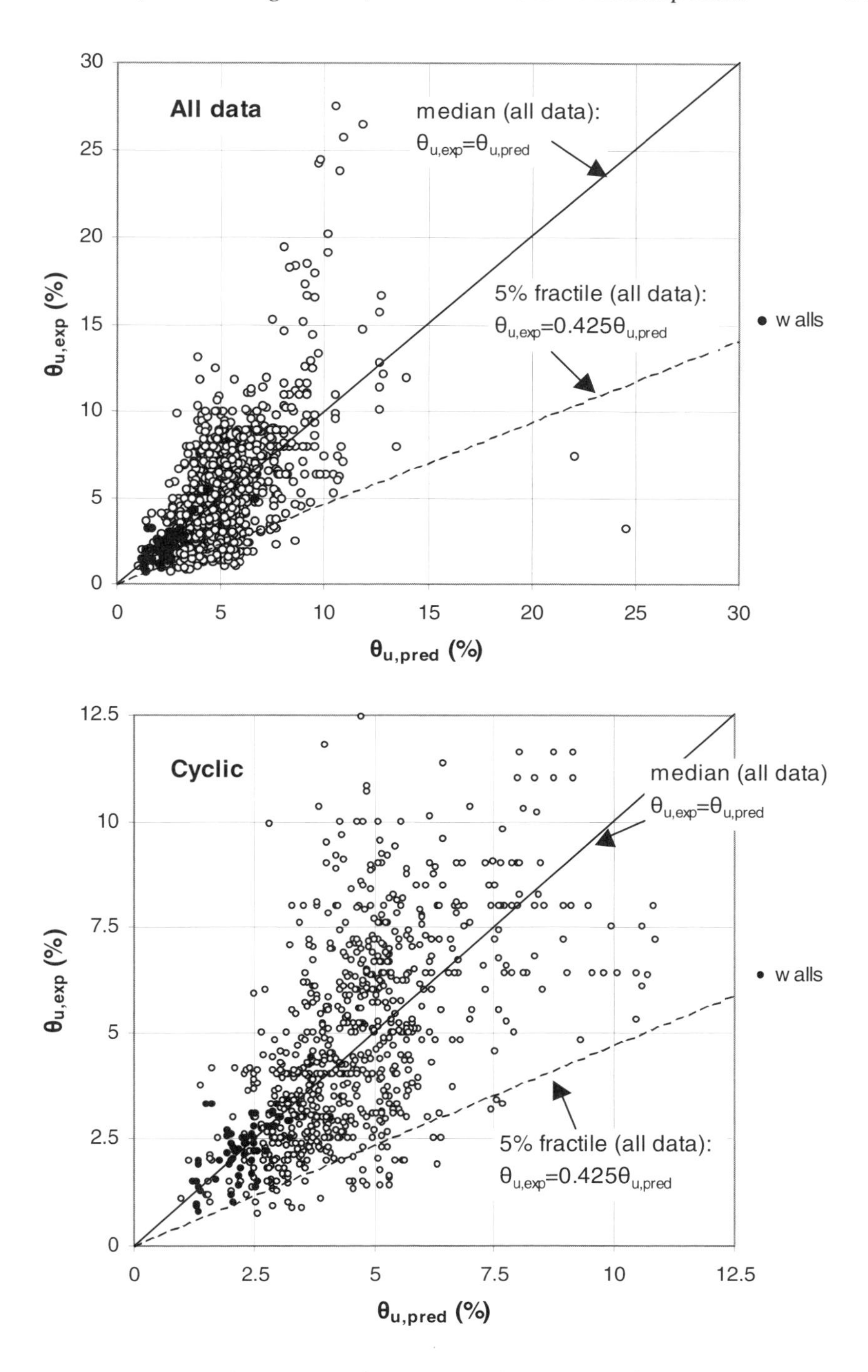

Figure 3. Comparison of predictions of Eq. (4) with the test results from which it was derived

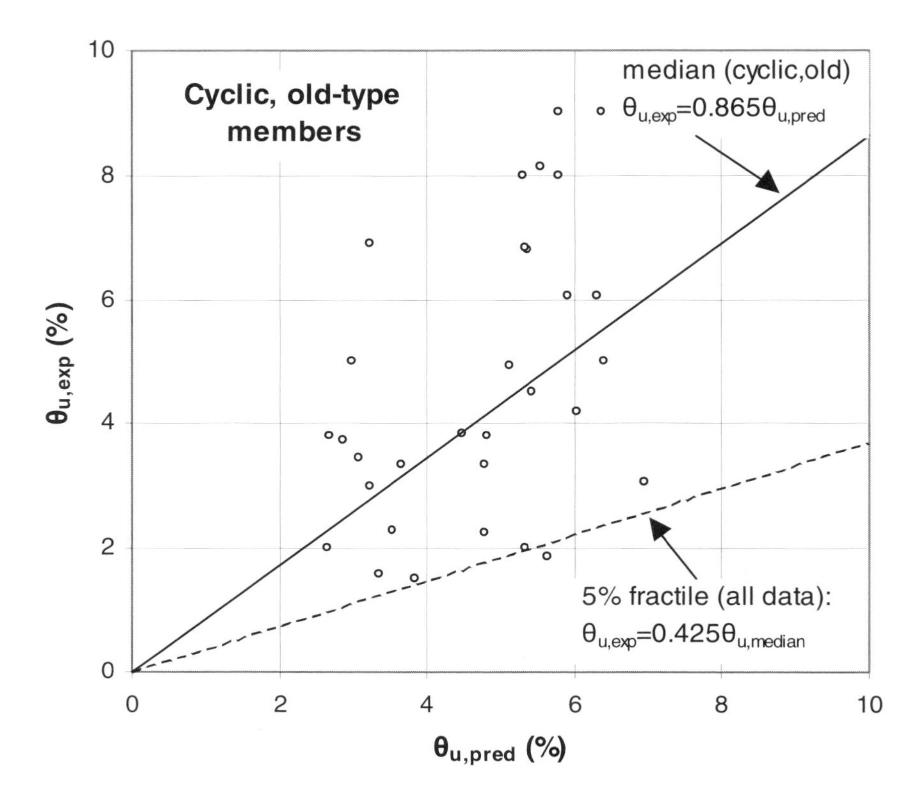

Figure 4. Comparison of the predictions of Eq. (4) with $\alpha=0$ to test results on members with old (i.e. non-seismic) type of detailing

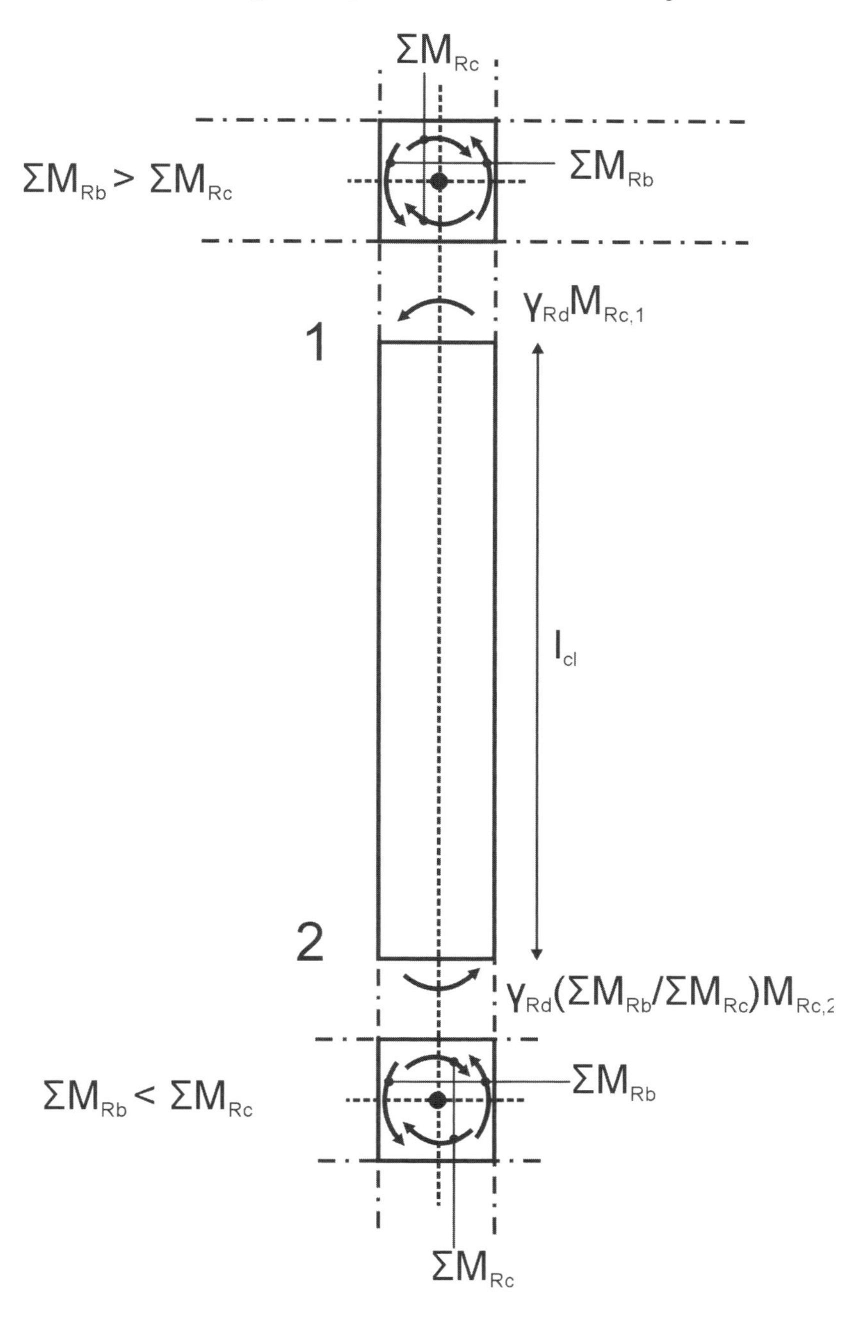

Figure 5. Bending moments at column ends for calculation of shear force $V_{E,max}$

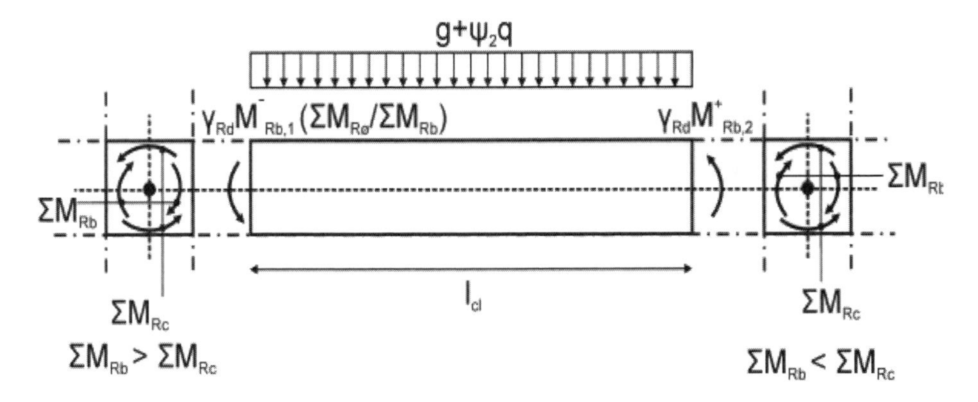

Figure 6. Bending moments at beam ends for calculation of shear force $V_{E,max}$

EXPERIMENTAL RESEARCH ON VULNERABILITY AND RETROFITTING OF OLD-TYPE RC COLUMNS UNDER CYCLIC LOADING

Stathis N. Bousias and Michael N. Fardis
Structures Laboratory, Department of Civil Engineering, University of Patras, Greece

Abstract: The results of an on going experimental campaign focusing on conventional and modern retrofitting techniques for reinforced concrete columns are presented. The parameters studied include the effect of jacketing (of reinforced concrete or fibre reinforced polymers) on deficient lap splices, the number of layers of the fibre reinforced polymers and the extent of the area of application of fibre reinforced polymers. Test results showed that jacketing is a very effective way of enhancing the deformation capacity of columns.

Key words: retrofitting, jackets, columns, fibre reinforced polymers, lap splices

1. INTRODUCTION

A large volume of existing structures worldwide has been constructed prior to the introduction of recent building codes for earthquake resistant design and, thus, are naturally vulnerable to seismic activity. Their inferior seismic performance often emanates from deficient design and execution processes: lack of capacity design philosophy, poor detailing, etc. The ductility deficit characterizing this class of structures leads to low deformation capacity systems, which, when subjected to deformation

S.T. Wasti and G. Ozcebe (eds.), Seismic Assessment and Rehabilitation of Existing Buildings, 245–268.

reversals well into the post-elastic region, respond with rapid loss of strength.

As a response to the pressing need for remedial measures enhancing the ductility of old-type members, conventional as well as less traditional techniques are currently employed. The first category comprises construction of reinforced concrete jackets made of cast-in-situ or sprayed concrete. Application of externally bonded FRP wraps for member rehabilitation belongs to the latter category, with its main contribution being the enhancement of the (limited) confining action provided by existing sparse stirrups. Despite the fact that the effectiveness, both in terms of structural response and cost, as well as the scientific documentation of this new technology have not been fully investigated and many issues remain unresolved, encasing RC members in FRP jackets is rapidly spreading.

Owing to their cost-effectiveness, concrete jackets have been, over the past two to three decades, by far the most widely-used technique for seismic upgrading of existing concrete members. The scarcity of experimental results on the cyclic behaviour of RC jacketed members is in contrast to the popularity of this retrofit measure. In particular, the limited data in the literature (Bett et al. [1988], Rodriguez and Park [1994], Gomez and Appleton [1998], Ersoy et al. [1993], Yamamoto [1992], Iliya and Bertero [1980]), do not include members with lap splices.

Unlike concrete jackets, the influence of FRP in enhancing column deformability is under intensive study worldwide, owing to certain advantages they offer over RC jacketing. The contribution of FRP-wrapping to the improvement of column deformation capacity through confinement has been investigated by several researchers (e.g. Ma and Xiao [2000], Zhang et al .[1999], Saadatmanesh et al. [1997], Chang et al. [2001], Ye et al. [2000], Seible et al. [1999]). Jacketing with FRP against lap-splice failures in RC columns has been studied by Ma and Xiao [1997], Saadatmanesh et al. [1997], Seible et al. [1997], Restrepo et al. [1998], Osada et al. [1999], Chang et al. [2001], Haroun et al. [2001], Saatcioglu and Elnabelsy [2001].

The results of an on-going experimental program on retrofitting of RC columns are presented herein. The campaign focuses on the following aspects: a) the effect of the number of FRP layers on the effectiveness of retrofitting, b) the size of the region over which they are applied on the member, and c) retrofitting of deficient lap-splicing of ribbed reinforcement through FRP wraps.

2. EXPERIMENTAL PROGRAMME

An experimental program comprising a total of 19 column specimens with dimensions, reinforcement detailing and materials typical of existing RC building structures, was undertaken to investigate the influence of important parameters of seismic retrofit. Both conventional (reinforced concrete) and innovative (fibre reinforced polymers, FRPs) jackets were considered.

The testing program includes two column geometries, representing non-seismically designed and detailed members (Figure 1):

- Type Q: a 250mm-square cross-section, reinforced longitudinally with four-14mm smooth bars (S220) (Figure 1(a)).
- Type R: a 250×500mm cross-section, reinforced longitudinally with four-18mm ribbed bars (S500) (Figure 1(b)).

The column height at which the lateral load is applied is the same for the two cases and equal to half a typical storey height, i.e. 1.6m.

The reasoning behind the selection of the aforementioned geometries was that, with a shear-span-ratio L/h = 6.4 in type Q specimens, cyclic behaviour and failure mode will be clearly dominated by flexure, whereas the behaviour of type R specimens (L/h = 3.2) may be affected by shear as well. The selection of a rectangular section with an aspect ratio of 2 in type R specimens was chosen to investigate the impact of confinement effected by the FRP wraps around the perimeter of the cross-section and hence over the compression zone.

Both types of specimens were reinforced in the transverse direction with 8-mm smooth (plain) stirrups of grade S220 at 200mm spacing, anchored by 135-degree hooks at one end and at 90-degree hooks at the other. This transverse reinforcement is typical in old structures and independent of column size or deformation demands. The 14mm-diameter smooth vertical bars have a yield stress of 313MPa and a tensile strength of 442MPa (average values from three coupons), while the corresponding values for the 18mm-diameter vertical bars are 514MPa and 659MPa. The yield and ultimate stresses for the mild steel used for ties are 425MPa and 596MPa. Concrete strength (measured on 150-by-300mm cylinders) at the time of testing ranges from 26 to 30 MPa (see Tables 1 and 2).

The specimens were cast into a heavily reinforced 0.6m-deep base within which ribbed vertical bars were anchored with 90-degree hooks at the bottom and smooth bars with a 180°-hook.

The behaviour of the columns was studied under cycling of horizontal displacements under the presence of constant axial force. The column specimens of both types were tested after being retrofitted.

To investigate the effect of FRP wrapping on the strength and deformation capacity of non-seismically designed columns, five type-Q specimens were constructed:

- Control specimen in its initial configuration (unretrofitted),
- Specimens with two layers of CFRP sheets wrapped around the lower either 600mm or 300mm,
- Specimens retrofitted with four layers of CFRP wraps around the lower either 300mm or 600mm.

Each layer of carbon FRP (CFRP) has a nominal thickness, t_f, of 0.13mm, an Elastic Modulus E_f = 230 GPa and a tensile strength of 3450MPa (failure strain ε_f = 1.5%) in the (main) direction of the fibres. A region above the base of the column was wrapped with CFRP, leaving a gap of about 10mm between the FRP wrap and the top of the column footing. The continuous CFRP sheets were attached to the column via epoxy resin after thoroughly cleaning the column surface from loose material and rounding the corners of the section. The latter has proven very important in earlier studies, as stress concentration at cross-section corners may lead to premature FRP rupture. At the end of wrap an anchorage length of one full side of the cross-section was provided.

A summary of the geometry and retrofitting schemes for type-Q specimens is presented in Table 1.

Tests on type R specimens focus on the effect of lap-splicing on the performance of columns in non-earthquake-resistant buildings and on the improvement of their behaviour through retrofitting with FRP jacket. The longitudinal reinforcement is either continuous at the connection of the column with the foundation, i.e. in the plastic hinge region, or, as usually is the case in existing (old) buildings, has straight ends lap-spliced near the bottom. The following specimens were tested:

- Unretrofitted specimen without lap-splices (as control)
- Three unretrofitted specimens, each with different lap-splice lengths (15-, 30-, and 45-bar diameters).
- Two specimens with different lap-splice lengths (15- and 30-bar diameters) but retrofitted with 2 layers of CFRP wraps.
- Three specimens with different lap-splice lengths (15-, 30-, and 45-bar diameters) but retrofitted with 5 layers of CFRP wraps.
- Two specimens without lap-splices, retrofitted with 2 or 5 layers of CFRP wraps as control specimen for those above.
- Three specimens with different lap-splice lengths (15-, 30-, and 45-bar diameters) but retrofitted with reinforced concrete jackets.

A summary of the geometry and retrofitting schemes for the specimens is presented in Table 2.

Horizontal loading was applied at the column head by a servo-hydraulic actuator at a distance of 1.6m from the base.

Testing was performed by cycling horizontal displacements at amplitudes increasing in 5mm steps along the testing axis. This load history with closely spaced single cycles was chosen over the usual protocols of 3 cycles at few displacement ductility levels, to monitor better the cyclic behaviour of the specimen up to failure. In type-Q specimens an axial load of approximately 750kN was applied through a jack at the top, along the member longitudinal axis. The corresponding value of axial load for type-R columns was 860 kN. The mean value of the normalized axial load, ν=N/Acfc, during the test is listed at the last columns of Tables 1 and 2. The jack acted against vertical rods connected to the laboratory strong floor through a hinge. With this setup the P-Δ moment at the base of the column is equal to the axial load, times the tip deflection of the column, times the hinge distance ratio from the base and the top of the column (i.e. times 0.5/1.6=0.3125). The P-Δ moments were removed from test results.

The rotation and axial displacement of two sections above the base (at 125mm and 250mm, for type-Q, and at 250 mm and 500 mm, for type-R) were also monitored through pairs of displacement transducers on opposite sides of the section.

3. TEST RESULTS

3.1 Type Q specimens

Results for type-Q specimens in the form of force-displacement loops, are presented in Figures 3 (for the control specimen, Q_0L0) and 5, 6 (for the retrofitted ones).

The behaviour of the control specimen (Q_0L0) during testing was in flexure, attaining a displacement ductility around 3 (Figure 3). The concrete cover and part of the core concrete at the lower 200mm of the column disintegrated and steel bar buckling was evident after concrete cover spalled off.

The effect of increasing the number of CFRP layers from 2 to 4 is shown by comparing the force-deformation loops in Figures 4(a), 6(a) to those of Figures 4(b), 6(b), respectively. The FRP wraps extend over the bottom 600mm of the column in the specimens of Figure 4, or over the bottom 300mm in those of Figure 6. The specimens with the two layers failed with FRP fracture and bar buckling (Figure 4b), while in those with 4 layers the test stopped after the peak cycle force dropped below 80% of peak resistance without fracture of the FRP. At that stage the concrete core inside the FRP had disintegrated. For the 600mm long FRP jacket, the increase of the number of FRP layers from 2 to 4 increases ultimate strength by less than 10% and ultimate drift, defined conventionally on the basis of the 20% drop in resistance rule, by about 30%. In the specimens with 0.30m-long FRP jacket, the improvement is below 5% in peak resistance and again about 30% in ultimate drift capacity. So, improvement is less than proportional to number of layers.

To investigate whether and to what extent a shorter length of confinement influences member response, two columns were tested after been retrofitted with 2 or 4 CFRP layers applied at the lower 300mm of column height (marginally longer than member structural depth). Comparing the force-displacement response curves in Figure 6 to those of Figure 4, no appreciable difference is observed, as damage in the unretrofitted specimen (Figure 4a) concentrated within the region between the first two stirrups (200mm).

Table 3 summarizes the ultimate resistance and deformation capacity of the type Q specimens, as well as their failure modes.

The envelopes of all force-displacement loops shown in Figure 7 testify the decisive contribution of FRP jackets in enhancing the deformation capacity of old-type members. Furthermore, the rather similar behaviour of specimens retrofitted with 2 and with 4 layers of fibre reinforced polymer wraps indicates that increasing jacket stiffness beyond a certain threshold does not increase proportionally the benefit to member deformation capacity. It is also seen that as long as confinement extends over the plastic hinge region, the response of the retrofitted specimens is the same in terms of strength and deformation capacity. Consequently, increasing the length over which FRP wrapping is applied is, on cost-benefit terms, not justified.

3.2 Type R specimens retrofitted with FRP

Specimen R_0L0 of the group of type-R specimens served as the unretrofitted control specimen with continuous longitudinal reinforcement. The specimen yielded in flexure but exhibited a mixed flexure-shear failure mode, with sudden drop in resistance at peak deflection of 45mm accompanied by bar buckling, inclined cracking and ultimate disintegration of the concrete core above the base. The deformation at failure was 40mm (2.5% drift ratio, Table 4) determined through the conventional rule of 20%-drop in lateral force resistance.

Three columns with lap-spliced longitudinal reinforcing bars, but otherwise identical to the control specimen above, were tested. They are characterized by different lap splice lengths: 15 bar-diameters (specimen

R_OL1), 30 bar-diameters (specimen R_OL3) and 45 bar–diameters (specimen R_0L4). The behaviour of the three specimens is shown in Figures 8(a) to (c), respectively. The response is conditioned by the presence and length of lap splices. The specimen with the shortest splices (R_0L1) displayed the lowest strength of all three, as reversed cyclic loading caused early spalling of the concrete cover and rapid degradation of bond. Specimens R_0L1 and R_0L3 did not reach the full flexural strength of the end section, while specimen R_0L4, with the 45-bar diameter lap, did. As a matter of fact, the experimental strength of specimens R_0L1 and R_0L3 was 80% and 95%, respectively, of the theoretical flexural capacity, whilst specimen R_0L4 as well as the control, R_0L0 reached 110% of the theoretical strength. The initial stiffness of all specimens with lap splices was similar to that of the member with continuous reinforcement, because during the early stages of loading slip between the lapped bars is minimal.

In Figure 9 the failure mode of specimens R_0L1 and R_0L3 is evident: damage appeared first by concrete splitting along the plane of lapped bars and progressed by crushing of concrete ahead of the end of the starter bars, due to high bearing stresses in that region. In the absence of dense stirrups, concrete crushing ahead of the rebar end is promoted by the fact that starter bars are usually (due to the sequence of construction) located at the corner of the stirrups and closer to the external surface than the continuing bars. During the subsequent cycles of increasing displacement amplitude and due to the sparse stirrups, shedding of concrete cover in the region of overlapping bars took place. Member lateral force capacity decreased rapidly due to insufficient force transfer between starter bars and member longitudinal reinforcement, soon after cover concrete spalled. The drift ratio at the conventionally defined failure of the column (i.e. point of reduction of peak cycle resistance below 80% of the maximum recorded lateral resistance in the direction of loading) was 1.5%, for both specimens R_0L1 and R_0L3, regardless of the length of the lap splice. Lap splice length affected peak resistance, which dropped by 30% in specimen with 15-bar diameters lap splice, or by 13% in that with the 30-bar diameters one, in comparison to the specimen with no lap splices or with 45-bar diameter ones. By comparing the magnitude of these reductions to the theoretical result derived from the simple rule often used for estimation of the flexural resistance of columns with insufficient lap splice length - namely using a yield strength of steel reduced by the ratio of available to required lap length - it is concluded that the lap length required for development of the same flexural capacity as specimens R_0L0 and R_0L4 is about 35 bar diameters.

Specimen R_0L4 (45-diameters lap-splice length) behaved much better than the other two lap-spliced specimens: its strength and deformation capacity was the same as that of the control column without lap-splicing.

Splitting cracks appeared also along the overlapping length, but the behaviour of the member afterwards was not conditioned by failure of the splice. The member sustained cycling of horizontal displacements in more or less the same way as the specimen with the continuous reinforcement and with similar rate of strength decrease after peak load.

A group of three specimens similar to the previous ones were tested after wrapping with 5 layers of CFRP. The response of the specimens is shown in Figure 10. In specimen R_P5L1 (15-diameters lap-splicing) retrofitting restored member strength to above 90% of that of the control specimen with continuous reinforcement, while member stiffness remained unaffected. In the two other specimens strength increased well above that of the unretrofitted column. The deformation capacity of all three columns was much higher than that of the unretrofitted column without lap splicing. Therefore, wrapping with 5 CFRP layers more than removes the deficiency in deformation capacity due to lap splicing and (with the exception of the column with the 15-bar diameter laps) that in strength as well. In the column with the 15-bar diameter laps, wrapping with 5 FRP layers did not fully re-instate flexural capacity and provided much lower energy dissipation (as evidenced by the narrow loops) than in the two other retrofitted columns. So, it may be concluded that, although improvement of such a lap-splice deficiency by FRP wrapping is possible, its full elimination is not.

Figure 11 shows force-deflection loops of two specimens with continuous reinforcement at the column base, for a column with the same geometry and retrofitting scheme (5 CFRP layers) as the three retrofitted ones above, but with lower concrete strength (18MPa). Although these results are not directly comparable to those of Figures 10(a) to (c) due to the lower concrete strength, they show that even when the lap splice length is of sufficient length to provide the full strength and deformation capacity of the unretrofitted column without lap splices (namely 45-bar diameters), FRP wrapping is not as effective in improving deformation capacity as in a column without lap splices.

An interesting observation on the behaviour of this latter retrofitted specimens was that no crushing of concrete in the area ahead of the starter bars occurred (as observed previously in specimen R_0L4), despite the fact that the region over which the CFRP was applied (lower 0.6m) was shorter than the lap splice length (0.81m), i.e. not confining the concrete at the end of the starter bar. It seems that the improvement in bond conditions along the 0.60m-long confined part of the lap splice reduces the force to be transferred by direct bearing of the head of the starter bar against concrete, and reduces the possibility of local crushing there.

Regardless of the lap-splice length, all columns tested after retrofitting sustained drift ratios of at least 5% at conventionally defined failure (Table

3). The modification of the response over that of the unretrofitted specimen is attributed to the confinement of concrete, which improved the conditions of force transfer from the longitudinal reinforcement to the starter bars. Frictional resistance along the splice length was maintained after concrete was split, owing to the increased confinement offered by the CFRP jacket. Despite cumulative lateral expansion of the compressed concrete inside the CFRP jacket, the jacket itself did not rupture in any of the specimens of the group, maintaining disintegrated concrete in place and contributing to dissipation of energy.

Figure 11 shows test results from specimens with 2 layers of CFRP (R_P2L1, R_P2L3), which is the practical minimum for retrofitting. Member response is improved with considerable increase in ductility supply and marginal enhancement in strength. The retrofitting effect of 2 layers is much less than that of 5 layers, but the reduction is not commensurate to that of the layers. For the specimen with 30-bar diameter laps, which had shown very satisfactory deformation capacity and energy dissipation when retrofitted with 5 FRP layers, wrapping with just 2 FRP layers is clearly not sufficient.

3.3 Type R specimens retrofitted with RC jackets

Three specimens identical to R_0L1, R_0L3 and R_0L4 were tested after been retrofitted with a 75-mm thick concrete jacket. The jacket was reinforced longitudinally with four 20-mm bars and transversely with 10-mm stirrups at 100mm centres. Shotcrete with a mean compressive strength of 36MPa was used for the jacket. No special measures were taken for improvement of the connection of the jacket to the existing member (roughening of interface, steel dowels, etc.). With the addition of the RC jacket the total cross-sectional dimensions become 650mm by 400mm leading to a shear span ratio in the direction of testing of 2.5.

The force-deformation loops obtained for these specimens are shown in Figure 12. In all three cases the addition of the jacket increased, as expected, the flexural strength of all columns. All three columns reached or exceeded their theoretical flexural capacity with the lap spliced bars assumed continuous. In other words, the RC jacket is quite effective regarding flexural resistance and in fully mobilizing the insufficiently spliced bars for it. Strength enhancement is roughly equal in the three columns, as shown in Table 4.

In specimens R_RCL1 (15-bar diameters lap) and R_RCL3 peak force was attained at 1.5% drift (double as that in the unretrofitted column, R_0L1) and the column yielded in flexure. After peak resistance inclined cracking developed with crack opening becoming wider with displacement cycling. Moreover, the jacket concrete disintegrated along a major part of the length of the corner bars, due to bond stresses. As a result, column strength dropped rather rapidly and the column ultimately failed by a combination of shear and bond at 4.2% drift. Shear- and bond-dominated behaviour led to limited energy absorption capacity, as shown by the narrow force-deflection loops in Figure 13(a) and (b). At the end of the test the column had disintegrated at the base. The only difference in behaviour between specimens R_RCL1 and R_RCL3 is the fact that the latter maintained peak resistance for a limited number of cycles and exhibited an abrupt reduction thereafter, while in the former resistance started dropping immediately after peak but at a lower rate. An immediate result of this behaviour is the marginally lower drift at conventionally defined failure of the latter specimen (3.8% compared to 4.2%).

Despite the longer lap splice length, the force-deformation response of specimen R_RCL4 (45-bar diameters lap splice length), did not deviate appreciably from that of the previous two retrofitted columns. The column yielded in flexure and, owing to the longer splice length, maintained its peak resistance for larger number of cycles. Shear cracking developed in this specimen, as well, but there was no bond failure and damage concentrated mainly at the lower part of the specimen. Fracture of the jacket concrete in compression near the base extended through the whole width of the jacket, accompanied by disintegration of the concrete in the initial column section and buckling of the bars, both in the jacket and in the original column (Figure 14(b)).

When the response of all three columns retrofitted with RC jacket is compared to that of the columns retrofitted with FRP jacket (Figures 15 to 17), especially when 5 layers of FRP are employed, it turns out that RC jacketing is less effective in improving deformation capacity than retrofitting with FRP jackets. This can be attributed to the fact that the shear strength enhancement provided by the transverse reinforcement of the RC jacket is not sufficient to balance the large increase in flexural capacity, shifting the behaviour from shear-dominated to a flexure-dominated one.

4. CONCLUSIONS

The paper presents the results of an on-going experimental program on retrofitting of RC columns. The program focuses on two aspects: a) retrofitting of columns with deficient lap-splicing of ribbed reinforcement through FRP wraps or concrete jackets, and b) the effect of the number of CFRP layers on the effectiveness of retrofitting.

Regarding FRP retrofitting of deficient lap-splices, it was found that with five FRP wraps over the bottom 0.6m of the column, deformation capacity is fully re-instated, even for very short lap splices within the FRP-confined length. Nonetheless, if the lap splice is too short (namely 15-bar diameters) energy dissipation capacity and post-peak strength degradation of the retrofitted column are not satisfactory. The same applies for longer lap splicing (30-bar diameters), if only 2 FRP layers are used. Interestingly, lap splices long enough not to adversely affect strength and deformation capacity of the unretrofitted column (45-bar diameters), prevent FRP wrapping from improving deformation capacity to the same level as in a column without lap splices.

Concrete jackets are very effective in removing the adverse effect of lap splicing on flexural capacity, even for very short lap splice lengths. Nonetheless, as: (a) the increase in column size reduces the shear span ratio and, (b) only a perimeter tie may be added to the jacketed column, the deformation capacity of the column may ultimately be controlled by shear, although the column initially yields in flexure. Therefore, the concrete-jacketed column has lower deformation capacity than the FRP jacketed one, but in the retrofitted structure it may be exposed to lower inelastic deformation demands due to its increased pre-yield stiffness and lateral resistance.

As far as the effect of the number of CFRP layers is concerned, it was found that, although two layers of CFRP are not equally effective as four layers of the same material, they are more cost-effective than 4 or 5 layers, in that the loss in performance is not commensurate to the reduction in FRP layers.

ACKNOWLEDGEMENT

The contribution of the NATO Science for Peace Program to the support of the research is acknowledged.

REFERENCES

1. Bett, J., Klingner, R. and Jirsa, J. 1988. Lateral load response of strengthened and repaired reinforced concrete columns. ACI Structural J., Vol. 85, No 5, pp.499-507.
2. Bousias, S. N., Triantafillou, T. C., Fardis, M. N., Spathis, L., Oregan, B., 2002. "Experimental behaviour of deficient rectangular columns with externally bonded FRP", 1st *fib* Congress, Osaka, paper W-189.
3. Chang, K. C.; Lin, K. Y. and Cheng, S. B., 2001. Seismic retrofit study of RC rectangular bridge columns lap-spliced at the plastic hinge zone, Proceedings of the International Conference of FRP composites in Civil Engineering, J.-G. Teng, ed., Hong Kong, China, 2001, pp. 869-875.

4. Ersoy, U., Tankut, T. and Suleiman, R. 1993. Behavior of jacketed columns. ACI Structural J., Vol. 90, No 3, pp.288-293.
5. Gomes, A.M., Appleton, J. 1998. Repair and strengthening of reinforced concrete elements under cyclic loading. Proceedings 11th European Conf. Earthquake Engineering, Paris.
6. Haroun, M. A., Mosallam, A. S., Feng, M. Q. and Elsanadedy, H. M., , 2001. Experimental investigation of seismic repair and retrofit of bridge columns by composite jackets, Proceedings of the International Conference of FRP composites in Civil Engineering, J.-G. Teng, ed., Hong Kong, China, pp. 839-848.
7. Iliya, R. and Bertero, V.V. 1980. Effects of amount and arrangement on wall – panel reinforcement on hysteretic behavior of reinforced concrete walls", Report No. UCB/EERC-80/04, University of California, Berkeley.
8. Ma, R., Xiao, Y., Li, K. N., 1997. Seismic retrofit and repair of circular bridge columns with advanced composite materials, Earthquake Spectra, Vol. 15, No. 4, pp. 747-764.Full-scale testing of a parking structure column retrofitted with carbon-fiber reinforced composites. Construction and Building Materials, Elsevier, Vol. 14, 2000, pp. 63-71.
9. Ma, R., Xiao, Y., Li, K. N., 2000. Full-scale testing of a parking structure column retrofitted with carbon-fiber reinforced composites. Construction and Building Materials, Elsevier, Vol. 14, 2000, pp. 63-71.
10. Osada, K., Yamaguchi, T. and Ikeda, S. 1999. Seismic performance and the retrofit of hollow circular reinforced concrete piers having reinforcement cut-off planes and variable wall thickness, in Transactions of the Japan Concrete Institute, Vol. 21, pp. 263-274.
11. Restrepo, J.I., Wang, Y.C., Irwin, R.W. and DeVino, B. 1998. Fiberglass/epoxy composites for the seismic upgrading of reinforced concrete beams with shear and bar curtailment deficiencies. Proceedings 8th European Conference on Composite Materials, Naples, Italy, pp. 59-66.
12. Rodriguez, M. and Park, R. 1994. Seismic Load Tests on Reinforced Concrete Columns strengthened by Jacketing, ACI Structural J., Vol. 91, No 2, pp.150-159.
13. Saadatmanesh, H., Ehsani, M.R. and Jin, L. 1997. Repair of earthquake-damaged RC columns with FRP wraps. ACI Structural J., Vol. 94, No. 2, pp. 206-215.
14. Saatcioglu, M., Elnabelsy, G., 2001. Seismic retrofit of bridge columns with CFRP jackets, Proceedings of the International Conference of FRP composites in Civil Engineering, J.-G. Teng, ed., Hong Kong, China, pp. 833-838.
15. Seible, F., Innamorato D., Baumgartner, J., Karbhari, V. and Sheng, L.H., 1999. Seismic retrofit of flexural bridge spandrel columns using fiber reinforced polymer composite jackets, in: Dolan CW, Rizkalla SH & Nanni A (eds.) Fiber Reinforced Polymer Reinforcement for Reinforced Concrete Structures, ACI Report SP-188. Detroit, MI, pp. 919-931.
16. Seible, F., Priestley, M.J.N. and Innamorato D., 1997. Seismic retrofit of RC columns with continuous carbon fiber jackets. Journal of Composites for Construction (ASCE), Vol. 1, No 2, pp. 52-62.
17. Yamamoto, T. 1992. FRP Strengthening of RC Columns for Seismic Retrofitting. Proc.10th World Conf. on Earthquake Engineering, Madrid, Balkema, pp.5205-5210.
18. Ye, L.P., Zhao, S.H., Zhang, K. and Feng, P., 2001. Experimental study on seismic strengthening of RC columns with wrapped CFRP sheets, Proceedings of the International Conference of FRP composites in Civil Engineering, J.-G. Teng, ed., Hong Kong, China, pp. 885-892.
19. Zhang, A., Yamakawa, T., Zhong, P. and Oka, T, 1999. Experimental study on seismic performance of reinforced concrete columns retrofitted with composite materials jackets,

in: Dolan CW, Rizkalla SH & Nanni A (eds) Fiber Reinforced Polymer Reinforcement for Reinforced Concrete Structures, ACI Report SP-188. Detroit, MI, pp. 269-278.

Table 1. Type-Q columns (250mm square section)

Specimen	Layers of FRP jacket	Height of FRP jacket (m)	Concrete strength (MPa)	Axial load ratio $\nu=N/A_cf_c$	Structural depth, d (mm)
Q_0L0 (control)	0	0	30.3	0.37	210
Q_P2H2	2	0.6	28.2	0.45	220
Q_P4H2	4	0.6	28.2	0.44	220
Q_P2H1	2	0.3	28.2	0.40	215
Q_P4H1	4	0.3	28.2	0.45	215

Table 2. Type-R columns (250mm by 500mm section)

Specimen	Specimen description	Lap splice	Layers of jacket	Jacket height (m)	Concrete strength (MPa)	Axial load ratio $\nu=N/A_cf_c$
R_0L0	Control, no splices	0	N/A	N/A	31.0	0.23
R_0L1	Unretrofitted, lap splices	15∅, 270mm	N/A	N/A	27.4	0.23
R_0L3	Unretrofitted, lap splices	30∅, 540mm	N/A	N/A	27.4	0.28
R_0L4	Unretrofitted, lap splices	45∅, 810mm	N/A	N/A	27.4	0.28
R_P2L0	FRP jacket, no splices	0	2	0.6	18.1	0.37
R_P5L0	FRP jacket, no splices	0	5	0.6	17.9	0.39
R_P2L1	FRP jacket, lap splices	15∅, 270mm	2	0.6	26.9	0.30
R_P5L1	FRP jacket, lap splices	15∅, 270mm	5	0.6	27.0	0.28
R_P2L3	FRP jacket, lap splices	30∅, 540mm	2	0.6	26.9	0.28
R_P5L3	FRP jacket, lap splices	30∅, 540mm	5	0.6	27.0	0.29
R_P5L4	FRP jacket, lap splices	45∅, 810mm	5	0.6	27.0	0.29
R_RCL1	Shotcrete jacket, lap splices	15∅, 270mm	N/A	1.50	36.7	0.21
R_RCL3	Shotcrete jacket, lap splices	30∅, 540mm	N/A	1.50	36.8	0.19
R_RCL4	Shotcrete jacket, lap splices	45∅, 810mm	N/A	1.50	36.3	0.19

Table 3. Summary of test results

Specimen	Peak force (kN)	Drift at failure (%)	Max. drift (%)	Comments
Q_0L0	44	2.2	2.5	Concrete crushing, rebar buckling
Q_P2H2	46	4.7	5.9	FRP fracture, bar buckling
Q_P4H2	50	6.9	6.9	Core concrete disintegration
Q_P2H1	50	5	6.25	FRP fracture, bar buckling
Q_P4H1	52	6.6	6.6	Core concrete disintegration

Table 4. Summary of test results of type R specimens

Specimen	Peak force (kN)	Drift at failure (%)	Max. drift (%)	Comments on failure mode
R_0L0	200	2.5	2.8	Bar buckling, concrete crushing, inclined cracks
R_0L1	145	1.9	2.8	Lap failure
R_0L3	170	1.9	3.1	Lap failure
R_0L4	200	2.5	2.8	Flexural failure, concrete crushing
R_P5L1	185	5.0	5.6	Core concrete disintegration
R_P5L3	210	5.6	5.6	Core concrete disintegration, bar buckling
R_P5L4	215	5.6	5.6	Concrete crushing, fracture of bar
R_P2L1	168	3.4	4.7	FRP fracture, Concrete crushing
R_P2L3	213	4.7	5.3	Concrete crushing
R_RCL1	345	4.2	4.8	Ductile shear failure, bond
R_RCL3	378	3.8	4.5	Ductile shear failure, bond
R_RCL4	357	4.7	5.1	Concrete crushing, bar buckling

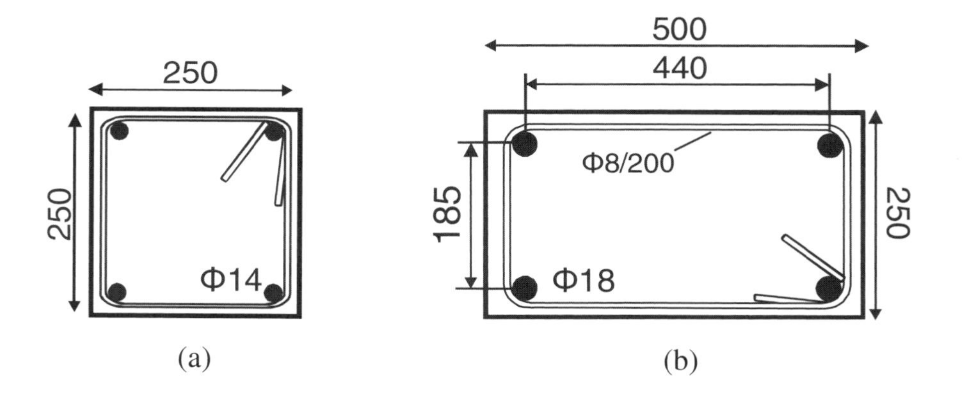

Figure 1. Specimen configuration (a) Type Q, (b) Type R

Figure 2. Test set-up

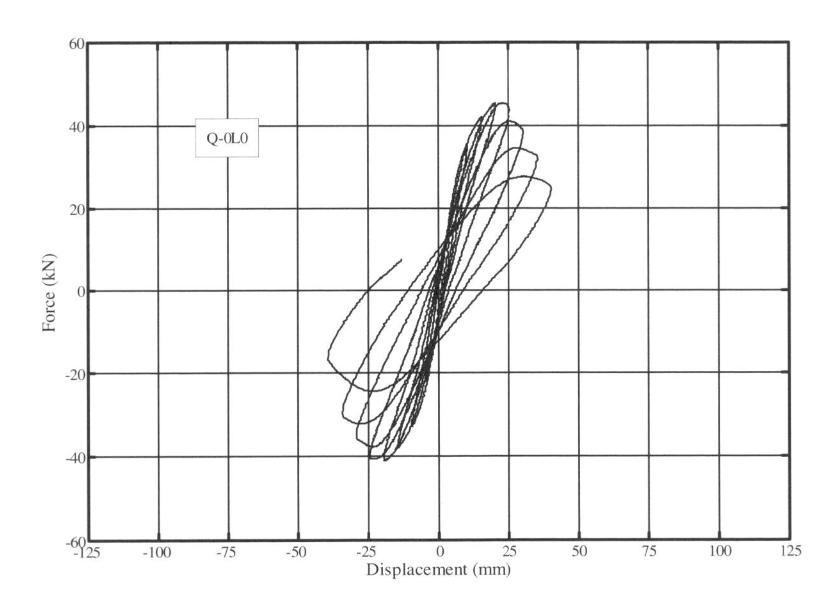

Figure 3. Unretrofitted column

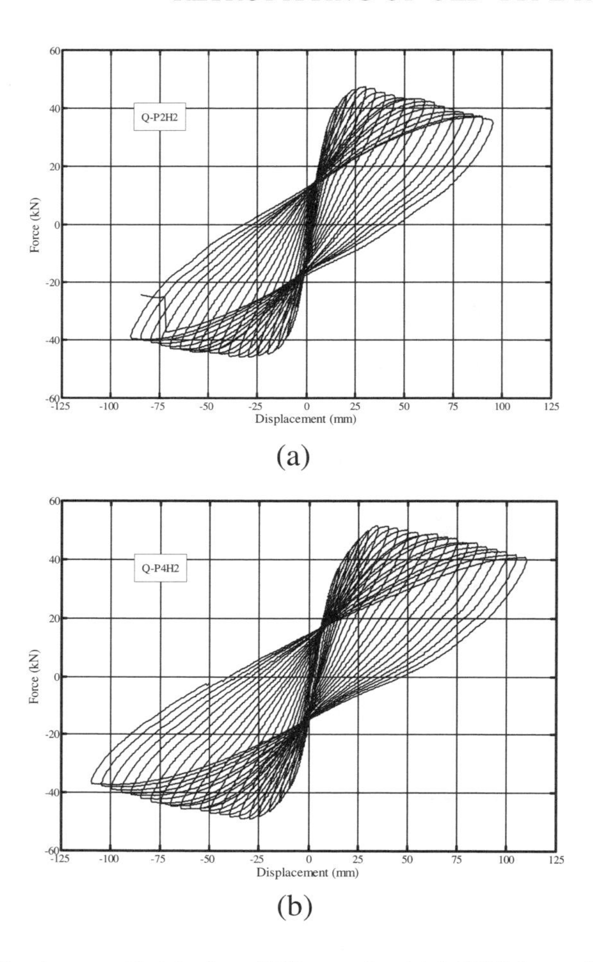

Figure 4. Specimens with 0.6m long FRP wrapping (a) 2-CFRP layers (Q_P2H2), (b) 4-CFRP layers (Q_P4H2)

(a) (b)

Figure 5. Failure of (a) unretrofitted specimen, (b) retrofitted specimen (2 CFRP layers)

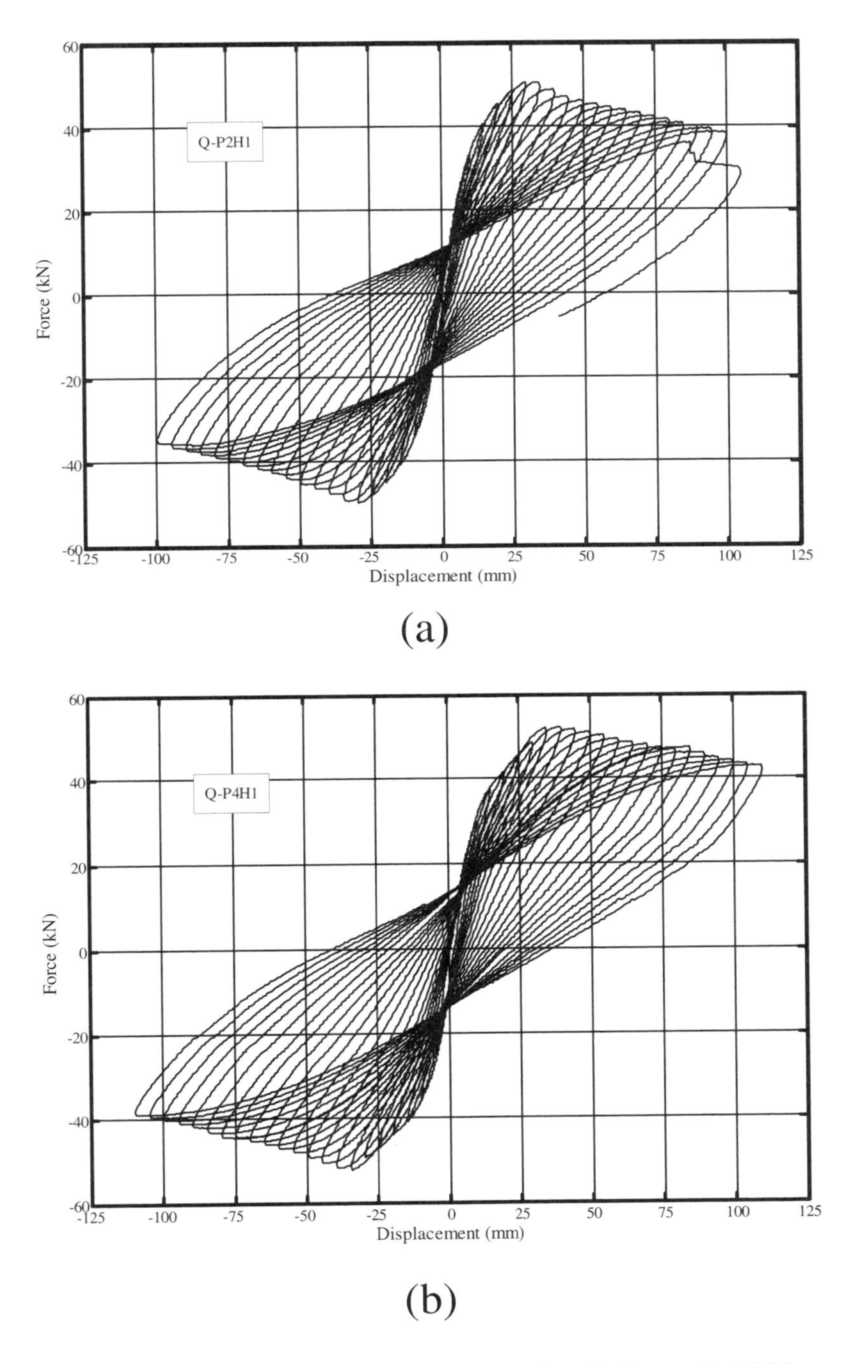

Figure 6. Specimens with 0.30m long FRP wrapping: (a) 2 layers (Q_P2H1), (b) 4 layers (Q_P4H1)

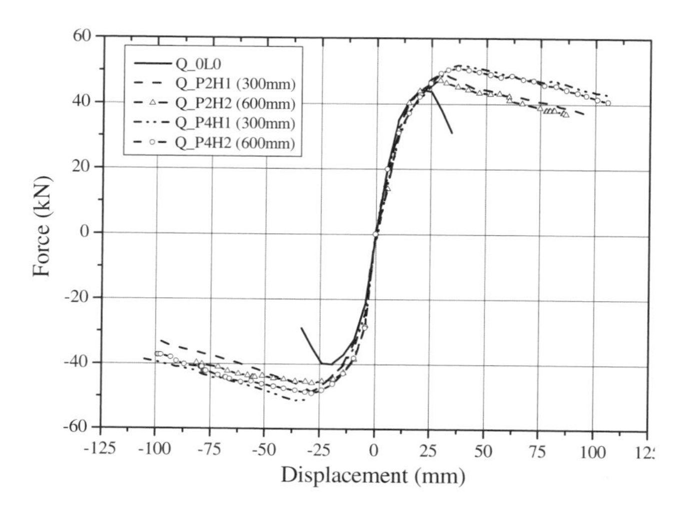

Figure 7. Comparison of envelope curves for type Q specimens

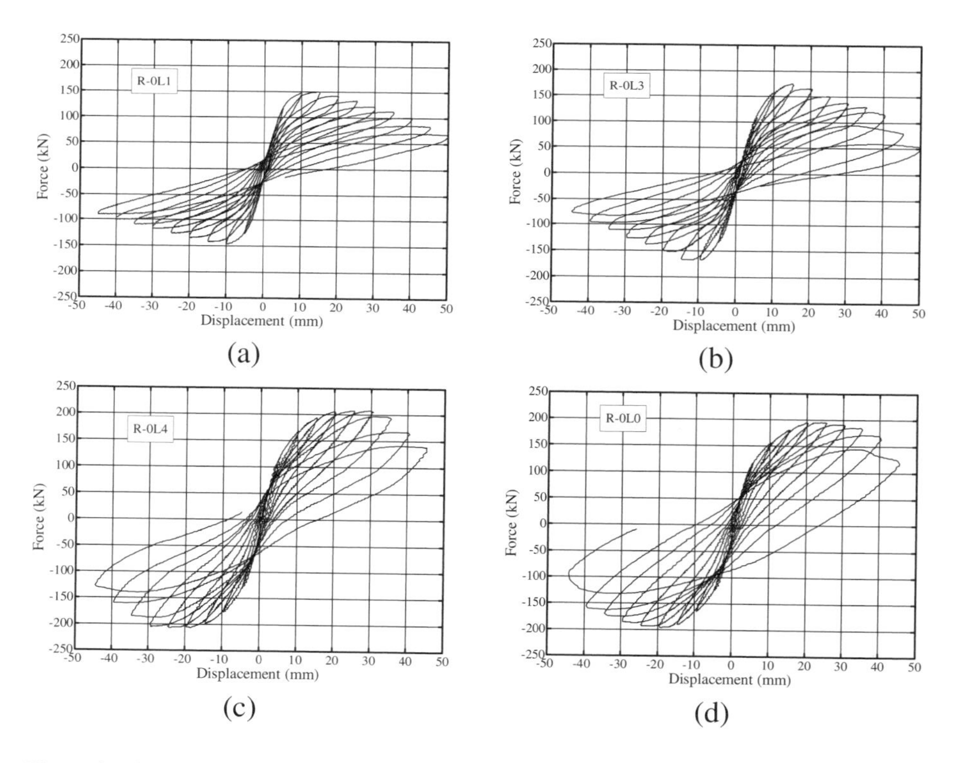

Figure 8. Effect of lap-splice length (un-retrofitted specimens): (a) R_0L1 (15-bar diameters), (b) R_0L3 (30- bar diameters), (c) R_0L4 (45- bar diameters), (d) no lap-splicing (R_0L0)

Figure 9. Failure of specimens with lap-splices: (a) R_0L1, (b) R_0L3, (c) R_0L4, (d) concrete crushing in front of rebar end face

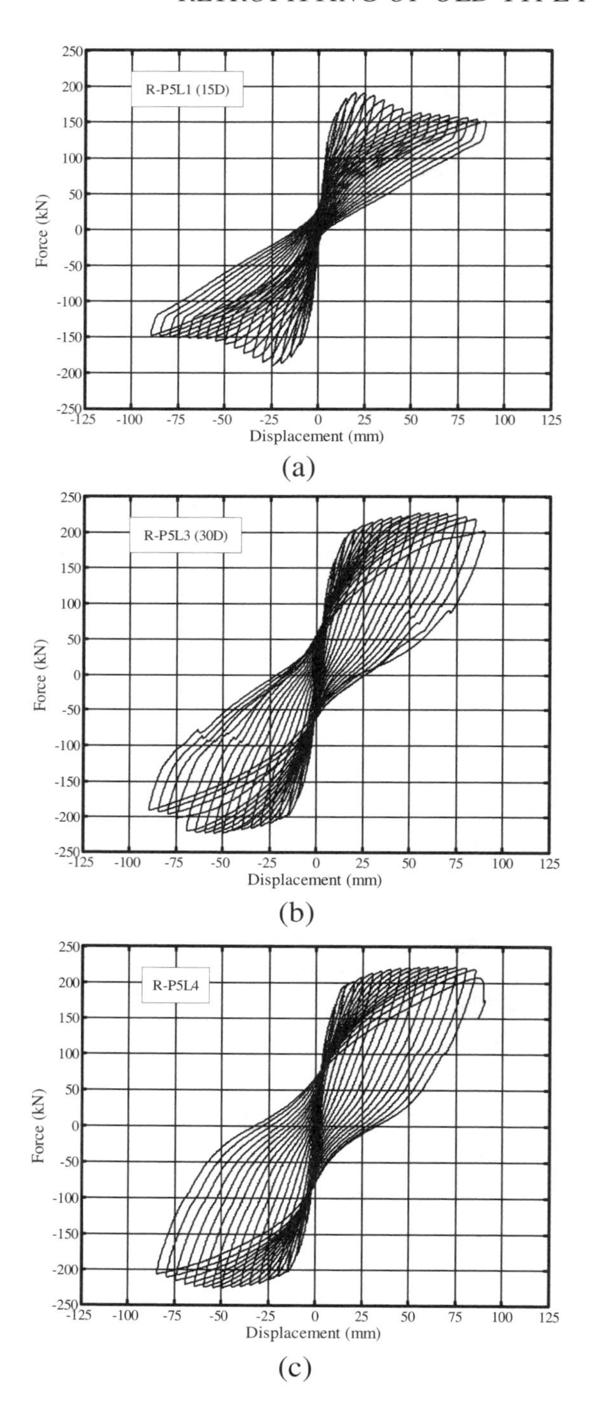

Figure 10. Columns retrofitted with 5 CFRP layers: (a) R_P5L1, (c) R_P5L3, (c) R_P5L4

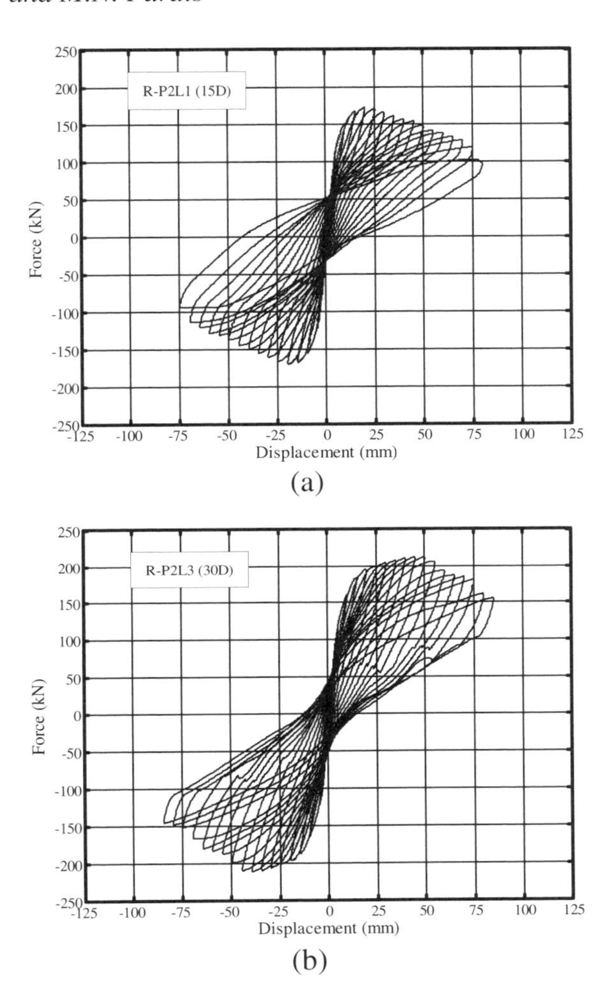

Figure 11. Specimens retrofitted with 2 CFRP layers: (a) R_P2L1, (b) R_P2L3

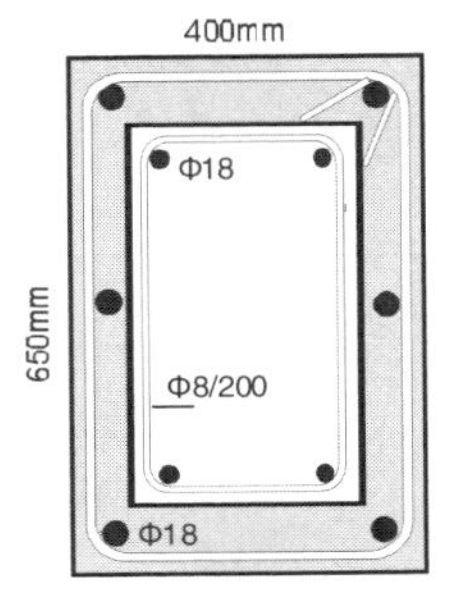

Figure 12. Cross-section of columns retrofitted with RC jackets

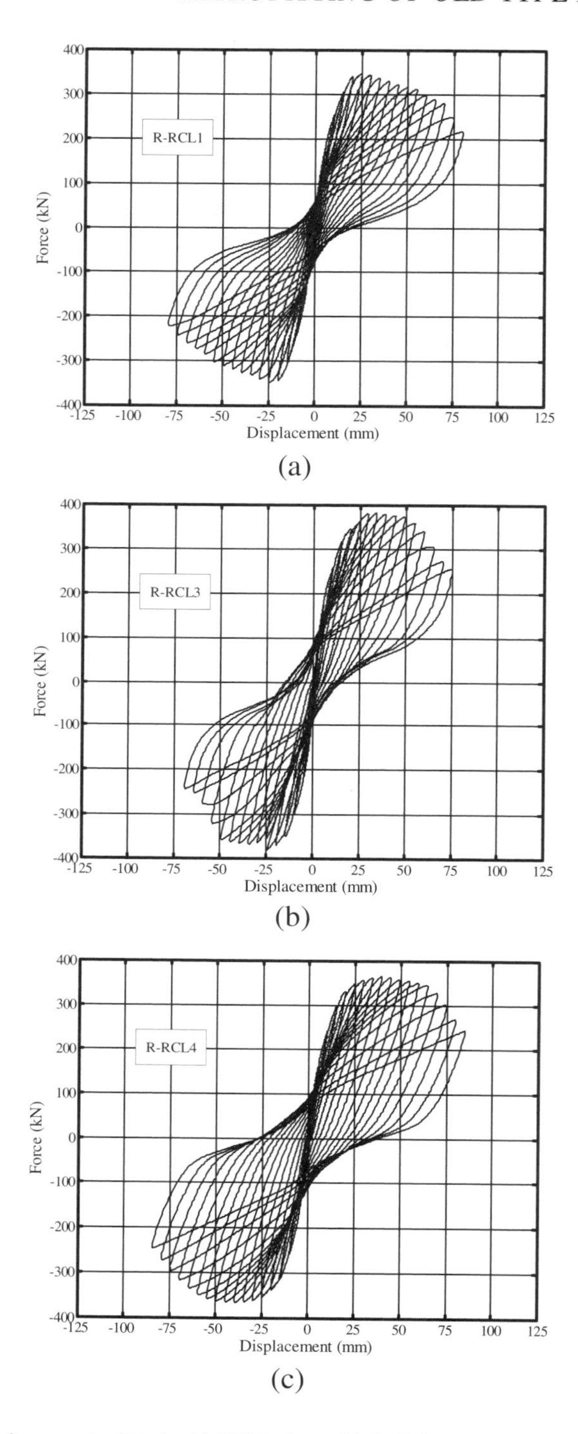

Figure 13. Specimens retrofitted with RC jackets: (a) R_RCL1, (b) R_RCL3, (c) R_RCL4

Figure 14. Failure of columns: (a) R_RCL1, and (b) R_RCL4

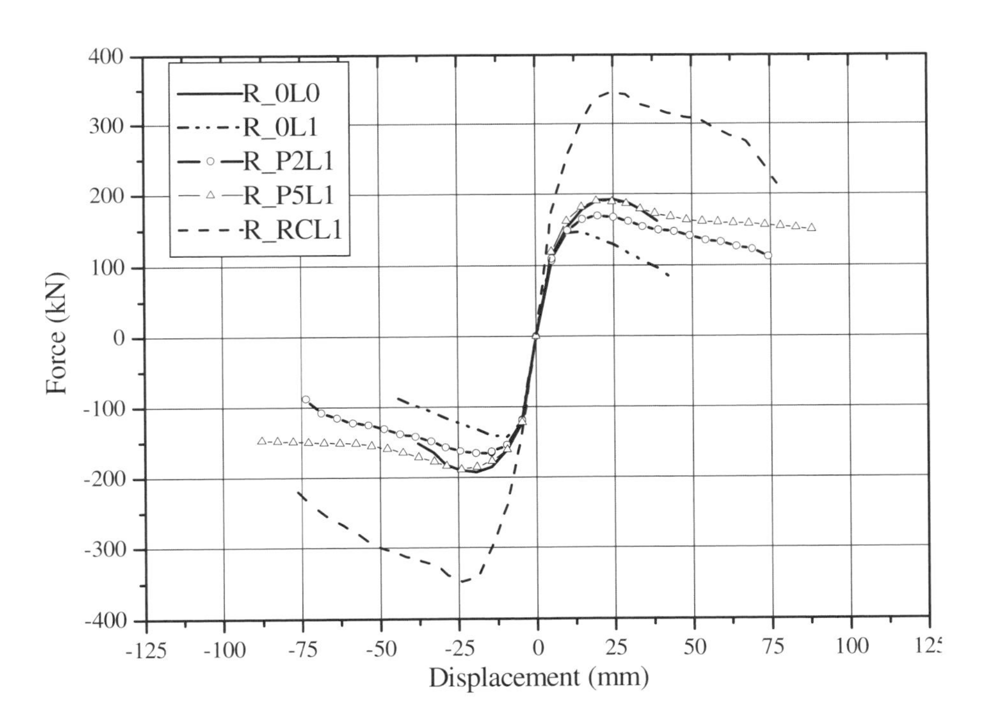

Figure 15. Comparison of envelope curves of type R columns with short lap splices (15D)

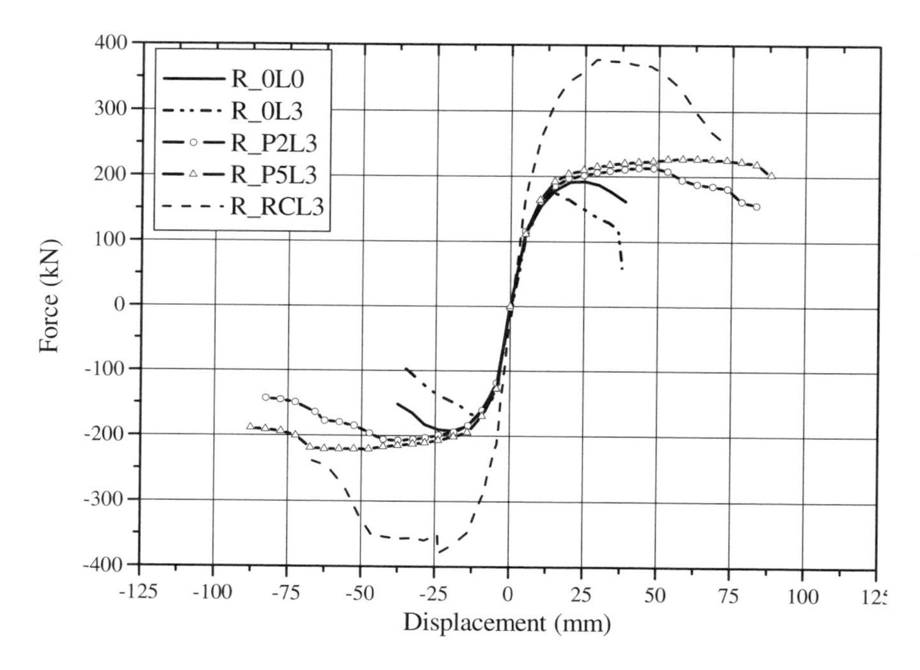

Figure 16. Comparison of envelope curves of type R columns with medium lap splices (30D)

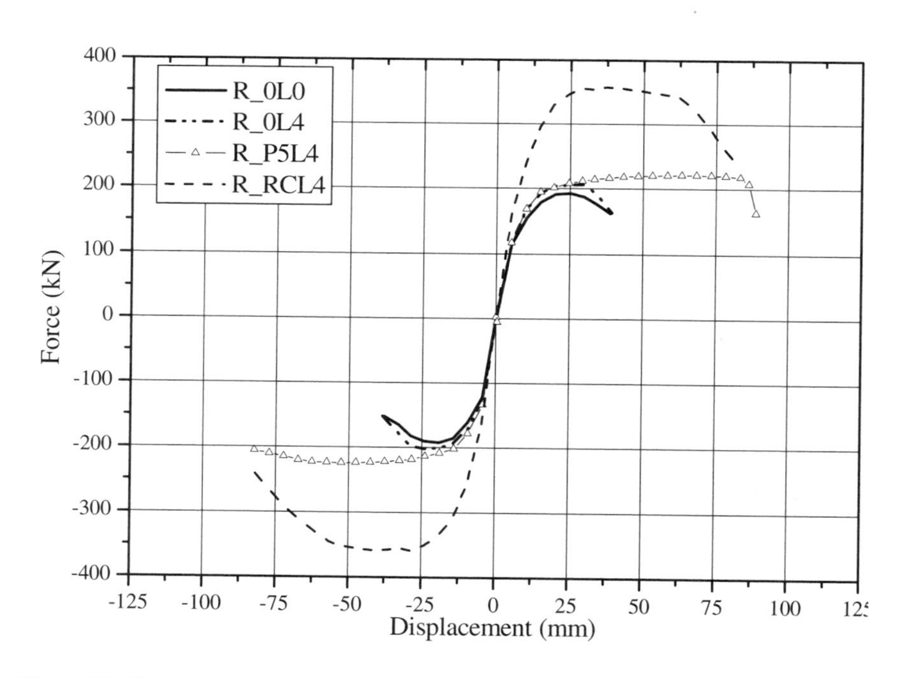

Figure 17. Comparison of envelope curves of type R columns with long lap splices (45D)

EARTHQUAKE ENGINEERING, SEISMIC VULNERABILITY ASSESSMENT AND SEISMIC REHABILITATION IN COLOMBIA

Luis E. Garcia
Universidad de los Andes, Bogota, Colombia

Abstract: The tectonics and seismicity of northwestern South America are presented. The construction types prevalent in Colombia are discussed. A brief description of the Colombian earthquake resistant regulations is given. The behavior of buildings during earthquakes previous to the Code enactment is presented. The accelerographic records obtained during the January 25 1999 earthquake are described and the building damage caused by the earthquake is discussed. Code compliance and enforcement are discussed. Issues associated with seismic vulnerability assessment and building retrofit in Colombia are presented.

Keywords: Colombian Seismic Code; Earthquake Engineering in Colombia, Field observations, Accelerographic Records, Building Damage, Construction Types, Response Spectra, Seismic Rehabilitation, Seismic Vulnerability Assessment

1. GEOGRAPHIC SETTING

Colombia, with an area of 1,138,300 km^2, is located in the Northwestern corner of South America. The country has coasts both in the Atlantic (Caribbean) and Pacific Oceans, to the north and west respectively, with bordering Panama between them. To the east Colombia borders with

S.T. Wasti and G. Ozcebe (eds.), Seismic Assessment and Rehabilitation of Existing Buildings, 269–304.

Venezuela and Brazil, and to the south with Ecuador and Peru (see Figure 2). As the northern Andes enter into the country from Ecuador, they divide into three cordilleras — eastern, central, and western — with the Magdalena River running between the first two with a south-north direction for 1550 km from a place just north of the border with Ecuador to the Caribbean. The Cauca River, a tributary of the Magdalena, runs between the western and the central cordilleras. Geologically the central cordillera is the oldest, and the eastern one the youngest. The active volcanoes in Colombia are located in the central cordillera. More than 85 per cent of the population of Colombia lives in the Andean region. The population of the country has increased more than three times since the 1950's (see Figure 3). Presently, sixty-eight per cent of the population lives in urban centers. The combination of these factors makes most Colombian cities vulnerable to natural disasters, and especially to seismic events, where the built environment plays such a large role with respect to the number of victims.

2. GENERAL TECTONICS OF NORTHWESTERN SOUTH AMERICA

The tectonics of the northwestern corner of South America is complex, to say the least. The fact that the Nazca, South American and Caribbean Plates converge in Colombian territory makes the tectonics of the region specially challenging (see Figure 4). The border between the Caribbean and South American plates is undefined. The Nazca plate forms a subduction zone under the South American plate in the Pacific Ocean coast. The direction and the convergence rate of the plates are shown in Figure 5 [Kellog and Vega, 1995].

The structural geology of the country has been studied with different degrees of detail. In general, a good mapping of large fault systems has been done for mining and petroleum exploration purposes. Special exploration has been done on a routine basis for the large hydroelectric projects with participation of leading world consulting firms. All this information was available for the identification of the main faulting systems [Paris, 1993]. They are shown in Figure 6. In general, the faulting in Colombia has a predominant N-S direction in coincidence with the three main cordilleras.

The main seismotectonic accident is the Subduction zone in the Pacific Ocean, being caused by the bending of the Nazca Plate as it subducts under the South American Plate. In the Colombian Pacific Coast, there is evidence of its existence from a point south of the Equatorial Line to 8° north. The ability to produce very large magnitude earthquakes of this subduction zone is known and the December 12, 1979 Ms = 7.9 earthquake certainly was produced by it. A Benioff zone develops with different dip angles that can be obtained from E-W sections of plots of focus of earthquakes. Its activity varies but earthquakes up to 120-130 km of depth can be assigned to it. Besides the subduction a large number of faults have been identified in the Andean part of the country.

3. SEISMICITY OF COLOMBIA

The first event of which a written record exists occurred in 1566, causing intensive damage in Cali and Popayán in the South West part of the country [Ramírez, 1975]. The instrumental seismicity of Colombia begins with the installation of the first seismographic station in 1922. The Geophysics Institute of the Javeriana University in Bogotá operated seven permanent seismographic stations scattered through the country, from 1957 to 1993. In 1993, the Colombian Seismological Network, administered by Ingeominas, a Colombian government agency, started operation.

Currently it has more than 20 permanent stations, linked via satellite to a main processing center located in Bogotá. Figure 7 shows the distribution of earthquakes, including historical and instrumental events, and covers data from 1566 to May 2002. From this figure, it is evident that the Nazca Plate subduction and the existing faults in the Andean region of produce most of the seismic activity in the country.

4. SEISMIC ZONING AND PEAK GROUND ACCELERATION

Current seismic design regulations, enacted in January 1998, contain the maps shown in Figures 9 and 10 [AIS, Ingeominas, and Uniandes, 1996]. The map in Figure 9 divides the country into High, Moderate and Low seismic hazard zones. The map in Figure 10 shows the effective peak ground acceleration in rock for a mean return period of 475 years.

5. TYPE OF CONSTRUCTION IN COLOMBIA

The construction industry in Colombia is one of the main contributors to the gross national product and is the source of employment of a significant percentage of the labor force, especially for moderately and low trained workers. From less than 4 million inhabitants at the turn of last century the population of the country has grown close to 40 millions today. Although the annual rate of growth declined in the last decade to 2% (see Figure 3b), the demand for housing is high and it explains why residential construction makes 75% of the production of the construction industry.

The main structural construction materials in Colombia are reinforced concrete and masonry. Although structural steel is used in large span roofs and in bridges, its use in the framing of buildings is limited. For reinforced concrete buildings, the moment resisting frame has been the traditional system and recently structural walls have appeared as an alternative in low-rise and high-rise buildings. Masonry is used for single-family dwellings, and low-rise and high-rise apartment buildings. Adobe construction is used in some rural areas but its use in Colombia is decreasing everyday.

Reinforced concrete was introduced in Colombia in the 1920's. The structures built in that time just followed what was common practice in North America and Europe. By the 1940's, a local evolution in the construction of slabs had occurred and left in place wood fillers made with "guadua" — a local type of bamboo — became popular and are still used in some regions of the country. During the 1950's, a development of slab-column frames occurred, and a special type of waffle slabs named "reticular celulado" became popular, its employment extending to other Latin-American countries. The use of reinforced concrete structural walls was a result of the requirements of the 1984 building code [MOPT, 1984], and the timid usage encountered in the late 1980's has increased steadily since then. Notwithstanding, moment-resistant building frames still make the majority of the reinforced concrete structural system of buildings.

Both non-engineered and engineered masonry is used in Colombia, with the former mainly in single dwelling low-cost construction, where the owner is the builder. Single-family dwellings are built of masonry as a rule. Unreinforced masonry was forbidden by the building code in 1984 [MOPT, 1984], but a large inventory of unreinforced masonry buildings still exists. The rest of masonry construction is engineered, with the following types being popular: (a) *Reinforced masonry* - This term locally means engineered masonry built using masonry units with vertical cells. It is used from one story homes to medium-rise buildings. It is mainly used in low-income housing projects that need fewer parking spaces. It was introduced to the country in the decade of 1970's; (b) *Confined masonry* - Consists of masonry walls surrounded by a light reinforced concrete frame. The use of confined masonry in Colombia dates from the 1930's and it is probably the most widely used structural masonry system. It is used from single story dwellings to apartment buildings up to five stories. It is engineered in most cases although the code includes a complete chapter on empirical design of this type of masonry.

With respect to non-structural elements, building partitions in apartment, and even office, buildings traditionally have been built using clay tile. The same is true for façades in apartment buildings where plastered clay tile and un-plastered solid clay brick are used. The seismic implications of this practice are discussed ahead.

6. BEHAVIOR OF BUILDINGS IN EARTHQUAKES PREVIOUS TO THE 1984 CODE

Three earthquakes occurring in the late 1970's and early 1980's were crucial in defining the scope of the first Colombian building Code enacted in 1984 [García, 1984, and MOPT, 1984]. These earthquakes were: (a) November 23 of 1979 ($\mathbf{M}_s = 6.4$) affecting the coffee-growing region; (b) December 12, 1979 ($\mathbf{M}_s = 7.8$) in the subduction zone in the Pacific Ocean coast; and (c) the March 31 of 1983 ($\mathbf{m}_b = 5.5$) affecting the city of Popayán.

The main features of the observed behavior were:

- Collapses and significant damage concentrated in low-rise buildings with less than five stories. A local prejudice that low-rise buildings did not require earthquake resistant design was the main culprit of this.
- Excessive flexibility under lateral loads of most buildings. Absence of seismic design or non-compliance of the drift requirements was the common reason for this behavior.
- A disproportionate large amount of column failures, mainly because of lack of appropriate lateral reinforcement combined with small element section area.
- Significant damage and even collapse of reinforced structural masonry buildings. Lack of appropriate building and supervision practice was the common factor. Absence of horizontal shear reinforcement was the culprit in some cases.
- In the Popayán earthquake, a large amount of damage to unreinforced masonry. The depth and large epicentral distance of the other two earthquakes probably prevented damage to unreinforced masonry.
- Non-structural damage in masonry partitions and façades. The importance of this fact was probably obscured by the more spectacular structural damage.

During the years between the enactment of the Code in 1984 and the occurrence of the January 25, 1999 earthquake, among several events four earthquakes with large to moderate magnitudes — October 18 of 1992 ($\mathbf{M}_s = 7.2$), June 6 of 1994 ($\mathbf{M}_s = 6.4$), January 19 of 1995 ($\mathbf{M}_s = 6.5$), and February 8 of 1995 ($\mathbf{M}_s = 6.4$) — were felt, with appreciable intensity, respectively in the Colombian cities of Medellín, Cali, Bogotá, and Pereira. Although the hypocentral distance to all of them, in each case, was more than 100 km, and no structural damage of importance was reported on buildings designed under the 1984 Code; they caused appreciable amounts of nonstructural element damage, especially in Medellín during the 1992 earthquake. These events increased the concern that the then current story drift requirements were insufficient and stricter limits were needed as a response to the continuing use of brittle unreinforced masonry façades and partitions.

7. COLOMBIAN EARTHQUAKE RESISTANT REGULATIONS

The need in Colombia of a modern mandatory building code was made evident by the March 31, 1983, Popayán Earthquake. Around the same time that the plan for having a Seismic Code enacted was being drafted the ATC-3 document [ATC, 1978] was published in the US. This document was deemed so important that the strategy was adjusted accordingly by giving it large local exposure — a translation into Spanish was made for this purpose — and by trying to establish contact with the drafters of ATC-3 so that key issues that required local adjustment could be trimmed. The result was spectacular. Help was enlisted, limited funding appeared from sources previously reluctant to contribute, and just two years after the ATC-3 document was published in the US the draft of what was to become the Colombian Seismic Code was already being used in a voluntary basis by numerous engineers.

The following quotation from "Confronting Natural Disasters - An International Decade for Natural Hazard Reduction" by the [Advisory Committee on the International Decade for Natural Hazard Reduction, 1987] summarizes what was accomplished then:

"... As stated earlier, technologies developed for application in one country are often applied in another without adaptation. The Applied Technology Council (ATC) developed recommended building practices for earthquake-resistant design for use in the United States. One of the first implementers — even before the United States — was Colombia. Building practices and materials in Colombia are somewhat different and the tectonic nature of Colombian earthquakes is different from that of California earthquakes, for which the ATC recommendations were principally formulated. Fortunately, contacts between Colombian and U. S. engineers involved in the ATC effort are strong. Colombian engineers were able to adapt the guidelines — with advice from U. S. developers — to their circumstances. Many other who have applied these findings have not had this advantage."

The Colombian seismic code was enacted in June of 1984 [MOPT, 1984]. The Code tried to take care of the problems brought out by earthquakes described previously. These earthquakes emphasized the deficiencies in the Colombian earthquake resistant building practice described. The absence of a code, and the lack of awareness of the problems associated with the excessive flexibility of frames responding to lateral loads, was the culprit of much of the bad behavior encountered. Because of these reasons, the 1984 Colombian code made strict story drift control one of its main objectives. The results were outstanding in the structural perspective: a survey conducted ten years ago among practicing structural engineers, assigned story drift as the controlling design parameter, as opposed to base shear when determining dimensions for a building structure [Meigs, Eberhard, and García, 1993]. Unfortunately, a corresponding evolution in nonstructural element building practice did not occur.

Because the 1984 Code was the first such a document to be made mandatory in Colombia, the Committee in charge of its drafting decided to include what was deemed critical at the moment — mainly life threatening aspects —, and postpone several important, but not critical, issues for future Code updates. Among the themes that were postponed were:

- A change in the structural systems permitted — the use of the Code as a vehicle for sponsoring a greater use of structural walls as opposed to moment-resistant frames was studied. The restrictions imposed on the allowable story drift were a result of these discussions.
- Limitations of building irregularities — Timid recommendations were given, but a full set of requirements for limiting irregularities was postponed.
- Non-structural elements — The draft of the 1984 Code had a chapter dealing with non-structural elements. It was suppressed, unfortunately, in the final version. The main reasons behind this were the lack of realistic alternatives for dealing with these elements, and the overwhelming pressure from the building and material industries on the high costs involved.
- Other structural materials — The 1984 Code had requirements for reinforced concrete, reinforced masonry and structural steel. The experience in the country with other structural materials such as wood or aluminum was minor; neither was there evidence of their wide spread use.

The Colombian Association for Earthquake Engineering initiated work on the update of the code in the early 1990's. The 1984 Code had been enacted by Decree of the President of Colombia, under special powers given by Congress, for a one-time issue only. In order to obtain new, and permanent, authorizations, special legislature was proposed in Congress which produced Law 400 of 1997 in August of that year creating among other things a Permanent Code Committee empowered to recommend to the President of Colombia updates of the Code when deemed necessary. The update of the Code was enacted in January 1998, using these authorizations from Congress [AIS, 1998] and its official denomination is "Reglamento NSR-98".

The seismic events that occurred from 1984 to 1997 increased the concern that the story drift requirements were insufficient and stricter requirements were needed, as a response to the continuing use of brittle unreinforced masonry façades and partitions. The update of the Code contains story drift requirements that are more conservative than those required by the 1984 version, plus several other schemes, to limit nonstructural element damage. The possible additional construction cost brought out by these new requirements motivated a formal evaluation of the type of structural systems used [García and Bonacci, 1994], of the drift design procedures [Sozen and García, 1992], and of their economic impact [Serna, 1994], [García, Pérez, and Bonacci, 1996], and [García, 1996]. The ultimate goal was to bring into a practical designer's perspective the issues involved; the rationale beyond selecting different member section dimensions; and, last but not least, the economic implications on the structural alternatives for the client to chose from, given a performance limit established by the story drift limit. The contents of the update of the Code are described in Table 2.

The three main features of the Code update, among others, are worth mentioning:

a) With respect to non-structural elements, a new Chapter was included for the design of these elements; special responsibilities were given to the architect, the builder, the supervisor, and the owner with respect to non-structural elements;
b) More strict drift limits are required, in order to incentive the use of more structural walls and restrict non-structural damage; and
c) All the critical facilities of the country are required to make seismic vulnerability assessments of their buildings and corresponding upgrades in no more than six years from the enactment of the new Code.

8. ACCELEROGRAPHIC RECORDS FROM THE JANUARY 25, 1999, EARTHQUAKE

Records from approximately 40 strong motion accelerographs were obtained for the main event of the January 25, 1999 earthquake that affected the coffee growing region of Colombia, and a corresponding number for the main aftershock that occurred in the evening the same day. The stronger records were obtained at the Universidad del Quindío, in Armenia on soft soil, just 14 km north from the epicenter (see Figures 11 to 13). Peak ground accelerations recorded in this station were 0.53**g** EW, 0.46**g** vertical, and 0.59**g** NS. The soil profile at this site consists of approximately 30 m of volcanic ash fill.

The two horizontal components of the Universidad del Quindío accelerographic record are plotted together in Figure 14, with the corresponding vertical acceleration is described using a color code. It is possible to observe that in some instances the two horizontal peaks occur simultaneously, and are accompanied by a very strong vertical acceleration. The acceleration response spectrum is shown in Figure 15. A high content of short period components can be observed. The EW component peaks at a vibration period of 0.25 seconds and the NS component at 0.56 seconds. The displacement response spectrum, shown in Figure 16, flattens at a period of approximately 1 second. The energy spectrum, Figure 17, shows characteristics periods of the order of 0.5 - 0.6 seconds for the horizontal components, and of 0.25 seconds for the vertical component. The Fourier spectrum, Figure 18, just confirms these observations.

In Filandia, a town located north of Armenia, 33 km from the epicenter, the peak ground accelerations recorded were 0.57**g** EW, 0.19**g** vertical, and 0.49**g** NS. The soil profile consists of approximately 20 m of volcanic ash fill. Figure 19 shows the acceleration response spectrum for this site. The Bocatoma site, located 10 km east of Pereira at the city waterworks sluice distant 48 km from the epicenter, is sited on rock, and the record obtained there corresponds to the rock register obtained closer to the epicenter. The peak ground accelerations recorded here were 0.084**g** EW, 0.028**g** vertical, and 0.050**g** NS. Figure 20 shows the acceleration response spectrum for this site.

Several records were obtained within the city of Pereira, located approximately 48 km from the epicenter. The Mazpereira station is located in a site with a 15 m deep man-made fill, in an area where the intensity of damage has been high in previous earthquakes and during this one. Peak ground accelerations recorded in this station were 0.25**g** EW, 0.10**g** vertical, and 0.30**g** NS. Figure 21 shows the acceleration response spectrum for this site. The Castañares station is located in a similar site with a 6 m deep man-made fill. Peak ground accelerations recorded in this station were 0.21**g** EW, 0.10**g** vertical, and 0.14**g** NS. Figure 22 shows the acceleration response spectrum for this site.

In Dosquebradas, a town that is part of metropolitan Pereira, 53 km from the epicenter, at the La Rosa site the peak ground accelerations recorded were: 0.18**g** EW, 0.07**g** vertical, and 0.19**g** NS. The soil profile consists of approximately 70 m of alluvial deposits of clay, silt and silty clay alternated with sand and gravel lenses. Figure 23 shows the acceleration response spectrum for this site. At the hospital of Santa Rosa de Cabal, a town located further north approximately 53 km from the epicenter the peak ground accelerations recorded were: 0.18**g** EW, 0.06**g** vertical, and 0.26**g** NS. The site has deep deposits of volcanic ash alternated with alluvial material.

9. OBSERVED BUILDINGS BEHAVIOR DURING THE JANUARY 25, 1999, EARTHQUAKE

In total 27 municipalities were affected by the January 25, 1999 earthquake. The official number of fatalities was 1185, plus 98 disappeared, 8523 people injured (83% of them in the department of Quindío and 17% in the department of Risaralda). The official figure for number of buildings affected is 79,446. Of these buildings 43,474 had minor to moderate damage, and 35,972 were severely damaged. The last figure is divided, in turn, in 17,551 buildings with partial or total collapse, and 18,421 condemned. Around 74% of the educational buildings of the region were damaged, affecting approximately 100,000 students. The official number of people left homeless is 160,397, of which 67,539 persons were living in temporary shelters after the earthquake. Unemployment in the coffee-growing region increased by 22%. Table 3 presents general data for Armenia and Pereira, the two largest cities affected.

Figures 25, 26, and 27 show the distribution by type of construction, building use, and age of construction, for a 12,454 buildings sample in the center part of the city of Pereira. It is important to note that 56.4% of the buildings correspond to unreinforced masonry construction, and that 91.7% were dwellings. Only 11% of the buildings were built after the 1984 code was enacted.

The results of the assessment of the buildings after the earthquake was performed using red tags for buildings either collapsed of with severe structural damage, orange tags for buildings with reparable structural damage or severe non-structural damage that required building evacuation while repair were performed, yellow tags for buildings with easily reparable damage that could be performed with the building occupied, and green tags for minor damage. Figure 28 shows the tagging results for the same building sample in Pereira.

The Earthquake Engineering Research Institute (EERI) Reconnaissance Team that studied the Quindío, Colombia Earthquake of January 25, 1999, stated:

"Many pre-1984 buildings collapsed or partially collapsed, since they were built before the adoption of more stringent seismic design standards in the 1984 Colombian Design Code."

The observed damage is consistent with age of construction, structural systems employed, and type of strong ground motion. The type of strong motion as depicted in Figure 14, with very strong accelerations in all three components, preclude any chance for unreinforced masonry to survive the earthquake. The short period waves containing the majority of the energy, accompanied by large vertical accelerations, are the worst combination for unreinforced masonry and low-height poorly constructed buildings. With 73% of the red tagged buildings being one and two-story buildings and 89.5% percent of the red and orange-tagged structures consisting of residential buildings: the unreinforced masonry is the culprit of most of the victims and observed damage. The sample of buildings four or more stories in height confirms that better quality construction was employed: 40% of the sample was green tagged, less than 5% of the sample was red tagged, and red tagged buildings higher than 6 stories comprise only 1% of the sample. Having 55% of the buildings four stories and taller tagged orange and yellow just points toward the problems of employing brittle unreinforced masonry partitions and façades.

10. CODE COMPLIANCE AND ENFORCEMENT

The building behavior during the 1999 Coffee Growing Region Earthquake was not very different from that observed during the earthquakes that inspired the enactment of the first code in 1984, as described previously in this document. The age on the inventory of buildings in the affected zone points toward the same patterns of damage observed previously for buildings constructed before 1984. The favorable impact of the first Colombian seismic code was made evident through the entire region affected by the earthquake. The amount of non-structural element damage points toward the need of changing the Colombian practice of building partitions and façades employing unreinforced masonry. Most of the damage observed in post-code buildings could have been avoided employing enhanced supervision and inspections practices. The Code update enacted in 1998, which tends to reduce most of the observed deficiencies in non-structural element usage and supervision and inspection, did not have a chance to be effective because few buildings had been built employing its requirements.

The sad confirmation that unreinforced masonry should not be used in seismic zones should encourage Colombia to implement a program for reducing the large inventory of existing unreinforced masonry buildings.

Earthquake strong motion amplification in sites with profiles of soils of volcanic origin and containing man-made fills correlate with the patterns of damage. City planning for the reconstruction is being influenced by their presence.

Enforcement of the Code has increased since 1998 when the issue of building permits was transferred from the municipalities to "curadores", architects and engineers that operate in a form similar to notaries. Building permit procedures were changed imposing large fines to government officials that allow construction of official building without a corresponding building permit; thus finishing a practice that was very common. Supervision of building construction has also improved. Critical facilities require a more conservative design through an Importance Coefficient (I = 1.3). Corresponding values of I are 1.2 for fire and police stations, 1.1 for school buildings, and 1.0 for other buildings.

11. SEISMIC VULNERABILITY ASSESSMENT AND REHABILITATION

The previous description of observed damage during the 1999 coffee region earthquake gives a clear picture of the general seismic vulnerability of the Colombian buildings. Many variables come into play, but several constants are applicable to the rest of the seismic regions of the country. The 1984 Code for several reasons was made applicable, when enacted, only for new buildings. The Colombian Association for Earthquake Engineering published in 1986 a document — AIS 150-86 — intended for applying the then new Code to existing buildings [AIS, 1986]. This document was made part of the 1998 update and still regulates seismic vulnerability assessment and rehabilitation issues.

The AIS-150 document established a procedure to evaluate the vulnerability based on two indexes: (a) an over-strength ratio, and (b) a flexibility ratio. The former compares the strength of the structure with respect to what is required for a new structure taking into account the type of detailing in the original structure, and the later compares the story drift expected for the design ground motions with the story drift limit of the code. Based on the results of the two indexes the structure is deemed vulnerable or not. For the rehabilitation of the structure, if vulnerable, the document gives guidance, but leaves most of the decisions on judgment of the rehabilitation designer. Under the update of the code it allows the use of corresponding ATC and FEMA documents, without making them mandatory.

The decision made in 1986 is still supported by most of the engineering community with respect to experience and judgment of the engineer as opposed to a very prescriptive procedure. The current requirements for vulnerability assessment give the responsibility to the designer of assigning an appropriate response modification factor, **R**, compatible with the type of detailing in the existing structure. The evaluation of the vulnerability performed using the over-strength and flexibility ratios is affected by is by special strength reduction factors, ϕ, depending on quality of the building design and construction, and scope of the maintenance given to the structure since its construction.

The main concern is viewed as a stiffness issue more than a strength problem because of the emphasis of the 1984 and 1998 codes with respect to story drift. This has led to a situation in which the preferred solution is associated with a stiffening of the structure that in most cases leads indirectly to strengthening using the same elements employed for increasing the rigidity of the lateral structure.

Law 400 of 1997 made mandatory the assessment and corresponding rehabilitation, if needed, of all critical facilities located in moderate and high seismic risk zones throughout the country as mentioned earlier. One of the initial respondents to this order given by law was the Public Health Ministry, and the main hospitals in the country were studied and many of them have been rehabilitated. A similar situation occurred with the Civil Aeronautics Administration, and all the main airports of the country have been evaluated and many of them are being rehabilitated. The three years given by Law 400/97 for the vulnerability assessment and the three extra ones for performing the rehabilitation were extended by congress to six and twelve years in 2001. The important fact is that the work is being performed.

One important issue that was taken seriously by Government was the way to approach the historic and architectural preservation buildings. The 1998 code update permits the use of lesser earthquake design forces in seismic rehabilitation of historical and architectural preservation buildings, if access to the public is somewhat restricted. Most of these buildings are not covered by the critical facilities scope. Notwithstanding, an effort was made to evaluate them. This has been a challenging task because the great majority corresponds to unreinforced masonry, adobe structures, or structures whose structural systems are difficult to be approached in modern earthquake resistant terms. This is the case of the National Capitol, and many important cathedrals and churches.

Table 4 lists several of the main structures that have been studied, or studied and rehabilitated throughout the country.

The experience has been similar to that in other countries that have similar programs of mandatory vulnerability assessment and rehabilitation. One of the main concerns has been the cost of rehabilitation of non-seismic features made simultaneously. The rational of not just making the seismic part but use it as a excuse to update the building in internal services and other features has overextended the assigned budgets in many cases. As experience has been gained, different government agencies have developed strategies to overcome the ensuing problems. To give an idea of the approach developed by some of these agencies, the Ministry of Public Health refused to provide funds for any kind of architectural changes in hospitals if the seismic rehabilitation issue is not solved at the same time.

Figure 29 corresponds to a building in the Universidad de los Andes campus. It was a four stories unreinforced masonry building whose rehabilitation was finished in mid 2002.

12. SCHOOL BUILDINGS SEISMIC SAFETY ISSUES IN COLOMBIA

Schools are predominantly built using reinforced concrete. The earthquakes listed in Table 1 produced intensive damage to school buildings. Reliable statistics exist for the 1999 Quindío earthquake. In total 74% of the educational buildings of the region were damaged, affecting approximately 100,000 students. Since the early 1990's several educational institutions, mainly universities and some private high schools, initiated programs of seismic vulnerability assessment of their physical plant and numerous building have been retrofitted as a result of these programs. Primary and secondary schools belonging to the public school system initiated programs mainly after the Quindío earthquake. It is worth mentioning here that fortunately the earthquake occurred during the Christmas vacation thus saving many lives because the earthquake occurred around 1 PM when the schools would have been fully occupied. When the school year started in February a program was started under which students were sent to other cities where they lived with host families and attended the schools of the children of the host family. This permitted the rebuilding of the schools and had the effect of being an eye opener of potential problems associated with seismic vulnerability for the school authorities of all the cities of the country.

All large cities located in moderate and high seismic risk zones initiated vulnerability assessment program with varying degrees of success. The largest such program was initiated by Bogotá, where a total of 2,518 Buildings belonging to 645 City Schools with an area of 1,006,000 m^2 and attended by approximately 400 000 students were evaluated. The activities for this project included:

- Definition of the methodology,
- Training of the personnel of the Education Department,
- Gathering of the information (building by building),

- Supervision of the gathering of the information,
- Approximate vulnerability assessment using several methodologies, including Hassan and Sozen method,
- Use of the seismic microzonation of Bogotá design response spectra,
- Development of a tool to assign priorities, and
- Detailed retrofit studies of key cases.

In order to assign priorities, a database utility was developed. This utility permits the combination of all features studied, such as:

- Vulnerability vs. number of students,
- Vulnerability vs. building area,
- Vulnerability vs. rehabilitation cost,
- Non-structural elements problems vs. any variable,
- Non-deferrable problems vs. number of students, and other.

These tools are being used by the city for budgeting, assignation of order of fix-ups, division between maintenance and rehabilitation issues, and other chores. The budget additions for school retrofitting assigned at this point by the City are: US$ 5 million for 2003, US$ 4 million for 2004, and US$ 3 million for 2005. This will cover 60 schools whose retrofit is deemed to be urgent. So far six large schools have been rehabilitated since the program started. Typical costs of retrofitting obtained from this project are of the order of US$ 15/m^2 to US$ 50/m^2 for the structural retrofit part with reposition of architectural features after the intervention being as high as the structural work.

One aspect that is being considered by the school authorities is the decline of population growth rates that have taken place in Colombia in the last decades. Figure 3 show the population growth in the last one hundred years, and the corresponding change in the growth rate. These trends affect number of schools and classrooms needed in the future, and influence the type of retrofit or substitution of existing schools.

13. COLOMBIAN EXPERIENCE IN BRIDGE SEISMIC VULNERABILITY ASSESSMENT AND REHABILITATION

Recent recognition of the social and economic impacts that result from severe damage or collapse of bridges during earthquakes has led the Colombian Government at national, departmental and city level to evaluate the vulnerability of important bridges and in some cases to perform complete seismic bridge rehabilitation. The awareness of the need of a mandatory seismic code for buildings came before an equivalent document for bridges.

The first Colombian seismic building code requirements were enacted in 1984, and an equivalent document for bridges was introduced in 1995.

Although important bridges traditionally were designed using seismic forces, since the introduction of the seismic requirements for bridges [Ministerio de Transporte, 1995] the use of modern earthquake resistant design and construction procedures is mandatory. When existing bridges are evaluated using these new requirements a series of inadequacies are brought to light. The problems encountered do not differ from those that have been encountered in other seismic countries, and that are highlighted everywhere by the occurrence of strong earthquakes. In the Colombian case the additional problems of hidden existing damage caused by scour, debris impact, settlement, heavy traffic, and other, makes the vulnerability assessment specially challenging. Limitations in the appropriate amounts of funds that can be devoted to the modernization of the rural road network and of the inner city intersections makes the rehabilitation of existing bridges a mandatory alternative in most cases.

Bridge design in Colombia traditionally have been made employing the American Association of State Highway and Transportation Officials, AASHTO, Specifications, although the Ministry of Transportation, and formerly the Ministry of Public Works, had required variations with respect to these Specifications in the design of bridges contracted by them, specially in the live loads employed. In 1991 the then existing Ministry of Public Works made a contract with the Colombian Association for Earthquake Engineering, AIS, for the development of a draft of a bridge design specification, with special emphasis in the seismic requirements. Within AIS, Committee AIS 200 was put in charge of leading the production of the draft. Committee AIS 200 studied several alternatives of the type of document to be produced. The use of the AASHTO Specifications as a model was decided. The Committee studied the applicability of the different requirements for the specific case of Colombia, and adopted several variations. The document is based on the 15th Edition of the AASHTO Specifications [AASHTO, 1992], and incorporates the 1994 Interim [AASHTO, 1994a]. The main variations with respect to the AASHTO Specifications are: (a) the design live load is heavier and span length dependent, (b) the approach to the live load distribution to the superstructure elements is different although leads to comparable results, (c) the seismic requirements were based in the ATC-6 document [ATC, 1986] with variations making them compatible with developments made after its publication and with the seismic hazard in Colombia, (d) includes requirements for reinforced earth. The Ministry of Transportation adopted the draft developed by Committee AIS 200 as the Colombian Bridge Design Specifications in 1995.

Halfway through the production of the Colombian requirements the members of Committee AIS 200 became aware of the publication of the draft of what was to become the AASHTO LRFD Specifications [AASHTO, 1994b]. This new document was studied in depth and a decision was made to translate it into Spanish and to permit its use as an alternative design procedure. Currently the AASHTO LRFD document is being studied as the basis for the update of the Colombian requirements that will be published in the near future. This new document will incorporate the ATC-32 recommendations [ATC, 1996] for seismic design, plus several features from recent documents and research. Although the current Colombian Bridge Design Specification does not address the seismic vulnerability assessment and rehabilitation directly, the procedures employed currently in Colombia by the consulting engineers engaged in assessment and rehabilitation of bridges follow the documents mentioned.

REFERENCES

1. [AASHTO, 1992], *Standard Specifications For Highway Bridges*, 15th Edition, American Association of State Highway and Transportation Officials, Washington, DC, USA, 470 p.
2. [AASHTO, 1994a], *Interim Specifications - Bridges - 1994 To Standard Specifications For Highway Bridges 15th Edition*, American Association of State Highway and Transportation Officials, Washington, DC, USA, 1994.
3. [AASHTO, 1994b], *AASHTO LRFD Bridge Design Specifications*, American Association of State Highway and Transportation Officials, Washington, DC, USA, 1006 pp.
4. [Advisory Committee on the International Decade for Natural Hazard Reduction, 1987], *Confronting Natural Disasters - An International Decade for Natural Hazard Reduction*, 1987, Commission on Engineering and Technical Systems, National Research Council, U. S. National Academy of Sciences and U. S. National Academy on Engineering, National Academy Press, Washington DC,
5. [AIS, 1986], Adición, Modificación y Remodelación del Sistema Estructural de Edificaciones Existentes Antes de la Vigencia del Decreto 1400/84 - Norma AIS 150-86, Asociación Colombiana de Ingeniería Sísmica, AIS, Bogotá, Colombia.
6. [AIS, 1998], Normas colombianas de diseño y construcción sismo resistente - NSR-98 (Ley 400 de 1997 y Decreto 33 de 1998), Asociación Colombiana de Ingeniería Sísmica - AIS, Bogotá, Colombia, 4 Vol.
7. [AIS, Ingeominas, and Uniandes, 1996], *Estudio General de Amenaza Sísmica de Colombia*, Comité AIS 300 - Amenaza Sísmica, Asociación Colombiana de Ingeniería Sísmica, Ingeominas, Universidad de los Andes, Bogotá, Colombia.
8. [ATC, 1978], Tentative Provisions for the Development of Seismic Regulations for Buildings — ATC-3-06, Applied Technology Council — ATC, Palo Alto, CA, USA, 505 p.
9. [ATC 1986], *ATC-6 - Seismic Design Guidelines for Highway Bridges*, 2nd Printing, Applied Technology Council, Redwood City, CA, USA, 204 pp.

10. [ATC, 1996], ATC-32 - Improved Seismic Design Criteria for California Bridges: Provisional Recommendations, Applied Technology Council, Redwood City, CA, USA, 1996, 215 pp.
11. [Committee ACI 318, 2002], Building Code Requirements for Structural Concrete (ACI 318-02) and Commentary (ACI 318R-02), American Concrete Institute, Farmington Hills, MI, 443 pp.
12. [García, L. E. and Bonacci, J. F., 1994], *Implications of the Choice of Structural System for Earthquake Resistant Design of Buildings*, paper presented at the Mete A. Sozen Symposium held during the 1994 Fall Convention of the American Concrete Institute, Tarpon Springs, FL, USA.
13. [García, L. E. and O. D. Cardona, 2000], *The January 25th, 1999, Earthquake in the Coffee Growing Region of Colombia - Introduction*, Twelve World Conference on Earthquake Engineering (12WCEE), Auckland, New Zealand, January 2000.
14. [García, L. E., 1984], *Development of the Colombian Seismic Code*, Proceedings of the Eight World Conference on Earthquake Engineering, Earthquake Engineering Research Institute, San Francisco, Ca., USA.
15. [García, L. E., 1996], *Economic Considerations of Displacement-Based Seismic Design of Structural Concrete Buildings*, Structural Engineering International, Volume 6, Number 4, International Association for Bridge and Structural Engineering, IABSE, Zurich, Switzerland.
16. [García, L. E., 2000], The January 25th, 1999, Earthquake in the Coffee Growing Region of Colombia - Accelerographic Records, Structural Response and Damage, and Code Compliance and Enforcement, Twelve World Conference on Earthquake Engineering (12WCEE), Auckland, New Zealand, January 2000.
17. [García, L. E., Pérez, A., and Bonacci, J., 1996], *Cost Implications of Drift Controlled Design of Reinforced Concrete Buildings*, Proceedings of the 11th World Conference on Earthquake Engineering, Acapulco, Mexico.
18. [Hassan, A. F., and Sozen, M. A., 1997], *Seismic Vulnerability Assessment of Low-Rise Buildings in Regions with Infrequent Earthquakes*, ACI Structural Journal, American Concrete Institute, Farmington Hills, MI, USA, (January-February), p. 31-39.
19. [Ingeominas y Universidad de los Andes, 1997], *Microzonificación Sísmica de Santa Fe de Bogotá*, Convenio Interadministrativo 01-93, Ingeominas - Instituto Nacional de Investigaciones en Geociencia, Minería y Química, UPES - Unidad de Prevención y Atención de Emergencias de Santa Fe de Bogotá DC., OND - Dirección Nacional para la Prevención y Atención de Desastres, Bogotá, Colombia, 130 p.
20. [Kellogg, J. N., and Vega, V., 1995], Tectonic Development of Panama, Costa Rica, and the Colombian Andes: Constraints from Global Positioning System Geodetic Studies and Gravity, Geol. Soc. Am., Spec. Paper 295.
21. [Meigs, B. E., Eberhard, M. O., and García, L. E., 1993], *Earthquake-Resistant Systems for Reinforced Concrete Buildings: A Survey of Current Practice*, Report No. SGEM 93-3, Department of Civil Engineering, University of Washington, Seattle, Washington, USA, November.
22. [Ministerio de Transporte, 1995], *Código Colombiano de Diseño Sísmico de Puentes*, Instituto Nacional de Vías, Bogotá, Colombia, 785 p.
23. [MOPT, 1984], Código Colombiano de Construcciones Sismo Resistentes, Decreto 1400 de Junio 7 de 1984, Ministerio de Obras Públicas y Transporte, Bogotá, Colombia.
24. [París, G., 1993], *Fallas Activas de Colombia*, Instituto de Investigaciones en Geociencias, Minería y Química, Ingeominas, Regional Pacífico, Cali, Colombia.

25. [Ramírez, J. E., 1975], *Historia de los Terremotos en Colombia*, Instituto Geográfico Agustín Codazzi, Bogotá, Colombia.
26. [Secretaría de Educación del Distrito Capital de Santafé de Bogota, 2000], Análisis de vulnerabilidad sísmica de las edificaciones de la Secretaría de Educación del Distrito y diseños de rehabilitación de algunas de ellas, Estudio Realizado por: Proyectos y Diseños Ltda., Bogota, Colombia.
27. [Seismology Committee, 1959], *Recommended Lateral Force Requirements*, Structural Engineers Association of California - SEAOC, San Francisco, USA.
28. [Serna, O. R., 1994], Implicaciones Económicas en el Cumplimiento de Límites de Deriva en Estructuras Aporticadas de Concreto Reforzado para Zonas de Amenaza Sísmica Alta e Intermedia, Master of Science Thesis, Report MIC-94-II-18, Departamento de Ingeniería Civil, Universidad de los Andes, Bogotá, Colombia.
29. [Sozen, M. A. and García, L. E., 1992], Earthquake Resistant Design of Reinforced Concrete Buildings Based on Drift Control - Examples, Proyectos y Diseños Ltda., Bogotá, Colombia.

Table 1. Main seismic Colombian events in the last 100 years

Year	Month	Day	Location	Magnitude	Depth (km)	Fatalities
1906	Jan	31	Pacific Ocean Coast, near Tumaco	m = 8.9	?	400
1967	Feb	9	Huila	$m_b = 6.3$	60	98
1967	Jul	29	Santander	$m_b = 6.0$	160	5
1979	Nov	23	Coffee growing region	$M_s = 6.4$	80	55
1979	Dec	12	Pacific Ocean Coast, near Tumaco	$M_s = 7.8$	40	500
1983	Mar	31	Popayán	$m_b = 5.5$	12	300
1992	Oct	18	Murindó, Chocó	$M_s = 7.2$	15	30
1994	Jun	6	Páez, limit Cauca and Huila	$M_s = 6.4$	< 20	500-1000
1995	Jan	19	Tauramena, Casanare	$m_b = 6.5$	15	10
1995	Feb	8	Calima, Valle	$m_b = 6.4$	90	5
1999	Jan	25	Quindio, coffee growing region	$m_b = 5.9$	10	1200

Table 2. 1998 Colombian Code Update Content

Title	Content	Remarks
A	General Earthquake Resistant Design Requirements	Updated
B	Loads	Updated
C	Structural Concrete	Updated
D	Structural Masonry	Updated
E	One and two-story dwellings	Updated
F	Metal Structures	Updated
G	Wood Structures	New
H	Geotechnical Requirements	New
I	Supervision	New
J	Fire Requirements	New
K	Other General Requirements	New

Table 3. January 25, 1999, earthquake – General characteristics for Armenia and Pereira

	Armenia	Pereira
Distance from epicenter	14 km	48 km
Population	320,000	460,000
No. of victims	700	40
No. of buildings	66,000	108,000
No. of heavily damaged buildings	9,430	9,870

Table 4. Vulnerability assessment and rehabilitation — Main cases

Building	Location	Structural system	Vulner.	Rehab.
Sede Sena Pereira	Pereira, Risaralda	Slab-Column system	yes	yes
Facultad de Ingeniería Universidad Nacional	Manizales, Caldas	Reinforced concrete frame	yes	yes
Centro Administrativo Distrital	Bogotá D. C.	Reinforced concrete frame	yes	yes
Museo Nacional	Bogotá D. C.	Unreinforced masonry	yes	yes
Capitolio Nacional	Bogotá D. C.	Unreinforced masonry	yes	yes
Biblioteca del Congreso	Bogotá D. C.	Adobe	yes	yes
Hospital Kennedy	Bogotá	Reinforced concrete frame	yes	yes
Universidad de Caldas	Manizales, Caldas	Unreinforced masonry	yes	yes
Hospital Simón Bolívar	Bogotá D. C.	Reinforced concrete frame	yes	
Hospital Tunal	Bogotá D. C.	Reinforced concrete frame	yes	
Hospital La Victoria	Bogotá D. C.	Reinforced concrete frame	yes	
Hospital San Blas	Bogotá D. C.	Reinforced concrete frame	yes	
Bloque B, Universidad de los Andes	Bogota D. C.	Unreinforced masonry	yes	yes
Bloque W, Universidad de los Andes	Bogota D. C.	Waffle slab	yes	
Gobernación del Quindío	Armenia, Quindío	Reinforced concrete frame	yes	yes
Museo de Antioquia	Medellín, Antioquia	Mixture reinforced concrete and unreinforced masonry	yes	yes
Aeropuerto El Dorado	Bogota D. C.	Waffle slab system	yes	
Aeropuerto José Maria Córdova	Medellín, Antioquia	Reinforced concrete frame	yes	
29 Airports	Different locations	Different	yes	some
Catholic Church Parishes	Bogota D. C.	Different	yes	
Barrancabermeja Oil Refinery Complex	Barrancabermeja	Different	yes	

Figure 1. South America

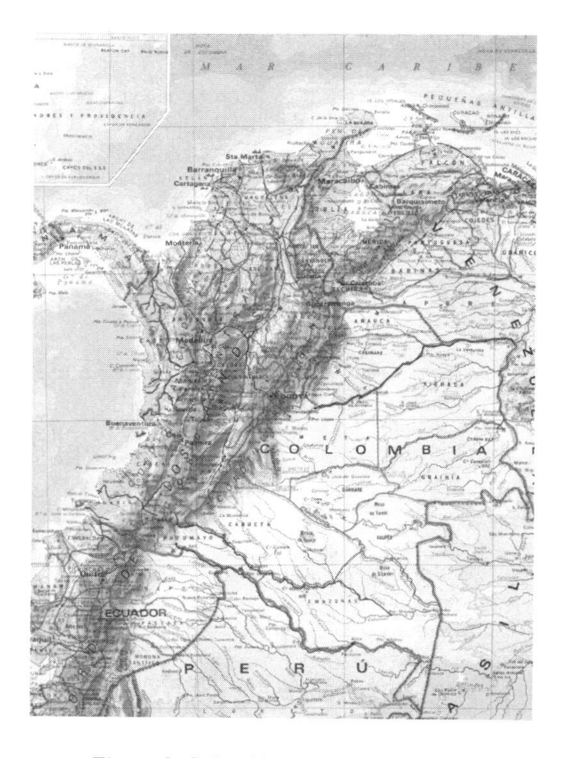

Figure 2. Colombia physical geography

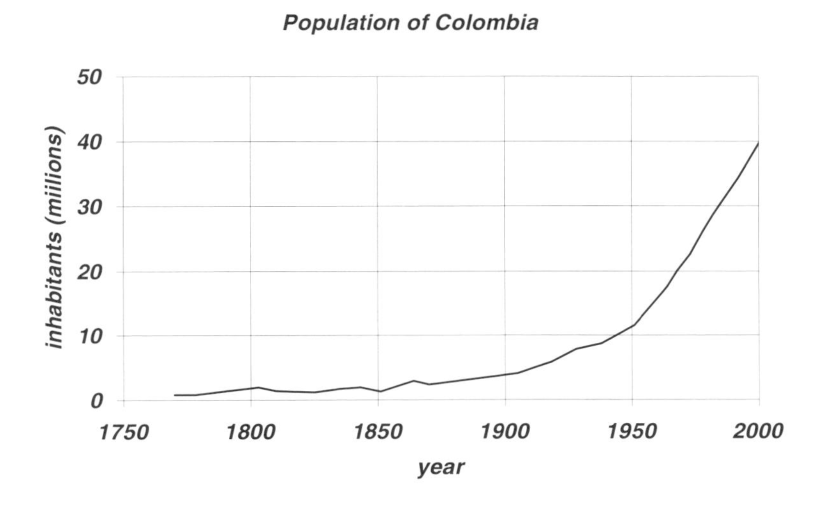

Figure 3. Colombia Population in the last 230 years

Figure 4. Tectonic plates that meet in Colombia

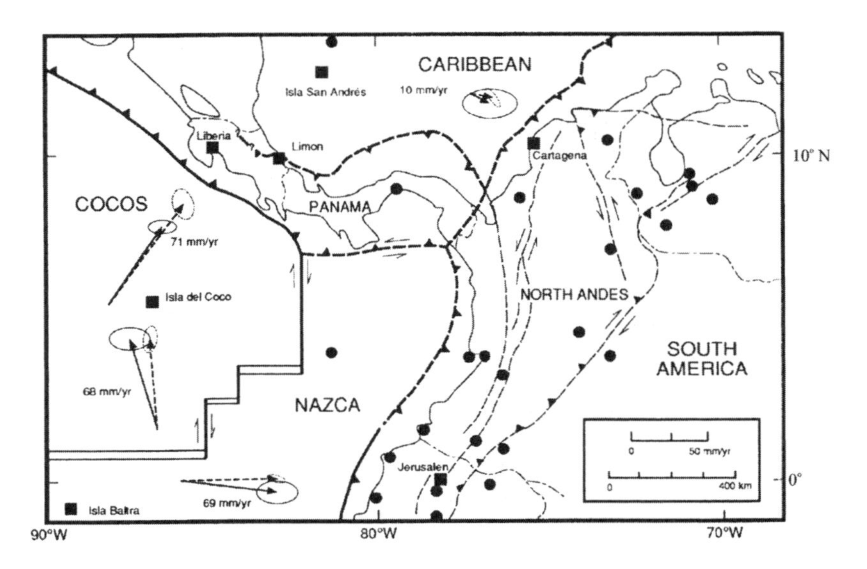

Figure 5. Direction and convergence rate of the tectonic plates in Colombia [Kellog and Vega, 1995]

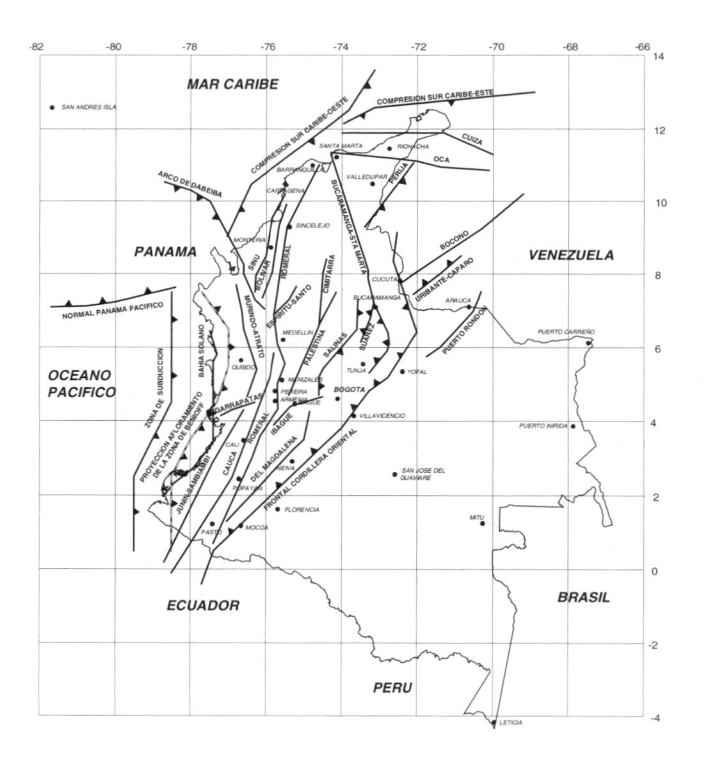

Figure 6. Main faulting systems in Colombia

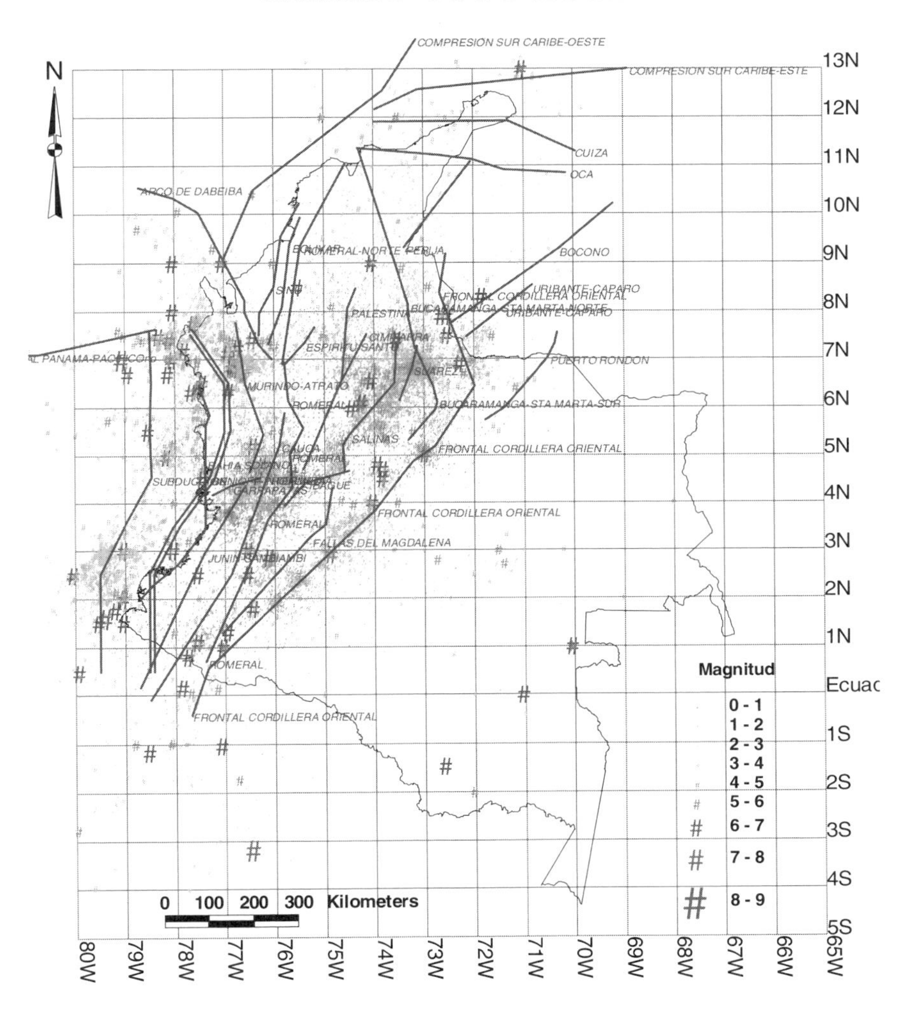

Figure 7. Location of earthquakes in Colombia 1566-2002

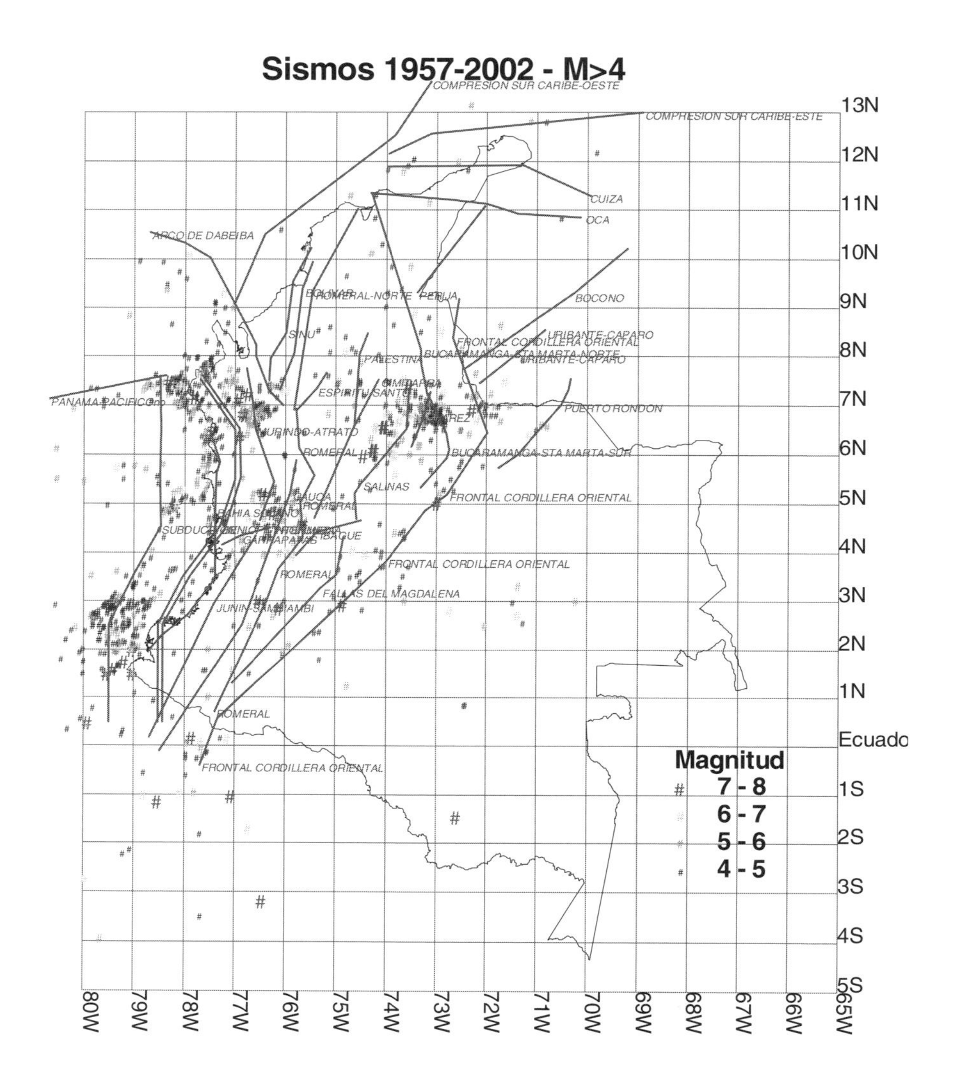

Figure 8. Location of earthquakes with Ms > 4 for 1957 to 2002

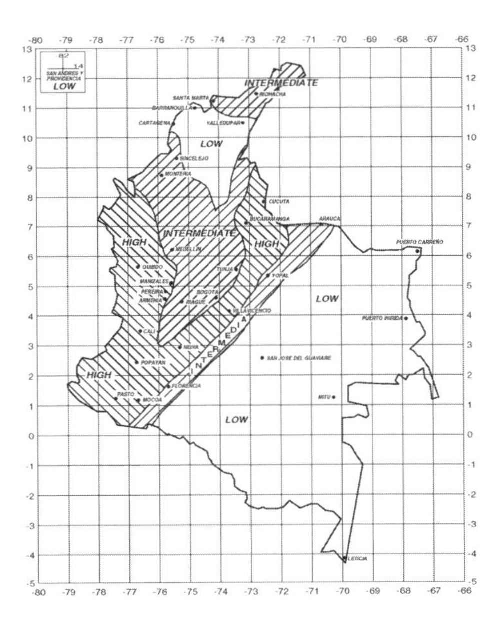

Figure 9. Seismic hazard zoning in Colombia

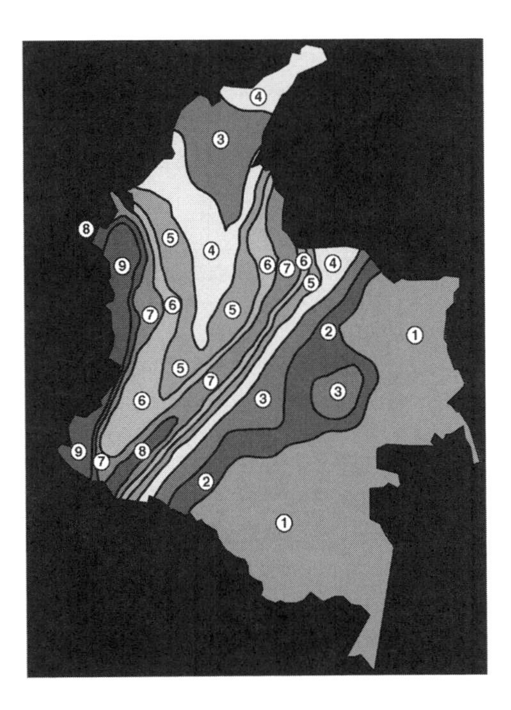

Figure 10. Effective peak ground acceleration

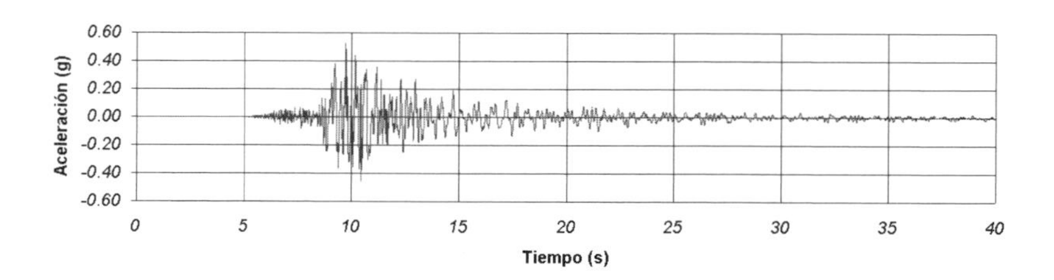

Figure 11. EW record at Universidad del Quindío in Armenia ($a_{max} = 0.53g$)

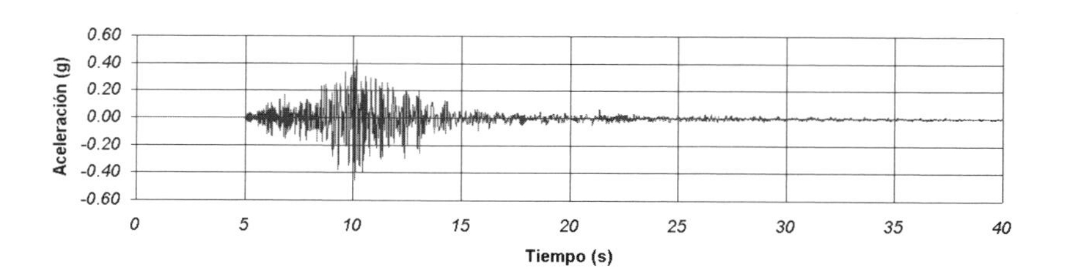

Figure 12. Vertical record at Universidad del Quindío in Armenia ($a_{max} = 0.46g$)

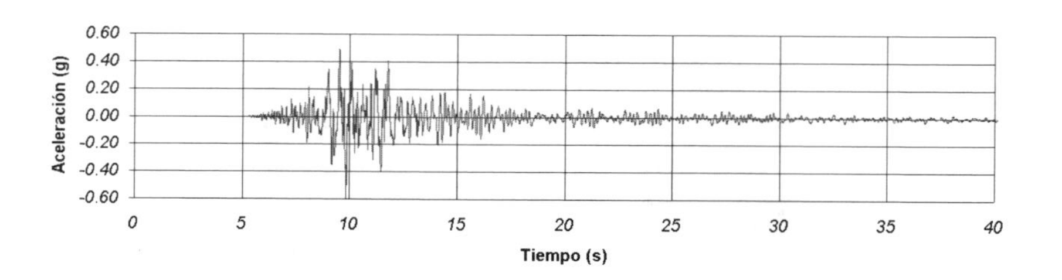

Figure 13. NS record at Universidad del Quindío in Armenia ($a_{max} = 0.59g$)

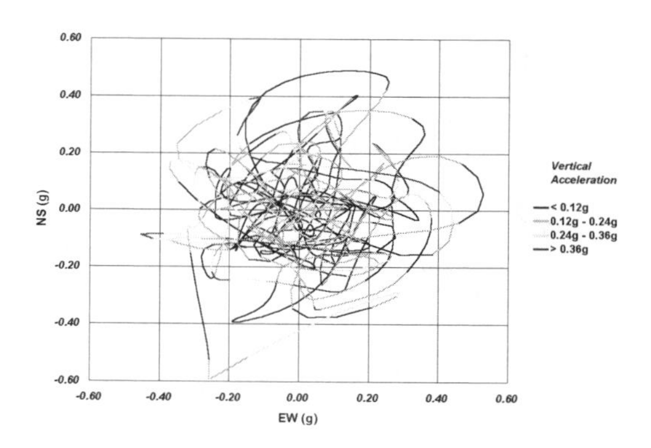

Figure 14. Universidad del Quindío (Armenia) site accelerograph shown in plan view

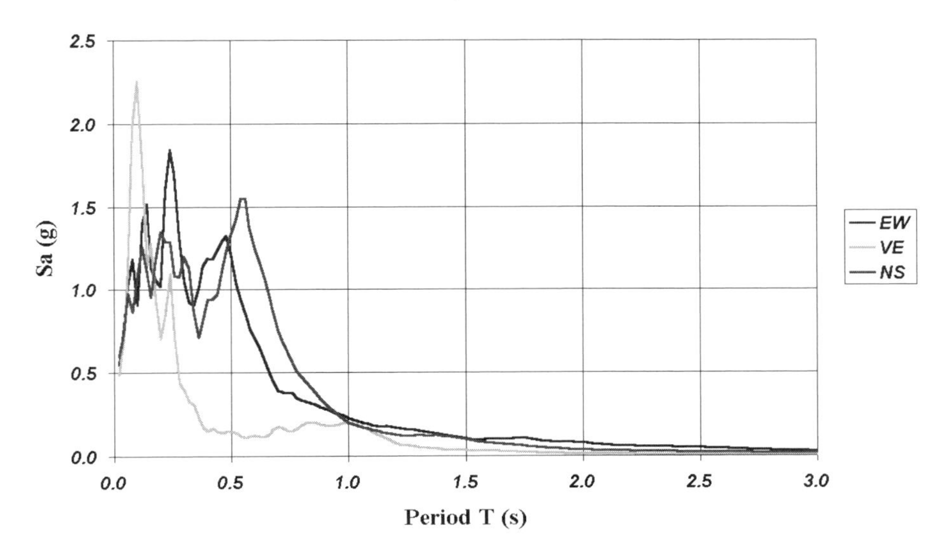

Figure 15. Acceleration spectrum at the Universidad del Quindío site in Armenia

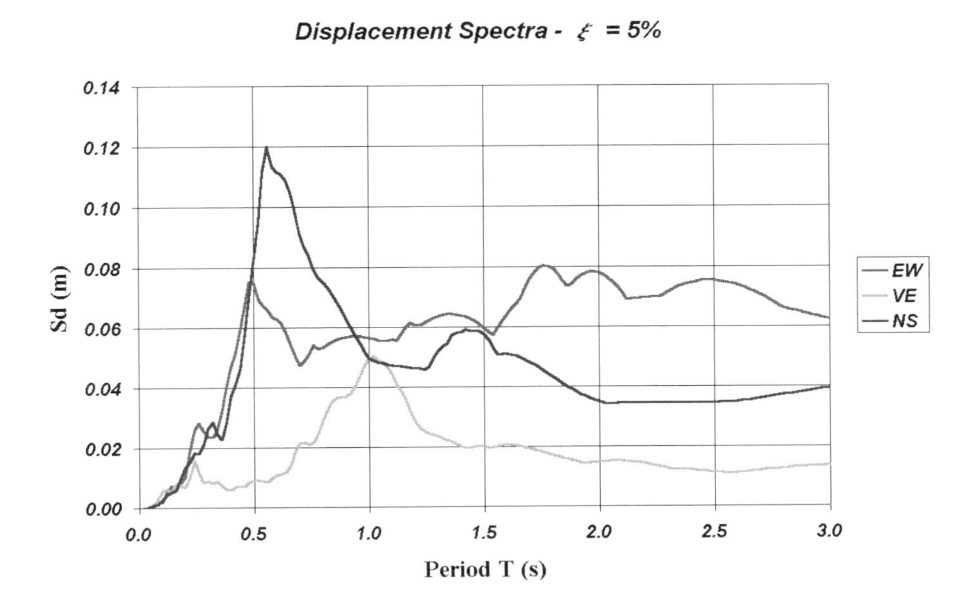

Figure 16. Displacement spectrum at the Universidad del Quindío site in Armenia

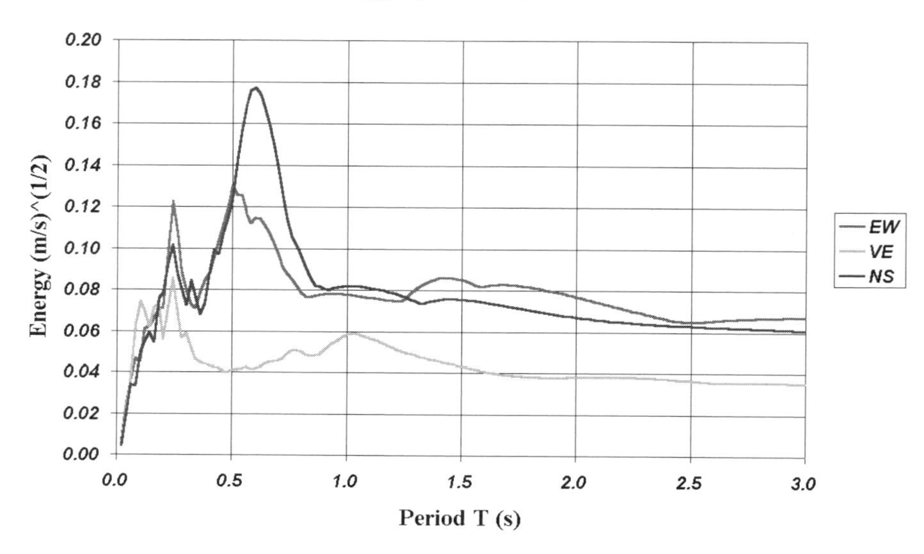

Figure 17. Energy spectrum at the Universidad del Quindío site in Armenia

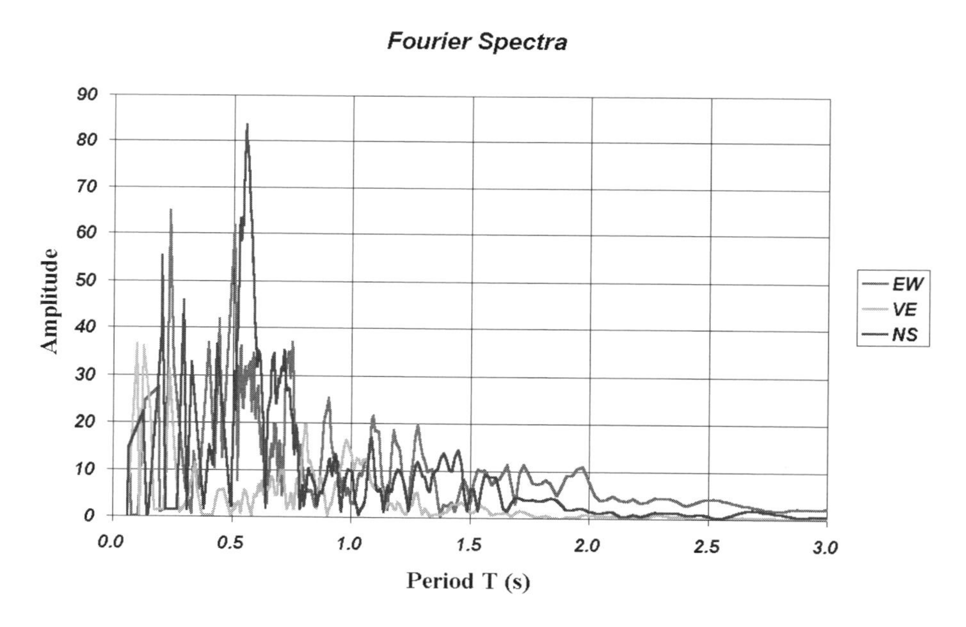

Figure 18. Fourier spectrum at the Universidad del Quindío site in Armenia

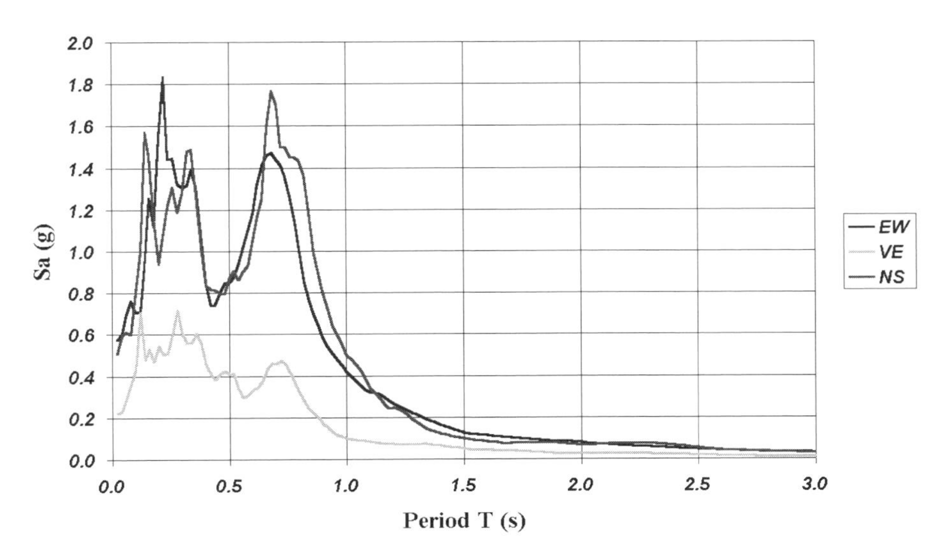

Figure 19. Acceleration response spectrum at the Filandia site

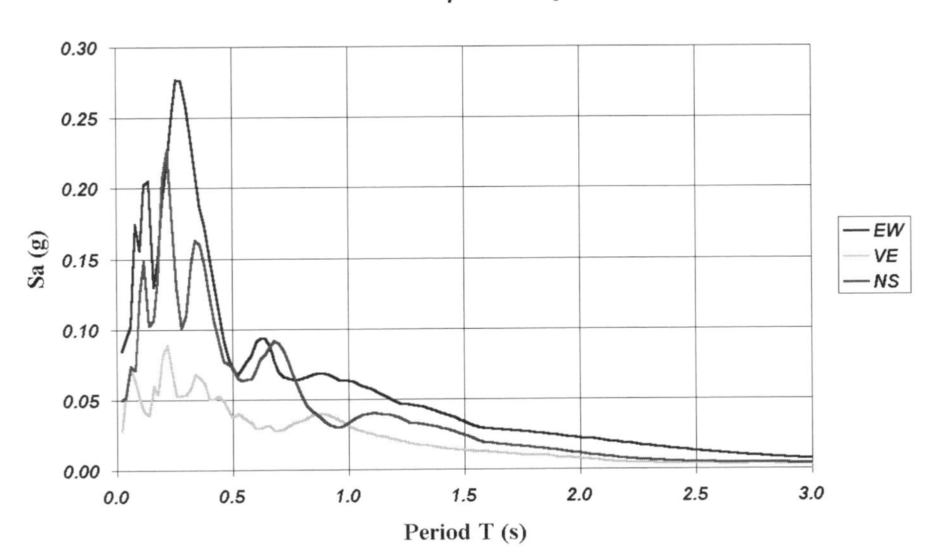

Figure 20. Acceleration response spectrum at the Bocatoma (Pereira) site

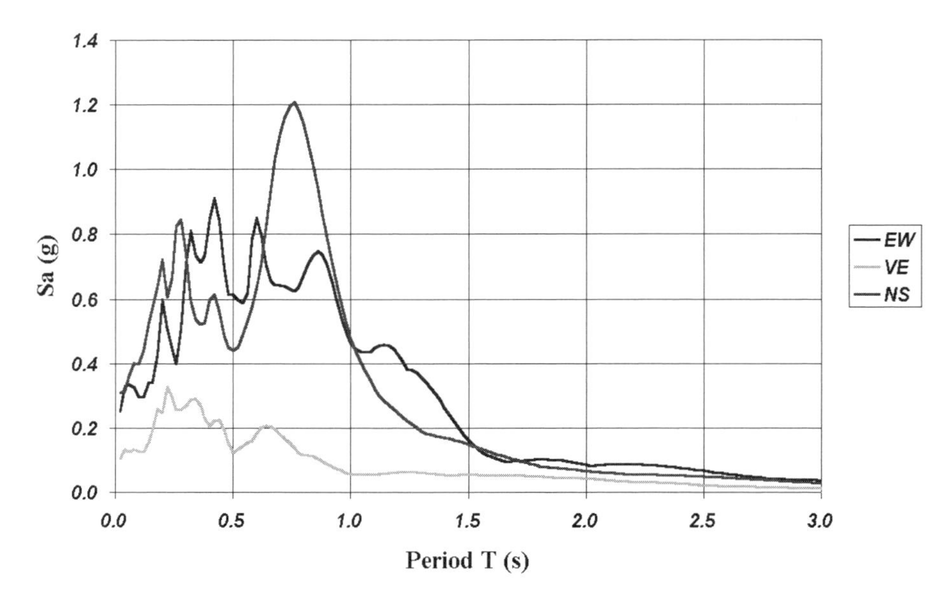

Figure 21. Acceleration response spectrum at the MazPereira (Pereira) site

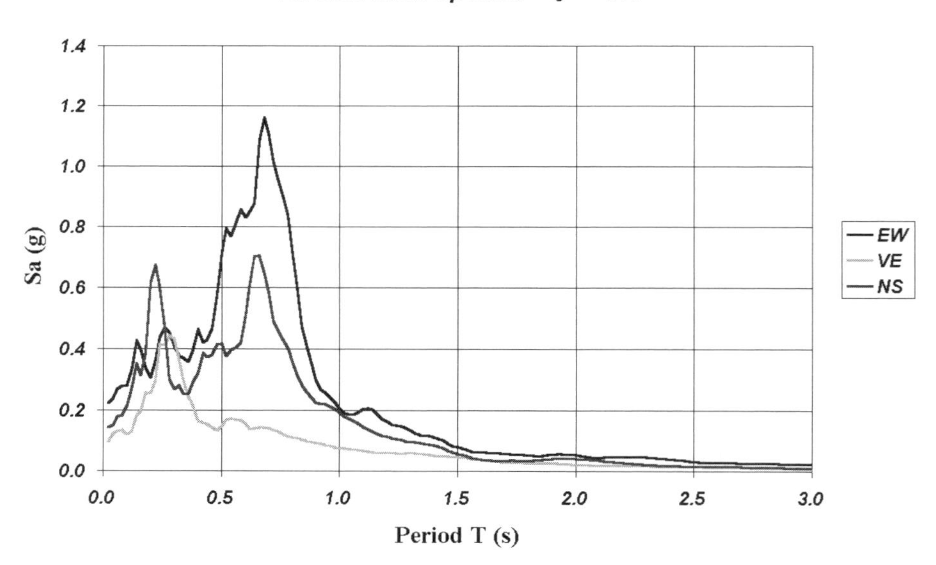

Figure 22. Acceleration response spectrum at the Castañares (Pereira) site

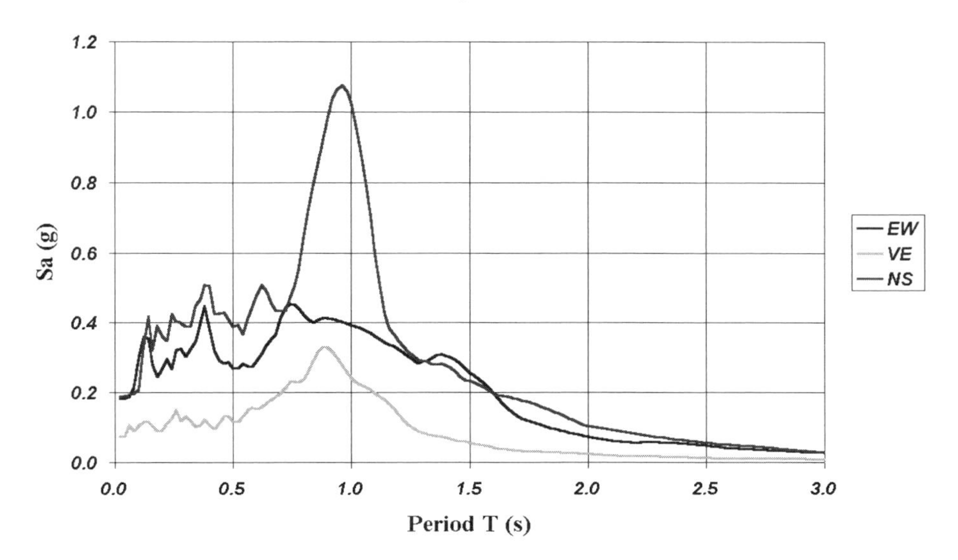

Figure 23. Acceleration spectrum at the La Rosa (Dosquebradas) site

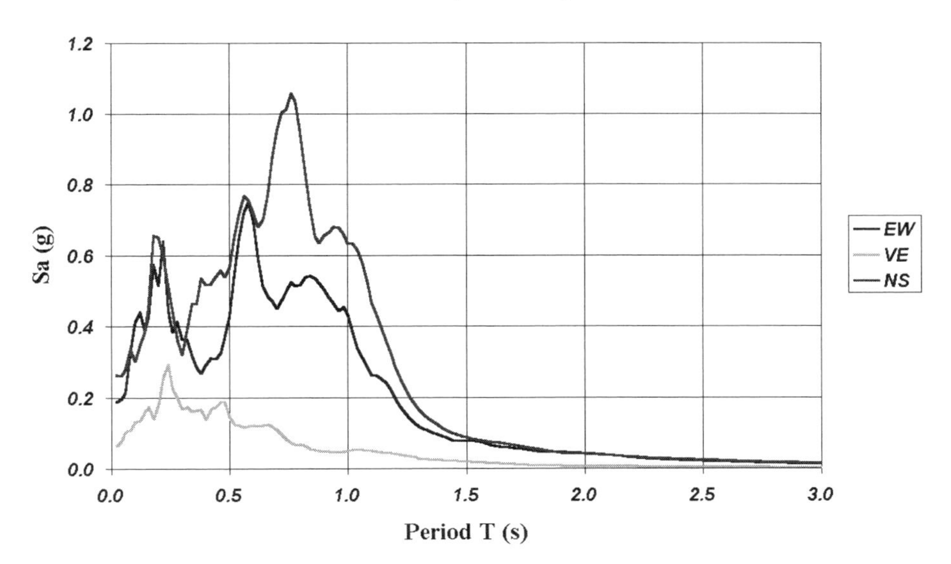

Figure 24. Acceleration spectrum at the Hospital of Santa Rosa de Cabal

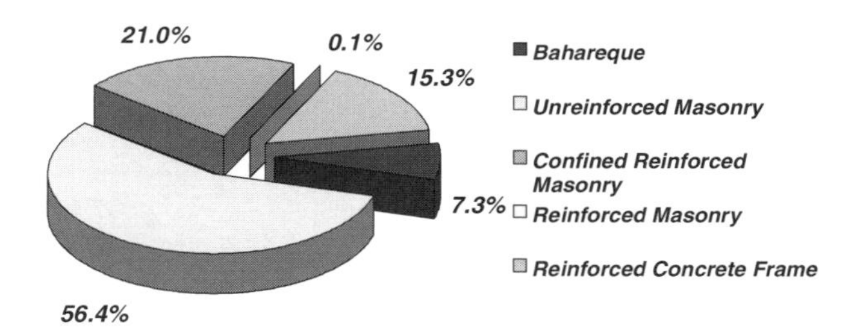

Figure 25. Type of construction for a 12,454 buildings sample in Pereira

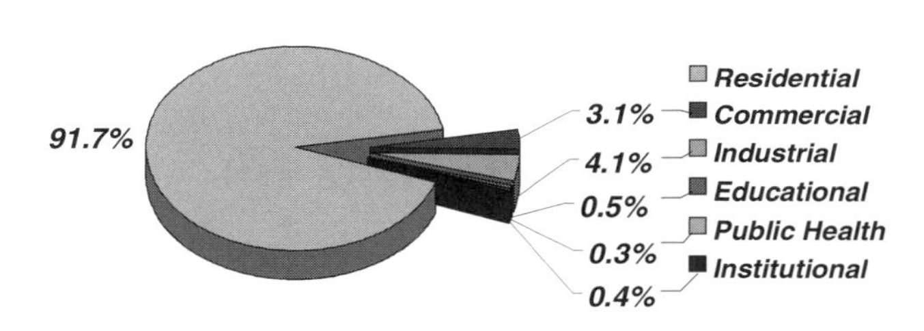

Figure 26. Building use for a 12,454 buildings sample in Pereira

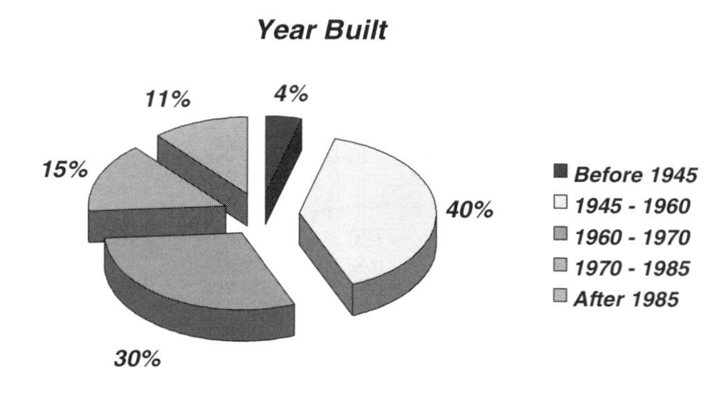

Figure 27. Age of construction for a 12,454 buildings sample in Pereira

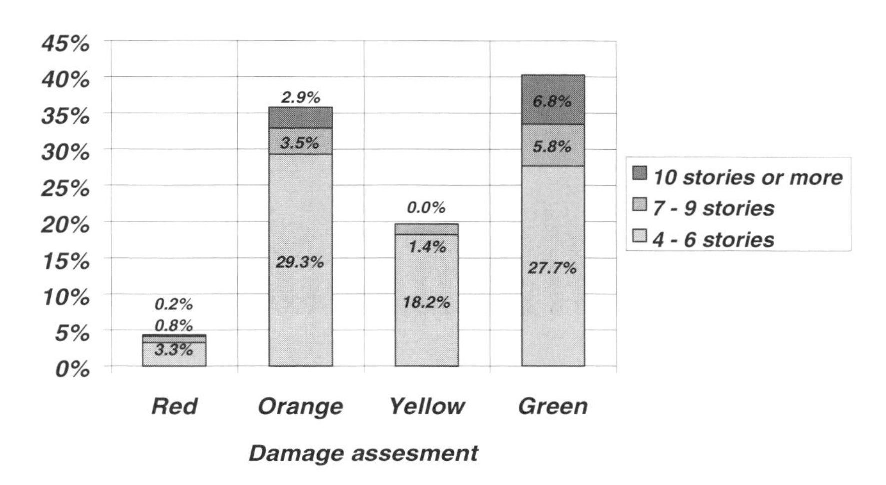

Figure 28. Tagging of the 12,454 buildings by height for buildings 4 stories and taller

Figure 29. (a) "Bloque B", Universidad de los Andes, Bogotá -Original unreinforced masonry construction

Figure 29. (b) "Bloque B", Universidad de los Andes, Bogotá- Rehabilitation work being performed

Figure 29. (c) "Bloque B", Universidad de los Andes, Bogotá- Rehabilitated building

Figure 29. (d) "Bloque B", Universidad de los Andes, Bogotá -Rehabilitated building

STRENGTHENING OF INFILLED WALLS WITH CFRP SHEETS

Ugur Ersoy, Guney Ozcebe, Tugrul Tankut, Ugurhan Akyuz
Emrah Erduran and Ibrahim Erdem
Structural Mechanics Division, Department of Civil Engineering
Middle East Technical University, Ankara, 06531 Turkey

Abstract: Introducing RC infills to selected bays in both directions is an effective approach for strengthening existing buildings having various deficiencies. However, if adopted for buildings in use, the building would need to be evacuated before starting the construction, which will last for several months. This would impair the practicality of the method. Hence, strengthening techniques applicable without interrupting the functioning of the building are urgently needed. In this study, it was intended to use existing partitions and non-load bearing exterior masonry infills as structural walls by strengthening them with CFRP strips. This paper reports test results related to frames having hollow clay tile infills strengthened by using CFRP conducted at METU.

Key words: Seismic strengthening, existing buildings, RC frames, brick masonry infill, CFRP, cyclic loading

1. INTRODUCTION

Strengthening of damaged reinforced concrete frames by introducing reinforced concrete infills to some selected bays has found wide application in Turkey and in other countries. The reinforced concrete infills increase the strength and the lateral stiffness of the framed structure significantly. This method has been found to be economical, effective and easy to apply.

S.T. Wasti and G. Ozcebe (eds.), Seismic Assessment and Rehabilitation of Existing Buildings, 305–334.

Surveys made in Turkey on existing reinforced concrete buildings have shown that a great number of these buildings do not satisfy the requirements of the current seismic code. The common deficiencies in these buildings are: (a) ends of beams and columns are not properly confined, (b) no lateral reinforcement is provided at beam-column joints, (c) ties are not anchored into the core concrete (90° hooks), (d) concrete strength is low and (e) lapped splices in column longitudinal bars are made at floor levels and usually the lap length is inadequate.

In addition to these deficiencies most of the framed structures inherit system weaknesses like soft stories, weak column-strong beam connections and short columns. In general, it is also observed that the lateral stiffness of these frames is inadequate and the drift requirements in the seismic code are not satisfied. Obviously, the deficiencies outlined here are not unique to Turkey and would apply to most of the existing old buildings in other countries facing seismic risk.

Strengthening of framed structures by introducing reinforced concrete infills to the selected bays in both directions would be an effective and acceptable approach for existing buildings having various deficiencies. However, once this technique is adopted for the strengthening of building in use, the building needs to be evacuated before starting the construction, which would last for several months. This would impair the practicality of strengthening by reinforced concrete infills. Hence, it can be said that strengthening techniques which could be applied without interrupting the functioning of the building are urgently needed.

2. REVIEW OF RESEARCH ON MASONRY REHABILITATION USING CFRP

Past research on strengthening of reinforced concrete structures can be classified in two groups. These are namely: (a) member strengthening and (b) system improvement. In member strengthening, certain members of the structure are strengthened to improve the seismic behavior of the structure. However, if the lateral rigidity of the structure is inadequate or if there are important system deficiencies like soft story, weak story, etc., and/or if the number of members that require rehabilitation is excessive, member strengthening would not be feasible. In such cases "system improvement" would be more feasible. During the last 40 years, studies related to rehabilitation of reinforced concrete structures by introducing reinforced concrete infills were carried out throughout the world. In this field the pioneering studies conducted at Middle East Technical University date back to the early seventies. In 1971, Ersoy and Uzsoy [1] made tests and concluded that the lateral load carrying capacity of a frame could be increased by 700 percent. At the end of their studies Altin et al. [2] and Sonuvar [3] concluded that the reinforced concrete infills, which were properly connected to the frame members, increased both the strength and stiffness significantly. Canbay et al. [4] stated in their work that the strength of a two-story, three-bay frame was increased by 400 percent as a result of the introduction of a reinforced concrete infill to the middle bay.

In addition to these experimental studies, many structures were repaired and strengthened using this technique after the earthquakes that hit different regions of Turkey in the last decade. However, the feasibility of this method is questionable as far as the pre-rehabilitation of large number of existing structures is concerned. As stated in the previous section, this procedure requires evacuation of the entire building under rehabilitation, which is generally neither feasible nor practical. Thus, alternative methods, which would not interrupt the use of the building, should be developed to strengthen large number of structures which do not possess adequate seismic safety. Externally bonded carbon fiber reinforced polymers (CFRP) could be the solution of this problem owing to their light weight, high strength and ease of application.

So far carbon fiber reinforced polymers (CFRPs) were mostly used in member strengthening such as wrapping of columns to increase ductility or shear strength of beams. Moreover, masonry structures with unreinforced walls had also been strengthened using CFRP. Within the framework of the latter application, Priestley and Seible [5] tested a full scale five story masonry building. The building was first loaded up to a certain damage level. The damaged building was then rehabilitated by using CFRP overlays, which were applied to the first and second story walls. After the testing of the rehabilitated building, it was observed that that even a very thin layer of CFRP (t=1.25 mm) resulted in 100 percent increase in the inelastic deformation capacity of the masonry structure. Moreover, measured shear deformations in the overlaid wall panels were reduced by 50 percent. The authors stated that CFRP strengthening of masonry structures was a feasible technique for the seismic rehabilitation of masonry structures.

Roko, Boothby and Bakis tested twenty five masonry prisms to determine the failure modes of sheet bonded FRP applied to brick masonry [6]. They used two types of masonry (high and low porosity) and two types of CFRP reinforcement (low and high modulus). They reported that all molded specimens with both low and high modulus had failed by shearing of the brick where FRP reinforcement had terminated, whereas extruded brick specimens failed by the debonding of FRP and thus did not yield significant moment resistance. At the end of these tests authors concluded that FRP could effectively be used for seismic rehabilitation of masonry structures.

In 1998, Triantafillou [7] tested twelve small masonry wall specimens. Specimens in the first group were tested in out-of-plane bending. Those in the second group were tested in bending. Each group consisted of six specimens, four of which were reinforced with epoxy bonded unidirectional CFRP. It was observed that the specimens tested in in-plane bending failed prematurely through debonding of the CFRP laminates. As a result of these observations, it was concluded that the achievement of full in-plane flexural strength depends on proper anchorage of the laminate to the masonry.

3. OBJECT AND SCOPE OF THIS STUDY

The main objective of the study reported here was to develop strengthening methods and techniques for buildings which could be applied with minimum disturbance to the occupants. It was intended to use the existing partitions and non-load bearing exterior masonry infills as structural walls by strengthening these infills. Two techniques were found to be feasible for strengthening the infills; (a) CFRP application to the surface of the infill with proper connection to the frame members and (b) precast concrete panels bonded to the infill and connected to the frame members.

Various techniques have been developed for strengthening the frames using CFRP and precast panels. Tests have been carried out at METU to see the efficiency of these techniques and to observe the behavior of strengthened frames using these materials. This paper covers only the tests related to frames having hollow clay tile infills strengthened using CFRP.

The specimens were one-bay, two story frames with hollow clay tile infills plastered on both faces and were tested under reversed cyclic loading.

In addition to these tests, three-bay two-story frames in which only the middle bay was infilled were also tested to see the sharing of lateral forces among the columns and the infilled frame. These tests have been briefly discussed herein as their details are given in another article in this volume.

4. EXPERIMENTAL PROGRAM

4.1 Test Specimens

In the first part of this study, seven one-bay, two-story reinforced concrete frames were tested. The test specimens consisted of two one-bay two-story frames connected to each other by a relatively rigid foundation beam. The main philosophy behind the design of this specimen was to prevent (or minimize) the rotations at the base of the first story columns. Reinforcement details and material properties of all the specimens were the same and all frames were infilled with hollow clay tiles. Six of these specimens were strengthened by using different CFRP arrangements. The remaining one was not strengthened and it served as a reference specimen.

The reinforced concrete frames were designed to simulate the deficiencies commonly observed in existing reinforced concrete residential buildings in Turkey. Therefore, they did not conform to the Turkish Seismic Code requirements [8]. The end zones of beams and columns were not properly confined, i.e. the ties were not closely spaced and the ends of the ties were bent 90°. No transverse reinforcement was used in the beam-column joints and the bottom reinforcement of the beams did not have sufficient anchorage length. Moreover, the concrete strength of the specimens was low. Column longitudinal bars were continuous in the first story (no lapped splice). The lap splice above the first story level had a lap length of forty bar diameters, 40db. The geometry and the reinforcement details of the frames are presented in Figure 1 and Figure 2, respectively.

The rigid foundation beam was heavily reinforced to prevent local failures. Moreover, the ties of the foundation beam were bent 135° into the core to ensure proper confinement.

After the construction of the reinforced concrete frames, the brick infills were placed and were plastered on both faces, Figure 1b.

The first specimen (SP-1) tested within the scope of this study was an unstrengthened specimen. This specimen was intended to serve as a reference specimen.

The second specimen (SP-2) was strengthened by applying two orthogonal layers of CFRP on both faces of the brick infills. The CFRP layers were not connected to the reinforced concrete frames. The CFRP was bonded to the infills by means of a special adhesive. No other connection was made between the CFRP layers. Both faces of the infill were fully covered by CFRP.

The exterior face of the third specimen (SP-3) was fully covered with two orthogonal layers of CFRP. The CFRP layers were applied on the brick infill and were extended to the reinforced concrete frame members. The CFRP layers were anchored to the frame members by special anchors to prevent the premature failure resulting from debonding. No special anchorage was used to fix the CFRP layers to the hollow clay tiles.

In the fourth specimen (SP-4), both faces of the specimen were fully covered by two orthogonal layers of CFRP. CFRP was extended to the reinforced concrete frame on the exterior side. Although the anchor dowels used in SP-3 contributed favorably to the behavior of the specimen, it was observed that the amount of dowels used was not sufficient. Thus, in SP-4, the number of anchor dowels used to connect CFRP to the frame members was increased. Moreover, CFRP layers were also anchored to the infill walls. Figure 3 shows the design details and the locations of the anchor dowels used in specimens SP-4. The manufacturing steps and the ways of application of the anchor dowels are given in Figure 4.

Although the lap splice length provided in specimens SP-1 to SP-3 above the first floor was 40db, these splices still caused problems due to the inadequate confinement in the lapped splice regions. To overcome this problem, in specimen SP-4, the lap splice regions were confined using two orthogonal layers of CFRP. The length of the confinement zone was 400 mm. Specimen SP-4 was far from being economical due to the amount of CFRP used, which fully covered both faces of the infill. To decrease the amount of CFRP used, in SP-5, it was decided to use only one layer of CFRP along the main diagonals of the hollow clay tile infill. The width of the CFRP layers used was 200 mm. As in the case of SP-4, the lap splice regions were confined using two layers of CFRP to eliminate premature failures. No confinement was provided at the foundation level, since there were no lap splices in this region.

Since the anchorage of the CFRP to the frame members and to the hollow clay tiles in specimen SP-4 proved to be effective, the same anchorage technique was used in SP-5.

Strengthening made in SP-5 increased both the strength and the amount of dissipated energy significantly. The failure of SP-5 was initiated by debonding of the CFRP strap at the foundation level. Moreover, yielding of longitudinal reinforcement and crushing of concrete were also observed at the base of the first story columns.

In SP-6, the bottoms of the first story columns were also confined using two layers of CFRP. The length of confinement zone was limited to the expected plastic hinge length (150 mm). This specimen displayed shear failures at the first story beam column joints. This failure was initiated by the failure of anchor dowels resulting in a sudden unloading in the tension strut.

In detailing the CFRP of the specimen SP-7, it was aimed to strengthen the first story joints for shear by using CFRP. For this purpose, the diameter of the anchor dowels used at the first story joints was increased. These dowels were expected not only to anchor CFRP struts to the specimen but also to increase the shear strength of the joints.

Table 1 summarizes the properties of the test specimens.

4.2 Test Procedure

The loading system and the test specimen that is used in the experimental investigation are shown in Figure 5. The test set-up is similar to one which was developed by Smith [9]. Twin specimens with a common foundation beam were constructed and laid upon the steel plates resting on ball bearings. The lateral load was applied through the foundation beam. Hence, the reaction force at each end of the twin specimens became the lateral force applied at the second story level. Axial load was applied to the columns by prestressing tendons as shown in Figure 5. The level of applied axial load in each test was about 25 percent of N0, where N0 is the nominal axial load capacity of the frame columns.

Test specimens were instrumented to measure the applied loads, lateral displacements, rotations in the foundation beam and diagonal strains of the infill. The instrumentation is shown in Figure 6.

Lateral displacements and support settlements were measured by means of linear variable differential transformer displacement transducers (LVDTs), which were mounted at each story and foundation levels. Two additional LVDTs were attached to the foundation beam to measure the rotations. The infills were further instrumented with electrical dial gages placed diagonally to monitor the shear deformations. After each test, story displacements were calculated by making corrections considering both support settlements and rigid body rotations. Since each specimen is made up of two specimens, experimental results were presented only for the specimen where failure was observed.

4.3 Test Results

The first specimen (SP-1) tested within the scope of this study was an unstrengthened hollow clay tile infilled reinforced concrete frame. The aim of this test was to provide a reference data for comparison with the results of the strengthened specimens. SP-1 showed an inferior, non-ductile frame behavior and failed at a maximum lateral load of 55 kN, (V=55 kN).

The second specimen was strengthened by the application of CFRP on the brick infills of the specimen on both sides, fully covering the faces. CFRP layers were neither extended nor anchored to the frame members. Since CFRP layers delaminated from the brick infills of SP-2, no significant increase in the lateral load carrying capacity could be achieved. However, the strengthening applied still resulted in an improvement in the behavior of the frame in terms of ductility. The inelastic displacement capacity of SP-2 was higher than that of SP-1. Moreover, the energy dissipated by SP-2 was almost 3.6 times the energy dissipated by SP-1.

SP-3 was strengthened by fully covering the external face of the infill with CFRP. No intervention was made on the interior face of the infill. Although CFRP was anchored to the frame members, CFRP layers on the second story panel of one of the twin frames completely delaminated at early stages of the test. No delamination was observed on the remaining three panels. Thus, the stiffness of the panel, where delamination took place, decreased drastically as compared to the others. Hence, in the remaining cycles, the reinforced concrete frame members bounding that panel underwent significant inelastic deformations. Due to this unfavorable behavior, the increase in the lateral strength of the specimen was insignificant.

It can be said that, the strengthening patterns applied in specimens SP-2 and SP-3 did not significantly improve the behavior of the frames.

In specimen SP-4, CFRP was applied to the brick infills on both faces. In contrast to SP-2, CFRP was extended to the reinforced concrete frame members. Also, anchor dowels were used to anchor CFRP to frame members and infills. As a result of this strengthening scheme, significant improvement in the seismic behavior could be achieved. The lateral load carrying capacity of SP-4 was 131.5 kN, which is almost 2.4 times that of SP-1. In addition to the increase in strength, SP-4 showed a far more ductile behavior than SP-1 as it dissipated 4.5 times more energy.

In deciding the strengthening pattern of SP-5, the main objective was to reach the strength increase achieved in SP-4 by using less CFRP. For this purpose, CFRP was applied as cross-struts on the infills. As in the case of SP-4, CFRP was both extended and anchored to the frame members. This strengthening pattern led to 75 percent reduction in the amount of CFRP used. Despite this reduction in the amount of CFRP used, SP-5 carried a lateral load of 118.8, which corresponds to 90 percent of the load carried by SP-4. In addition, SP-5 dissipated 3.1 times more energy than that dissipated by SP-1. A photograph of specimen SP-5 after the test is given in Figure 7.

The strengthening pattern used in SP-5 was more economical and resulted in a considerable improvement in the behavior of the frame. Thus, the strengthening patterns used in the remaining specimens were slight modifications of this pattern. In SP-6 the only difference was the confinement of the plastic hinge zones at the base of the first story columns by CFRP. However, this modification had an adverse affect on the behavior of the frame. The failure mechanism changed from a ductile flexural failure to a brittle shear failure in the first story beam-column joints.

In SP-7, additional efforts were made to prevent the shear failure of the beam-column joint observed in specimen SP-6. These zones were strengthened by CFRP strips. However, this modification did not improve the behavior. Both the behavior and the lateral strength of SP-7 were very similar to those of SP-6.

The hysteretic load displacement curves for all specimens are given in Figures 8 through 14.

5. DISCUSSION OF TEST RESULTS

In this section, the test results are evaluated considering strength, stiffness, and energy dissipation characteristics of the test specimens.

5.1 Strength

In evaluating the effectiveness of a strengthening technique, the strength increase attained is considered to be one of the most important parameters. The lateral load capacity of each specimen is given in the second column of Table 2. In addition, the response envelope curves are given in Figure 15 to enable comparison of behavior. Response envelope curves were developed by connecting the maximum values at each cycle.

Table 2 and Figure 15 show that the CFRP applied to specimens SP-2 and SP-3 did not increase the strength of the structure significantly due to the debonding of CFRP at the early stages of testing. It should be recalled that in SP-4, CFRP was applied on both faces of the infill, in a manner to fully cover the entire surfaces. Also the CFRP was anchored to the infill and the frame members by special anchor dowels. Therefore, SP-4 was the extreme case for strengthening using CFRP. The strength of SP-4 was more than twice when compared to SP-1, in which no strengthening was carried out. However, due to the fact mentioned above, the economic feasibility of the approach used in strengthening of SP-4 can be questioned.

In specimens SP-5, SP-6, SP-7, CFRP strips were placed as cross-bracings. Each strip consisted of one layer instead of two as in the case of SP-4. The strength increase in these three specimens was not as high as the one observed in SP-4. However, the strength of the specimen was doubled when compared to that of the reference specimen SP-1. The amount of CFRP used in SP-5, SP-6 and SP-7 was much less than the one used in SP-4. The CFRP configuration used in these three specimens led to an economical solution for strengthening the existing buildings with minimum disturbance to the occupants.

5.2 Stiffness

Strengthening of infills by CFRP sheets did not result in significant changes in the stiffness of the test specimens. As can be seen from the third column of Table 2, the initial stiffnesses of the test frames were very close to each other. As the strength demand of a structure is directly related to the stiffness characteristics, it can be concluded that the proposed strengthening method increases the capacity of the structure without causing a significant increase in the strength demand. This is an important advantage of strengthening by using CFRP.

5.3 Energy Dissipation

Energy dissipation capacity of a structure is an indicator of ductility and is one of the important factors that determine the survival of the structure in a major earthquake.

The dissipated energy for each cycle was computed as the area under the load deformation curve. In Figure 16, values of cumulative energy dissipated by each specimen in terms of total drift ratio (roof displacement/total height) are presented. The total energy dissipated by each specimen is also given in Table 2.

From Figure 16 and Table 2, it can be seen that SP-3, SP-4, and SP-5 dissipated more energy than the remaining specimens. It should be noted that the high energy dissipation in SP-3 is due to the significant damage in the second story columns. Thus, the behavior of SP-3 in terms of high energy dissipation cannot be considered satisfactory due to the unfavorable failure mechanism observed in this specimen.

The difference between the energy dissipation capacities of SP-5, SP-6, and SP-7 is mainly due to the failure mechanisms of these specimens. Failure of SP-5 was initiated by the formation of the plastic hinges at the base of the first story columns and completed by the crushing of concrete in these regions. In other words, the failure mechanism of SP-5 was relatively ductile. However, both SP-6 and SP-7 failed by a far more brittle mechanism. Since the bases of the first story columns were confined using CFRP, the failure zone moved to the next weak link which was the first story beam-column joints. Both specimens failed due to the sudden shear failure at these joints.

5.4 Interstory Drifts

Interstory drift is defined as the relative displacement between two consecutive floors divided by the floor height. Interstory drift is generally accepted as a measure of the non-structural damage. The Turkish Earthquake Code limits the interstory drift to 0.0035 based on elastic analysis [8].

In Table 2 the maximum first story interstory drift ratios for each specimen are given. In case of SP-3 and SP-4, infills sustained significant damage during the test. However, although interstory drift value for first floor of SP-5 in the last cycle increased up to 0.017, no significant damage was observed in the infill walls. This can be considered as a significant contribution of CFRP.

As a summary of the tests on two-story, one-bay specimens, it can be concluded that SP-4 and SP-5 displayed superior response in terms of lateral load capacity, stiffness and ductility. Although SP-4 exhibited a better response than SP-5, the amount of CFRP reinforcement used in this specimen was about four times that of SP-5. Thus, the CFRP detailing used in SP-5 seems to be the most efficient one.

6. TESTS ON 2-STORY, 3-BAY SPECIMENS

Tests on two-story, one–bay frames, with masonry infills reinforced with CFRP sheets were very helpful in order to understand the behavior and observe the strength increase due to the presence of CFRP. Special techniques and anchors developed were tested and improved in this experimental study. Also, design criteria were developed considering the test results obtained.

In the system improvement approach, the aim is to replace the existing lateral load resisting system which is composed of frames with various deficiencies by a new lateral load resisting system which consists of structural walls. Structural walls can be formed by infilling the selected bays of the frames by reinforced concrete infills, properly connected to the frame members. As was mentioned previously, seismic rehabilitation of damaged framed structures by introducing reinforced concrete infills has been applied to hundreds of buildings after the major earthquakes in Turkey in the last decade.

In the earlier sections of this paper it was shown that CFRP sheets anchored to the masonry infill (hollow clay tile) and connected to the frame members transformed the masonry wall into a structural wall.

Whether the rehabilitation is made by introducing reinforced concrete infills to some selected bays of the structure or by strengthening the masonry infills using CFRP strips, the main idea is to have adequate structural wall area in both directions so that the total lateral load can be carried by these walls.

The researchers at METU were curious to see the relative distribution of the lateral load to the walls and the frames in a structural system where only the selected bays are converted into structural walls.

To observe the distribution of lateral load between the columns and the structural wall (infilled frame), particularly in the post-elastic stages, a three-bay, two story test specimen was designed, Figure 17. Only the middle bay of the specimen shown was filled to convert it into a structural wall. In Specimen S1, the infill was cast in place concrete, properly connected to the frame members by dowels. The thickness of the reinforced concrete infill was 70 mm.

In Specimen S2, the middle bay was filled with hollow clay tile type of masonry wall, plastered on both faces. Before the test, CFRP sheets arranged as cross bracing were bonded and anchored on both faces of the infill, Figure 18. The thickness of the hollow clay tile wall, including the plaster was about 80 mm.

As can be seen in Figure 18, only the middle bay was infilled. The two interior columns were connected to the infill and therefore were considered to be an integral part of the infilled frame. The two exterior columns were the only frame columns. A special force transducer was designed to measure the axial load, bending moment and shear. Two such special transducers were placed at the bottom of the exterior columns as shown in Figure 18. The lateral load applied to the specimen minus the shear measured at the bottom of two exterior columns would give the base shear taken by the wall, i.e. the infilled frame. A photograph of the specimen before the test is shown in Figure 19.

As shown in Figure 18, the axial loads on the columns were provided by two concrete blocks placed on the top of the second story (total weight to four columns was 18 kN). The reserved cyclic lateral load was applied only at the second story level.

The load-top displacement envelopes of the two infilled frames (S1 and S2) together with that of the bare frame are shown in Figure 20.

As can be seen in Figure 20, the lateral load capacity of Specimen S2 (hollow clay tile infill, strengthened by CFRP cross-strips) almost reached the capacity of S1 in which the infill was a reinforced concrete wall. The stiffnesses of S1 and S2 were not very different from each other, the stiffness of S2 being somewhat lower than S1. The basic difference between these two specimens was observed beyond the point where the specimens reached their capacities. While the displacement of S1 beyond the peak point increased with small reductions in strength, the strength drop in S2 was very significant once the maximum load was reached. It should also be noted that Specimen S1 had a much greater deformation capacity as compared to S2.

The readings obtained from the special transducers placed at the bottom of exterior columns revealed that up to almost the maximum load, the infilled frame took about 98 percent of the lateral load applied. Near the end of the test when the infilled frame suffered considerable damage, this share dropped to about 90 percent. The tests clearly showed that the contribution of the frames to lateral load carrying capacity is very small if adequate infill area is provided.

In Figure 21, the first story drift ratios are plotted against the lateral loads for specimens S1, S2 and the bare frame. As can be seen from this figure, the drift limit becomes very important for buildings rehabilitated by strengthening the masonry infills with CFRP strips. Beyond a certain drift limit (in this case 0.0035) the capacity drops drastically.

7. ANALYTICAL STUDIES

Analytical studies are evidently required to determine the amount of hollow clay tiles to be reinforced by using CFRP strips to bring the building to a satisfactory seismic performance level. Preliminary studies are underway. In these studies CFRP reinforced infills were modelled as diagonal tension struts. The strut model was developed by carrying out a trial and error procedure using the test results. The strut model developed was applied to a sample building and the results of the analyses were used to develop design criteria for strengthening of reinforced concrete structures using CFRP. Initial findings indicated that satisfactory seismic performance can be achieved by the strengthening of masonry infills with CFRP of twice the area of cast-in-place reinforced concrete infills required.

8. SUMMARY AND CONCLUSIONS

In Turkey, most of the residential buildings were designed and/or constructed with various deficiencies mainly due to lack of inspection. This fact results in an enormous number of structures to be strengthened. Upgrading all of these structures to a level that they could be in use after a major earthquake does not seem to be feasible. Thus, the objective in strengthening these structures should be collapse prevention.

Conventional strengthening techniques such as introduction of reinforced concrete infills do not seem feasible since they require the evacuation of the structure during the rehabilitation period. In this study, owing to the favorable mechanical properties and ease of application, Carbon Fiber

Reinforced Polymers (CFRP) was proposed as a solution to this engineering problem.

In this study, 7 one-bay, two-story, and 2 three-bay, two-story hollow clay tile infilled reinforced concrete frames were tested to evaluate the effectiveness of CFRP reinforcement in system strengthening. Six out of seven one-bay, two-story frames were strengthened using CFRP reinforcement. The other specimen was tested without any strengthening. This specimen served as a reference specimen. All frames were constructed with the common deficiencies seen in practice.

The conclusions given here are mainly based on the test results reported by the authors. Considering the type, configuration and size of the specimens tested, location of the lateral load and the load history, one should not generalize the conclusions without making careful judgments. Further experimental and analytical studies are needed to confirm some of the conclusions given below.

a) Tests on 1-Bay 2-Story Frames

– The effectiveness of the strengthening of infills by CFRP layers primarily depends on the anchorage of CFRP to the infill and frame members.
– CFRP reinforced masonry infill walls with CFRP overlays or CFRP strips anchored to the frame members and the masonry infill, improved the earthquake response of reinforced concrete frames having severe various deficiencies.
– CFRP strengthening did not increase the lateral stiffness significantly. The maximum increase in initial lateral stiffness was about 34 percent in specimen SP-5.
– The drift limit becomes very important for buildings rehabilitated by strengthening the masonry infills with CFRP strips. Beyond a certain drift limit the capacity drop is drastic. For specimens SP-5, SP-6 and SP-7, severe strength degradation started at a drift limit of 0.5 percent.
– It was observed that the energy dissipation capacity was increased in specimens which were strengthened by CFRP.
– Test results showed that local strengthening made at the base of the columns could lead to inferior behavior. Local strengthening made in specimen SP-6 by confining the base of the column at the foundation level by wrapping with CFRP changed the failure mechanism and resulted in a brittle shear failure. In the specimen where this local strengthening was not applied (specimen SP-5), the failure mode was flexure and therefore the behavior was more ductile.
– Although strengthening by CFRP strips applied on the infill was not as effective as the rehabilitation made by introducing reinforced concrete infills, the fact that CFRP strengthening can be applied without disturbing the occupancy makes this technique very attractive.

– Test results indicated that the failure of test specimens strengthened by applying CFRP layers to the infills is more brittle than those strengthened by introducing RC infills.

b) Tests on 3-Bay 2-Story Frames

– The lateral load capacity of Specimen S2 (hollow clay tile infill, strengthened by CFRP cross-strips) reached the capacity of S1 in which the infill was a reinforced concrete wall.
– The stiffnesses of S1 and S2 were not very different from each other, the stiffness of S2 being somewhat lower than S1.
– The basic difference between these two specimens was observed beyond the point where the specimens reached their capacities. The specimen S1 with reinforced concrete infill wall displayed a ductile and stable response beyond the peak point; the specimen S1 demonstrated a rather brittle behavior once the maximum load was reached. It should also be noted that Specimen S1 has a much greater deformation capacity as compared to S2.
– Drift control is very important for buildings rehabilitated by strengthening the masonry infills with CFRP strips. Beyond a certain drift limit the load drops very suddenly. This limit was 0.0035 for specimen S2.

ACKNOWLEDGMENT

The research work presented in this study is supported in part by the Scientific and Research Council of Turkey (TUBITAK) under grant: YMAU-ICTAG-1574 and by NATO Scientific Affairs Division under grant: NATO SfP977231. The contributions of TUBITAK and NATO are gratefully acknowledged.

REFERENCES

1. Ersoy, U. and Uzsoy, S., "The Behavior and Strength of Infilled Frames", Report No. MAG-205 TUBITAK, Ankara, Turkey, 1971. (In Turkish)
2. Altin, S., Ersoy, U., Tankut, T., "Hysteretic Response of Reinforced Concrete Infilled Frames", ASCE, Journal of Structural Engineering, Vol. 118, No. 8, August 1992, pp. 2133-2150
3. Sonuvar, M. O., "Hysteretic Response of Reinforced Concrete Frames Repaired by Means of Reinforced Concrete Infills", Ph.D. Thesis, Department of Civil Engineering, Middle East Technical University, June 2001
4. Canbay, E., Ersoy, U. and Ozcebe, G. "Contribution of RC Infills to the Seismic Behavior of Structural Systems", accepted for publication in ACI Structural Journal
5. Priestley, M. J .N., Seible, F, "Design of Seismic Retrofit Measure for Concrete and Masonry Structures", Construction and Building Materials, Vol.9, No.6, 1995, pp. 365-377
6. Roko, K., Boothby, T. E., Bakis, C.E., "Failure Modes of Sheet Bonded Fiber Reinforced Polymers Applied to Brick Masonry", ACI SP 188-28, 2000
7. Triantafillou, T. C., "Strengthening of Masonry Structures Using Epoxy-Bonded FRP Laminates", Journal of Composites for Construction, ASCE, May 1998, pp.98-104
8. ___________, "Specification for Structures to be Built in Disaster Areas", Ministry of Public Works and Settlement, Government of Republic of Turkey, Ankara, 1998 (in Turkish)
9. Smith, B. S., "Model Test Results of Vertical and Horizontal Loading of Infilled Frames", ACI Journal, pp.618-624, August 1968.

Table 1. Properties of test specimens

			SPECIMEN						
			SP-1	SP-2	SP-3	SP-4	SP-5	SP-6	SP-7
Longitudinal Reinforcement	Yield Strength (MPa)		388	388	388	388	388	388	388
	Ultimate Strength (MPa)		532	532	532	532	532	532	532
	Diameter (mm)		8	8	8	8	8	8	8
	# of Bars	Columns	4	4	4	4	4	4	4
		Beam	6	6	6	6	6	6	6
Transverse Reinforcement	Yield Strength (MPa)		279	279	279	279	279	279	279
	Ultimate Strength (MPa)		398	398	398	398	398	398	398
	Diameter (mm)		4	4	4	4	4	4	4
	Spacing (mm)		100	100	100	100	100	100	100
	End Hooks		90°	90°	90°	90°	90°	90°	90°
Compressive Strength of Concrete (MPa)			19.5	15.3	12.9	17.4	12.0	14.7	17.5
Compressive Strength of Mortar (MPa)			4.3	4.3	3.1	2.9	4.1	4.2	4.3
CFRP	Infill	Application Face	None[1]	Both[2]	Ext.[3]	Both	Both	Both	Both
		Type	-			Blnkt[4]	Strip[5]	Strip	Strip
		Anchorage	-	-	Yes	Yes	Yes	Yes	Yes
	RC Frame	Application Face	None	None	Ext.	Ext.	Ext.	Ext.	Ext.
		Type				Blnkt.	Strip	Strip	Strip
		Anchorage			Yes	Yes	Yes	Yes	Yes

[1] No CFRP is applied on either face
[2] CFRP is applied on both faces
[3] CFRP is applied on exterior face only
[4] CFRP Blanket that fully covers the infill surface
[5] CFRP strips that extend along the main diagonals of the infill

Table 2. Summary of the test results

Specimen	Maximum Lateral Load (kN)	Initial Stiffness (kN/m)	Total Dissipated (kN-m)	Maximum Interstory Drift (percent)
SP-1	55.8	29,660	2.5	1.70
SP-2	64.6	29,520	6.1	1.49
SP-3	65.4	21,820	8.7	1.14
SP-4	131.5	36,430	11.1	2.68
SP-5	118.8	39,604	7.8	1.75
SP-6	100.4	32,624	4.2	0.88
SP-7	105.7	24,392	4.0	0.86

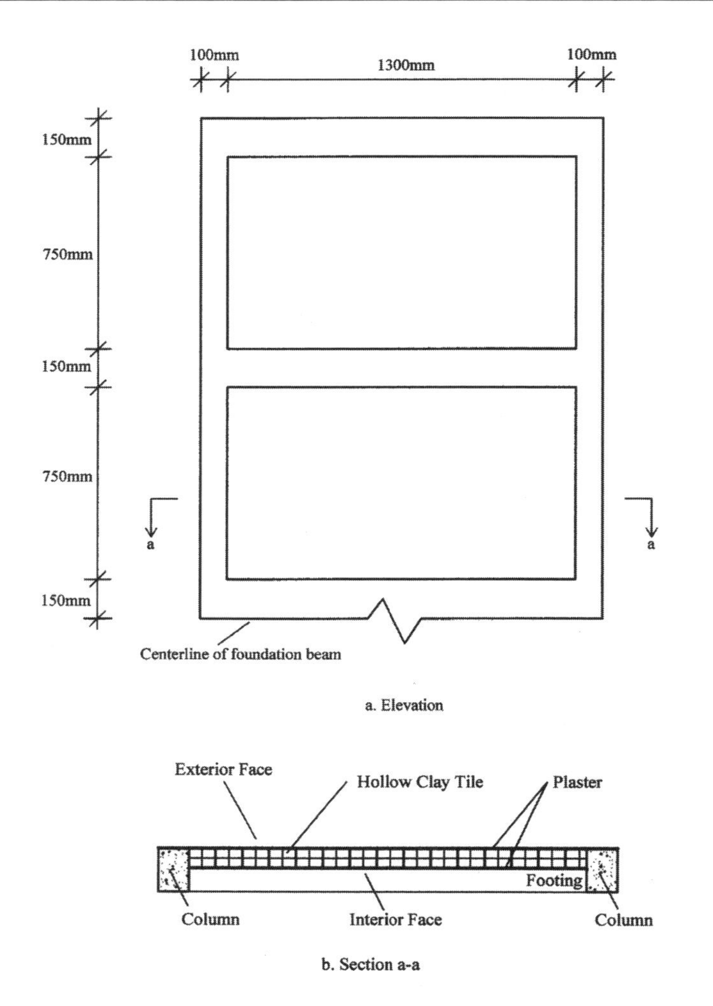

Figure 1. Geometry of the one-bay, two-story test specimens

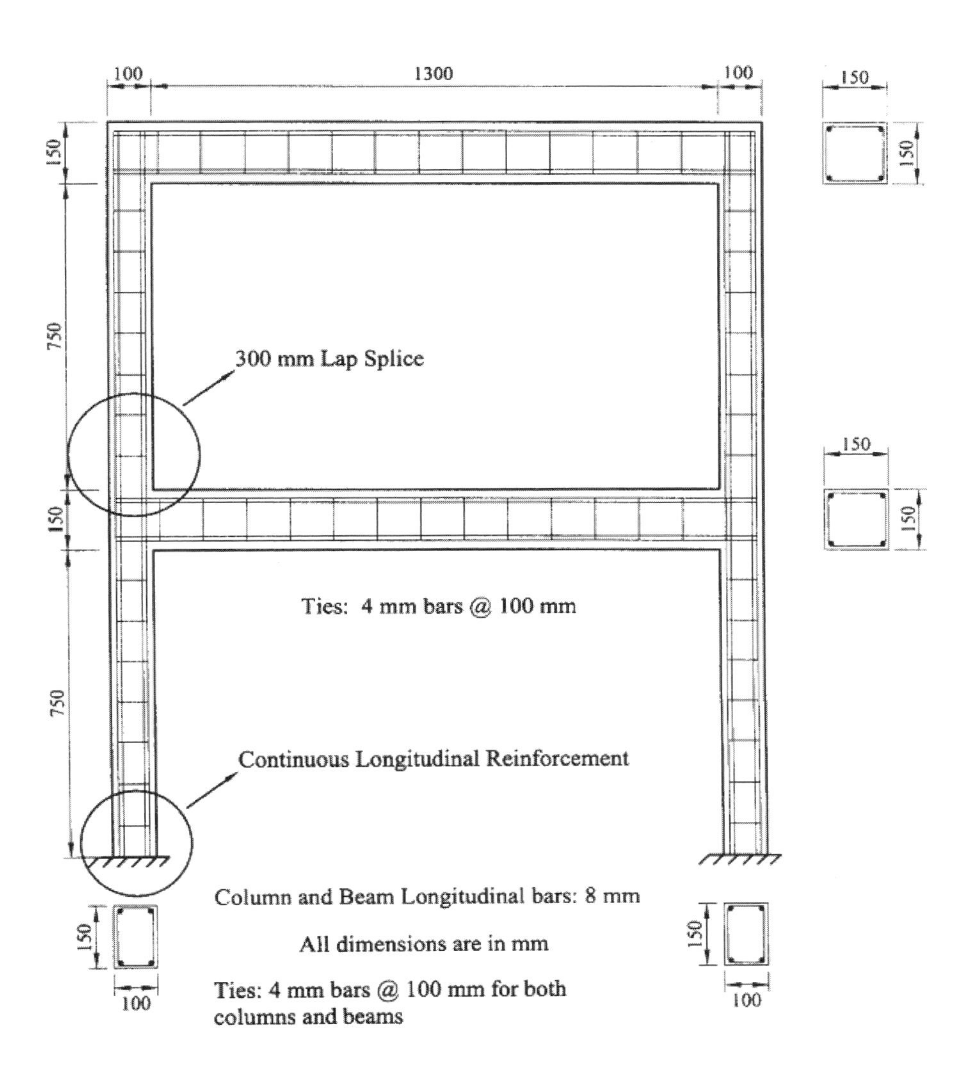

Figure 2. Geometry Reinforcement details of the one-bay, two-story test specimens

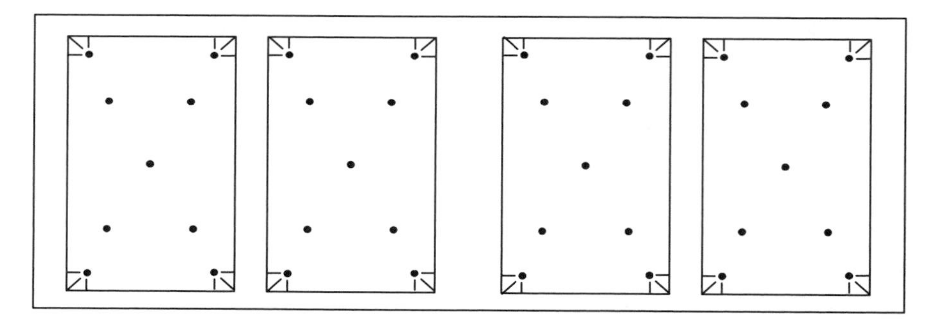

Interior Face

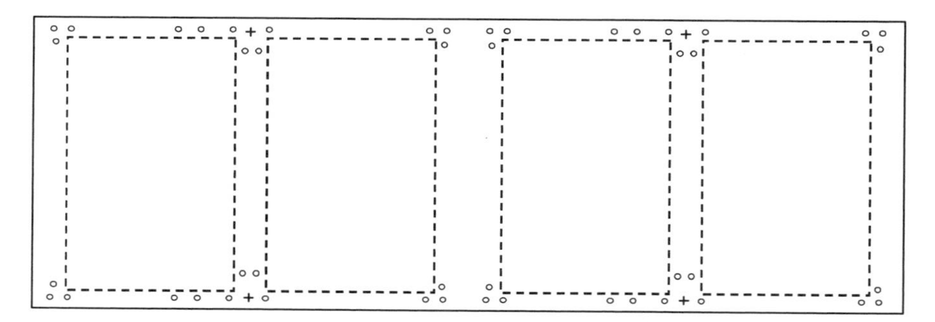

Exterior Face

Anchor		Applied to	Depth in RC (mm)	Width of Strip (mm)	Diameter of Hole (mm)
TYPE A	•	Infill		50	10
TYPE B	-	Concrete	50	25	10
TYPE C	o	Concrete	60	25	10
TYPE D	+	Concrete	60	30	12

Figure 3. The location and design details of anchor dowels in specimen SP-4

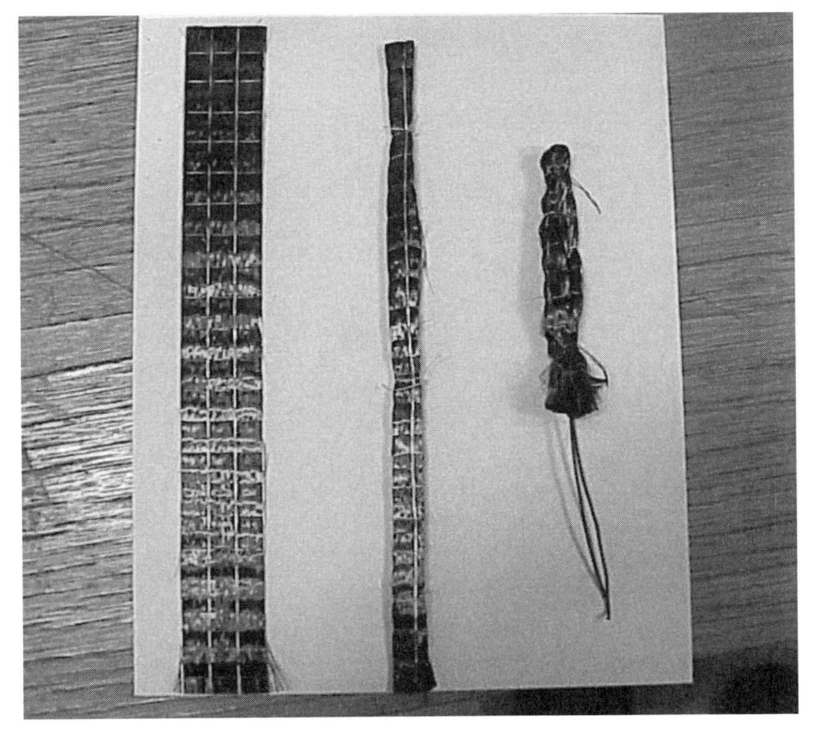

(a) Manufacturing process of anchor dowels

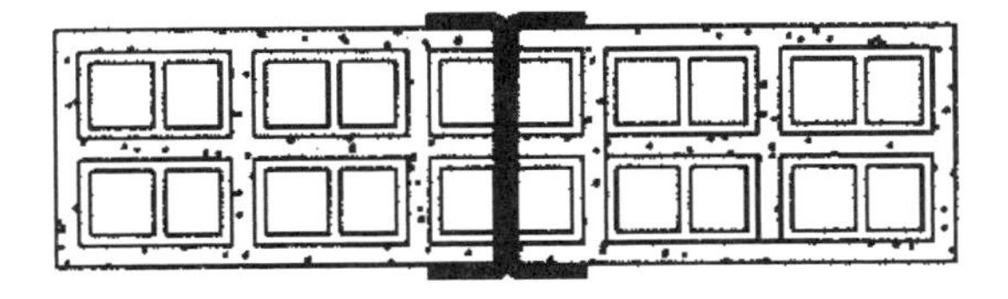

(b) Type A – applied to infill

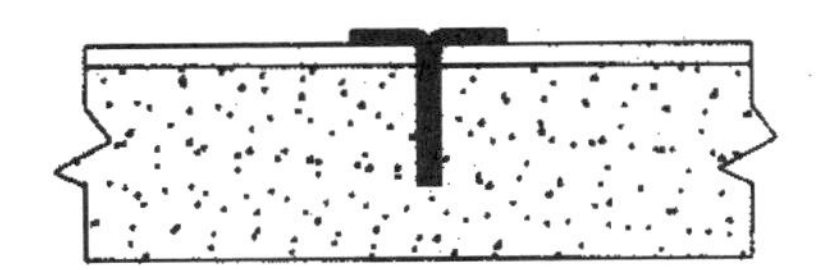

(c) Types B, C, D – applied to frame members

Figure 4. Manufacturing and application of anchor dowels

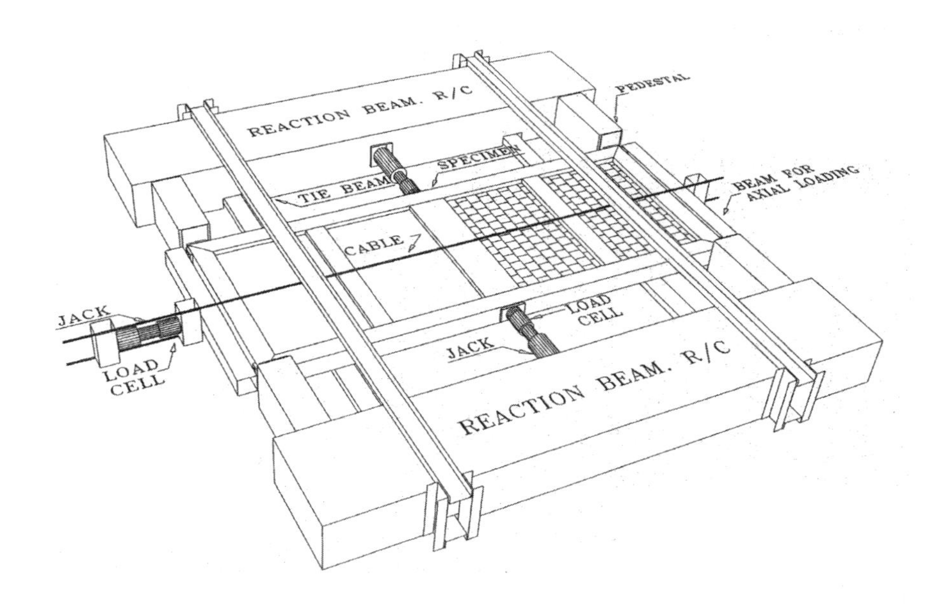

Figure 5. Test Setup

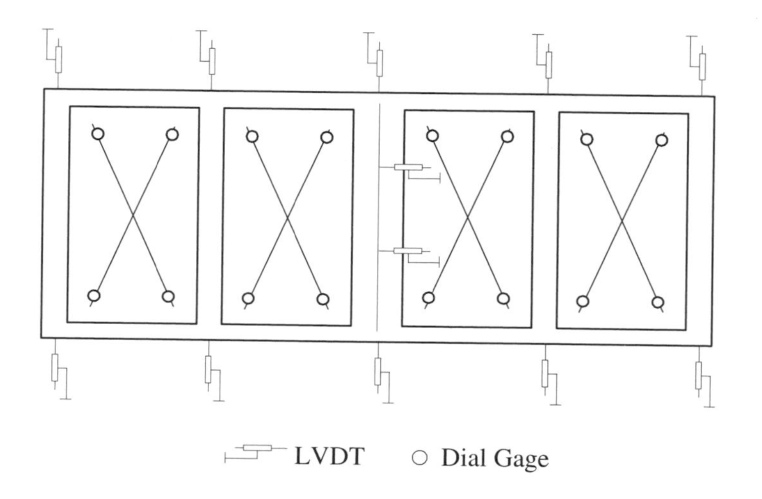

Figure 6. Instrumentation

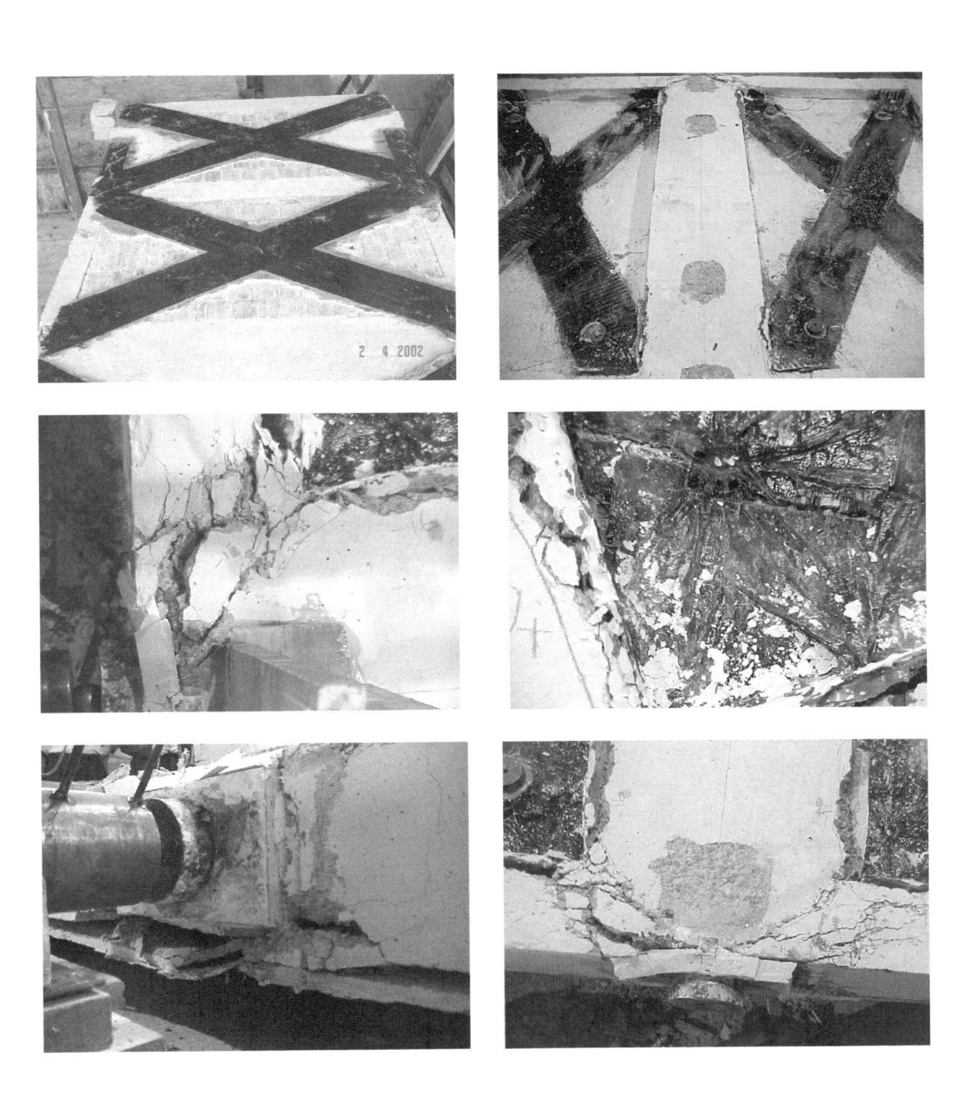

Figure 7. Specimen SP-5 after the test

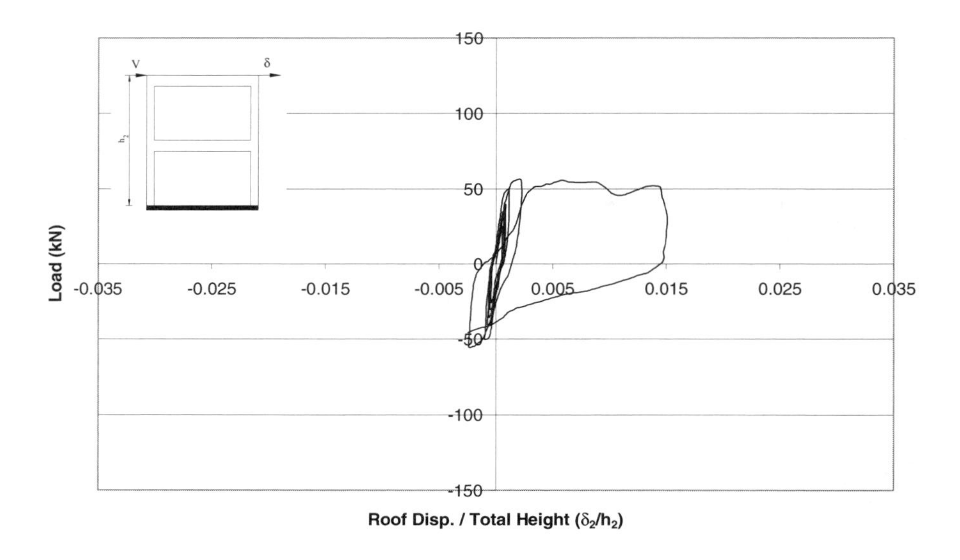

Figure 8. Lateral load versus roof drift ratio hysteresis curve for SP-1

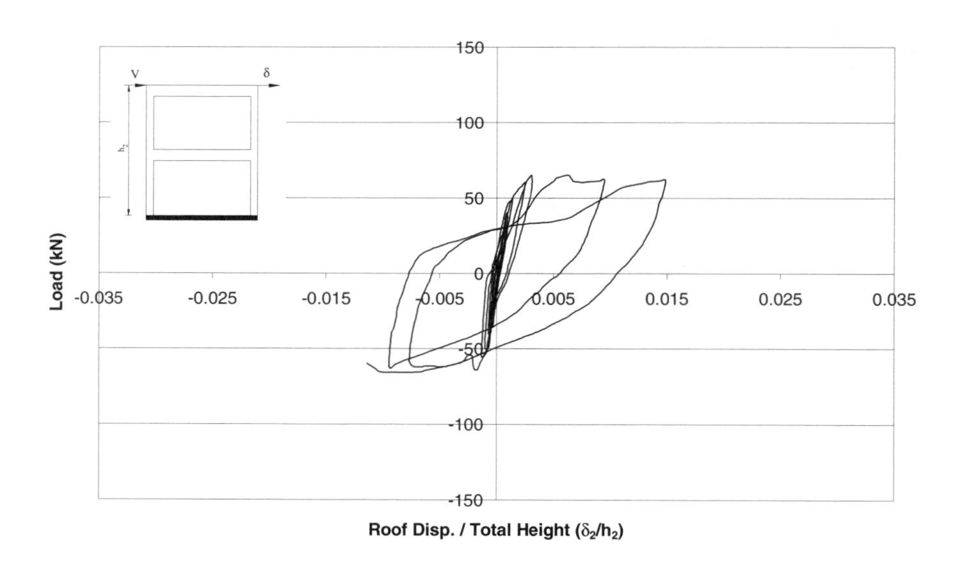

Figure 9. Lateral load versus roof drift ratio hysteresis curve for SP-2

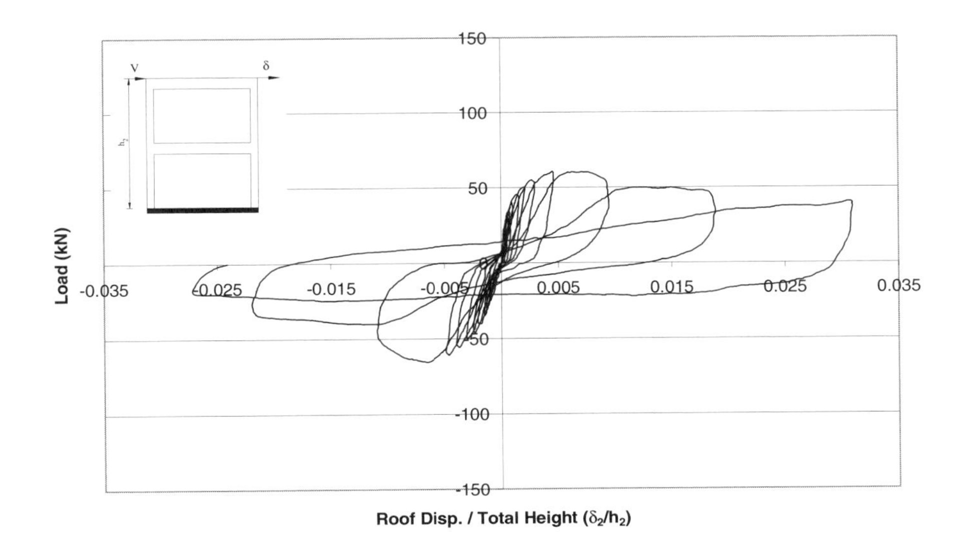

Figure 10. Lateral load versus roof drift ratio hysteresis curve for SP-3

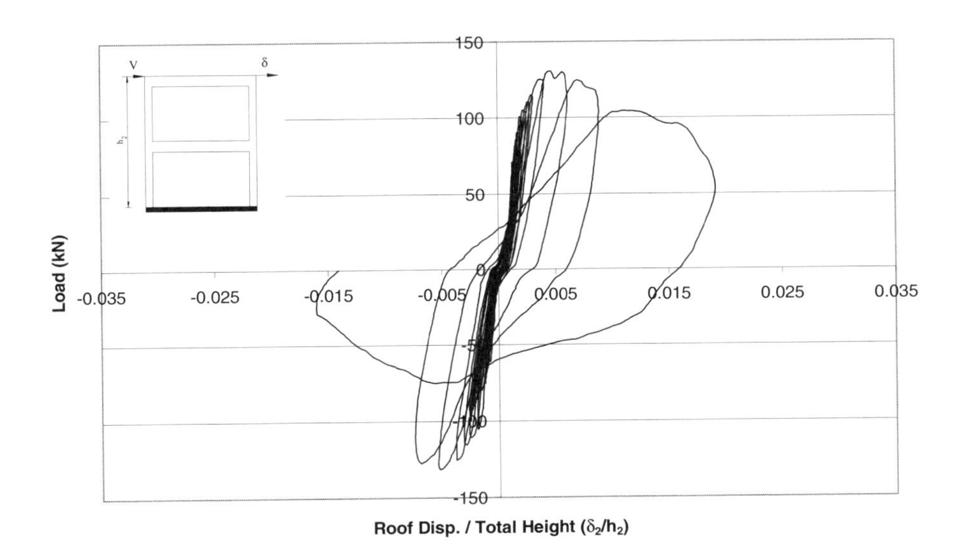

Figure 11. Lateral load versus roof drift ratio hysteresis curve for SP-4

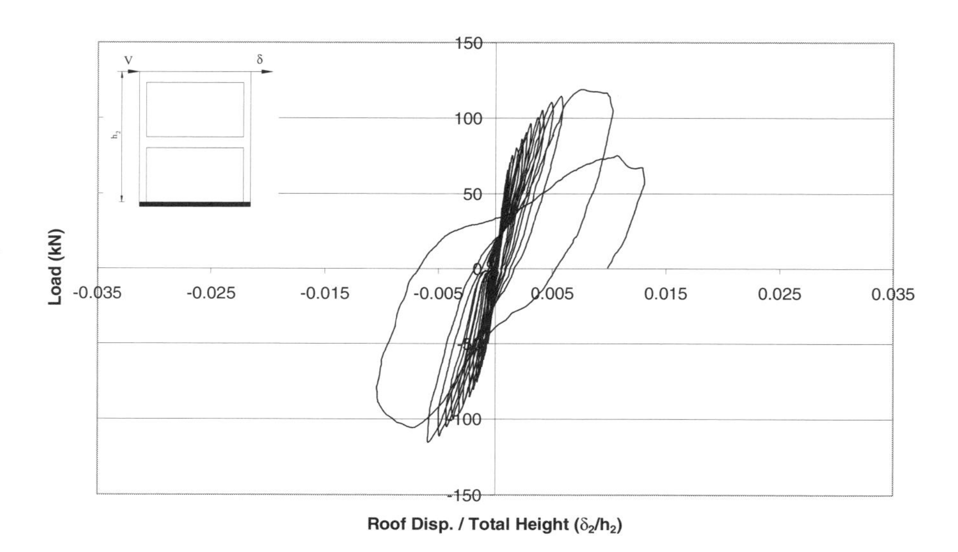

Figure 12. Lateral load versus roof drift ratio hysteresis curve for SP-5

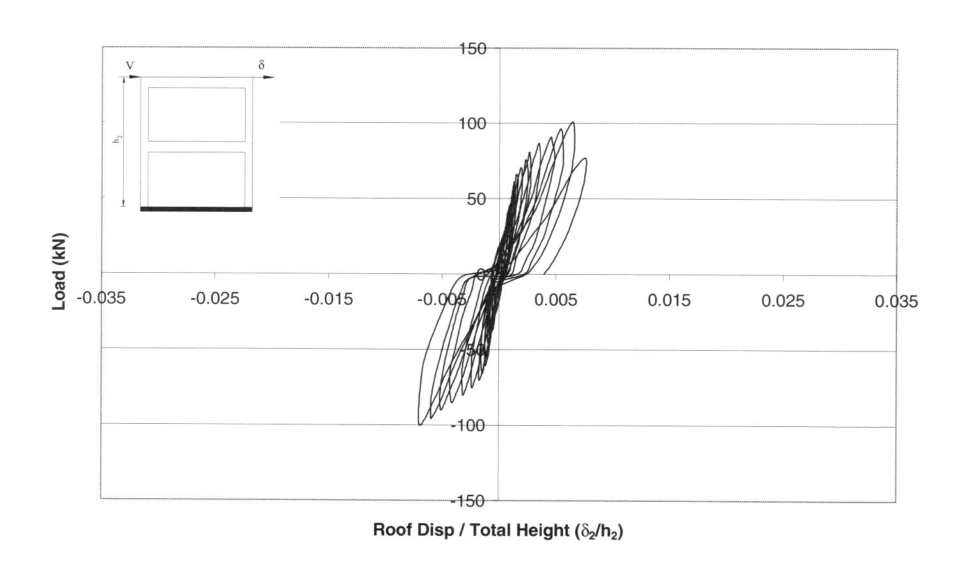

Figure 13. Lateral load versus roof drift ratio hysteresis curve for SP-6

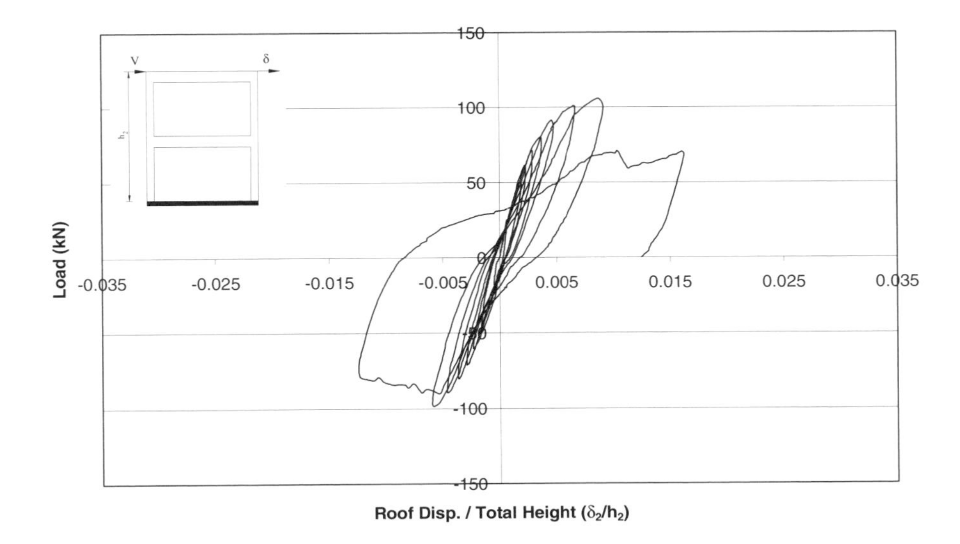

Figure 14. Lateral load versus roof drift ratio hysteresis curve for SP-7

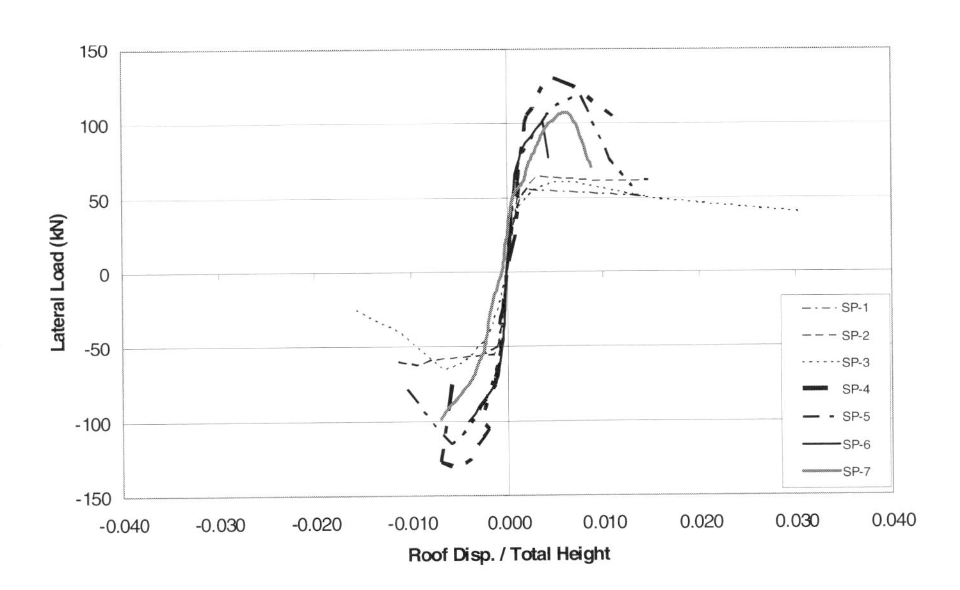

Figure 15. Response envelope curves

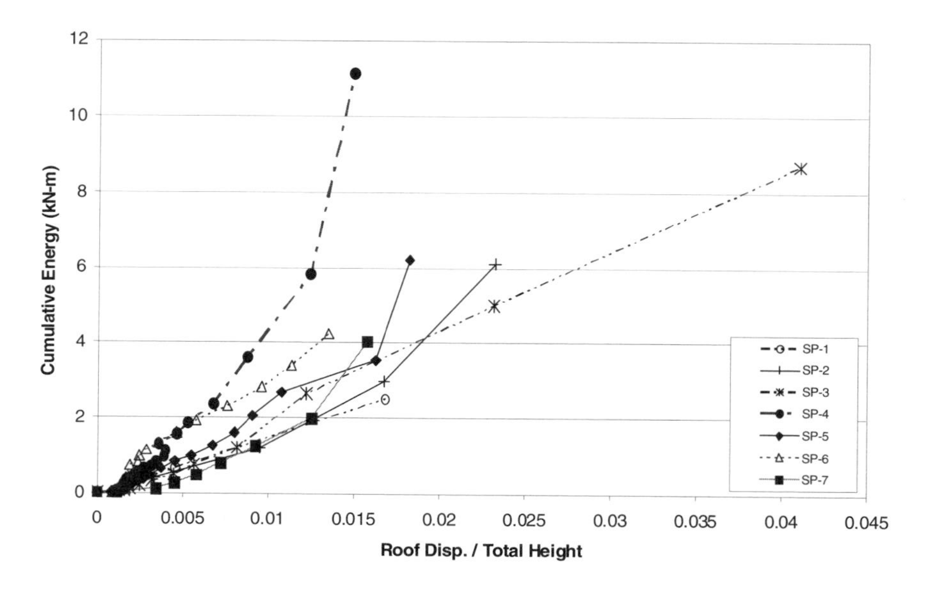

Figure 16. Cumulative energy dissipation for test specimens

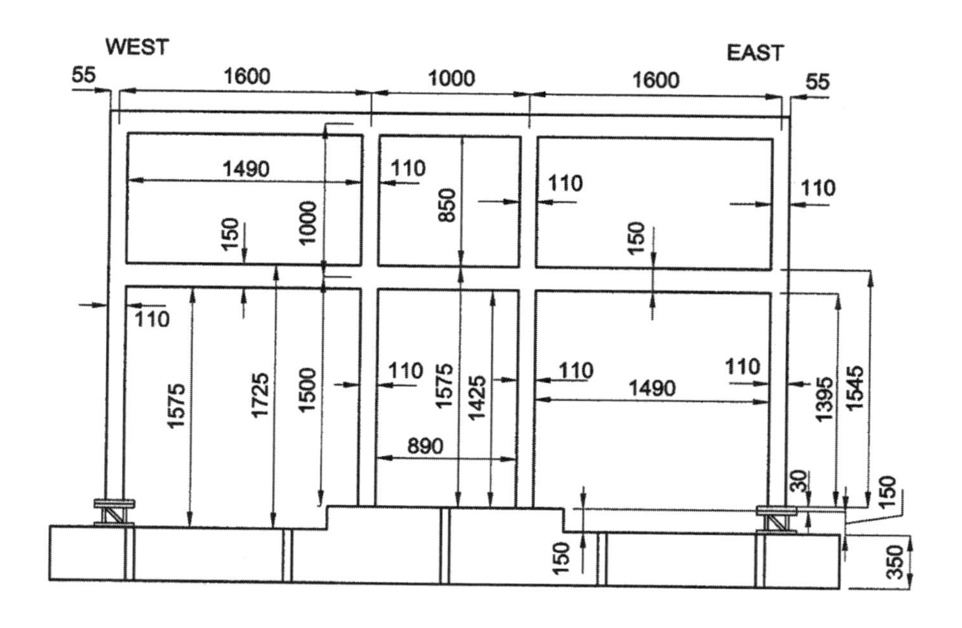

Figure 17. Dimensions of the test specimen

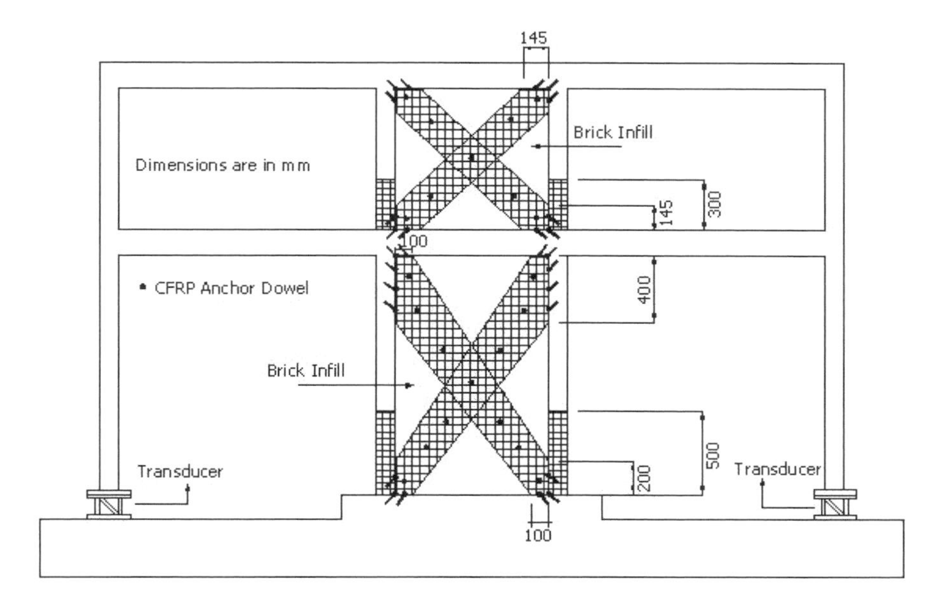

Figure 18. CFRP strips used to strengthen the infill, specimen S2

Figure 19. Specimen S1 before test

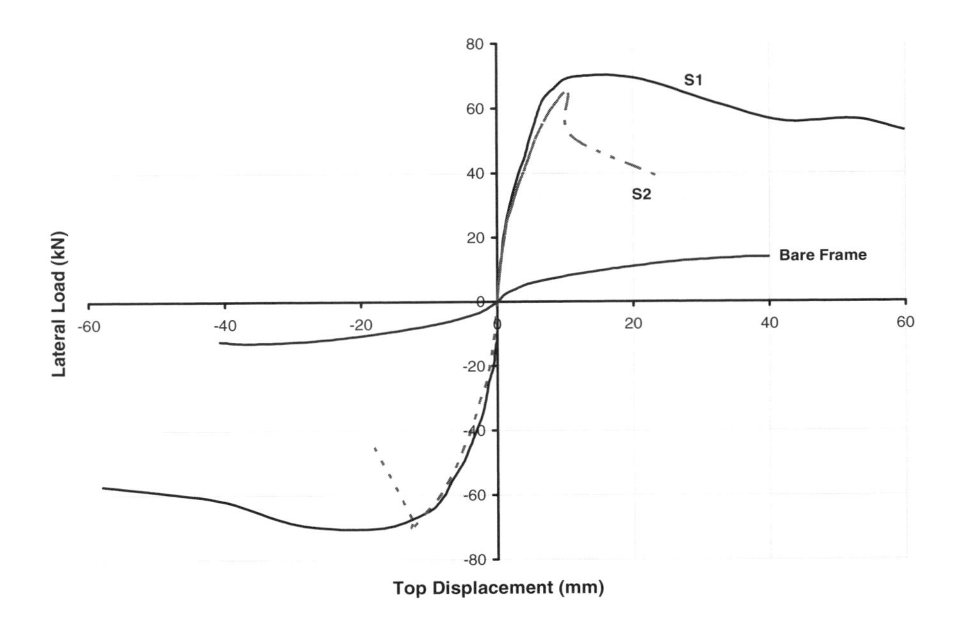

Figure 20. Load versus top displacement envelop curves

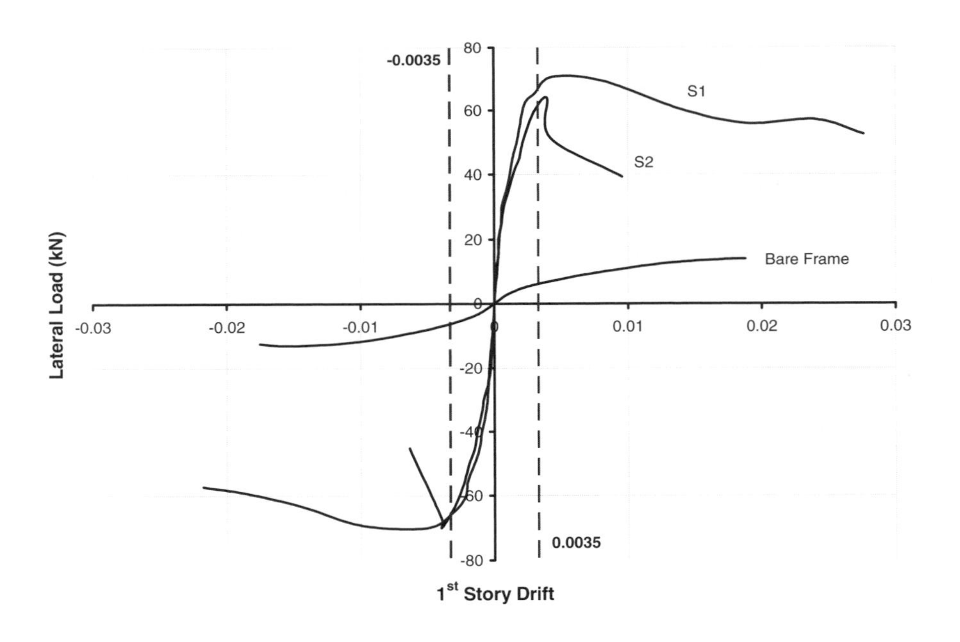

Figure 21. First story drift ratios

STRUCTURAL BEHAVIOUR OF ORDINARY RC BARE AND BRITTLE PARTITIONED FRAMES WITH AND WITHOUT LAP SPLICE DEFICIENCY

Faruk Karadogan, Ercan Yuksel and Alper Ilki
Structural And Earthquake Engineering Laboratory of Istanbul Techical University

Abstract: In the framework of an experimental program supported by a NATO project, it is aimed to examine the performance of a new strengthening technique for existing buildings using FRP sheets. For this purpose, three preliminary ½ scale tests were carried out in Istanbul Technical University. Since lap splicing deficiency in columns has been selected as one of the major factors causing early collapse of buildings, two similar sets of one bay-two story specimens have been fabricated with and without lap splicing deficiency. The early results achieved at the end of tests, and the theoretical predictions are presented in this paper.

Key words: brittle partitioning, infill wall, reinforced concrete frames, strengthening

1. PURPOSE

It is very well known that an effective rapid evaluation technique is urgently needed especially for existing low rise reinforced concrete buildings in seismically active regions of the country. The general structural deficiencies and the material characteristics of these buildings, which have

S.T. Wasti and G. Ozcebe (eds.), Seismic Assessment and Rehabilitation of Existing Buildings, 335–356.

similarities to the buildings damaged or collapsed during the severe earthquakes in the last decade in Erzincan, Dinar, Adana and Kocaeli-Golcuk, are all known. It is not easy to have an analytic way of being sure to what extent the code requirements are satisfied by these ordinary buildings. There is no doubt that the observation of progressive damage accumulation in realistically prepared complementary experiments will have a tremendous amount of contribution not only to develop evaluation techniques which consider all the local parameters but also to propose new, effective and even cheaper strengthening techniques. Actually, these are the two major goals of the experimental and theoretical research program, which is going on in Istanbul Technical University (ITU) in connection with other institutions under the general sponsorship of NATO. The results compiled in this paper are the results of three tests carried out on one bay-two story RC frame specimens. Those are the first specimens detailed, fabricated and tested as scheduled. They consist of two bare frames and one infilled and plastered frame as is done in practice. Two of the specimens have been fabricated so that no lap splicing deficiency exists. However, the other one has lap splicing deficiencies at critical column sections. One of the specimens of each group is infilled with brittle brick masonry.

The possible differences between the structural behaviors of the three specimens have been observed carefully to collect information for the deformations or accumulated damage levels, ultimate strengths and especially for the participation of each component, namely the frame and the wall, to the combined ultimate strength of the infilled frame.

2. SPECIMENS

The geometry and expected ultimate strength of specimens, loading protocol, the details of the specimens have been decided in a common meeting supervised by counterpart researchers from Middle East Technical University (METU) mainly considering: *i-* the room and loading capacities of the laboratory, *ii-* the structural features of earlier tests carried out in different institutions. Depending on these decisions the prototype shown in Figure 1 has been prepared. Two frames have been cast first with continuous column longitudinal bars, specimens BC-0-1-11 and IC-0-1-5.8. 13.5×20×20 cm brittle bricks used widely in practice have been cut in half in the transverse direction to the holes and used as infill material in specimen IC-0-1-5.8. At the bottom sections of the third and fourth specimens, BL-0-1-19.5 and IL-0-1-19.5, only 20ϕ development length for bars 4ϕ12 in columns have been used to represent the lap splice deficiencies encountered frequently in practice. Although mild steel has been used, longitudinal bars had no hooks in the critical regions of the columns.

Specimen IL-0-1-19.5 has been infilled with the same bricks used for the specimen IC-0-1-5.8 and been plastered in a similar way. Different stages of fabrication are given in Figure 2.

During the stages of concrete casting several standard cylinder specimens have been taken and cured in the same way as the frames, and tested for compressive strength of concrete, whether mixed in laboratory or obtained from concrete producers. For the mortar used in the construction of brick walls and at the stage of plastering many standard cube and cylinder specimens have again been taken and tested in the laboratory. The results are tabulated in Table 1; stress and strain relationships obtained for mild steel are shown in Figure 3.

3. TESTING SETUP

The general view of a bare frame at the stage of testing is shown in Figure 4 together with the necessary adapter pieces and devices used for applying lateral and vertical loads on to the specimen.

More details can be found in Figures 5 and 6, where the fixation of the specimens on the adapter foundation, which has been specially cast for these tests, the points where the transducers are located for displacement measurements and the locations of strain gauges can be followed up better.

The point load created by the actuator is applied to the specimens through a thick plate, which is approximately centered at the center of rigidity of the specimens to prevent the possible secondary torsional effects. In the case of eccentrically placed infill walls this had more importance and extra provisions were taken for safety and reliability of the test. All out of plane displacements were recorded to observe the magnitude of the inevitable torsion.

Axial loads acting on the columns have been kept constant as much as possible during the whole course of test by checking the intensities and pumping oil to the jacks when necessary, Figure 7. The intensity of column axial forces was at the order of 15% of the axial capacity of the columns. Doing that theoretically it has been assumed to have no important frame action in the prototype. In other words the expected changes of axial force due to lateral loads are neglected or the rotations at joints are omitted theoretically. Up to more or less 60% of the expected ultimate load, *the force control mode of actuator system* and afterwards *the displacement control* has been preferred with the displacement protocol shown in Figure 7.

4. THEORETICAL PREDICTIONS FOR POSSIBLE STRUCTURAL BEHAVIORS

The computer program DOC2B developed in the Civil Engineering Faculty of ITU has been utilized first [1] to predict the nonlinear behavior of bare frames with different compressive strengths of concrete.

Briefly, it can be said that DOC2B takes into consideration not only the geometrical nonlinearities but also the distributed plastic deformations due to material nonlinearity. The linearization technique employed by DOC2B is based on the secant modulus. If it is required, the moment-curvature (M-χ) diagrams of critical sections can be updated at each step of the load increments considering the current axial force on the section. The stress-strain relationships of confined concrete and steel with strain hardening are used to obtain the M-χ relationships. The structures subjected to any kind of monotonic load combinations can be analyzed by DOC2B.

The base shear – top displacement curves obtained for the specimens BC-0-1-11, BL-0-1-19.5 and the bare frame of the specimen IC-0-1-5.8 are presented in Figure 8, together with M-χ diagrams of the specimen BC-0-1-11 referred to in the analysis for column and beam sections.

DOC2B has been modified so that it is possible to take into consideration any kind of nonlinear model for the equivalent struts of the infill walls, [2], [3], [4]. The one proposed to be used as the default model of the program is given in Figure 9.

The material characteristics, which are needed in this model have been derived by referring to the early results of other current research going on in the Structural and Earthquake Engineering Laboratory of ITU. One of the so-called pure shear tests proposed by ASTM and carried out in ITU is shown in Figure 10, together with the associated shear stress-shear strain diagram obtained experimentally, [5].

The analytical base shear – top displacement curve for the specimen IC-0-5.8 obtained using the modified version of DOC2B is given in Figure 11.

5. TEST RESULTS

Experimentally found base shear versus top displacement curves of two bare frames, namely the specimens BC-0-1-11 and BL-0-1-19.5 are presented in Figure 12 and 13, respectively. Experimentally found base shear versus top displacement curves of the infilled frame, namely the specimen IC-0-1-5.8 is presented in Figure 14.

Envelopes of base shear – top displacement curves for all of the specimens are put together for having better comparisons in Figure 15.

If the theoretical predictions and experimentally obtained base shear-top displacement curves are presented in the same diagram, the results given in Figure 16 are obtained.

Damage accumulated in the beam-column connections is presented in Figure 17.

Damage accumulated in the support regions for the displacement ductility level around 4, is presented in Figure 18.

At the bottom sections of the columns, base shear – curvature relationships have been obtained evaluating the strains measured on the longitudinal reinforcement bars, Figure 19.

The average diagonal shortening of upper infill wall is given in Figure 20 for specimen IC-0-1-5.8.

After having reached an approximate displacement ductility ratio of 4, the general appearances of the specimens are as shown in Figure 21.

Crack widths at critical sections are shown in Figure 22 and Table 2 for different levels of top displacements.

The out of plane displacements of infilled frame measured during the test are presented in Figure 23.

6. GENERAL OBSERVATIONS AND EARLY RESULTS

1. The general observations on the overall behavior of three specimens tested for the time being, as part of the envisaged total of 12 specimens for the experimental program in ITU, indicate that:
2. Infill walls placed eccentrically into the frame to represent the exterior walls of real buildings, may cause difficulty in testing of the specimens, but a properly chosen eccentricity given on purpose for loading, significantly reduces the problem. Only a limited amount of torsional effects namely, out of plane displacements, on the infill frame specimens were recorded.
3. Testing bed and fixation details that have been specially prepared for these tests are good enough to be used for the rest of the program. However the way of vertical load application to the specimens should be reviewed and had better be changed so that *frame action* be maintained during the whole course of test. Since recently two lateral loads are planned to be applied on two story levels of specimens, new adapter pieces should be designed and be fabricated for that purpose. The theoretical work carried out to compare the possible nonlinear behavior of specimen BC-0-1-11 subjected to two different loading patterns shows some important differences in structural behavior as expected. Lateral

load capacities and distribution of plastic deformations are affected because of the loading patterns.

4. The displacement ductility of approximately 4 is achieved without having serious instability or local or premature failure such as crushing of the concrete in the core of column sections or having buckled reinforcement etc. However severe damage has been observed in the beam-column connections of the first story, which has not been introduced into the nonlinear analysis.
5. Bond between the peripheral RC elements and infill walls has been lost even at the lower level of displacement cycles. Hence, the small gap between the wall and frame became very effective on the overall behavior of the infilled frame. It is interesting to observe also that the infill wall had no diagonal crack even at the end of the test, but a certain extent of crushing on one corner only, which is not similar to the common or generally expected X type failure mode of brittle brick walls. The infill walls have almost their own vertical and lateral load-bearing capacities at the end of the test. This has to be kept in mind and if a very slight improvement in connection is provided between frame elements and wall, this will cause substantial increments in the lateral load carrying capacity of the compound system as has been shown in the theoretical analysis. This result is in parallel with the results of earlier tests carried out in ITU on different types of integrated infill walls, [6] [7]. Strips of carbon or glass fibers over the boundaries of walls must be tried within the framework of experimental investigation for strengthening of specimens planned in laboratory.
6. Theoretical predictions are encouraging not only for the work done for the bare frame but also for the work initiated to take into account the nonlinear behavior of the infill walls. It should not be forgotten during the evaluation stage of the results that the total capacity of wall has not been utilized in the tests but only in theoretical work.
7. It is clearly observed that the contribution of the frame to the lateral load capacity of the infilled frame is smaller than the lateral ultimate capacity of the bare frame. This is just because of the big difference between the displacement ductilities of the two structural systems. This has to be considered not only in the evaluation stage of testing the lateral load bearing capacity of buildings but also at the stage of strengthening.
8. Although during the test of lap-spliced specimen (BL-0-1-19.5), several cracks indicating the initiation of loss of bond were observed, the overall behavior of this specimen was not worse than the specimen with continuous longitudinal reinforcement (BC-0-1-11). This may be attributed to the differences in the concrete compressive strengths of these specimens.

The results to be obtained at the end of similar tests on specimens strengthened by carbon or glass fibers will be presented elsewhere.

ACKNOWLEDGEMENTS

This study is financially supported by NATO and The Scientific and Technical Research Council of Turkey. The assistance of H. Saruhan, C. Demir, G. Erol, A. Tezcan and E. Yilmaz are acknowledged.

REFERENCES

1. Yüksel, E., "Nonlinear Analysis of 3D Structures with Certain Irregularities", PhD Thesis Submitted to Istanbul Technical University, 1998, (In Turkish).
2. Combescure, D., "Modélisation du Comportement sous Chargement Sismique des Structures de Bâtiment Comportant des Murs de Remplissage en Maçonnerie", 1996.
3. Madan, A., Reinhorn, A., Mander, J., Valles, R., "Modeling of Masonry Infill Panels for Structural Analysis", Journal of Structural Engineering, Vol.123, No.10, 1997.
4. Saneinejad, A., Hobbs, B., "Inelastic Design of Infilled Frames", Journal of Structural Engineering, Vol.121, No.4, 1995.
5. Strengthening of Brittle Brick Walls using Carbon or Glass Fibers, An Experimental Study Being Conducted in Structural and Earthquake Engineering Laboratory of ITU.
6. Yüksel,E., Teymür, P., "Bölme Duvarlarına Yönelik Çalışmalar", Yapı Mekaniği Laboratuvarları Toplantısı, TÜBİTAK, Kasım 2001.
7. Yüksel,E., İlki,A., Karadoğan,F. "Strengthening of Reinforced Brittle Masonry", 11th European Conference on Earthquake Engineering, Paris,1998.

Table 1. Features of the specimens

Specimen	Testing date	Frame type	Lap-splice	Characteristic concrete compressive strength, MPa	Characteristic mortar compressive strength, MPa
BC-0-1-11	23-11-2002	Bare	Continuous	11.3	-
IC-0-1-5.8	15-03-2003	Infilled	Continuous	5.8	2.05 (wall),3.79 (plaster)
BL-0-1-19.5	01-04-2003	Bare	20 ϕ_l	19.5	-
IL-0-1-19.5		Infilled	20 ϕ_l	19.5	-

Table 2. Crack widths

	Top displecement (mm) **Push**					Top displecement (mm) **Pull**			
Crack no	22	44	66	88	Crack no	-22	-44	-66	-88
1	0.1	0.3	1.8	2.0	1'	0.3	0.7	1.6	2.5
2	0.3	0.8	1.6	2.0	7'	0.1	0.3	0.7	1.8
3	0.2	0.5	1.4	3.0					
25	0.1	0.3	0.8	1.2					

	Top displecement (mm) **Push**				Top displecement (mm) **Pull**		
Crack no	14	22	33	Crack no	-11	-22	-33
41	0.1	0.5	1.0	41'	0.8	2.0	2.5
42	0.7	0.9	1.5	44'	0.8	2.0	6.0
44	0.1	0.9	3.5				

	Top displecement (mm) **Push**					Top displecement (mm) **Pull**			
Crack no	22	44	66	88	Crack no	-22	-44	-66	-88
1	0.4	1.8	2.7	3.5	2'	0.4	1.4	2.0	3.6
4	0.6	1.6	1.4	3.0	3'	1.0	1.8	1.6	3.5
5	0.7	1.4	2.5	3.5	7'	0.2	1.2	2.0	2.2
18	0.1	0.4	0.8	2.5	17'	0.1	0.1	0.5	1.6

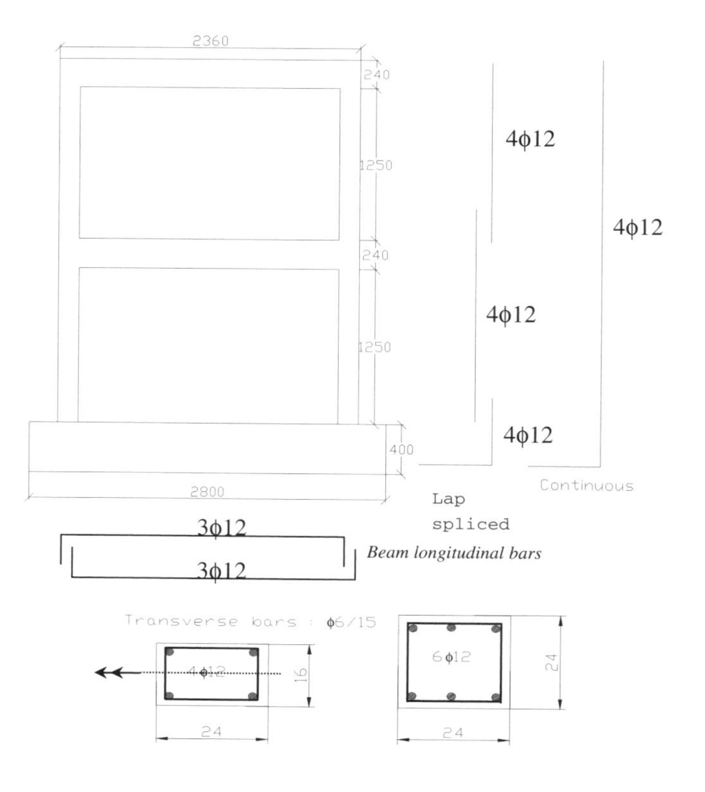

Figure 1. General view of the specimens

Figure 2. Different stages of fabrication

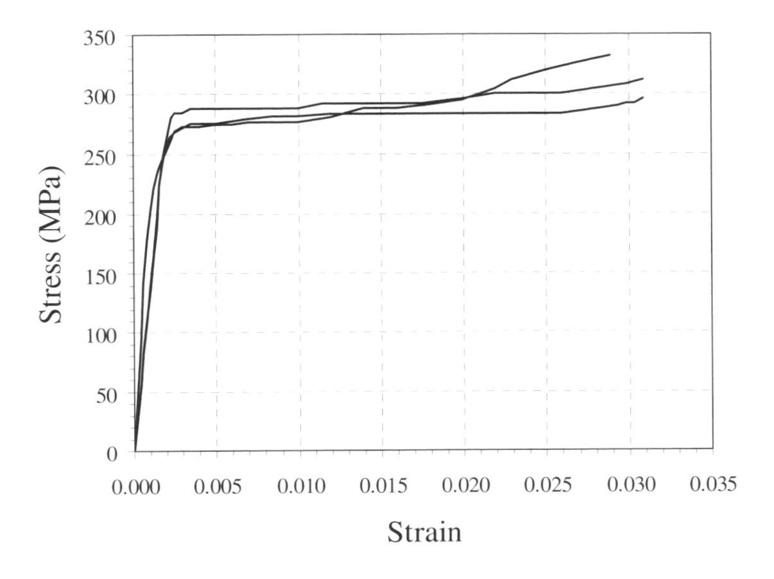

Figure 3. Stress-strain relationships of longitudinal reinforcement

Figure 4. The general view of a bare frame specimen at the stage of testing

Figure 5. Lateral and vertical loading pieces

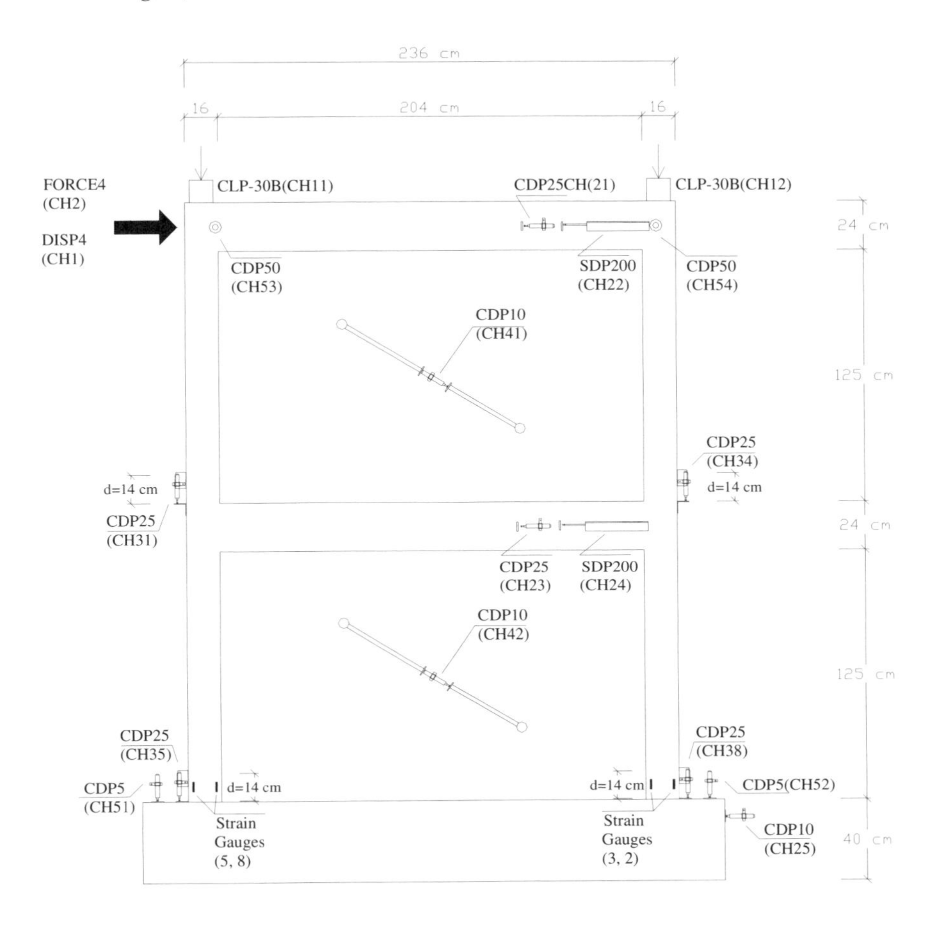

Figure 6. Locations of measuring instruments

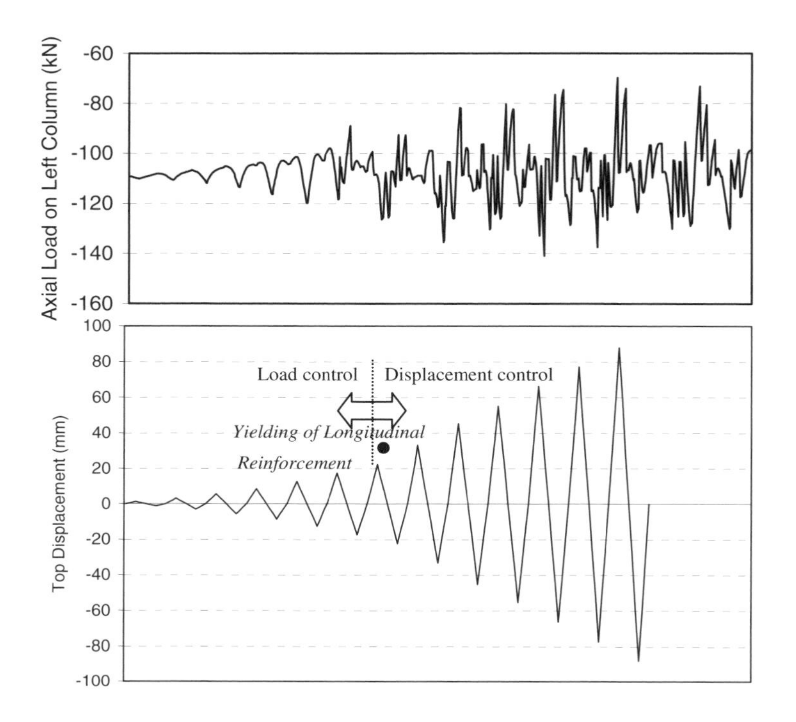

Figure 7. Variation of axial load and loading protocol for specimen BL-0-1-19.5

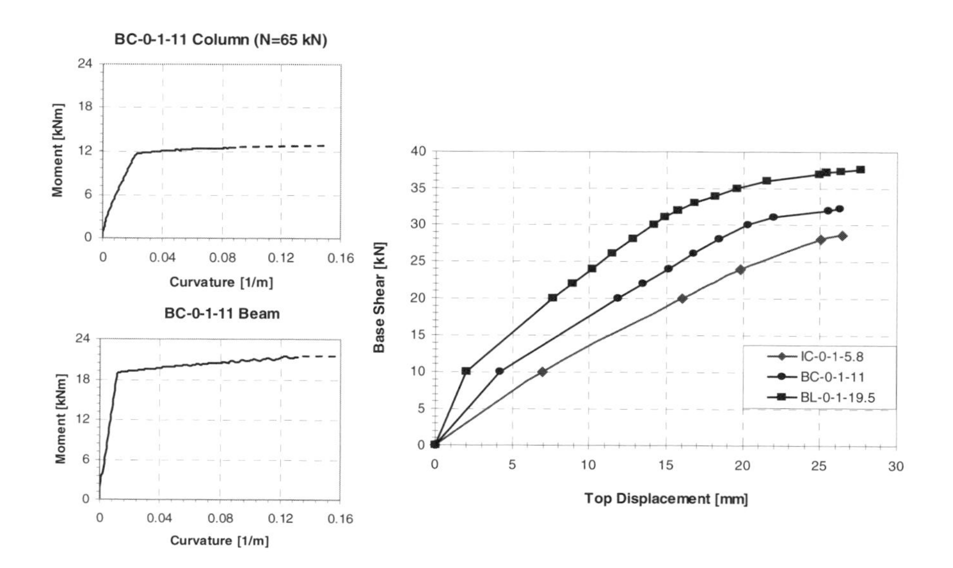

Figure 8. Prediction of base shear versus top displacement of bare frames

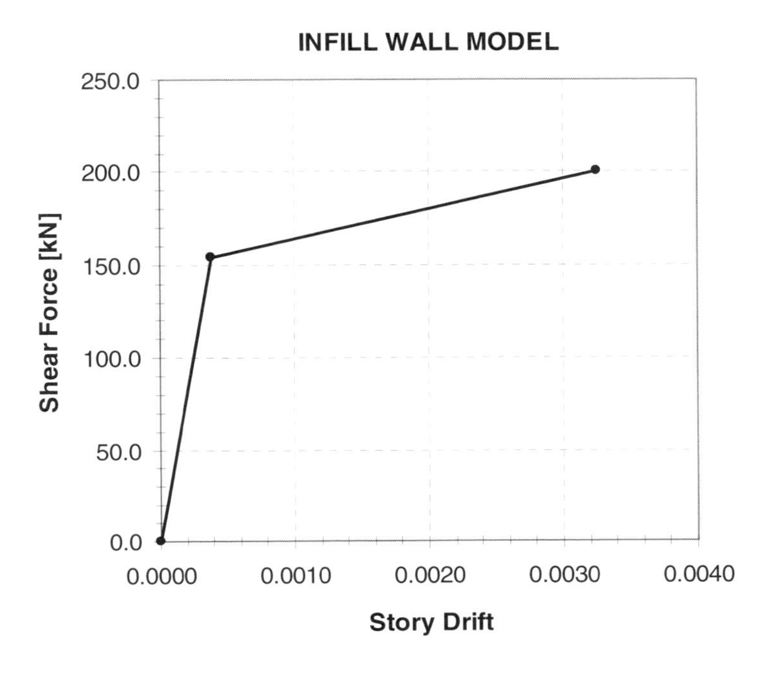

Figure 9. Default equivalent strut model used in DOC2B, [1], [2]

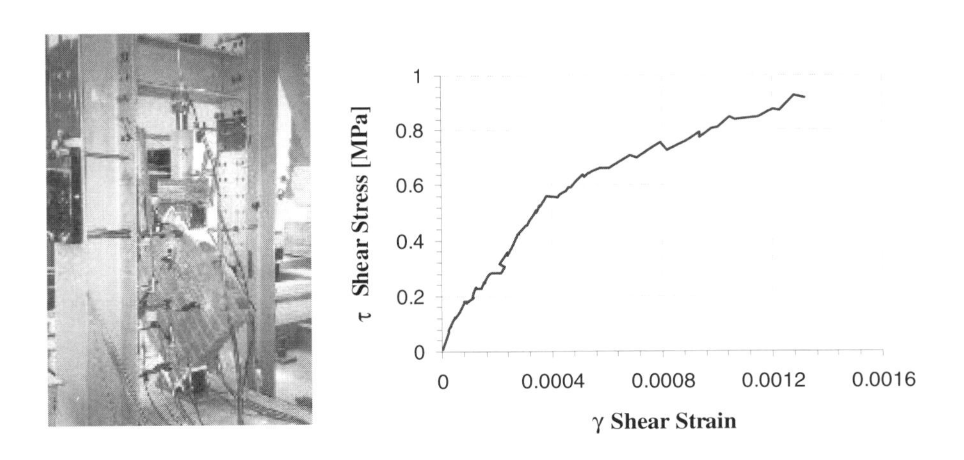

Figure 10. Testing setup for shear test and experimentally obtained shear stress-shear strain diagram of brittle brick wall

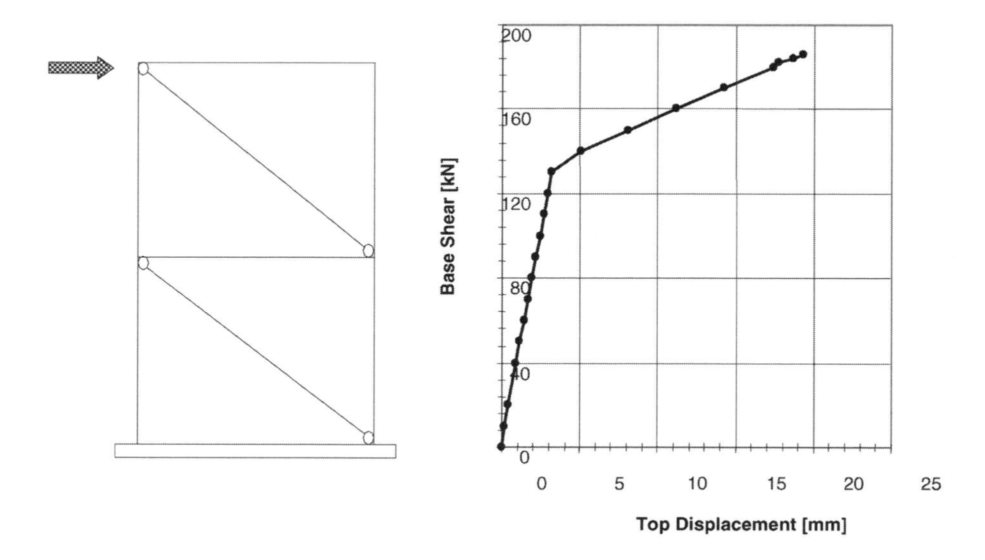

Figure 11. Analytical base shear versus top displacement curve for the infilled specimen IC-0-1-5.8

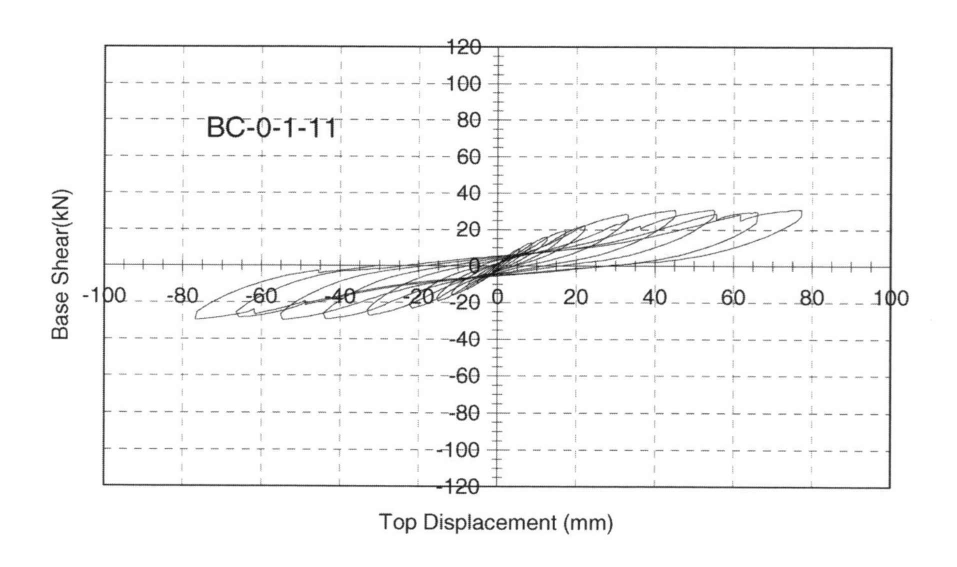

Figure 12. Experimental base shear versus top displacement curves for BC-0-1-11.

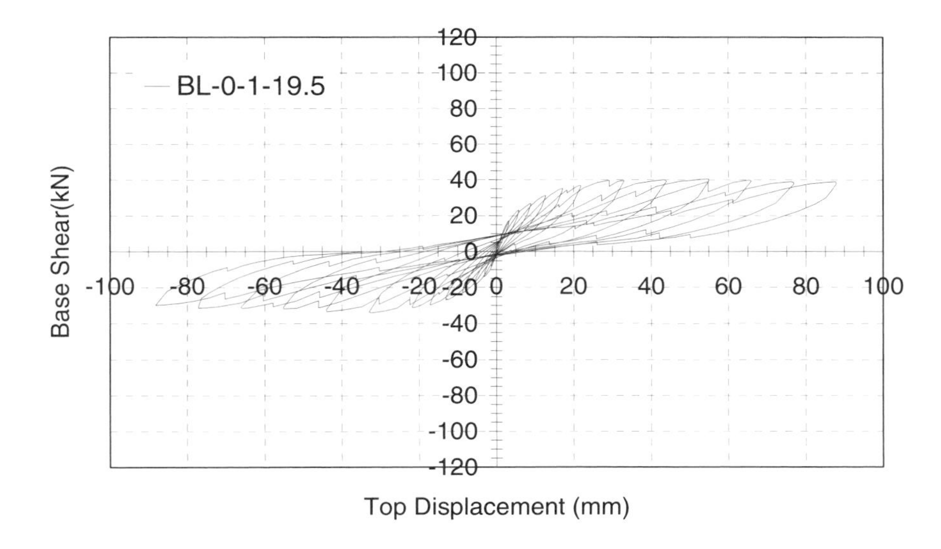

Figure 13. Experimental base shear versus top displacement curves for BL-0-1-19.5.

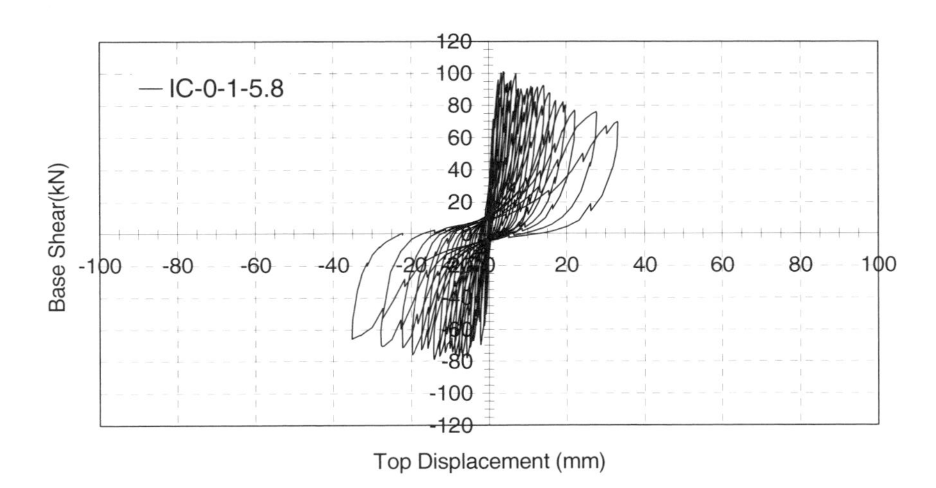

Figure 14. Experimental base shear versus top displacement curves of IC-0-1-5.8

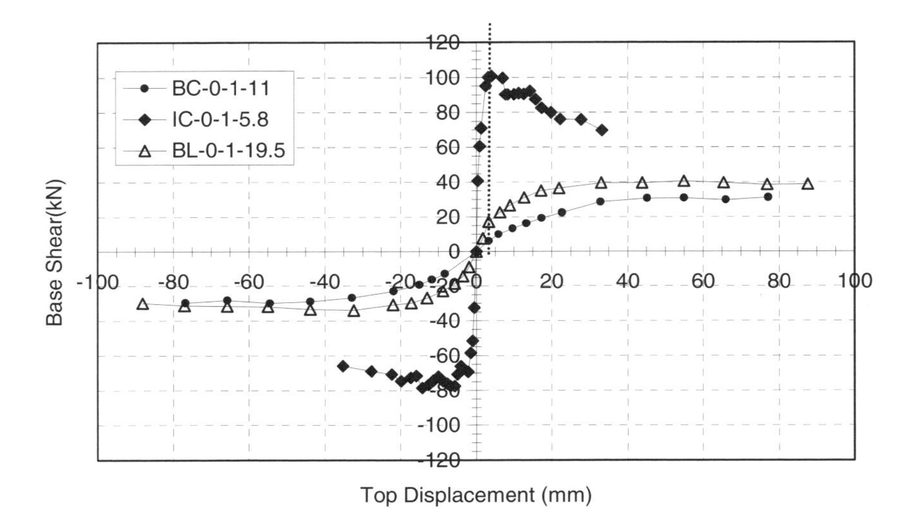

Figure 15. Experimental envelope curves for all specimens

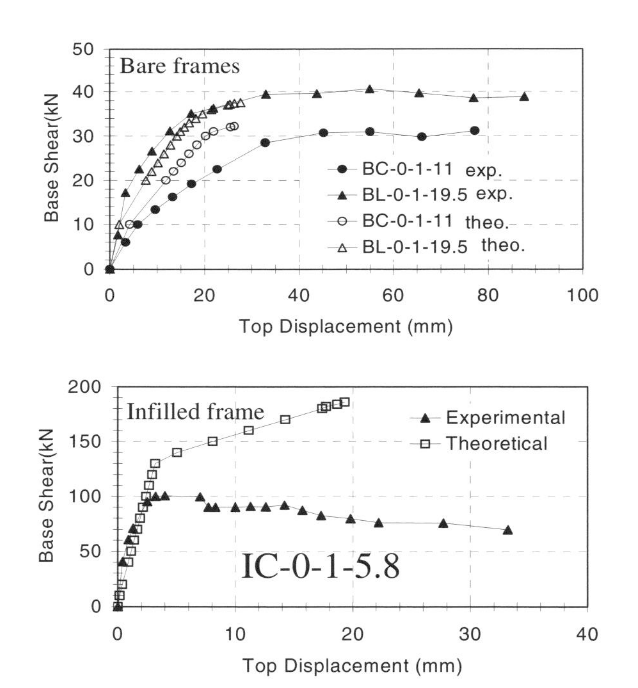

Figure 16. Theoretically and experimentally obtained envelope curves of cyclic P-□ curves of all of the specimens

BC-0-1-11

BL-0-1-19.5

IC-0-1-5.8

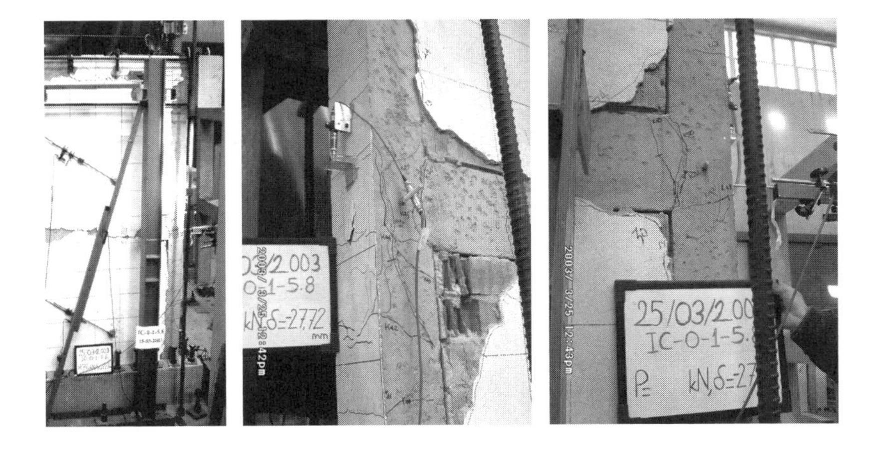

Figure 17. Beam-column connections of bare frames

BC-0-1-11

BL-0-1-19.5

IC-0-1-5.8

Figure 18. Damages accumulated at support regions

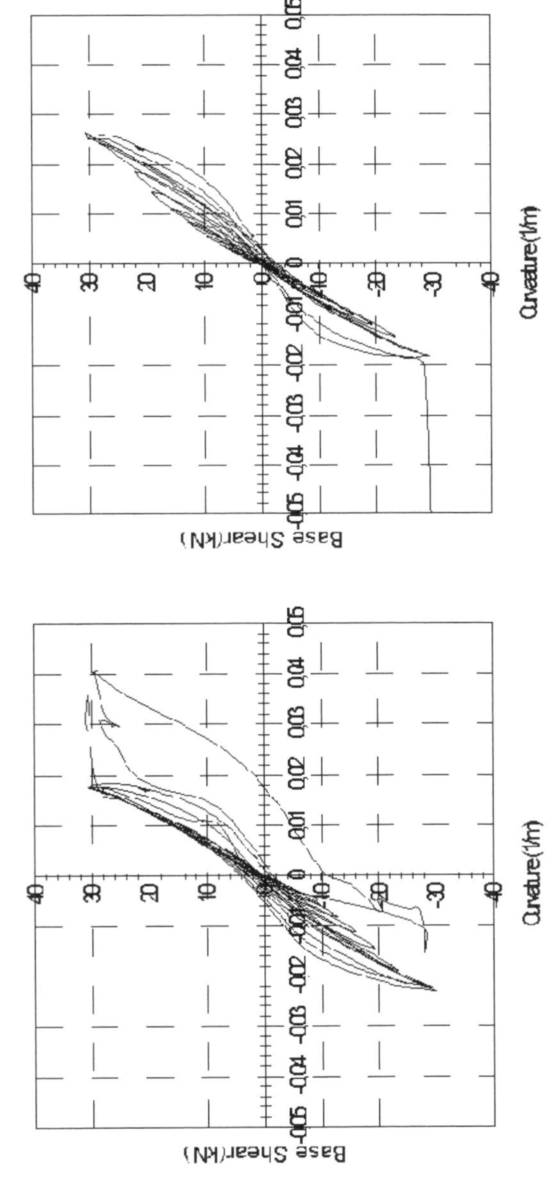

Figure 19. Experimental base shear-curvature relationships

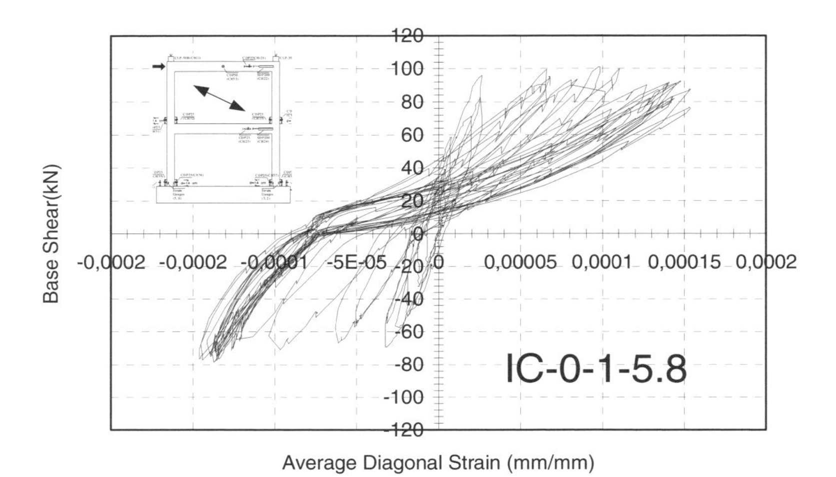

Figure 20. Experimentally observed base shear versus diagonal strains

BC-0-1-11 BL-0-1-19.5

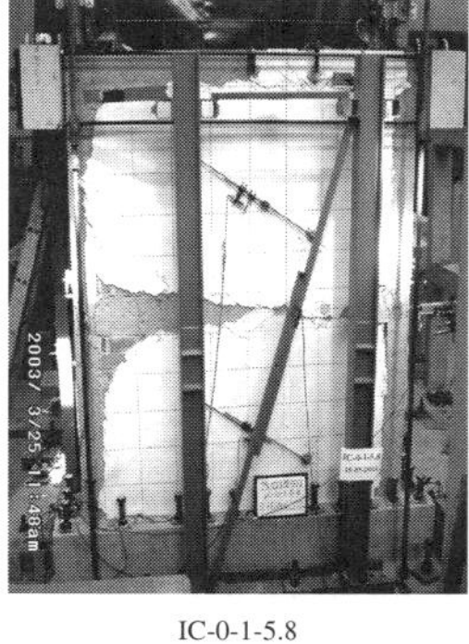

IC-0-1-5.8

Figure 21. General appearances of specimens after tests having reached a displacement ductility ratio of approximately 4

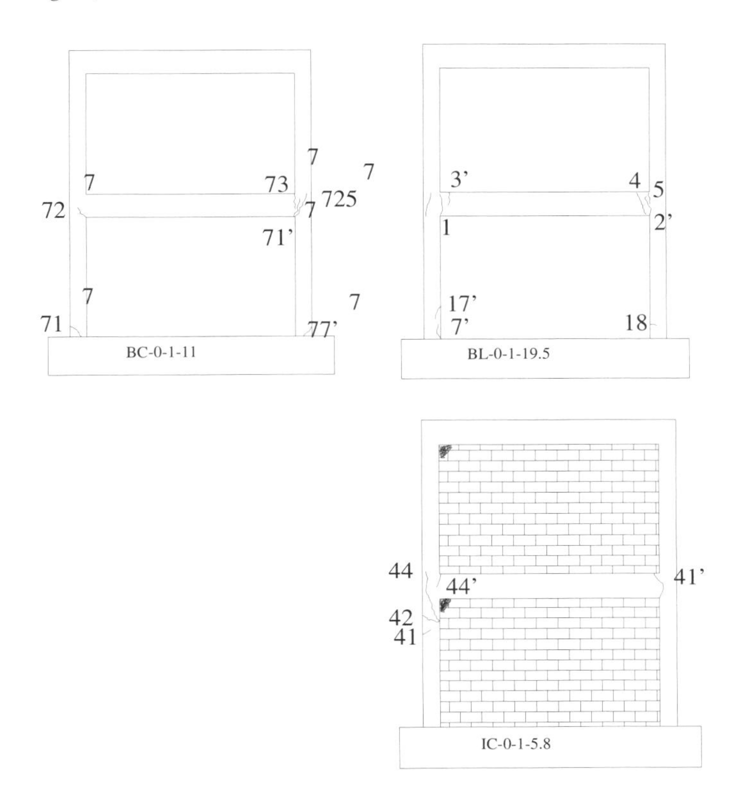

Figure 22. Major cracks at the end of the tests

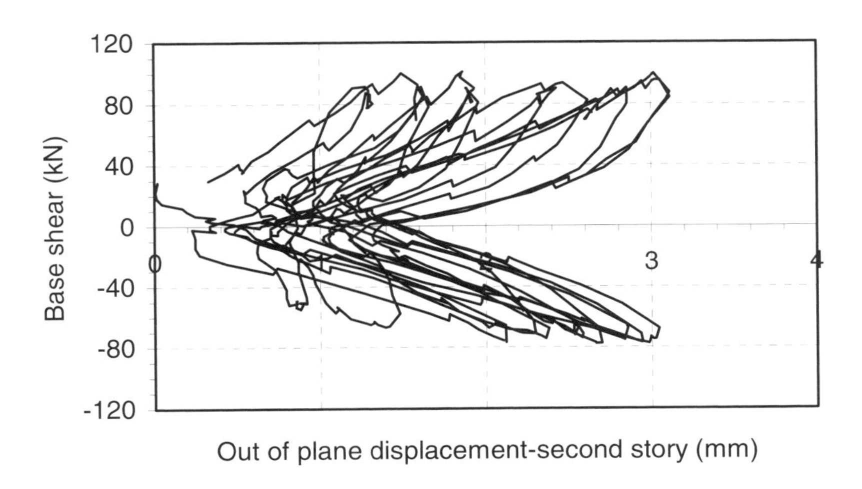

Figure 23. Base shear force versus out of plane displacement relationships

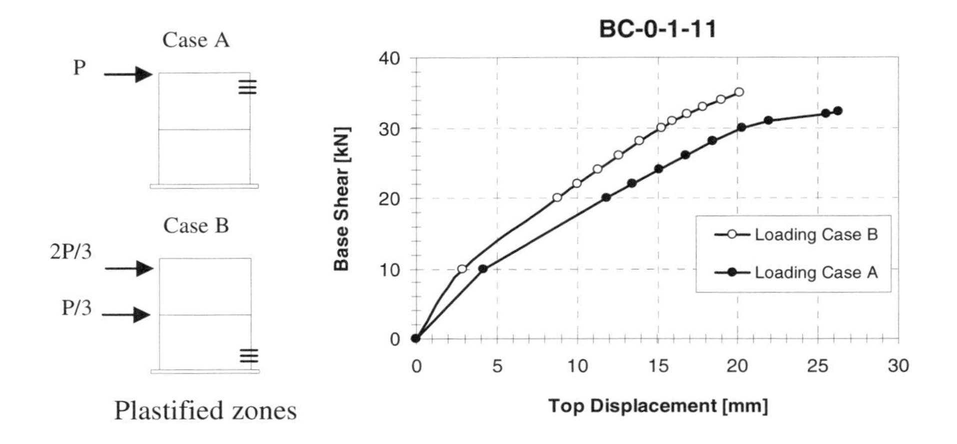

Figure 24. Effect of lateral load application setup

SEISMIC RETROFIT OF R/C FRAMES WITH CFRP OVERLAYS

Experimental Results

Sevket Ozden, Umut Akguzel and Turan Ozturan
Dept. of Civil Engineering, Kocaeli University, Kocaeli, Turkey
Dept. of Civil Engineering, Bogazici University, Istanbul, Turkey

Abstract: This study is part of a research program that is being carried out as a NATO Project 977231 "Seismic Assessment and Rehabilitation of Existing Buildings", led by Middle East Technical University. Within the framework of the research program a new seismic retrofitting methodology is experimentally under investigation. The behavior of hollow brick infilled reinforced concrete frames strengthened by Carbon Fiber Reinforced Polymers (CFRP) has been studied. The main deficiencies of the one-third scale one-bay, two-story test frames were insufficient column lap splice length, poor confinement, and inadequate anchorage length of beam bottom reinforcement. In all specimens beams were stronger than columns and no joint shear reinforcement was used.

Key words: Seismic Retrofit, CFRP, R/C Frame Structure, Lap Splice, Anchorage Length, Brick Infill

1. INTRODUCTION

Although the design philosophy of the Turkish Earthquake Code ensures strength, stiffness and ductility for reinforced concrete frame structures, a wide spread of heavy earthquake damage is observed in such structures during the recent earthquakes due to lack of inspection at the design and construction stages. Frequently encountered deficiencies of the existing

S.T. Wasti and G. Ozcebe (eds.), Seismic Assessment and Rehabilitation of Existing Buildings, 357–382.

buildings may be listed as; weak column-strong beam construction, inadequate anchorage length of beam bottom reinforcement, improper detailing and spacing of hoop reinforcement in columns, beams and joints, and insufficient lap splice length for column longitudinal reinforcement. As a result of these deficiencies the structural behavior deviates from a ductile mode and such buildings cannot resist major earthquakes.

Seismic retrofit of structures may either be "member strengthening" or "overall system upgrading". In the member strengthening technique the individual structural elements such as beams, columns or shear walls are strengthened to overcome their existing deficiencies. Such a retrofit approach may not be feasible in the case when numerous members are to be retrofitted. Moreover, member strengthening may usually be ineffective for structures having insufficient lateral stiffness. For such buildings, system upgrading turns out to be the most goal oriented and cost effective seismic retrofit technique.

By system upgrading, it is aimed to increase the lateral load bearing capacity and the stiffness of the overall structure. Addition of new reinforced concrete shear walls for system upgrading has been intensively studied by different universities and applied during recent earthquakes. In this technique, the brick infill walls are demolished before the reinforced concrete shear wall is cast in place. The time of construction for this application is relatively long and the disturbance to the occupants is mostly beyond the comfort limit especially for a pre-earthquake seismic strengthening process. Externally bonded carbon fiber reinforced polymers (CFRP) applied directly on the existing hollow brick walls may be a good solution for this problem due to the rapid application, light weight, and high strength capacity of the material.

2. PREVIOUS RESEARCH

The use of fiber reinforced polymer (FRP) composites in civil engineering applications has grown very rapidly in recent years. Nowadays, civil engineers are interested in exploiting composite materials in structural applications by taking advantage of their high strength-to-weight ratio, corrosion and fatigue resistance, and relatively low cost.

It is anticipated that the use of FRP on masonry will involve walls resisting in-plane and out-of-plane loads and, possibly infilled panels. Indeed, the majority of the work conducted to date has been on the out-of-plane capacity of walls with externally applied FRP.

The use of FRP as a repair and strengthening material may be an alternative to the conventional techniques. Schwegler (1994) and Weeks et al. (1994) performed various tests on concrete and masonry elements reinforced with FRP. Results showed a marked improvement in the ductility and load carrying capacity of the elements tested [1,2].

Ehsani et al. (1997) studied the effects of shear strengthening with epoxy-bonded FRP overlays applied to the exterior surfaces of brick masonry. The studied parameters were the strength of FRP, fiber orientation and anchorage length. The results showed that both the strength and ductility of tested specimens were significantly increased with this technique [3].

Triantafillou (1998) presented the results of his research related to the use of FRP composites as strengthening materials against shear for concrete, masonry and wood members. He stated that the technique is highly effective and the design of members strengthened with FRP could be based on the classical truss analogy [4].

Triantafillou (1998) also studied the short-term strength of masonry walls strengthened with externally bonded FRP laminates on 12 identical small wall specimens. He developed design models for the dimensioning of FRP reinforcement in masonry walls under monotonic out-of-plane bending, in-plane bending and shear forces, all combined with axial load. It was concluded that highly reinforced locations near the highly stressed zones gave considerable strength increases and the shear capacity of strengthened walls could be very high especially for low axial loads [5].

Young-Joo Lee (1999) investigated the force transfer mechanism between FRP and brick specimens. Glass and carbon fiber reinforced polymer (GFRP, CFRP) sheets were used on brick wall specimens. It is reported that the specimens reinforced with wider FRP sheets showed premature failure modes such as FRP sheet breakage, de-bonding and brick failure. Brick prism specimens reinforced with CFRP showed de-bonding of sheet and shear failure in brick, while the specimens reinforced with GFRP sheet underwent sheet rupture and mortar joint failure [6].

Tumilan et al. (2001) published a paper on Strengthening Masonry Structures with FRP composites and concluded that FRP strengthening increased the shear capacity of the wall specimens and more ductile behavior was observed. It is stated that although the FRP application increased flexural and shear capacities of masonry walls, the effectiveness of this technique is highly dependent upon the FRP end anchorage to the surrounding beam, column or foundation [7].

Marjani (1997) studied the behavior of brick infilled reinforced concrete scaled frames. Effect of plastering on the brick was also investigated. It was found out that hollow clay infill increased both strength and stiffness of the reinforced concrete frames significantly. It was also reported that plastering both sides of the infill improved the behavior of the specimen and improved the ductility by delaying the infill diagonal cracking [8].

Altin (1990) investigated the behavior of reinforced concrete frames strengthened with reinforced concrete infill. Fourteen one-third scale, one-bay, two-story specimens were designed and detailed in accordance with the code and infill walls were introduced to undamaged frames. The main test variables were the pattern of the infill reinforcement, the connection of the infill to the frame, the effect of axial load and the strength of the columns. The loading was reversed cyclic. It was concluded that the reinforced concrete infill walls, which were properly connected to frame members, increased both strength and the stiffness significantly; besides, the column strength and axial load on the columns improved the overall behavior and lateral load capacity [9].

Sonuvar (2001) repaired the heavily damaged one-third scale, one-bay, two-story reinforced concrete frames with reinforced concrete infill walls. Test specimens were constructed with common deficiencies in Turkey such that lack of confinement, poor concrete quality, strong beam-weak column, inadequate lap splice length, poor detailing of beam bottom reinforcement and ineffective ties [10].

Canbay (2001) tested a two-story, three-bay reinforced concrete frame. After the structure was significantly damaged, reinforced concrete infill was introduced in the middle bay. It was reported that the strength, stiffness and energy dissipation capacity of the repaired specimen increased significantly and the newly introduced reinforced concrete infill carried 90 percent of the lateral load [11].

Mertol (2002) tested two one-third scale, one-bay two-story reinforced concrete frames with hollow tile brick infill under reversed cyclic lateral load with constant axial load, one being with CFRP overlay. Test results indicated the importance of CFRP anchorage to the surrounding beams and columns [12].

Keskin (2002) studied the behavior of brick infilled reinforced concrete frames strengthened with CFRP reinforcement on two similar specimens tested by Mertol. In the first test, only one side of the specimen was strengthened with CFRP overlay and it was extended to the frame members. The second specimen was with CFRP overlay applied on either side of the infilled frame. It was concluded that the CFRP anchorage via CFRP dowels play an important role in the behavior and capacity of the strengthened frames [13].

Erduran (2002) also studied the behavior of brick infilled reinforced concrete frames strengthened with CFRP reinforcement on similar specimens tested by Keskin. The CFRP was applied as X-braces on to the brick wall and connected to the surrounding reinforced concrete members. It was pointed out that the tested CFRP detail resulted in a significant increase in the energy dissipation and lateral load carrying capacity. On the other hand, the failure of CFRP strengthened specimens was relatively brittle once the ultimate load capacity was reached [14].

3. TEST PROGRAM

Four specimens were tested to highlight the effect of brick infill and CFRP overlays on the strength and behavior of poorly constructed reinforced concrete frames (Table 1). All four specimens, namely Pilot, U1, U2, and U3, reported herein, are one-third scale, one-bay and two-story reinforced concrete frames tested under reversed cyclic lateral load. The vertical load on the frames kept constant at a level of approximately 10 percent of the column axial capacities. The main deficiencies of the test frames may be listed as follows:

a) Insufficient lap-splice length of column longitudinal reinforcement.
b) Poor confining reinforcement for both columns and beams.
c) Inadequate anchorage length of beam bottom reinforcement.
d) No joint shear reinforcement.
e) Strong beam – weak column design.

The first set of the specimens, namely Pilot and U1, were tested to observe the bare frame behavior. The difference between bare frame specimens were the concrete compressive strength, anchorage detail of the beam reinforcement to the beam-column connection, and the lap splice length of the column longitudinal reinforcement. Specimens U2 and U3 were tested with hollow tile brick infill in both stories. The brick infill walls were plastered on both sides. The difference between specimen U2 and U3 was the existence of CFRP cross-overlay on the subsequent specimen U3. Frame reinforcement detail for specimens U1, U2 and U3 was kept the same, while the concrete compressive strength was alike for these specimens. Brick infill details for specimens U2 and U3 are given in Figure 1.

3.1 Loading Setup

Bare and infilled frame specimens were fixed at the base and loaded horizontally with a deformation controlled 250 kN capacity hydraulic actuator (Figure 2). For bare frame specimens, namely Pilot and U1, the horizontal cyclic loading was applied to the second story beam level only, while the load was divided into two via a steel spreader beam and applied both at the first and second story levels for brick infilled specimens U2 and U3, such that two thirds of the applied load goes to the upper story level.

Out of plane deformation of the bare frames was not restrained during testing. On the other hand, a steel frame was constructed around the brick infilled specimens to restrain the out of plane deformations. In either cases, deformations were measured and recorded throughout the test (Figure 3).

3.2 Instrumentation

An electronic data acquisition system with control feedback was used to measure the level of applied load, displacements and rotations. In all the specimens, the reversed cyclic load level and the frame top displacement were monitored to apply the predetermined loading pattern. Curvature measurements on bare frame columns were made to highlight the effect of inadequate lap splice length on the behavior. Two different measurement lengths were used on one of the first story columns to emphasize the effect of the gauge length on curvature readings (Figure 4).

Out of plane displacements were continuously monitored and recorded both for the bare frame (Pilot, U1) and infilled frame (U2, U3) specimens. However, the infilled frames were restrained against such deformations by means of a steel frame constructed in the test rig. For specimens U2 and U3 shear deformations on the brick infill, horizontal base slip, and frame base rocking were also measured (Figure 5). The measurements were relative to the frame foundation in all specimens.

3.3 Reinforcement Detail

Reinforcement detail is the same for all specimens except specimen Pilot; differences being the lap splice length of column reinforcement (l_p=200mm) and the anchorage detail of the beam reinforcement to the beam-column connection as given in Table 1. The longitudinal reinforcement both for beams and columns were 8 mm diameter plain bars with a yield strength of f_y=312 MPa. The yield strength of 4 mm diameter transverse plain bars was f_y=277 MPa. The reinforcement detail of the specimens is given in Figure 6.

4. TEST RESULTS

The reversed cyclic lateral load behavior of the specimens is discussed below. Loading histories, load versus roof drift ratio curves and moment curvature graphs are given for each specimen. In addition, photographs taken after the test are given to show the damage level attained.

4.1 Specimen: Pilot

Specimen Pilot was a reinforced concrete bare frame with a column rebar lap splice length of l_b=$25D_b$. Concrete compressive strength measured on 150x300 mm cylinders was f_c'=22.4 MPa. The specimen was tested under the lateral load history given in Figure 7. The lateral load was applied as a point load to the top story beam level. The lateral load versus roof drift ratio curve of the specimen is given in Figure 8. The moment curvature behavior of the column sections with lap splice length deficiencies was monitored by two curvature-meters mounted on the same column, with different gauge lengths (Figure 9 and 10). The axial load on each column was 30 kN throughout the test.

During the test, no visible cracks were observed until the end of the first full load cycle where the load level was 5.8 kN. A constant lateral load of approximately 8 kN, was applied for the next three cycles, and during these load cycles flexural cracks were observed at both ends of the first story beam. The flexural cracks on the columns began to widen at the 7th load cycle where the lateral load level was 12.5 kN. Until this load cycle, no visible cracks were observed in the beam column joint since the beam longitudinal bars were not anchored to the joint effectively; hence the transferred joint shear was small.

Between load cycles 8 and 11, the load level was in the range of 13-14 kN and the flexural damage was concentrated on the first story beam ends and both ends of the first story columns. During the load cycle 10 spalling of the concrete cover was observed at the upper end of the first story column to the west. The crack widths at this load level were approximately 3 mm on either first story column bases and approximately 1.5 mm on both ends of the first story beam. The first visible x-cracks on the connection were observed at load cycle 12. The flexural cracks at the upper end of the first story columns were located right at the beam bottom level and widened concurrently with the widening of the flexural cracks at the base of the same members. Minimal cracking was observed on the second story beams and columns.

4.2 Specimen: U1

Specimen U1 was the second of the bare frame specimen set. The reinforcement detail of U1 was exactly replicated for the specimens with brick infill, namely U2 and U3. Concrete compressive strength was f_c'=15.4 MPa. The reversed cyclic lateral load history, lateral load versus roof level drift ratio, and first story column moment-curvature graphs of specimen U1 are given in Figures 11, 12 and 13 respectively. The lateral load was applied at the second story beam level as a point load. There was a constant axial load of 30 kN on each column throughout the lateral load history.

The first visible cracks were observed concurrently on the base of the lower east column and at the west end of the first story beam in load cycle 3, where the lateral load level was in the range of 7 kN. Hairline cracks on the beam column connection were first observed in 6th load cycle where the load level was 9.5 kN.

During load cycles 7 and 8, new cracks started to form on the lower column ends and on the beam-column connections. Beyond the load cycle 8, the lateral load capacity of the specimen stabilized under increasing lateral displacements. At the end of load cycle 13, the measured crack widths on the first story beam ends and column bases were in the range of 1 and 2 mm respectively (Figure 14).

Damage due to lateral loading was accumulated in the first story columns and the beam. Heavy cracking was observed on the first story beam column joints compared to the specimen Pilot (Figure 15). Two individual flexural cracks on the lower column bases were observed, one being right at the base and the second being at a height of 50 mm (Figure 16).

4.3 Specimen: U2

Specimen U2 was the first specimen tested with brick infill. Frame concrete strength was f_c'=14.7 MPa, while the mortar strength used between the bricks and used as plaster was f_m'=5.5 MPa. Specimen U2 was tested under the lateral load history given in Figure 17, and the lateral load roof drift ratio curve for this specimen is given in Figure 18.

The brick infill was constructed such that it was flush with the outer surface of the reinforced concrete frame in specimen U2. Due to such an eccentric location of the brick infill, specimen U2 was prone to out-of-plane deformations under in-plane lateral loading. Therefore it was restrained against out-of-plane deformations with a steel frame mounted in the test rig (Figure 19). Roller bearings mounted at the second story beam level had no restraining effect in the direction of lateral load. The axial load on each column was 30 kN throughout testing. The lateral load was divided into two via a spreader beam as shown in Figure 2 such that 66 percent of the actuator load goes to the second story level.

The first visible crack on specimen U2 was observed at the second story brick wall to top beam interface at the 5th load cycle, where the lateral load level was 40 kN. In the next cycle, these hairline cracks were extended down through the column-wall interface. Subsequently, new hairline cracks were observed at the first story wall-to-beam interface.

In load cycle 7 horizontal cracks were observed on the first story columns located approximately 200 mm from the base. Concurrently, first story brick walls were cracked at the frame corners and the crack orientation was almost perpendicular to the wall diagonal axis. The load level in this cycle was 50 kN.

In load cycle 8, where the load level was 55 kN, no new cracks were observed. It was seen that the plaster on the frame started to spall off in load cycle 9 and the lateral load level started to decrease. In load cycle 12, the cracks observed in the previous cycles started to widen suddenly and the load capacity degradation was more pronounced (Figure 20). The accumulated damage on specimen U2 at the end of the test is shown in Figure 21.

4.4 Specimen: U3

Specimen U3 was a brick infilled frame specimen strengthened with CFRP (*SikaWrap Hex-230C*) X-overlays. Manufacturer specified "fiber tensile strength" was 4100 MPa, and the tensile E-Modulus was 231,000 MPa for the CFRP. 200 mm width CFRP diagonals were applied on both sides of the specimen U3 (Figure 22). X-overlays were anchored to the reinforced concrete frame and brick infill through CFRP anchors at locations shown in Figure 22. The anchorage depth of CFRP anchors on the reinforced concrete frame was 50 mm (Type A and B) (Figure 23). Type C anchors pass through the brick wall and were bonded to the overlays. In either case CFRP anchors were bonded to overlays over a circular area with a radius of 50 mm.

Concrete compressive strength was f_c'=16.0 MPa and the mortar strength was f_m'=5.1 MPa. The applied lateral load history is given in Figure 24. Load was applied in increments of 5 kN at each load cycle until failure, and displacement controlled cycles were applied beyond that point. Lateral load was applied through a steel spreader beam, through which the story loads were shared as 2 to 1 for the second and the first stories. The specimen was restrained against out-of-plane deformations with the same steel frame used for specimen U2 as shown in Figure 19. Axial load level was kept constant at 30 kN on each column throughout the test.

No visible cracks were observed until load cycle 5 was reached, where the load level was 40 kN. The first crack was observed at the interface between the second story brick panel and the upper beam. The length of the crack was approximately 300 mm. Concurrently, horizontal flexural cracks were observed on the first story columns, at a height of 145 mm from the foundation level (crack number 5 in Figure 26).

In load cycle 8, where the load level was 55 kN, the above mentioned flexural cracks started to be continuous around the column indicating a shift from flexural behavior to an axial tensile behavior of columns. The column base also failed under tension at a further load level (cycle 10, P=65 kN) (crack number 20 in Figure 26). The specimen lateral load capacity was reached when the horizontal cracks on the columns started to widen suddenly and the first story brick panel started to separate from the foundation.

Damage until the ultimate load level was concentrated to the first story columns in the form of flexural and flexural/tensile cracks. The number of cracks on the first story columns was bigger than that of the specimen U2. No cracks were observed on the second story beam and columns throughout the test. Furthermore, no cracking was observed on the first story beams. Damage was concentrated in the first story columns and the brick infill panels throughout the testing (Figure 27).

The following observations were made relative to the CFRP overlay behavior:

1. No tensile failure of X-overlays was observed.
2. x-overlays peeled off under compression,.
3. x-overlays close to the frame corners buckled under compression and cracked.
4. CFRP anchors at the foundation level on the inner side of the frame were pulled out while the ones on the outer side were ruptured.
5. The pull-out-with-concrete-cone type of CFRP anchor failure indicated the deficiency of anchorage depth.
6. Capacity loss may be associated with the failure of CFRP anchors. The first drop in load level is observed concurrently with the pull-out failure of the anchors at the foundation level on the inner side of the frame. Consequently, the lateral load resistance increased again and the second drop in load is associated with the rupture of the CFRP anchors at the foundation level on the outer side of the frame.

Beam-column connections remained uncracked throughout the testing. The dimension of the mostly damaged brick infill area located close to the column-foundation connection was 35x35cm as shown in Figure 27. Failure of the brick in this region made the CFRP overlay buckle under compressive deformations. Besides, first story columns were overstressed at this level.

5. DISCUSSION OF TEST RESULTS

Specimens tested within the framework of this study did not conform to the requirements of the Turkish Earthquake Code [15]. Detailing deficiencies of the specimens may be listed as: inadequate hoop detailing in beams and columns, the lack of ties in the beam-column connection, insufficient lap splice length of column longitudinal and beam bottom reinforcement. In addition, the concrete strength was around f_c'=20 MPa. Plain bars were used as longitudinal reinforcement and the workmanship was poor.

The lateral strength and stiffness of the frames tested were closely related to the reinforcement detail, concrete compressive strength and the existence of hollow clay brick infill and CFRP x-overlays.

Beam longitudinal reinforcement anchorage detail is different between specimens Pilot and U1. As a result of this difference the overall crack pattern is significantly affected. Flexural cracking was concentrated on the first story beam ends in specimen Pilot while heavy cracking in the beam column joint was observed in specimen U1. This may be due to the smaller shear force transferred to the joint in specimen Pilot in comparison with U1.

The first flexural cracking in bare frames was observed on the both ends of east column at the first story during a westward load cycle. In further load cycles beyond the ultimate capacity, the flexural cracks on the first story columns just beneath the first story beam were widened and led the specimen to sudden failure. The reason may be the insufficient lap splice length of the column re-bars at the first story level.

On the other hand, first cracking was observed between the brick infill panel and the frame members in the case of the infilled specimens. Significant capacity loss and relative slip between panels and the frame elements was observed concurrently in the case of specimen U2. Consequently, infill wall compression struts started to fail resulting in rapid degradation in strength and stiffness.

The specimen with CFRP x-overlay U3, experienced a behavior different than U2. Initial cracking was observed in the first story columns at a height of 145 mm, where the lap splice length was 160 mm. Significant loss in lateral load capacity was observed associated with the sudden widening of these cracks on either of the first story columns and the pull-out failure of the anchors at the foundation level on the inner side of the frame. Consequently, the lateral load resistance increased again and the second drop in load is associated with the rupture of the CFRP anchors at the foundation level on the outer side of the frame. At further load cycles, overall frame base experienced slip and base rocking in specimen U3. The negative slope after the peak load was comparable in specimens U2 and U3.

The difference in behavior due to the existence of brick infill walls and the CFRP x-overlay can be seen in Figure 28 clearly. The initial stiffness of brick infilled specimen and the specimen with CFRP x-overlay were similar. Peak displacements and lateral loads are given in Table 3.

The peak loads were attained at 1.0 0.2 and 0.3 per cent roof level lateral drift ratios in specimens U1, U2 and U3 respectively. The same ratio was 1.7 for specimen Pilot. It should be noted that U2 and U3 were infilled specimens, latter being strengthened with CFRP overlays. The lap splice length of specimen U1 was smaller than that of the Pilot specimen.

6. CONCLUSIONS

The conclusions given here are mainly based on the test results reported by the authors.

a) The improper lap splice length in the columns mainly governed the ultimate load level and the failure mode of the specimens.
b) Existence of brick infill reduced the drift level attained at the peak lateral load, and the application of CFRP x-overlays seemed ineffective in increasing the drift levels at failure load.
c) Initial stiffness values were comparable in specimens with brick infill and with CFRP x-overlay.
d) Under tensile forces CFRP x-overlays remained intact, but under compressive forces they peeled off and buckled in the post-peak region.
e) The specimen with CFRP x-overlay failed due to the insufficient CFRP anchorage detail provided. Pull-out and rupture type of CFRP anchor failures were observed on the same specimen.

ACKNOWLEDGEMENT

The authors gratefully acknowledge the technical support of Bogazici University Structures Laboratory where all the experiments were conducted, and financial support from METU-Civil Engineering Department Projects (NATO: SfP977231, TUBITAK: IÇTAG-I 575).

NOTATION

D_b : reinforcing bar diameter (mm)
f_c' : concrete compressive strength (150x300mm cylinders), (MPa)
f_m' : mortar compressive strength (50x50x50mm cubes), (MPa)
P : corrected total lateral load (kN)
l_p : lap slice length (mm)
Δ : roof level displacement under lateral load (mm)

REFERENCES

1. Schwegler, G. (1994). "Repair and upgrading techniques of unreinforced masonry structures utilized after the Friuli and Campania/Basilicata earthquakes." Earthquake Spectra, Earthquake Engrg. Res. Inst., Berkeley, Calif., 10(1), 171-185.
2. Weeks, J., Sieble, F., Hegemier, G., and Priestley, M.J.N. (1994). "the US-TCMMAR full-scale five story masonry research building test: Part V-Repair and retest." Rep. SSRP-94/05, Struct. Sys. Proj., University of California, San Diego.
3. Ehsani, M.R., Saadatmensh, H., Al-Saidy, A., "Shear Behaviour of URM Retrofitted with FRP Overlays", ASCE Journal of Composites for Construction, Vol. 1, No. 1, February 1997, pp. 17-25.
4. Triantafillou, T.C., "Composites: A New Possibility For The Shear Strengthening of Concrete, Masonry and Wood", Composites Science and Technology, Vol.58,2, 1998, pp. 1285-1295.
5. Triantafillou, T.C., "Strengthening of Masonry Structures Using: Epoxy Bonded FRP Laminates", ASCE Journal of Composites for Construction, Vol. 2, No. 2, May 1998, pp. 96-104.
6. Young-Joo Lee "Masonry Structures Strengthening with FRP Sheet", M.S. Thesis, Department Of Architectural Engineering, Pennsylvania State University, P.A., 1998.
7. Tumilan, J.G., Micelli, F., Nanni, A., " Strengthening of Masonry Structures with FRP Composites", A Structural Engineering Odyssey Conference Proceeding, Section 52 Chapter 5, May 2001.
8. Marjani, F., "Behaviour of Brick Infilled Reinforced Concrete Frames Under Reversed Cyclic Loading", Ph.D. Thesis, Department of Civil Engineering, Middle East Technical University, September 1997.
9. Altin, S., "Strengthening of Reinforced Concrete frames With Reinforced Concrete Infills", Ph.D. Thesis, Department of Civil Engineering, Middle East Technical University, February 1990.
10. Sonuvar, M.O., "Hysteretic Response of Reinforced Concrete Frames Repaired by Means of Reinforced Concrete Infills", ", Ph.D. Thesis, Department of Civil Engineering, Middle East Technical University, June 2001.
11. Canbay, E., "Contribution of the RC Infills to the Seismic Behaviour of Structural Systems", Ph.D. Thesis, Department of Civil Engineering, Middle East Technical University, December 2001.
12. Mertol, H.C., "Carbon Fiber Reinforced Masonry Infilled Reinforced Concrete Frame Behaviour", M.Sc. Thesis, Department of Civil Engineering, Middle East Technical University, July 2002.
13. Keskin, R.S.O., "Behaviour of Brick-Infilled Reinforced Concrete Frames Strengthened by FRP Reinforcement: Phase I", M.Sc. Thesis, Department of Civil Engineering, Middle East Technical University, July 2002

Erduran, E., "Behaviour of Brick-Infilled Reinforced Concrete Frames Strengthened by FRP Reinforcement: Phase II", M.Sc. Thesis, Department of Civil Engineering, Middle East Technical University, July 2002

Turkish Earthquake Code for Structures to be Built in Disaster Regions, Ministry of Public Works and Settlement, Ankara, 1998.

Table 1. Specimen Properties

Specimen	Column Re-bar Splice Length	Beam Re-Bar Anchorage Detail	Concrete Strength f_c' (MPa)	Mortar Strength f_m' (MPa)	Total Vertical Load, N (kN)	Lateral Load Applic. Point
Pilot	25×D_b (200mm)	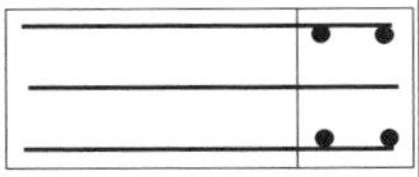	22.4	N/A	60	 P
U1	20×D_b (160mm)	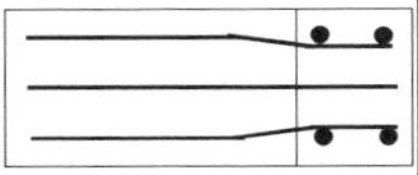	15.4	N/A	60	
U2			14.7	5.5	60	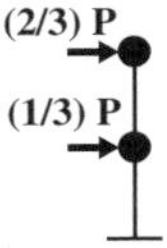 (2/3) P, (1/3) P
U3			16.0	5.1	60	

Table 2. Roof Displacement and Lateral Load Peaks in Positive Quadrant

Pilot		U1		U2		U3	
H (kN)	Δ (mm)	H (kN)	Δ (mm)	H (kN)	Δ (mm)	H (kN)	Δ (mm)
6.03	1.69	4.05	0.72	20.50	0.01	19.19	0.06
7.77	3.20	6.04	1.99	26.10	0.03	25.21	0.14
8.08	4.10	7.55	3.95	30.00	0.07	32.08	0.32
7.70	4.07	7.48	4.43	34.10	0.10	35.05	0.39
9.52	5.98	7.24	4.54	39.90	0.26	40.36	0.52
11.24	10.55	9.48	6.66	43.90	0.78	44.00	0.61
13.18	14.02	9.89	7.87	49.10	1.64	48.70	0.74
13.75	17.84	11.00	10.98	53.40	2.77	54.99	1.03
14.02	20.81	11.16	17.20	59.30	3.93	59.52	1.25
14.05	29.21	11.07	20.38	44.20	6.14	64.87	1.70
13.80	36.68	10.28	22.24	39.37	6.64	70.01	2.17
13.24	43.75	10.42	25.71	40.55	7.88	73.90	3.19
11.42	52.13	9.64	35.35	38.55	10.49	75.15	4.93
9.96	59.48	7.93	46.02	31.88	15.78	64.14	6.37
		6.16	56.53			58.73	7.73
						51.64	9.53
						55.77	12.41
						62.72	15.72
						70.42	25.21
						44.64	31.81

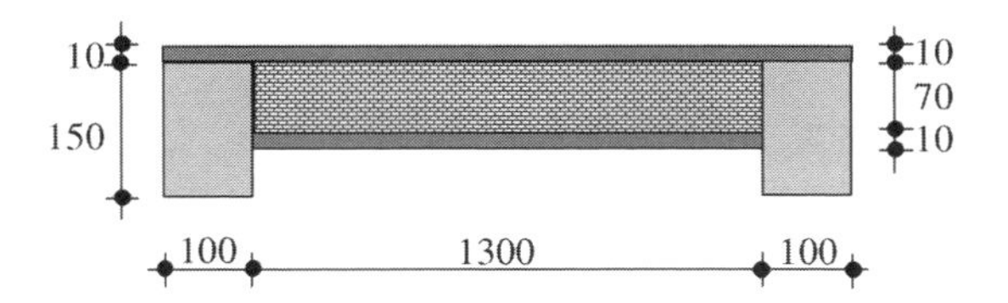

Figure 1. Brick Infill and Plastering Detail for Specimens U2 and U3

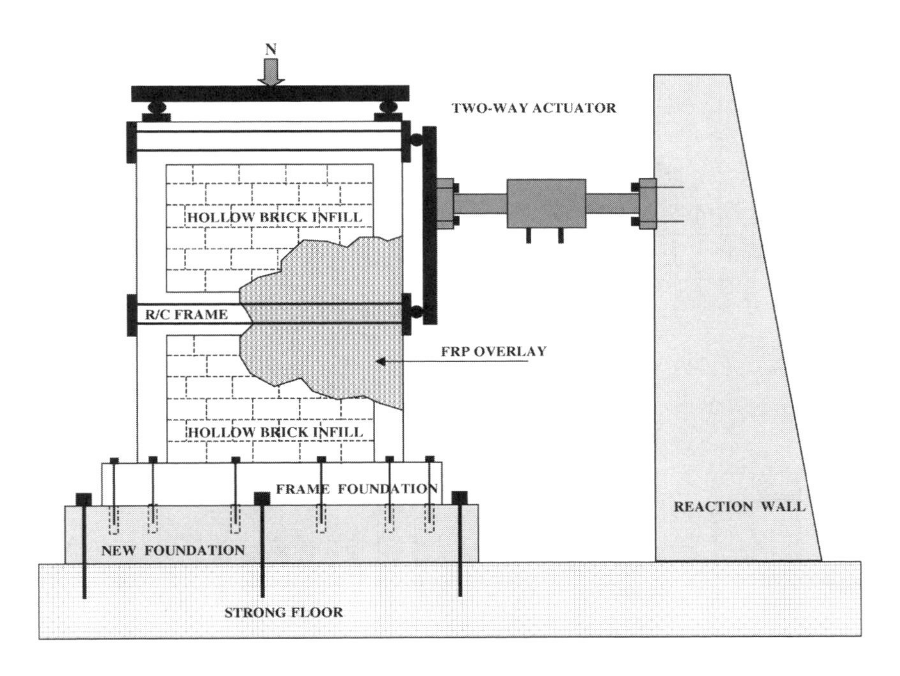

Figure 2. Test Setup

Figure 3. Loading and Data Acquisition Systems for Specimen Pilot

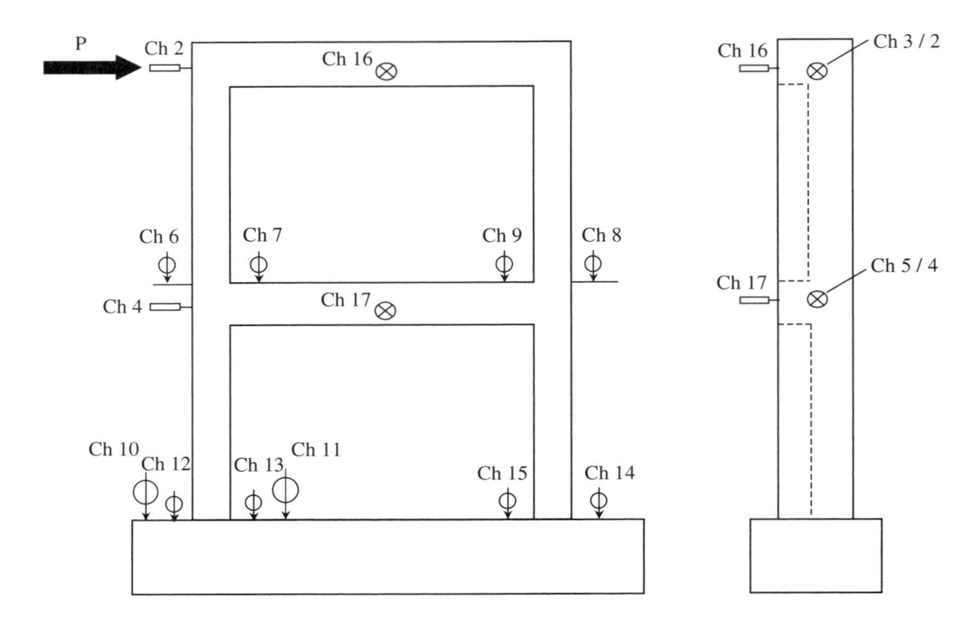

Figure 4. Measurement Points for Bare Frames (Pilot and U1)

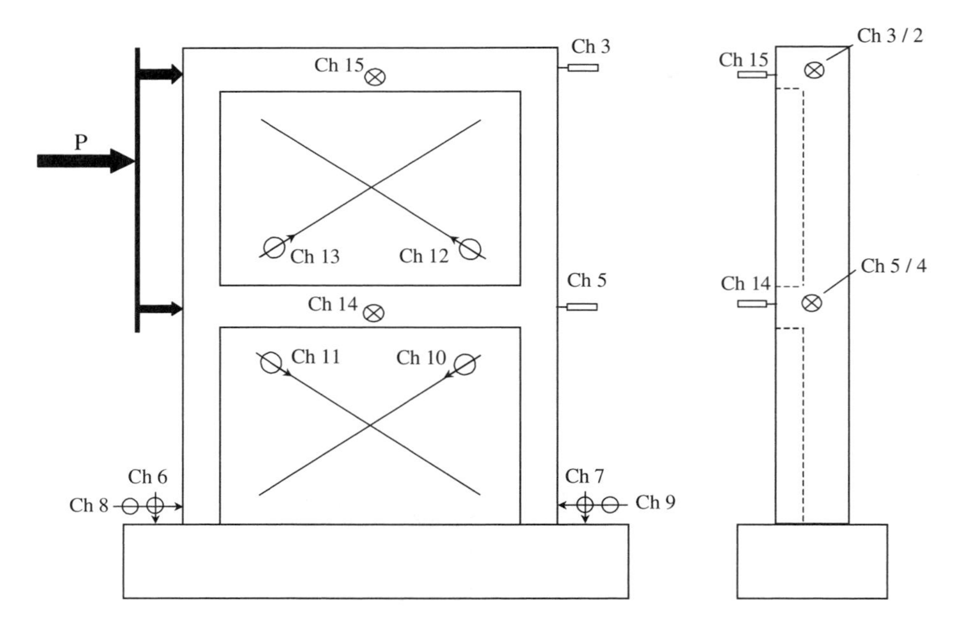

Figure 5. Measurement Points for Infilled Frames (U2 and U3)

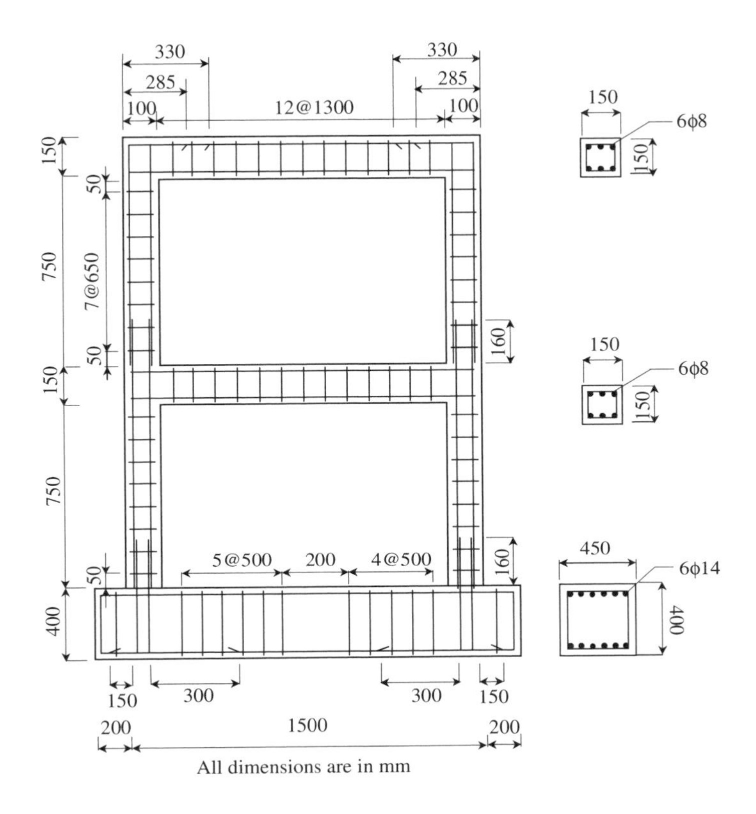

Figure 6. Reinforcement Detail of Specimens

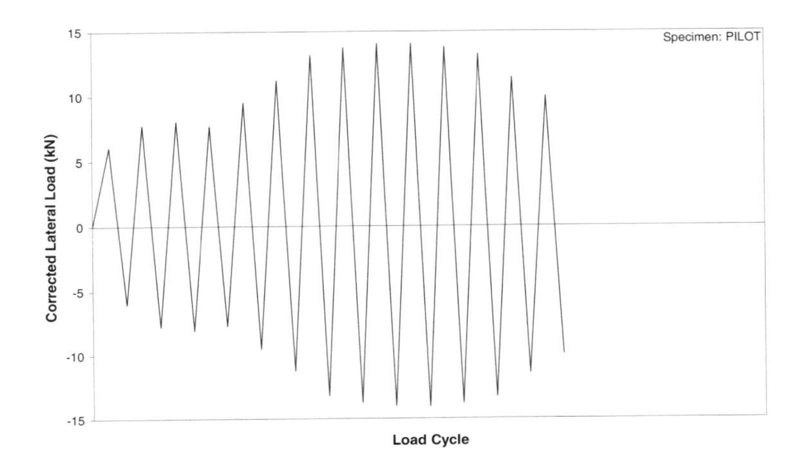

Figure 7. Loading History of Specimen Pilot

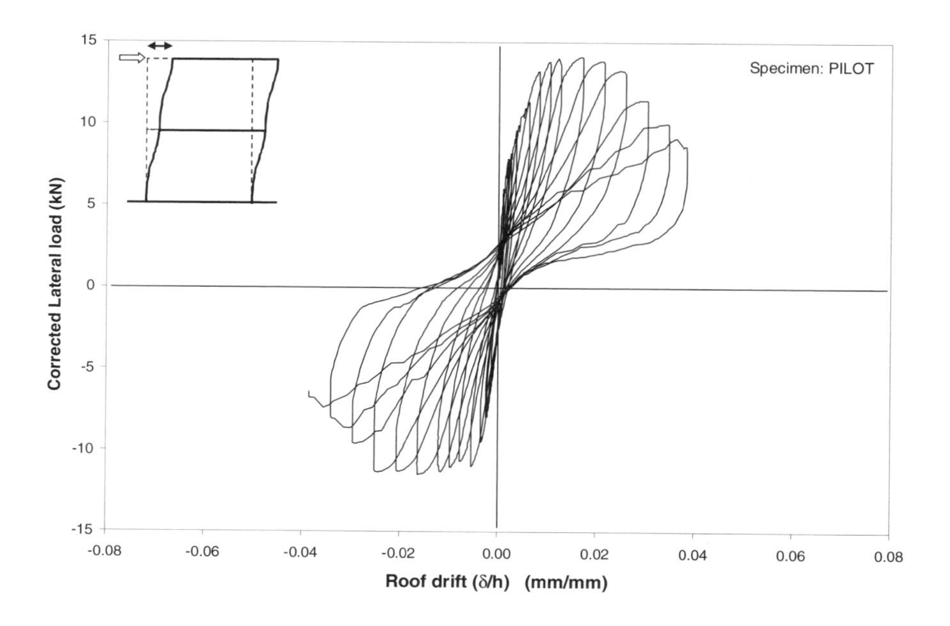

Figure 8. Load–Roof Drift Curve of Specimen Pilot

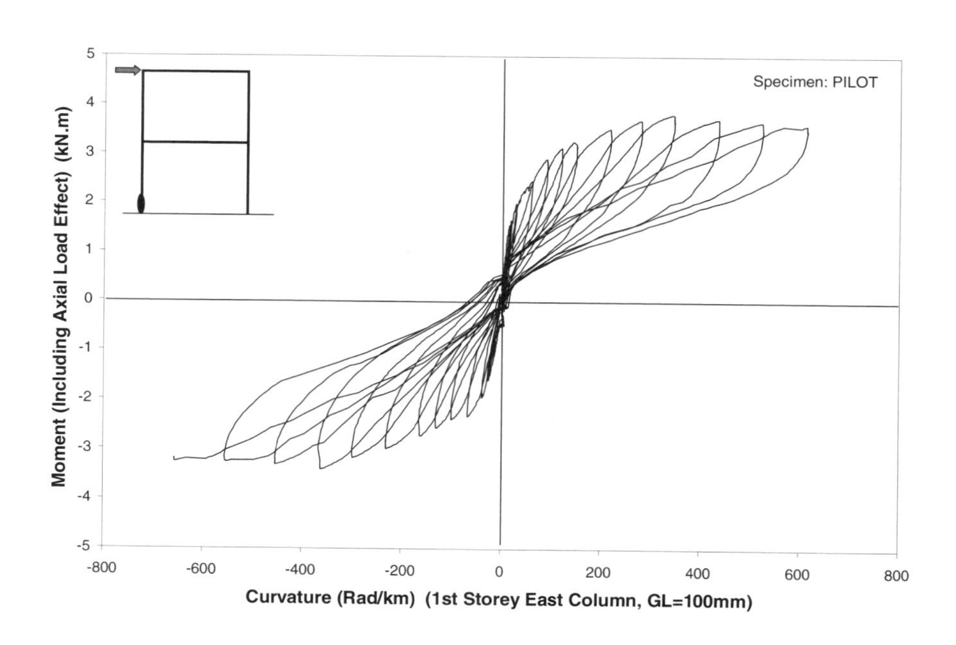

Figure 9. Column Moment-Curvature of Specimen Pilot (GL=100mm)

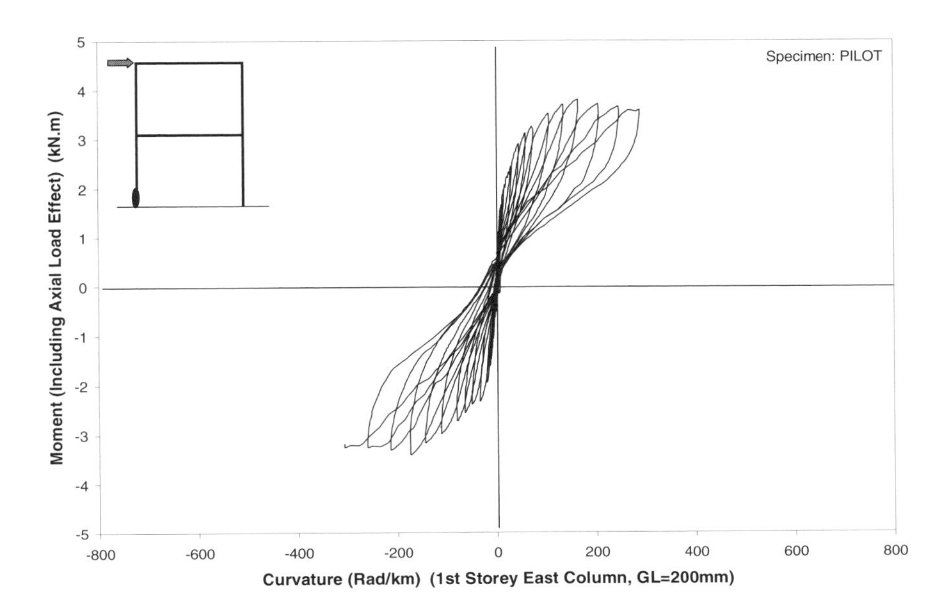

Figure 10. Column Moment-Curvature of Specimen Pilot (GL=200mm)

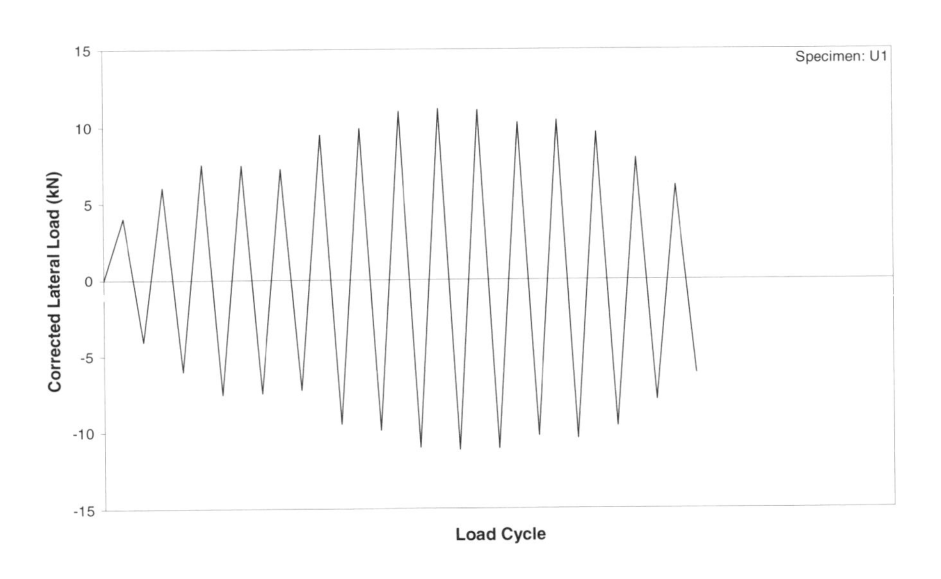

Figure 11. Loading History of Specimen U1

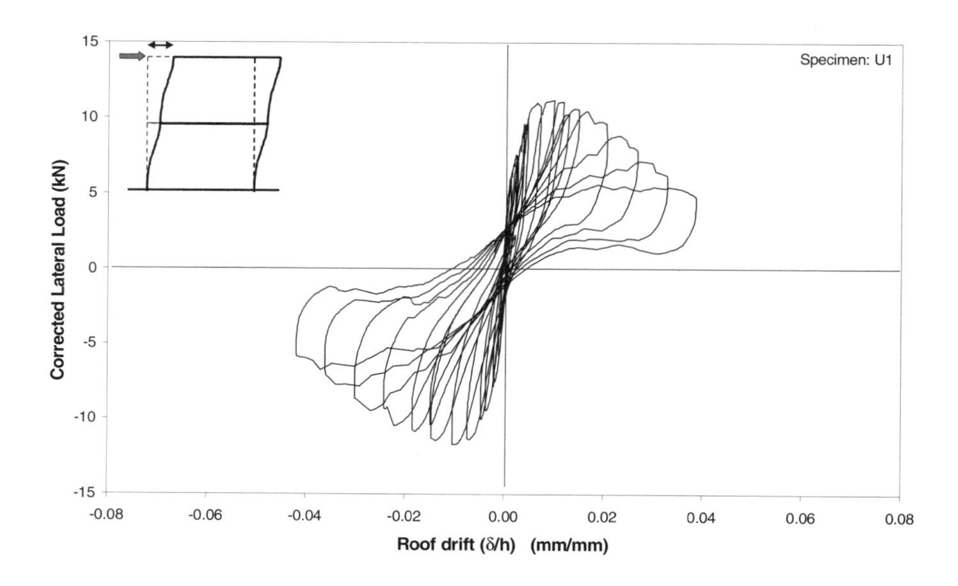

Figure 12. Load–Roof Drift Curve of Specimen U1

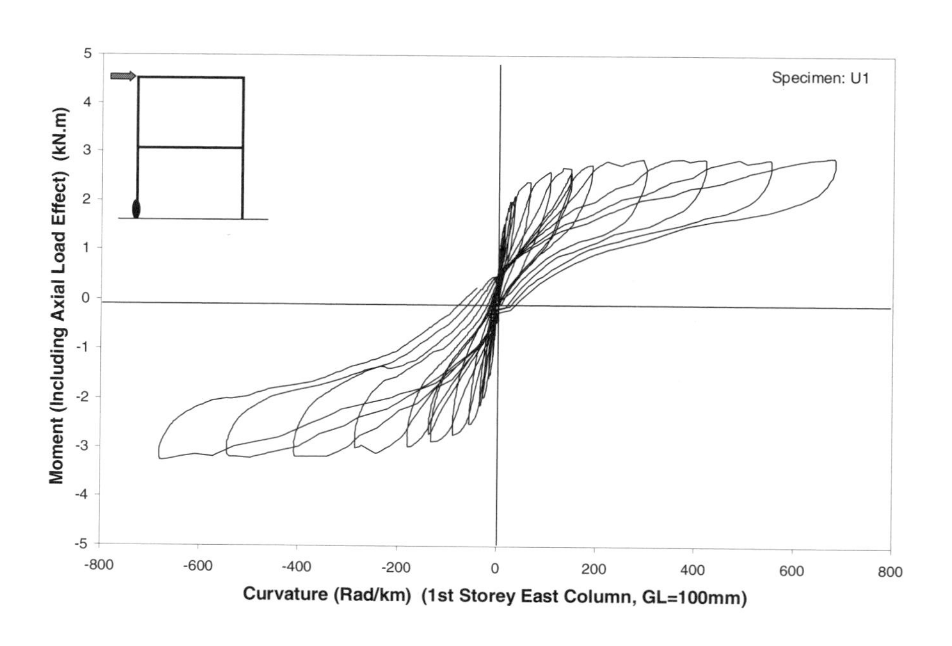

Figure 13. Column Moment-Curvature of Specimen U1 (GL=100mm)

Figure 14. Specimen U1 after the Test

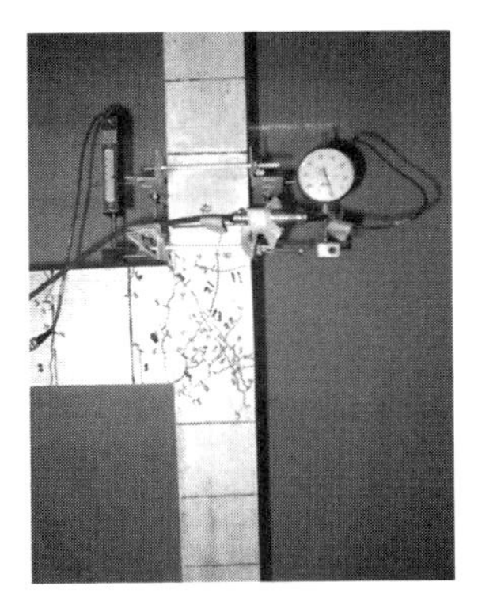

Figure 15. Connection Damage on Specimen U1

Figure 16. Column Base Damage on Specimen U1

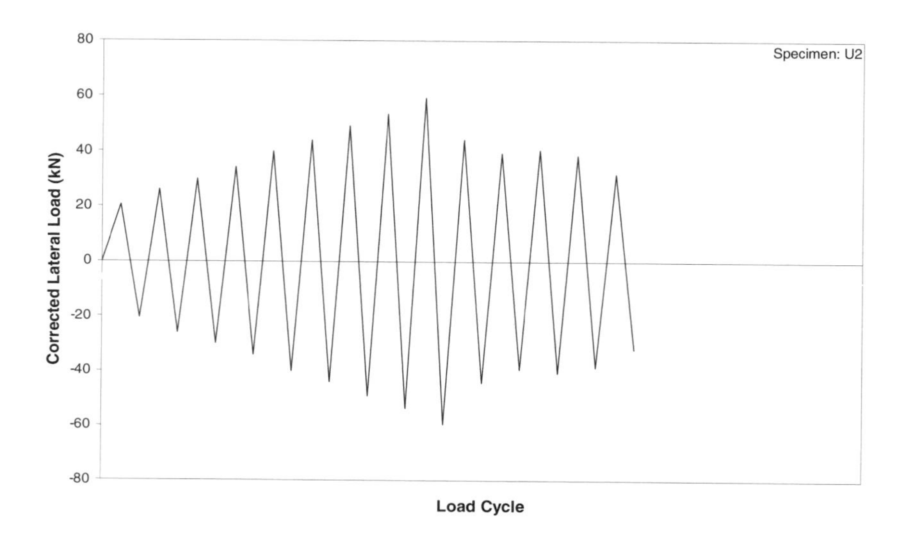

Figure 17. Loading History of Specimen U2

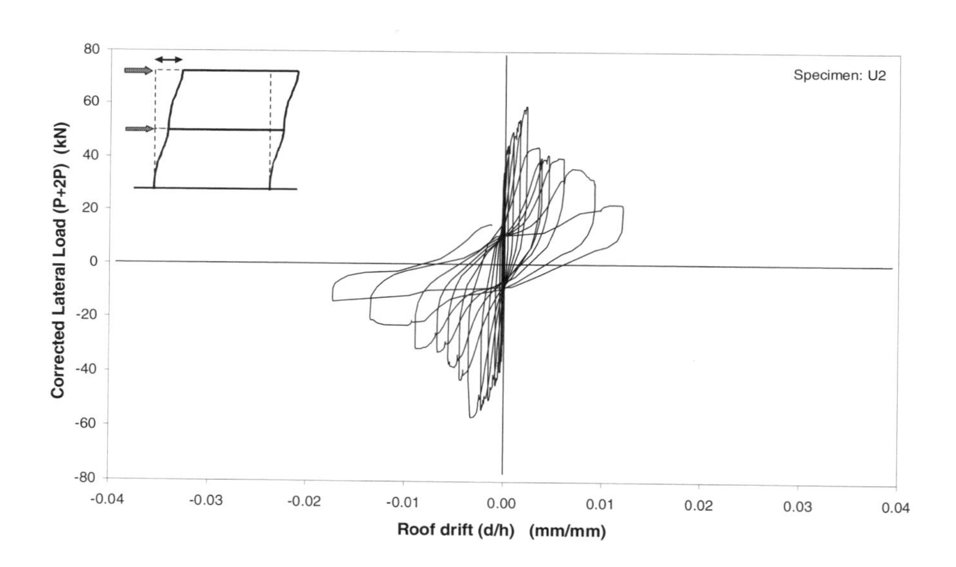

Figure 18. Load-Roof Drift Ratio of Specimen U2

Figure 19. Restraining Steel Frame and Specimen U2

Figure 20. Crack Propagation on Specimen U2 at Load Cycle 12

Figure 21. Specimen U2 after Test

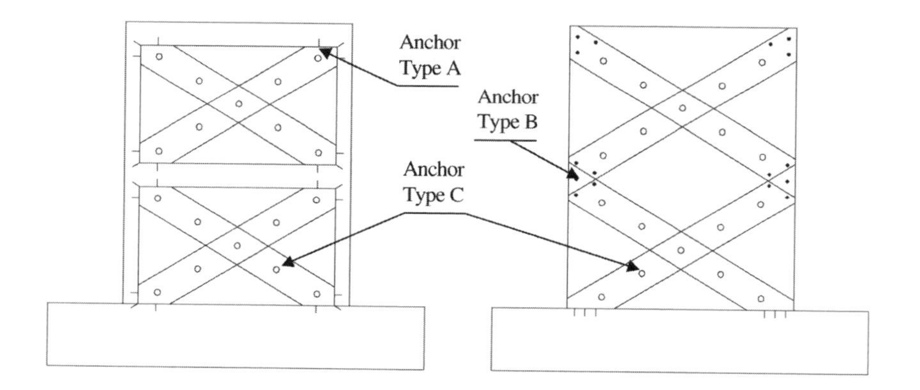

Figure 22. CFRP X-Brace Detail in Specimen U3

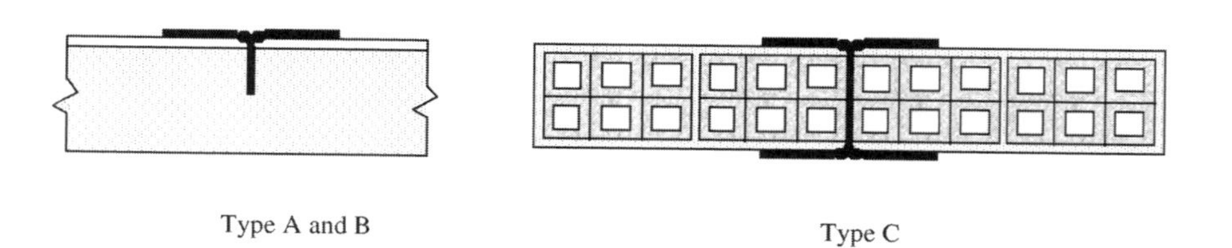

Figure 23. CFRP Anchor Detail for Specimen U3

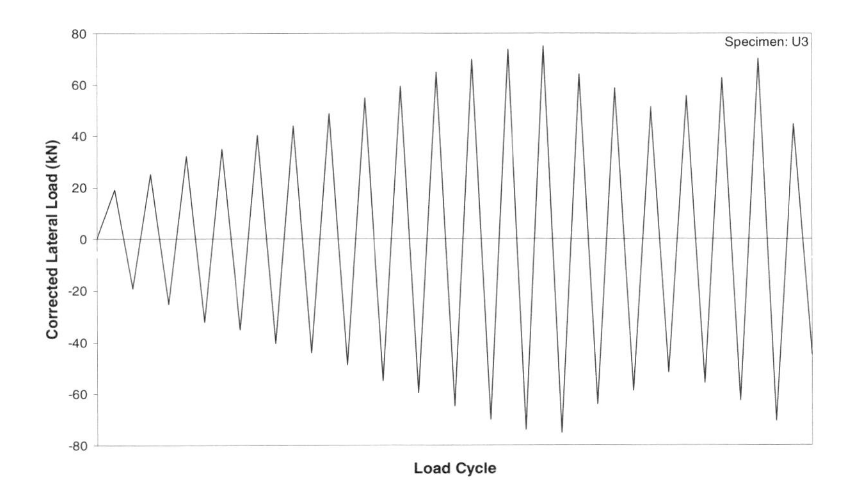

Figure 24. Loading History of Specimen U3

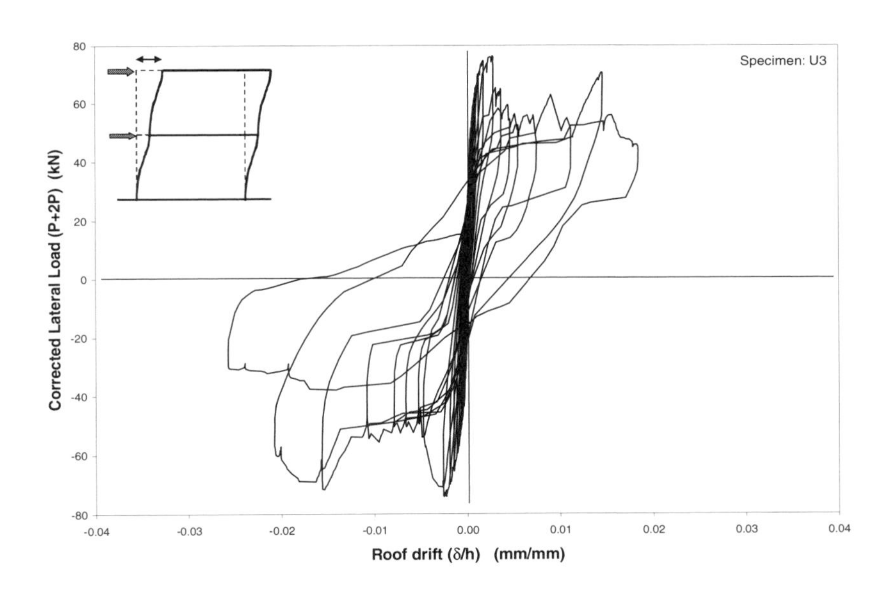

Figure 25. Load-Roof Drift of Specimen U3

Figure 26. Column Cracking Due to Inadequate Lap-Splice Length – Specimen U3

Figure 27. Specimen U3 after Testing

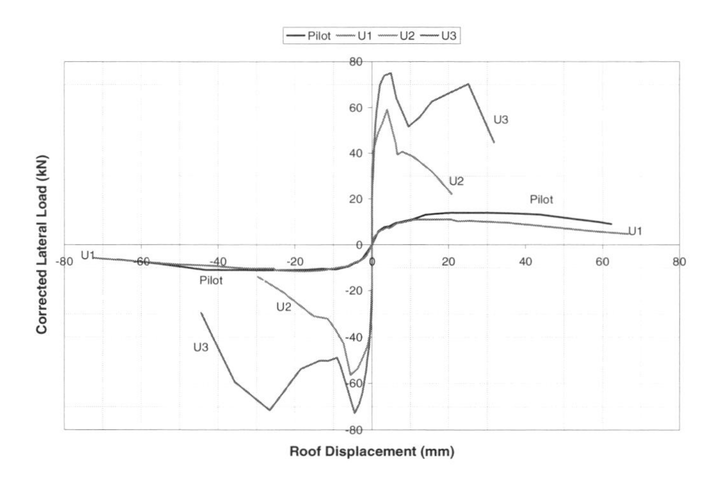

Figure 28. Load-Roof Displacement Envelope Curves

EXPERIMENTAL RESPONSE OF A PRECAST INFILL WALL SYSTEM

Robert J. Frosch, James O. Jirsa and Michael E. Kreger
Purdue University, University of Texas at Austin, University of Texas at Austin

Abstract: Many reinforced concrete moment resisting frame buildings in seismic zones lack strength and ductility. A precast infill system may provide an economic advantage for the retrofit of these structures. In order to evaluate the performance of this type of system, a two-story nonductile frame was constructed and rehabilitated with a precast infill wall and post-tensioning system. Three tests were performed on the rehabilitated model structure to evaluate the overall system behavior, verify performance of connection details, and investigate the performance of post-tensioning used to provide column tensile capacity at the boundary elements of the wall. The behavior of the structure through its loading history is discussed and compared with design assumptions.

Keywords: infill wall, reinforced concrete, seismic rehabilitation, shear wall

1. INTRODUCTION

Many reinforced concrete moment resisting frame buildings in seismic zones lack strength and ductility. A large number of these buildings were designed and constructed during the 1950's and 60's and do not possess the deformation capacity required for the level of lateral strength provided.

S.T. Wasti and G. Ozcebe (eds.), Seismic Assessment and Rehabilitation of Existing Buildings, 383–406.

According to the recognized methods of the day, the lateral forces prescribed were much lower than that required for elastic behavior. Therefore, ductility must be present in the structure to sustain the imposed deformation. The building codes, however, did not specify detailing requirements to achieve inelastic deformation capacity and structural toughness under significant cyclic lateral deformations.

The principal deficiency in concrete moment frames is inadequate flexural ductility or shear capacity in beams or columns and lack of confinement, frequently in the joints[1]. In addition, since structures were primarily designed for gravity loads, compression lap splices were typically used. Typical nonductile features (weak links) are summarized below.

Beams and Columns:

- Inadequate flexural capacity
- Inadequate shear capacity
- Lack of confinement
- Lack of column tension lap splices

Beam-Column Joints:

- Lack of confinement
- Inadequate joint shear capacity
- Strong beam-weak column

2. PRECAST INFILL SYSTEM

A precast panel system was developed that enables precast panels to be assembled into an infill wall.[2,3] A precast system can be constructed rapidly without the need for extensive formwork and the relatively cumbersome and sometime difficult procedures associated with moving and placing large quantities of fresh concrete within an existing building. The precast panels can be brought into an existing structure through the use of elevators and light forklifts. The panels have shear keys along the sides to allow for force transfer and are connected to one another through the use of a reinforced grout strip. Panels are connected to the existing frame through the use of steel pipes (shear lugs) that eliminate the need for interface dowels. The existing structure is cored in selected locations to allow for insertion of pipes and continuity of the wall vertical reinforcement. A schematic of the precast wall system is presented in Figure 1.

The tensile capacity of the existing frame columns must be increased to provide overturning capacity to the infill wall. Typical nonductile structures were constructed with only compression lap splices which do not provide for tensile yielding of column reinforcement. A post-tensioning system located adjacent to the existing frame columns (boundary elements) was used to improve the column tensile capacity without using conventional jacketing. This system is illustrated in Figure 2.

3. SPECIMEN DESIGN

To evaluate the performance of the precast infill wall system along with the post-tensioning system, a large-scale test was performed. The test specimen was designed in two stages. First the building frame was designed according to the building codes governing design during the period of construction of the original structure. Secondly, a rehabilitation scheme was designed utilizing present design codes and based on research dedicated to the behavior of the precast panel-to-panel connections[4] and the precast panel-to-frame connections.[5]

3.1 Existing Frame

The model structure was constructed with details typical of 1950's and 60's construction. The test structure had a column spacing of 13 ft. - 4 in. and floor heights of 8 ft. which corresponded with a 2/3 scale for typical moment frame buildings investigated. The specimen selected represents a "cut-out" section of the building and is shown in Figure 3.

The main nonductile features typical of 1950's and 1960's design contained in the test structure are presented below:

Column Lap Splices: Column splices were provided at the footing and at each floor level. They were designed as compression splices with a lap of 15 inches (20 d_b) for the #6 bars.

Column Ties: The columns were reinforced with #3 ties with 90° hooks spaced at 12 in.

Beam Reinforcement: The positive moment bars were discontinuous at the beam-column joint where a 4 in. embedment (6 in. in the prototype) into the joint was provided. Negative moment steel was also discontinuous in the midspan regions where positive moment controls design. Shear reinforcement in the form of U stirrups was provided.

3.2 Infill Wall

The precast infill wall system was designed to strengthen and stiffen the model test structure. A general layout of eight precast panels per floor as shown in Figure 4 was selected. The resulting panel size was consistent with handling requirements for ease in transport and placement. In addition, a grid layout of 2 rows by 4 columns permitted shear lugs to be distributed in the closure strips around the frame boundary with three lugs along the top and bottom boundaries of the wall and one lug at the midheight of the column.

The specific philosophy used in design was to provide for monolithic behavior of the precast infill wall system. Additionally, the infilled wall should have sufficient shear strength to permit the development of flexural yielding at the base of the wall. The shear strength of the wall should be controlled by the shear lugs provided around the interface since yielding of the lugs will provide some ductility. Since the column splices were not modified, a tension failure would be inevitable during a major seismic event. Therefore, shear resistance from the columns was neglected as was column tensile capacity in considering the collapse limit state design.

The post-tensioning steel was designed to provide precompression forces to the wall to improve wall behavior during different stages of lateral loading in addition to moment capacity. For a service limit state design, the post-tensioning was designed to provide for uncracked wall behavior so that the gross sectional properties of the wall could be used to reduce lateral drift. The high stiffness provided at this design level should maintain displacements at small values which may be necessary to meet building performance requirements based on continuity of operations. For a damage limit state design, the post-tensioning was designed to delay the column splice failure. The precompression force allows for an increase in the lateral loads before splice failure occurs. Both the amount of post-tensioning steel and initial prestressing force were varied during testing to ascertain their effect on the behavior of the structure. Additionally, changing the amount of post tensioning steel allowed for evaluating both flexural and shear failure modes of the structure.

4. TEST SERIES

4.1 Existing Frame

Existing structures are typically found in a cracked condition due to forces the building frames have experienced in service. Therefore, the model structure was laterally loaded with a triangular load distribution applied at the two floor levels (Figure 5) to produce cracking in the frame elements simulating the likely condition of a structure prior to rehabilitation. The existing frame was also tested to determine its overall behavior and to provide comparison of experimental results with those obtained analytically.

As shown in the load-displacement response in Figure 6, the initial stiffness was substantially reduced at first cracking. As cycling and loading increased, however, the stiffness continued to decrease. The effect of cracking can be observed from the slopes and the hysteresis exhibited.

Overall, the frame behaved as expected. The lateral force resistance of the frame was low. At the maximum load applied to the specimen, the stress in the positive moment steel was approximately 21 ksi on the extremely short development length (4 in.) which indicated that pullout failure would occur at a low base shear. Pullout failure of bottom beam bars would occur suddenly and would be nonductile. Subsequently, forces would be redistributed to the columns leading to failure of the splices. The structure in this condition could not be expected to resist an earthquake producing significant ground motions. Structural rehabilitation, therefore, would be recommended to seismically upgrade the structure.

4.2 Infill Wall System

The infill wall system was tested to establish the behavior of the rehabilitation technique. The specimen was tested cyclically using a triangular load distribution applied at the two floor levels as conducted for the existing frame to simulate first vibration mode effects for the structure. Flexural and shear strength were investigated in order to examine the precast panel connections, the infill wall design procedure, and the post-tensioning system.

4.2.1 Test 1 (Flexural Hinge)

Test 1 was performed to demonstrate that the infill wall system can be designed to achieve a ductile mechanism through the formation of a flexural hinge at the base of the wall. This mechanism allows the structure to sustain significant deformations while maintaining its lateral load capacity. The overall goal of the test was to load the structure until the limit load of the structure was reached as determined by initial yielding of the post-tensioning bars.

Since the structure was tested cyclically in both directions, two different system characteristics could be determined in one test. Different post-tensioning bar sizes were used on opposite ends of the wall to assess the effect of the amount of steel. Two 1 in. post-tensioning bars were used adjacent to the east column while two 1-1/4 in. bars were used adjacent to the west column. The amount of reinforcement was selected so that the wall shear strength would be higher than the flexural strength, thereby, allowing the formation of a flexural hinge. To investigate the effects of post-tensioning, an initial prestressing load of 56 kips per bar was selected. The same load was applied to all bars to produce uniform axial compression on the wall (moments eliminated).

The load-displacement relationship is presented in Figure 7. All loads are presented as base shear, and test cycles are designated by the load or displacement target level during testing superscripted by the cycle number. For example 0.3^1 is the first cycle to 0.3% drift.

The column reinforcement strain gage measurements indicated that a zero tension stress state occurred in the push direction at 47.4 kips. In the pull direction, zero tension was indicated at 59.7 kips. In addition, the pre-existing crack at the base of the column was observed open during an inspection at 90 kips. The structural response, however, as loading increased beyond the zero tension load remained the same; the same initial stiffness was maintained up until approximately 90 kips in the push direction and approximately 100 kips in the pull direction (Figure 8). The load at which the linear-elastic behavior changed was termed the decompression load as will be discussed later.

As loading increased beyond the decompression load, there was a gradual decrease in the stiffness. Cracking occurred on the bottom existing frame columns with the first crack occurring at the top of the column splice. As loading further increased, additional cracking was noted at intervals along the height of the column that progressed from the bottom of the column and moved upward.

When load in the push direction reached 255 kips (255^1), splitting cracks along the spliced bars were noted in the east column splice region indicating that failure was imminent. Therefore, it was decided to cycle at this load level to investigate behavior immediately prior to splice failure. Load in the pull direction to the same level did not produce splitting cracks in the west column. An additional cycle (255^2) was completed at the same load level to further investigate the splice failure. Although cracking patterns did not change, significant changes in the hysteresis were noted. Additionally, the response beyond the decompression load in cycle (255^2) was linear and did not exhibit the gradual change in stiffness that was observed in cycle 255^1. This reduction in stiffness was expected since flexural cracking occurred in the first cycle (255^1).

Following cycling at 255 kips, the load was increased to 259 kips in the push direction (0.2^1) when splice failure occurred in the east column. The structure was pushed to approximately 0.2 percent drift and unloaded. Loading was increased to 0.2 percent drift (270 kips) in the pull direction. Splice failure, however, did not occur because the initial post-tensioning in this column was slightly higher which provided increased splice capacity.

The structure was subsequently unloaded. Further testing was conducted in only the push direction to avoid failing the west column splice in this test.

In Cycle 0.3^1, the structure maintained good initial stiffness matching approximately the loading cycle to splice failure as shown in Figure 7. As loading increased beyond 120 kips, there was a slight decrease in stiffness; however, the load-displacement response was close to the original behavior with only a slight loss in load capacity. This response suggested that the splice bars were contributing to the tensile resistance of the structure and the overall stiffness. The structure was cycled at this displacement level to assess the contribution of the splice steel. The response in the second cycle (0.3^2) was much different than the first (0.3^1) as shown in Figure 9 which presents the post-splice failure response. It is clear that the splice bars were contributing to tensile capacity during cycle 0.3^1, but the damage in the splice was sufficient to eliminate any significant tensile capacity for later cycles.

The structure responded in all cycles with the same initial stiffness (K_1) up to 60 kips (zero tension load). There was a gradual transition in stiffness until approximately 120 kips (slightly beyond the decompression load of 90 kips). Following this level, behavior in all cycles was fairly linear with approximately the same stiffness (K_2). There was a trend, however, of gradually degrading stiffness with increasing displacement cycles. In cycles up to 0.45 % drift, the post-tensioning steel remained elastic.

In the last cycle (0.55^1) of this test, the behavior was similar to the previous cycles. It is significant to note, however, that yielding of one east (north side) post tensioning bar was reached while the other bar remained elastic. The overall drift at this level was measured at 0.575 % as shown in Figure 9. After unloading, a larger residual drift remained since one of the post-tensioning bars yielded.

4.2.2 Test 2: Shear Test (500 kips Post-Tensioning)

Test 2 was designed to investigate the shear strength of the infill wall. Additionally, the effect of the prestressing force and the amount of post-tensioning steel were further investigated. In order to produce a shear failure in the wall, an increase in the flexural capacity was required. Therefore, two 1 in. post-tensioning bars were added to each side of the wall. The bar configuration consisted of four 1 in. bars on the upload side and of two 1 in. and two 1-1/4 in. bars on the download side. An initial prestressing load of 65 kips per bar was selected to further investigate the effects of the post-tensioning. The response curve is presented in Figure 10.

Test 2 commenced by loading in the pull direction, the direction in which splice failure in the column had not occurred. The zero tension stress state could not be determined from the column strain gage readings, but the slope of the response curve changed slightly at approximately 100 kips in both directions (point A) indicating that the zero tension load had been reached. As load was increased beyond point A, there was a gradual change in stiffness up to point B (approximately 200 kips) where a distinct change in the load response was observed. Loading beyond point B occurred at a new slope and was fairly linear in both the push and pull directions. The distinct change in behavior occurring at point B was termed the decompression load as in Test 1. Loading beyond point B produced different behavior in both loading directions. The specimen was stiffer in the pull direction since the tension column splice was intact. In the push direction, a lower stiffness was expected since only the unbonded post-tensioning bars provided tensile capacity.

Loading in the pull direction (450^1) was continued until a base shear of 421 kips was reached when splice failure occurred in the west column. Loading was further increased beyond splice failure to 450 kips after which there was a slight reduction in stiffness, but the base shear was still increasing very slightly. In cycle 450^2 in the pull direction, the difference in the hysteresis loops following splice failure can be noted. There was a distinct change in stiffness from the pre-splice behavior because the unbonded post-tensioning system was required to carry the tensile forces that previously had been carried by the tensile reinforcement in the column.

As noted from the overall response, there was some hysteresis in the loading loops due to cracking and column damage in the splice region. After unloading, however, the structure returned to approximately its original position with only small residual drifts since the post-tensioning steel remained elastic throughout testing. The load-displacement response formed an S shape, characteristic of non-linear elastic behavior.

4.2.3 Test 3: Shear Test (No initial Post-Tensioning)

Test 3 was designed to investigate the shear strength of the wall in the absence of initial prestressing load. Since compressive stresses in the wall can enhance the shear strength, Test 3 was intended to obtain a lower bound shear capacity for the wall. The same bar configuration used in Test 2 was maintained. The prestressing force, however, was released in all post-tensioning bars. The anchor nuts were reinstalled in the snug-tight position to permit loading of the structure. Monotonic loading to failure in the pull direction was used, and the response is presented in Figure 11.

The structure was loaded monotonically in the pull direction since a localized failure of the loading system prevented full testing in the push direction. The initial stiffness was maintained up to approximately 300 kips.

The stiffness gradually decreased with further loading as increased shear cracking occurred in the wall; the increased cracking increased the shearing deformations. The trend of stiffness degradation continued until approximately 460 kips where a distinct change in the slope of the curve was noticed. Fairly large deformations occurred with only a small increase in load.

At 491 kips, concrete spalled in the wall at the top of the main compression strut. The spalling occurred within the grout strip under the existing beam at the location of the west post-tensioning anchorage. Cracking patterns indicated that a primary compression strut directed from the "toe" of the wall to the post-tensioning anchorage had formed. It was within this strut that the wall spalled. From the flatness of the response curve at 491 kips and the crushing that occurred in the compression strut, it was likely that the shear capacity of the wall was reached. Unloading occurred at approximately the initial loading stiffness.

5. CRACKING PATTERNS

Web shear cracking initiated at approximately a 45 degree angle in the middle first story wall (V_{base}=248 kips). A similar crack was also noted in the second story wall at slightly higher loads. As loading increased, shear cracks extended, and there were additional cracks in both stories. The general orientation of all cracks was along a line connecting the column base to the top post-tensioning anchorage. The cracking patterns of the first and second floors following testing are shown in Figure 12 and Figure 13, respectively.

Diagonal cracks were continuous through panels and grout strips. There was no indication of distress along grout strips. The cracking pattern indicated that the wall behaved in a monolithic fashion even though it was constructed of smaller, individual units. A maximum crack width of 0.8 mm was measured during testing. After reaching this width, it was noticed that additional cracks would open so that the initial crack would close slightly.

Web shear cracking was predicted using a Mohr's circle elastic stress analysis which included the effect of compressive stresses from the post-tensioning system. Cracking was assumed to initiate when the principal tensile stress reached $6\sqrt{f_c'}$ (psi). The cracking load was calculated to be 158 kips which was in excellent agreement with the test results.

6. CAPACITY ANALYSIS

6.1 Zero Tension

The applied force required to produce zero tension (Figure 14) at the extreme fiber of the wall was measured as previously discussed. The load required to produce this stress condition was calculated using basic flexural mechanics and gross sectional properties of the flanged wall (I_g=3,468,000 in^4, A_g=1176 in^2). The equation for this stress state is presented below.

A comparison of the experimental results with the theoretical calculations is presented in Table 1. The calculations were made assuming only axial compression. In reality, the post-tensioning system varied slightly from one side of the structure to the other producing applied moments. These moments could also be accounted for in the analysis which would provide for the different zero tension loads measured in the two directions in Test 1.

6.2 Decompression Load

There was an abrupt change in stiffness at a load approximately double the zero tension load. Prior to this point, the structure maintained approximately the same stiffness even beyond the zero tension load. This behavior was not expected as a transition in behavior was anticipated when the structure reached the zero tension stress state.

It was discovered that the compressive stresses remaining in the structure at the zero-tension condition provide clamping forces to maintain the stiffness. The applied moment must fully overcome the pre-compression force or decompress the section before the stiffness of the section changes significantly. Before decompression, no significant tensile stresses are required to be resisted by the column tensile steel as shown in Figure 15 for Test 1 and Figure 16 for Test 2. A very small change in behavior was noticed after the zero tension load.

The system can be viewed by considering the applied moment as an eccentric axial load, P, where the axial load is the post-tensioning force as illustrated in Figure 17. As the axial load eccentricity increases, pre-compression across the interface is reduced. As P approaches the edge of the structure, the initial compression is eliminated and the moment on the section at this stage is termed the decompression moment in this paper. The decompression moment can be calculated as follows.

$$M_{DC} = Pe$$

Here, e = Wall Length / 2

For this analysis, it was assumed that the section cannot resist tension. The column base was precracked prior to wall construction, and the wall was connected along an unroughened surface. In Table 2, calculated loads are compared with experimental results. The values were calculated assuming pure axial compression and neglecting any initial eccentricity of the applied pre-compression. The analytical results compare extremely well with those measured.

6.3 Splice Failure

The existing frame column acted as a boundary element of the infill wall; therefore, tensile stresses developed from the overturning moments were required to be resisted by the column reinforcement. A splitting tensile splice failure occurred in both columns due to the short compression lap length.

The reason that failures occurred in Test 1 and 2 at different lateral loads was the different initial post-tensioning forces applied. In general, as the initial post-tensioning increased, the lateral load to produce splice failure increased. The test results are presented in Table 3.

Stresses in the splice reinforcement and ties were determined from measured strains immediately prior to failure. In the east column, the column bars reached a stress of 59 ksi. From the load-displacement response, however, it appears that the bars probably reached yield as indicated by the load plateau. This seems highly likely since the bars had a yield strength of 61 ksi, and there was variability in the strain gage readings. Stresses in the two column ties in the splice region increased significantly as the splice failure was approached, but did not reach yield. A maximum tie stress of 16 ksi was measured. The download column provided similar findings; two column bars apparently reached yield according to strain gage readings while the column ties reached 35 ksi. Even though several of the bars may have reached initial yield, the embedment lengths were too short to allow for any significant deformation capacity of the bars and any increase in stresses due to strain hardening.

6.4 Flexural Strength

The ultimate flexural capacity of the structure following splice failure was achieved through yielding of the post-tensioning steel. The capacity can be computed from strength design procedures recommended by ACI 318.[6] Since the high strength bars have a yield plateau, the yield strength of the material can be used to determine the moment capacity. The yield strength for the post-tensioning bars can be estimated as 90 percent of ultimate.

$$F_y = 0.9F_u \qquad T = A_s F_y \qquad a = \frac{T}{0.85 f_c' b} \qquad M_n = T\left(d - \frac{a}{2}\right)$$

A flexural strength of 273 kips was reached in Test 1 while the computed strength using the actual yield strength of the post-tensioning bar was 245 kips.

6.5 Shear Strength

The approximate shear strength of the wall was 491 kips. As previously mentioned, the load-displacement response for Test 3 and the initiation of compressive strut failure indicated that the shear strength had been reached. The test was not continued beyond this load level because the capacity of the loading system was reached.

The wall shear strength was controlled by concrete crushing in the primary compression strut. No interface or diagonal shear failures occurred that limited the shear capacity in either the first or second story wall. Both walls (6 in. and 4 in.) reached a shear stress of $7.1\sqrt{f_c'}$ (psi) which was calculated based on the lowest strength wall panels (4500 psi) and the entire length of the wall (L_w= 172 in.) as recommended by ACI 318[6] Chapter 21. The shear strength (492 kips) greatly exceeded the design strength of 146 kips that was calculated based on the combination of the pipe shear capacities at the interface.

By using a simple strut and tie model of the structure (Figure 18), prediction of the compression strut failure was possible. The crushing occurred in the wall below the top floor beam due to the reduction in cross sectional area created by the 4 in. wall. It was assumed that the bearing stresses from the post-tensioning anchorage spread out along a 45 degree angle as shown in Figure 19. By using this model and the bearing strength of concrete ($0.85f_c'$) as recommended by ACI[6], compression strut failure was estimated at approximately 445 kips which compares reasonably well with the test result of 491 kips.

7. OBSERVED BEHAVIOR VERSUS DESIGN ASSUMPTIONS

The infill wall was designed for a base shear of 146 kips assuming that the shear lugs were to carry the entire base shear. In reality, a large portion of the base shear was carried by a combination of friction from the compressive force acting at the "toe" of the wall and the shear resistance from the existing column at the failed splice. Even if it was assumed that the shear lugs were at their peak capacity, a large portion of the base shear would remain unaccounted.

The precast wall panels and the panel connections were designed to be consistent with the design assumption made for the frame connection. They were designed to transfer shear to the shear lugs without failure originating in either the panels or panel connections. The design of the panels worked according to design assumptions since no distress of the panels could be seen adjacent to the shear lugs. The cracking pattern, however, does not indicate struts forming adjacent to the shear lugs as was implicit in the design assumption. Instead, a primary compression strut was directed at the "toe" of the wall indicating that the entire wall was effective in transferring shear.

The infill wall reached a base shear of 491 kips which was 3.4 times greater than the design level.

8. SUMMARY AND CONCLUSIONS

Three tests were performed on the rehabilitated model structure to evaluate the system behavior of the precast infill wall and the post tensioning system used to provide column tensile capacity. Primary variables included the amount of post-tensioning steel and the initial post-tensioning loads. Loads were cyclically applied to the floor slabs in a triangular distribution to simulate earthquake effects. Testing allowed examination of design recommendations for the precast infill wall, post-tensioning system, and panel connections.

The precast wall system performed exceptionally well. Diagonal cracks were continuous through panels and grout strips. There was no indication of distress along grout strips. Shear cracking extended along a line from the "toe" of the wall to the top post-tensioning anchorage. There was some cracking concentrated around the pipe shear lugs between the wall and frame, but failure did not appear to be imminent. Overall, crack patterns indicated that the wall behaved monolithically even though it was constructed of multiple units, and both the panel and frame connections performed according to design. At ultimate, the shear strength of the structure was controlled by concrete crushing of the primary compression strut; no interface or diagonal shear failures occurred that limited the shear capacity in either the first or second story wall.

The post-tensioning system performed according to design. Testing demonstrated that the splice capacity of the existing frame could be controlled with the initial post-tensioning force. Following splice failure, a large crack formed at the base of the wall with the wall/frame system rotating nearly as a rigid body around the "toe" of the wall in the direction of applied load due to the unbonded tendons. The flexural capacity of the structure during this stage of behavior was determined by the yield strength of the post-tensioning steel.

The precast infill wall system developed in this research may eliminate many of the costly and time-consuming procedures currently used in cast-in-place infill wall construction. Based on the research conducted, design and detailing guidelines were also developed to use the precast infill wall system as an alternative to cast-in-place construction. These guidelines are presented in Reference 3. The precast system should provide engineers with a technique that offers the potential to reduce overall costs of rehabilitating existing structures while allowing the rehabilitation to be tailored to the requirements of the owner. Furthermore, the system can be used to decrease nonstructural damage and costs associated with the damage and to increase life safety.

ACKNOWLEDGEMENTS

This study was conducted at the Phil M. Ferguson Structural Engineering Laboratory and was sponsored by the National Science Foundation (Grant No. BCS-9221531). The authors would like to thank Loring Wyllie, Jr. of Degenkolb Engineers and Tom Sabol of Engelkirk and Sabol Consulting Structural Engineers for their practical advice and assistance with the research. Finally, thanks are extended to Wanzhi Li and Michael Brack for their help during the experimental phase of this research.

CONVERSION FACTORS

1 in. = 25.4 mm 1 kip = 4.448 kN 1 ksi = 6.895 MPa

REFERENCES

1. FEMA-172, *NEHRP Handbook for Seismic Rehabilitation of Existing Buildings*, Building Seismic Safety Council, Washington, D.C., 1992.
2. Frosch, R.J. (1996), "Seismic Rehabilitation Using Precast Infill Walls, *Ph.D. Dissertation*, University of Texas at Austin, 234 pp.
3. Frosch, R.J., Li, W., Jirsa, J.O., and Kreger, M.E. (1996). "Retrofit of Non-Ductile Moment-Resisting Frames Using Precast Infill Wall Panels," *Earthquake Spectra*, 12(4), pp. 741-760.
4. Frosch, R.J. (1999). "Panel Connections for Precast Concrete Infill Walls," *ACI Structural Journal*, Vol. 96, No. 4, July-August 1999, pp. 467-472.
5. Frosch, R.J. (1999). "Shear Transfer Between Concrete Elements Using a Steel Pipe Connection," *ACI Structural Journal*, Vol. 96, No. 6, November-December 1999, pp. 1003-1008.
6. ACI Committee 318, "Building Code Requirements for Reinforced Concrete (ACI 318-02) and Commentary (ACI 318R-02)," American Concrete Institute, Farmington Hills, MI, 2002.

Table 1. Loading to Reach Zero Tension

	Experimental (kips)	Calculated (kips)
Test 1		
Push	47	51
Pull	60	51
Test 2		
Push	100	108
Pull	100	108

Table 2. Decompression Moment Analysis

Test	Experimental (kips)	Calculated (kips)
1	90	85
2	190	186

Table 3. Splice Failure Test Results

Column	Splice Failure, V_{Base} (kips)	Initial Post-Tensioning (kips)
East	259	237.2
West	421	507.2

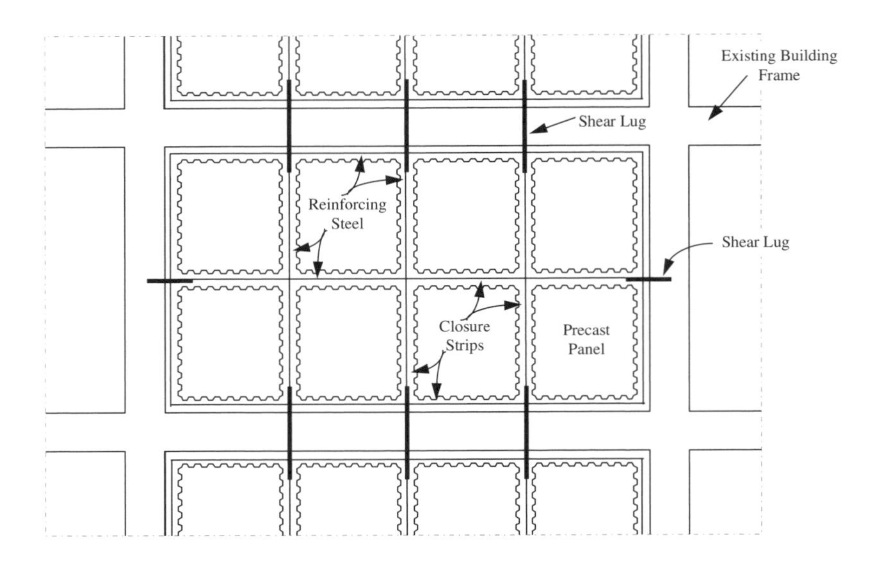

Figure 1. Precast Infill System

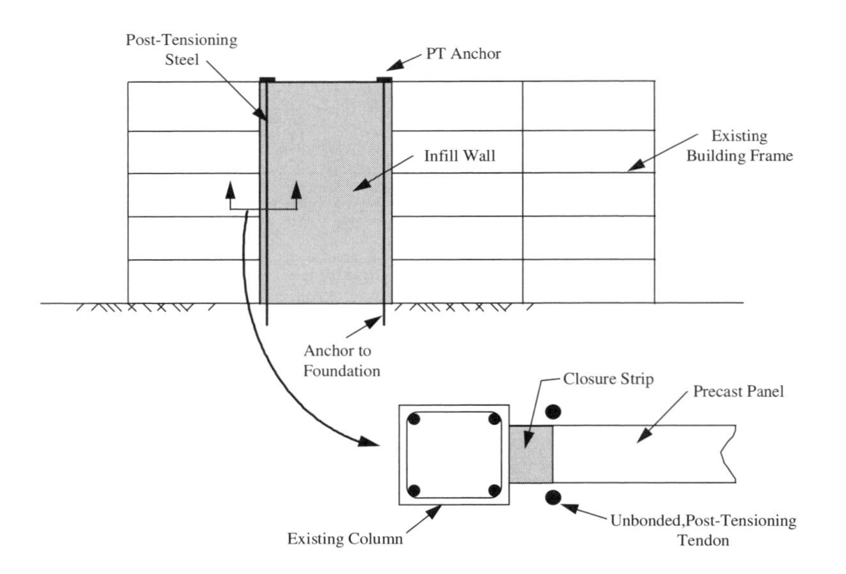

Figure 2. Post-Tensioning System

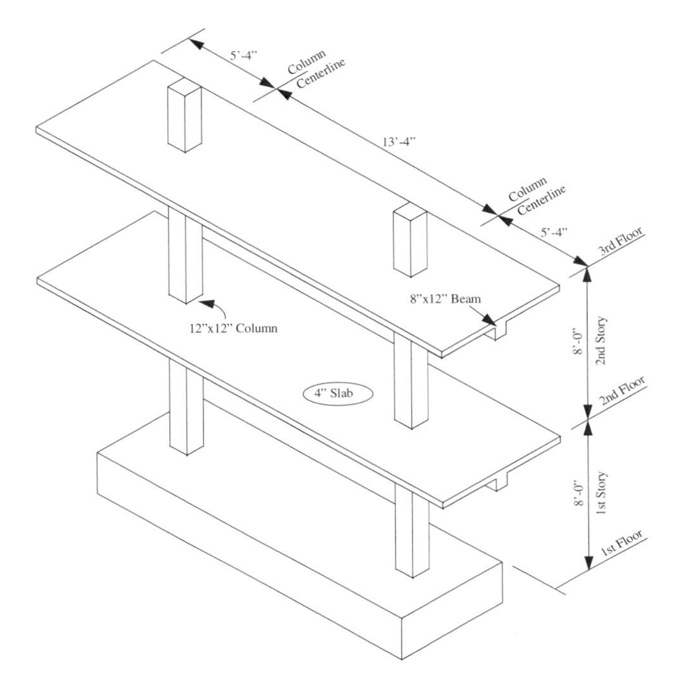

Figure 3. Existing Frame

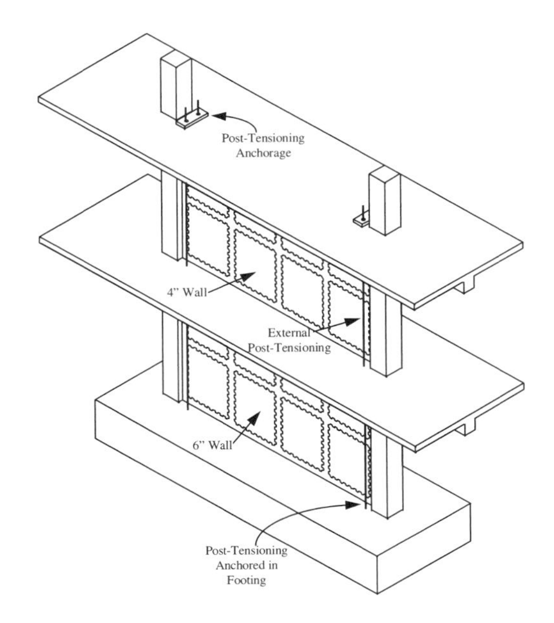

Figure 4. Infill Wall

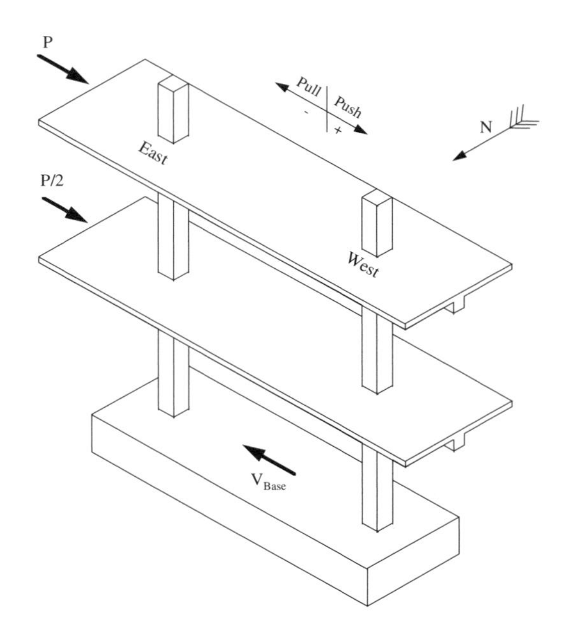

Figure 5. Loading Distribution

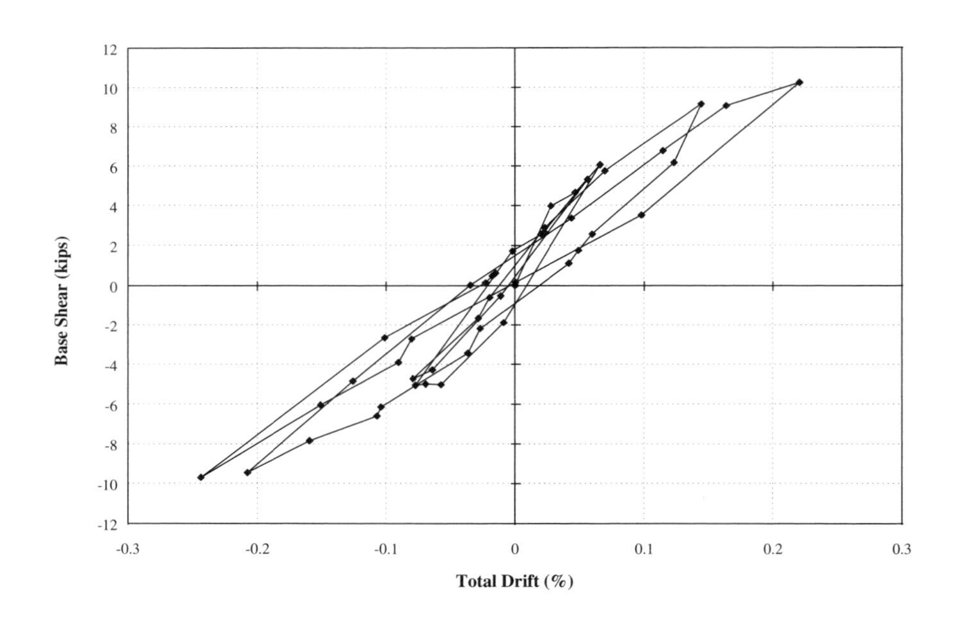

Figure 6. Existing Frame Response

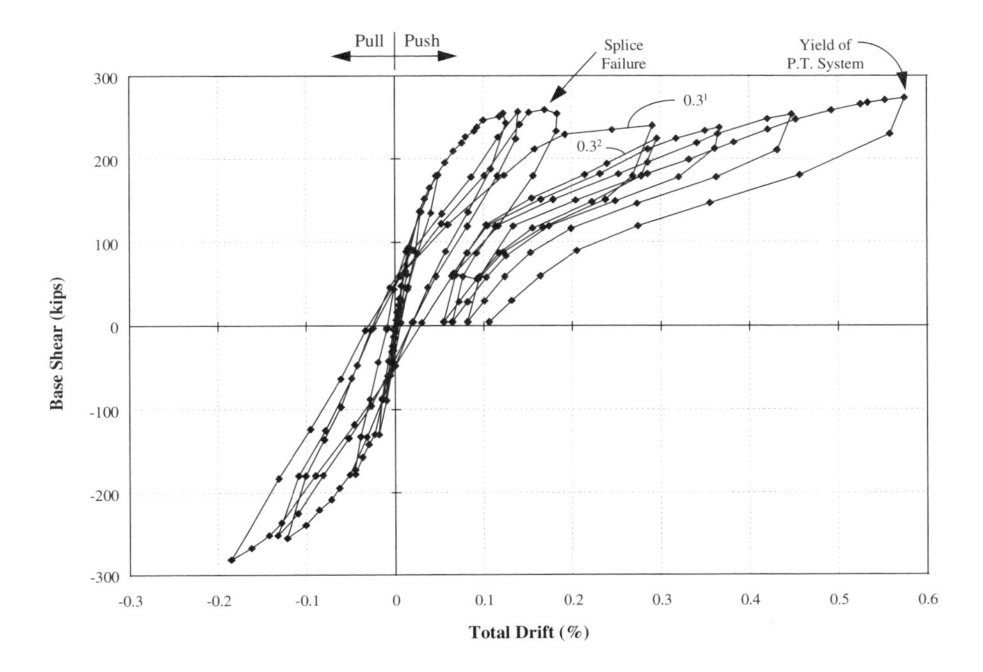

Figure 7. Test 1 Response

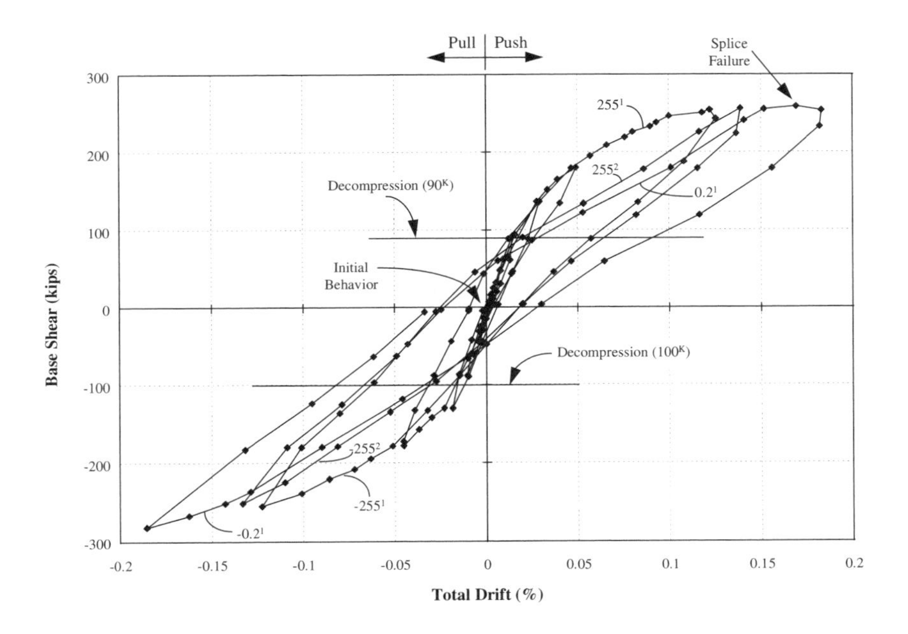

Figure 8. Test 1 Response Prior to Splice Failure

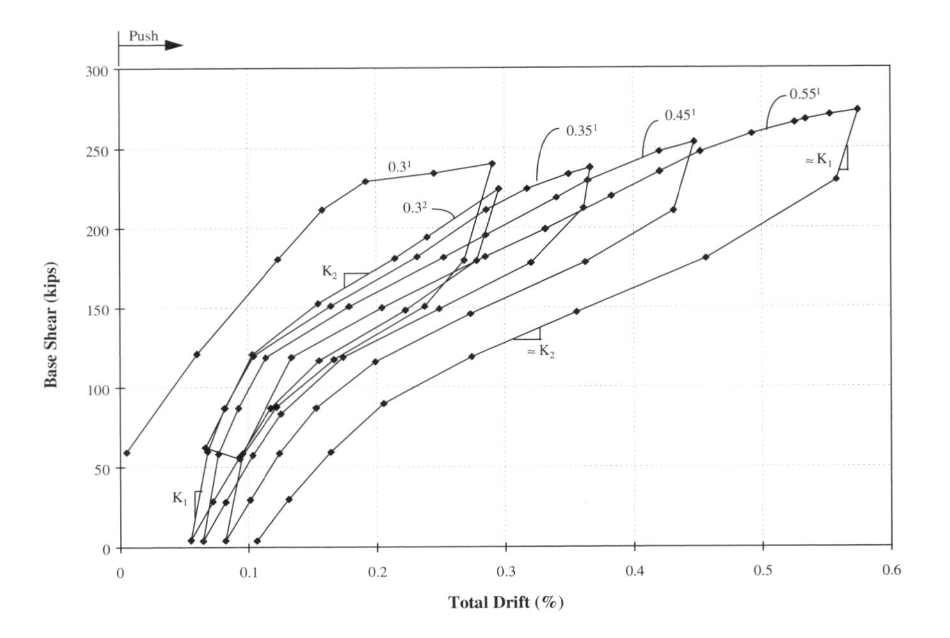

Figure 9. Test 1 Response After Splice Failure

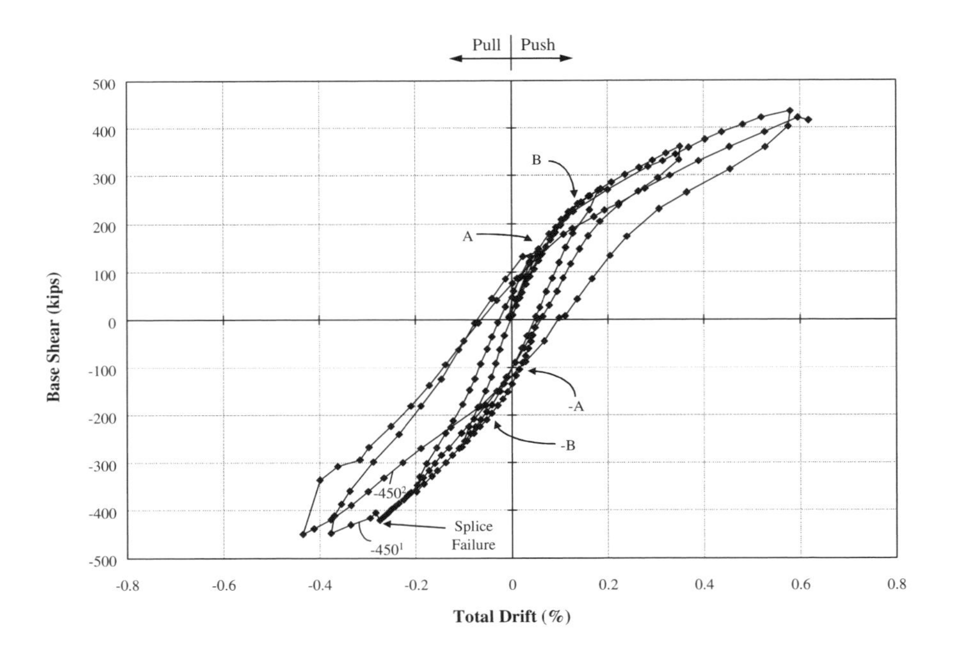

Figure 10. Test 2 Response

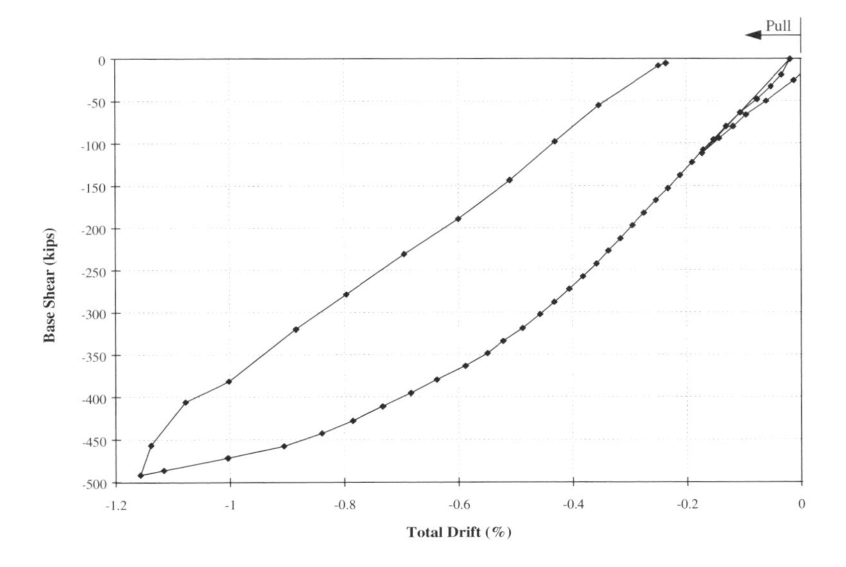

Figure 11. Test 3 Response

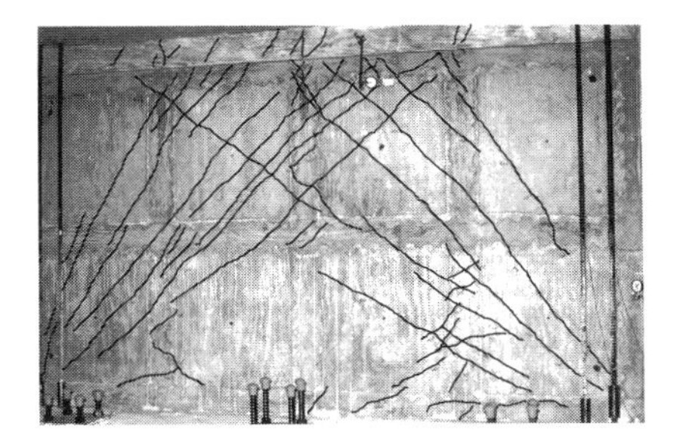

Figure 12. First Floor Cracking Pattern

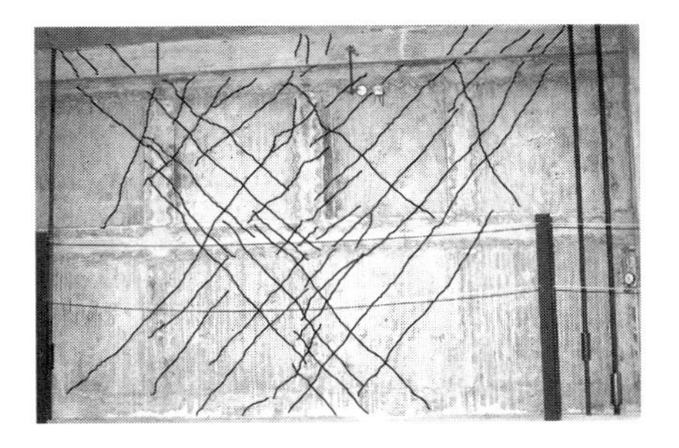

Figure 13. Second Floor Cracking Pattern

$$\frac{P}{A} + \frac{Mc}{I} = 0$$

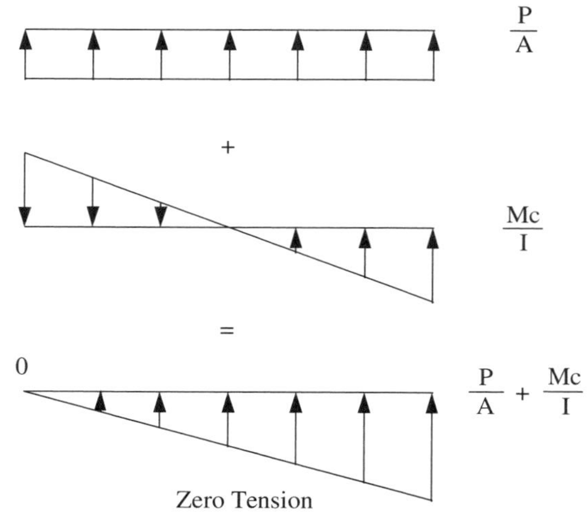

$\frac{P}{A}$

+

$\frac{Mc}{I}$

=

0

$\frac{P}{A} + \frac{Mc}{I}$

Zero Tension

Figure 14. Zero Tension Stress State

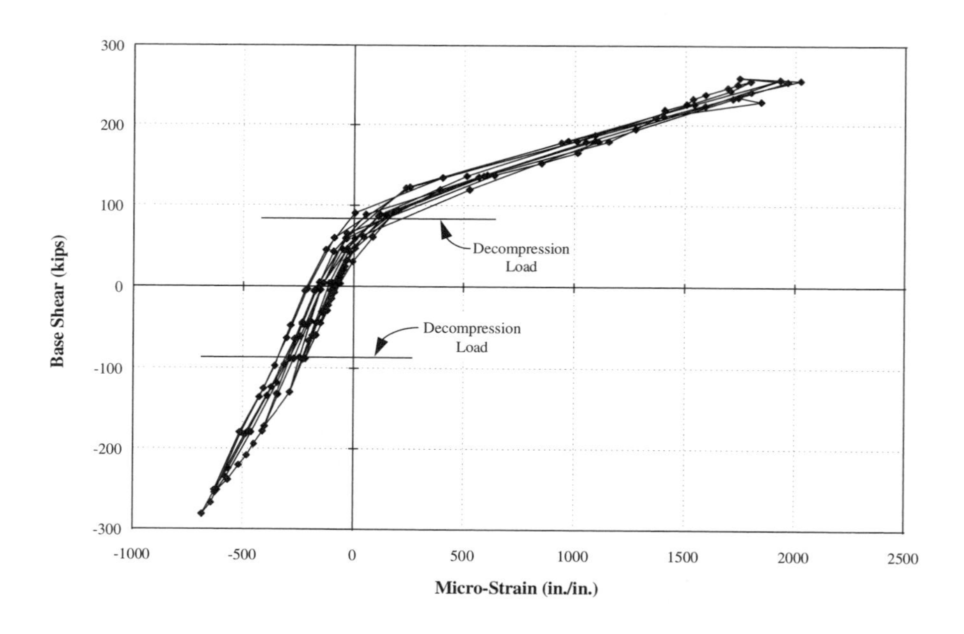

Figure 15. East Column Tension Steel (Test 1)

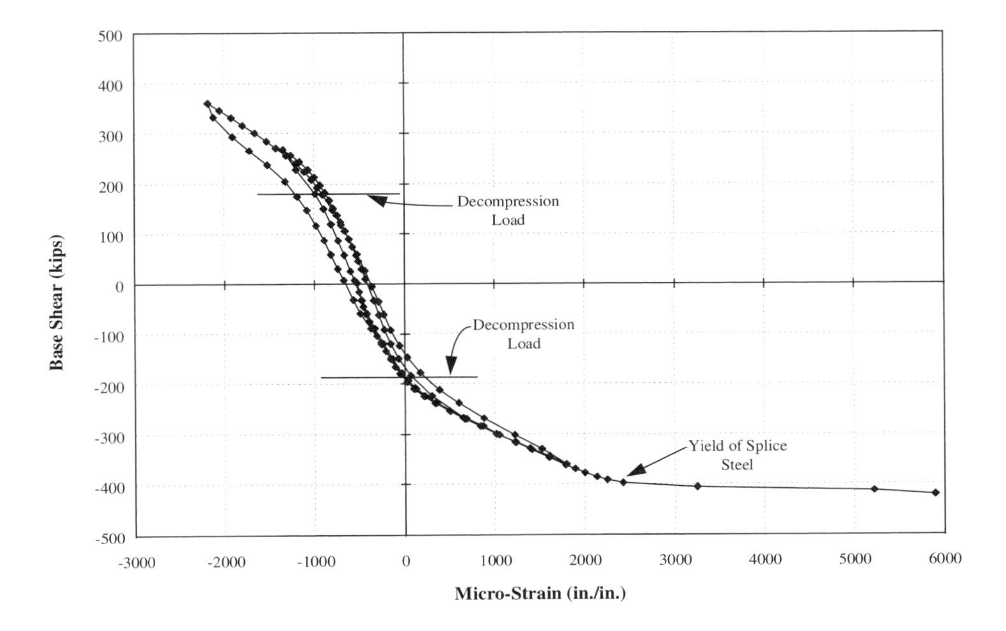

Figure 16. West Column Tension Steel (Test 2)

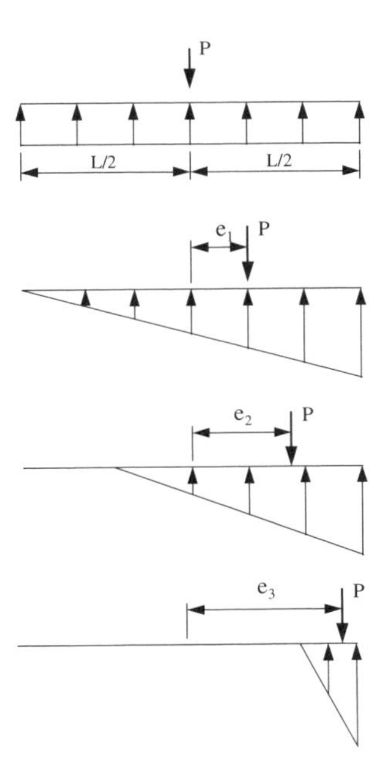

Figure 17. Analysis of Decompression Load

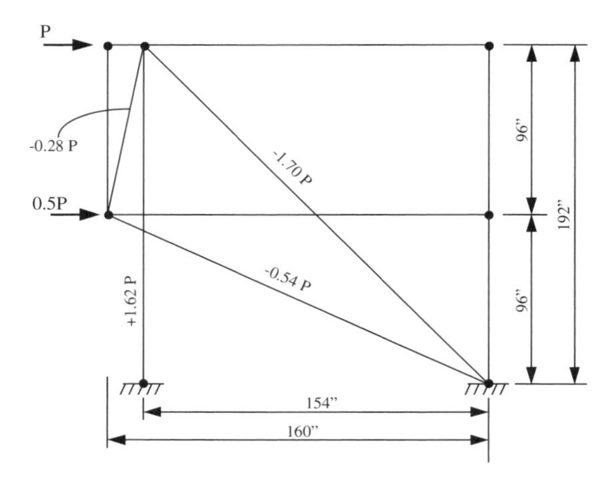

Figure 18. Strut and Tie Model

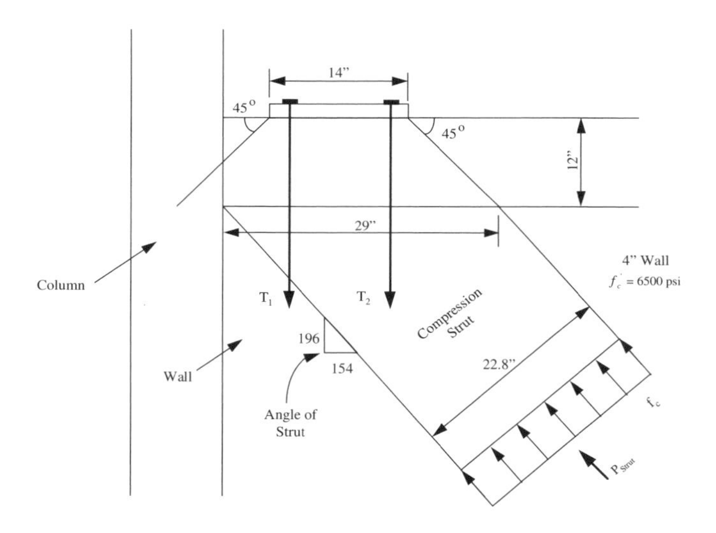

$$P_{Strut} = A_{Strut}(0.85 f_c')$$
$$= (22.8)(4)(0.85)(6.5) = 503.9 \text{ kips}$$

Strut force from analysis

$V_{Base} = 1.5$ kips

$P_{Strut} = 1.7$ kips

$$V_{Base} = 503.9\left(\frac{1.5}{1.7}\right) = 445 \text{ kips}$$

$$V_{Base} = 445 \text{ kips} \quad \approx \quad 491 \text{ kips from Test 3}$$

Figure 19. Nodal Point Analysis

A COMPARATIVE STUDY ON THE STRENGTHENING OF RC FRAMES

Ibrahim Erdem, Ugurhan Akyuz, Ugur Ersoy and Guney Ozcebe
Structural Mechanics Division, Civil Engineering Department,
Middle East Technical University, 06531 Ankara, Turkey

Abstract: In order to improve the behavior of buildings and in order to prevent total collapse, necessary amount of strengthening must be provided. The frame of the test specimen (1/3 scaled, 2-story, 3-bay) is detailed such that it has the common deficiencies of existing buildings in Turkey. Two types of strengthening techniques, namely introducing an infill RC wall, and CFRP strengthened hollow clay tile wall, are investigated. The test specimens are subjected to reversed cyclic quasi-static loading. By means of special transducers, axial force, shear force and bending moment at the base of the exterior columns are measured. Strength, stiffness, and story drifts of the test specimen are determined.

Key words: Earthquake, reinforced concrete (RC), strengthening, earthquake resistance, carbon fiber reinforced polymer (CFRP)

1. INTRODUCTION

Deficiencies introduced during design and/or construction may cause catastrophic results when the structure is subjected to high lateral loads. Due to inadequate lateral stiffness, many reinforced concrete buildings were highly damaged or collapsed in Turkey during the major earthquakes in the last decade. In order to improve the behavior of such buildings and to

S.T. Wasti and G. Ozcebe (eds.), Seismic Assessment and Rehabilitation of Existing Buildings, 407–432.

prevent total collapse, the necessary amount of strengthening must be provided. Various strengthening methods have been developed to reduce the damage caused by earthquakes. These methods aim to increase the capacity of the some individual members or that of whole structural system.

Many researchers have investigated the strengthening methods that can be used for the seismically deficient structures. The most commonly used methods are stated by Sugano [(1)] as: providing back-up structure, infilling existing frames, bracing existing frames, placing side walls, jacketing of existing members.

Seismic repair and/or strengthening can be accomplished either by system behavior improvement or member repair/strengthening. When the existing structure is safe under gravity loads, to compensate the inadequacy of the structure against seismic loads, a new lateral load resisting system, which will increase the lateral strength and the lateral stiffness of the existing system, is introduced. This is called system improvement. Various techniques based on this principle have been developed and applied in the past. Among them, the most widely used technique is the formation of new stiff walls through infilling some bays of the existing frame with reinforced concrete infills. The use of RC infill walls as a method of seismic behavior improvement for moderately damaged structures is commonly used in Turkey. This approach was first proposed following pilot tests made by the METU staff on infilled frames (Ersoy and Uzsoy [(2)]).

Later, Altin, Ersoy and Tankut [(3)] performed 14 tests to investigate the behavior of RC infilled undamaged frames. They designed and constructed two-story one bay frames according to the Turkish Reinforced Concrete Code. The reinforcement pattern of the RC infill wall, the compressive strength of the boundary columns, the connection between the wall and the frame (anchorages) and the amount of axial load on the columns were the main parameters of this study. The specimens were tested under reversed cyclic quasi-static loading. The authors concluded that the connection of the RC infill wall to the frame is very important. Strength and stiffness of the specimens are highly affected by the quality of the connection. It has been also observed that an increase in the compressive strength of the column increases the lateral strength of the specimen. In addition, a high axial load level on the column increases the capacity of the specimen.

Sonuvar [(4)] constructed five two-story, one bay, one-third scale RC frames. The frames had common deficiencies found in buildings in Turkey. The bare frames were first damaged and then strengthened with RC infill walls. Specimens were tested under reversed cyclic quasi-static loading. The author also carried out an analytical study to compare the test results. In the analytical model the infill walls were represented as equivalent struts. The study revealed that the RC infill wall improves the behavior of the damaged frame in terms of strength, and stiffness.

Canbay [5], also, studied the effects of the infill on the behavior of the damaged RC frame. He constructed a one third scale, two-story, three-bay frame. Two force transducers were developed by the author to measure the forces under the exterior columns. The bare frame was moderately damaged under reversed quasi-static loading. The damaged frame was rehabilitated by introducing an RC infill wall in the middle bay of the frame. The strengthened frame was also tested under reversed cyclic quasi-static loading. The seismic behavior of the bare frame and that of the strengthened frame was compared in terms of strength and stiffness. The author also carried out an analytical study. He performed a push-over analysis using the package program DRAIN-2D. In this analytical study, the infill wall was represented as a column with rigid arms at the story levels. Limit analysis results for both the bare frame and for the infilled frame were reported. The author concluded that strength, stiffness and energy dissipation capacity of the RC frame increased significantly by introducing the RC infill wall. The initial stiffness of the strengthened damaged frame was about 15 times that of the undamaged bare frame and the lateral load capacity of the frame increased up to 4 times with rehabilitation. He also found out that the infill wall carried 90% of the applied lateral load until yielding, and after yielding, the wall carried 80% of the lateral load. Canbay also showed that results of the limit analysis and push-over analysis were in good agreement with the test results. On the other hand, the moment-curvature relationship of the columns was calculated for the bare frame satisfactorily, while that for the damaged columns could not be predicted because of the past loading history.

Inserting RC infills to the existing structure is applicable at the cost of a certain discomfort to the occupants. Therefore, the application of this technique in the strengthening of undamaged existing buildings is not practical. The scope of this study is to develop a new strengthening technique, namely the strengthening of hollow clay tile walls with CFRP, without evacuating the occupants. For this purpose experimental research programs were initiated and the uses of new materials in system and member rehabilitation were studied.

Seven one-bay two-story hollow brick infilled twin frames were produced in the structural mechanics laboratory at the Middle East Technical University [6-9]. These were strengthened with CFRP and tested under reversed cyclic load applied in the horizontal closed loading frame. The results showed that the stiffness of the CFRP strengthened members increases by only 20% as compared to unstrengthened infilled frames, which was considered as not being significant. Therefore one can conclude that the strength demand is not increased. On the other hand the base shear capacity increased up to 200-250%, which showed the success of the strengthening technique.

Triantafillou [10] studied the short-term strength of masonry walls which were strengthened with externally bonded fiber reinforced polymer (FRP) laminates. He constructed 12 small wall specimens. Six of them were tested in in-plane bending and the other six were tested in out-of-plane bending. The walls tested in in-plane bending failed due to debonding of the FRP laminates in the tension zone. The others which were tested in out-of-plane bending failed by crushing of the masonry in the compression zone. Design models for dimensioning of FRP reinforcement were prepared by the author. They were compared and verified with the experimental results. The following conclusions were drawn by the author: (a) when the wall is under out of plane bending, the increase in bending capacity is quite high, which is the case of upper story levels of masonry structures where the level of axial load is low, (b) FRP reinforcement increases the strength considerably in the case of in-plane bending, (c) special care should be taken to prevent the debonding of the FRP laminates, (d) under low axial loads, FRP laminates increase the in-plane shear capacity.

In the present study two system improvement techniques, namely strengthening by introducing RC infill walls and strengthening with CFRP, are compared. As already mentioned, the first one is a simple and effective method which has been widely used for many years in Turkey. Proper application of RC infills provides considerable lateral strength, stiffness and ductility. Strengthening with CFRP is a new approach. It is a promising technique since its application is fast with minimum disturbance to the occupants. Application of CFRP, with care, provides significant increase in lateral strength but it does not provide as much ductility as the RC wall provides, because of the brittleness of the material.

2. EXPERIMENTAL PROGRAM

In this study, the behavior of two strengthened reinforced concrete frames was investigated. Test specimens developed by Canbay [5] were used. These were two-story three-bay 1/3 scale undamaged RC frames. Only the middle bay of the frames was infilled. The first frame was strengthened with an RC infill wall and the second one was strengthened with CFRP applied on the existing hollow clay tile infill wall in the interior bay.

The RC frames were detailed in such a way that they included common deficiencies of residential buildings in Turkey. Some of these deficiencies can be stated as:

- Poor lateral strength

- Lap splices made at floor levels with inadequate lap splice length
- Unconfined joints
- Inadequately confined member end zones,
- Poor detailing of the ties (instead of 135°, the hooks were bent at 90°)
- Improper detailing of the beam bottom reinforcement
- Poor concrete quality.

Since the infill was introduced only to the interior bay, the two interior columns were integral parts of the infilled frame. Therefore the structural system consisted of an infilled frame with two boundary columns and two frame columns (exterior columns). In order to observe the distribution of the lateral force between the frame columns and the infilled frame, special force transducers developed by Canbay (5) were mounted at the base of the frame columns (exterior columns).

2.1 Test Specimens

The test specimens consisted of two-story, three-bay reinforced concrete frames. They were cast vertically. The width of the middle bay was 1000 mm and that of the other two bays was 1600 mm. The first story height was 1500 mm and the second story height was 1000 mm. All columns were 110×110 mm and all beams were 110×150 mm in size. The infill wall in the first story had dimensions of 890×1425 mm and the second story wall had dimensions of 890×850 mm. The schematic view of the frame is given in Figure 1. To simulate real construction, the second story was cast 15 days after the first story concrete was cast. For both stories low strength concrete was aimed at. Reinforcement details of the RC frame have been given in Figures 2-3. To have a 1.7% steel ratio, 4-eight millimeter diameter plain bars were used as longitudinal reinforcement in the columns. Due to the usage of the force transducers at the foundation level, the exterior column longitudinal reinforcement was welded to four L-shaped bars which were welded to the base plate. 30x70 mm L-shaped bars were used. The 30 mm leg of these bars was welded to the steel plate of the transducer and the 70 mm leg was welded to the reinforcement of the column in order to prevent the slip of the longitudinal bars. In Figure 4 the details of the welded region are given.

2.1.1 Reinforced Concrete Infill, Specimen S1

The thickness of the RC infill, which was used in the first test, was 70 mm. The connection of the frame to the RC infill was achieved by steel dowels (10 mm diameter deformed bars). The arrangement of the dowels is given in Figure 5. The length of the extension of the dowels from the columns and from the beams into the wall was 200 mm (20 times the diameter of the dowel) and that from the foundation to the wall was 300 mm.

Two layers of infill reinforcement meshes were prepared using 6 mm diameter plain bars. The reinforcement ratio in the vertical direction (ρ_v) was 0.00545 for both stories. The horizontal reinforcement ratios (ρ_h) for first story and for the second story were 0.00567 and 0.00570, respectively. Figure 6 shows the mesh reinforcement in detail.

2.1.2 Hollow Clay Tile Infill, Specimen S2

CFRP strips were used to strengthen the infill of this specimen. CFRP was applied on the hollow clay tile infill constructed in the middle bay of the both stories. The hollow clay tile used in the second test was 1/3 scale. The details of the clay tile are presented in Figure 7.

2.1.3 Carbon Fiber Reinforced Polymer (CFRP), S2

Previous studies carried out in the METU Structural Engineering Laboratory showed that the configuration and detailing of the CFRP is very important to obtain satisfactory results. These studies revealed the following results:

- To increase lateral strength, CFRP should be extended and anchored to frame members,
- It is not needed to cover whole brick panel with CFRP, because diagonally placed CFRP sheets are almost as effective as covering the whole bay,
- A firm anchorage should be provided between CFRP and the panel,
- If there is a lap splice problem in the frame, that portion of the frame should be strengthened by wrapping of CFRP,
- Anchor dowels should be used to connect CFRP to the infill and to the frame members.

The most effective and yet economical configuration of CFRP has been tried to be determined in order to limit the lateral displacement and to provide lateral rigidity to the system. A model was developed using the structural analysis program, SAP-2000. Two struts were placed diagonally to cover the area of CFRP and one diagonally placed strut was used to represent the brick wall in one story one bay frame whose dimensions were the same as that of the middle bay of the first story of the test frame. The model was analyzed under a constant lateral load. The area bounded by the struts (representing the CFRP) changed in each analysis. The bounded area and the lateral displacement for each different configuration were checked. After comparing the results, the most efficient CFRP configuration for the first story was determined. In the second story, CFRP sheets with constant thickness were placed diagonally. The configuration of the CFRP application is given in Figure 8. 8 mm diameter and 70 mm deep holes were drilled on the inner faces of the frame, on the foundation and from one side to the other side of the brick infill. The locations of the anchor dowels can also be seen in Figure 8. The corners of the columns at the lap spliced regions were smoothed to prevent stress concentrations in CFRP and then wrapped with short CFRP sheets of 500 mm length in the first story and 300 mm length in the second story. Since the CFRP used is unidirectional, these confinements were made by using orthogonally placed two sheets, to provide both lateral and transversal confinement. Figure 9 shows the test specimen prior to testing.

2.2 Materials

The reinforcing steel used in this study comprised plain bars. The results of tensile tests are given in Table 1. Low strength concrete was aimed for the frame and higher strength concrete was aimed at for the RC infill wall. The average compressive strengths obtained are presented in Table 2.

A number of hollow clay tiles were tested to determine the compressive strength. The specimens were loaded parallel to the holes. The average failure load was found to be 46 kN. The compressive strength of the hollow clay tile is calculated as 16.3 MPa if the net area is considered. If the gross area is taken into consideration then the value is reduced to 7.8 MPa.

CFRP is composed of an epoxy-based matrix and carbon fibers. One-directional high strength carbon fiber (C1-30) was used in this study. The CFRP has only tensile strength along the direction of carbon fibers. The mechanical and physical properties of the carbon fiber used in this study are presented in Table 3. More than the strength of the carbon fiber, it is the strength of the CFRP system that is important. To determine the tensile strength of CFRP, coupon tests were conducted at METU and University of Texas at Austin. The samples had two orthogonal layers similar to the ones used on the specimens. The average tensile strength was found to be 800 MPa, and the ultimate strain obtained was 1.25%.

2.3 Test Procedure and Measurements

The test setup and instrumentation developed by Canbay [5] were used with minor modifications. In order to obtain response envelope curves of the specimens three LVDT (Linear Variable Differential Transformer) displacement transducers, to obtain the member forces two force transducers, and to evaluate lateral deformation due to shear and to calculate the curvatures 12 dial gages were mounted on the specimen. A load cell was used to measure the applied horizontal load. The instrumentation and load setup is given in Figure 10.

Both test specimens were tested in the vertical position. A reaction wall was used for the loading. The lateral load was applied at the second story level. The load was quasi-static and reversed cyclic type. Out of plane motion of the test specimen was prevented. Additional 9 kN constant vertical loads, which correspond to an axial load level of $0.055f_{ck}A_c$, were applied on top of the specimens during the tests.

The loading history curves of the test specimens are given in Figure 11, and the displacement history curves, are given in Figure 12.

2.3.1 Specimen Strengthened with RC Infill, S1

The purpose of the test was to evaluate the behavior of the frame strengthened by applying an RC infill to the middle bay and to compare the results with the other strengthening method. At cycle number 13, the lateral load reached was 70 kN with a corresponding top displacement of 11 mm. After this cycle the loading was continued by controlling the top displacement. The yield displacements for the positive and for the negative directions were determined as 10.5 mm and 12 mm respectively. The drift ratio for each story was calculated by dividing the relative displacement of that story by the height of the story. At the yield stage, the first story had 0.44% drift and the second story had 0.50% drift. Maximum drift ratios for the first and second story were obtained as 2.83% and -2.56% respectively. The drift ratio curves for the first and second story are given in Figures 13-14, respectively.

At the final cycle, the strength decrease in the positive direction was 24% and the displacement ductility corresponding to this stage was 5.7. Pinching of hysteresis loops was observed towards the end of the testing due to high deformations causing slip.

2.3.2 Specimen Strengthened with CFRP, S2

The purpose of the second test was to compare the behavior of the frame strengthened with RC infill with that of the one strengthened with CFRP. The test started with 5 kN lateral load and increased by 5 kN every cycle up to the failure of the carbon fiber anchor dowel at the foundation level. The exact yield points for the positive and for the negative directions could not be determined because of this failure.

The specimen showed almost no inelastic deformations up to 4th cycle. After this cycle, due to formation of the cracks, stiffness degradation started and inelastic deformations occurred. This phenomenon continued until the failure of the anchorages. When the strength of the specimen was reduced considerably (40%) the test was terminated. In the final cycle (16th cycle) the lateral load was 39 kN in the positive direction and 45 kN in the opposite direction. Corresponding displacements were +23 mm and -18 mm respectively. The ratio of the maximum displacement to the displacement corresponding to the maximum load was calculated as 2.22 in the positive direction and 1.43 in the negative direction.

Drift ratio curves for the first and the second stories are given in Figures 15 and 16, respectively. The maximum drift ratio for the first story was obtained in the final cycle as 0.98% and that for the second story was calculated as 0.93%. Because of the story height difference drift ratios for the first story were slightly higher than for the second story. As can be seen from the figures drift ratios increased rapidly after the failure of the carbon fiber anchor dowels and the failure of the CFRP. The increase in the drift ratio in the positive direction and in the negative direction was about 100% and 50% respectively after the failure of the CFRP.

3. DISCUSSION OF TEST RESULTS

The first test specimen which was strengthened with the RC infill wall exhibited a behavior very similar to the behavior of a cantilever beam because almost all lateral loads (≈90%) were carried by the infill frame. The failure of the specimen (considerable strength decrease and/or large drift) was followed by yielding of the infill wall and the heavy damage of the first story exterior beam-column joints with wide cracks in the first story beams. The RC wall yielded not at the foundation level but at the end of the lap splice region where the capacity of the wall was about half of the capacity of the wall at foundation level.

The behavior of the second specimen strengthened with CFRP was similar to that of a diagonally braced frame. The diagonally placed CFRP strips mostly provided the lateral stability of the frame. The failure of the CFRP strips was due to their weak points which were the connections to the frame and to the foundation. The connection of the strips to the frame was achieved with the carbon fiber anchor dowels. Only a limited amount of carbon fiber could be used for the purpose. It is well known that the bond strength of the dowel depends mostly on the strength of the concrete, which was very low in this specimen. For these reasons the failure of the second specimen was due to the bond failure of the CFRP anchor dowels to the frame.

In order to compare the behavior of the specimens tested in this study with the bare frame tested by Canbay [5], Table 4 was prepared. In this table S1 represents the first specimen strengthened with RC infill, S2 is used for the second specimen strengthened with CFRP.

3.1 Strength

The lateral strength of the specimen is defined as the maximum lateral load carried by the specimen. The loading history of S1 is given in Figure 11. The strength of this specimen was 70 kN in both the positive and in the negative directions. The strength of S1 was 4.7 times greater than the strength of the bare frame. The lateral strength of S2 was 4.3 times greater than that of the bare frame.

3.2 Stiffness

The lateral stiffness is the force required to impose unit displacement to a specimen, which is the slope of the load-displacement curve. Because the load-displacement curve was not linearly elastic, a single stiffness could not be described. The stiffness of the test specimens at each cycle is given in Figure 17, so that the change in the stiffness during the test can be observed clearly. The stiffness in each cycle was calculated by the sum of the maximum positive load and maximum absolute negative load divided by the sum of the absolute corresponding displacements. Stiffnesses of the two specimens were not very different while initial stiffness of S1 was larger than that of S2. After the 4th cycle the stiffness values of both specimens were nearly the same. The stiffness of S2 was slightly larger than that of S1 after the 12th cycle. This is because in specimen S1 the damage observed at this cycle was more severe than that of specimen S2 at the same cycle.

As expected, the improvement in the stiffness of the frame due to strengthening is very significant. The initial stiffness of S1 was 14.6 times that of the bare frame. Second specimen S2 did not have an initial stiffness as large as the first specimen had but it was 11.0 times the initial stiffness of the bare frame and 0.75 times of that of first specimen strengthened with an RC infill wall. At the end of the first test, stiffness degradation of the first specimen was 94% while stiffness degradation of the second specimen was 91%. Stiffness values at important stages are also presented in Table 4.

3.3 Response Envelopes

Response envelope curves were prepared to evaluate the strength and the stiffness characteristics of the test specimens. They were obtained by connecting the maximum points of the lateral load-second story displacement curve of the specimens. Response envelope curves of specimens S1 and S2 are given in Figure 18 with the response envelope curves of the bare frame [5]. As can be seen from this figure, strengthening made by infilling the middle bay improved both strength and stiffness significantly. The stiffness and strength of S1 were somewhat higher than those of S2.

Although the behavior of S1 and S2 were similar up to the maximum load, beyond this point the behavior of S2 was much more brittle.

3.4 Drift Ratio Envelopes

Drift ratios of each story and the corresponding loads at each half cycle was plotted to form the envelope curves, Figures 19-20. On the same figure the drift limit for elastic analysis specified by the Turkish Seismic Code [11] is also given in order to enable comparison. Every code provides similar limits to prevent extensive structural and non-structural damage and to minimize the second order effects. The maximum drift ratio is limited to 0.0035 or 0.02/R where R is the behavior factor. For a normal ductility system composed of frames and shear walls, and for a normal ductility system composed of frames only, the R factor is 4. S1 and S2 are considered to be normal ductility systems. Therefore the limit for S1 and S2 was determined as 0.0035. For the comparison of the drift ratio of each specimen, the limit 0.0035 is given in these figures.

The specimens had small strength and stiffness degradation till the specified limit and after that point, drift ratios increased more rapidly with the applied load. Both specimens showed considerable inelastic deformation subsequently. The necessity of the limit can be understood from the mentioned figures. Specimen S1 had a better behavior than S2 in terms of drift ratio as expected. When Figure 19 is compared with Figure 20 it is seen that the second story showed a better drift envelope for both specimens. It is natural to have this result since the second story height was less than the first story height. It should be noted that the strength reduction in S1 beyond the peak was much less compared to S2.

4. ANALYTICAL PREDICTION OF MOMENT – CURVATURE RELATIONSHIPS OF THE COLUMNS

The analytical study for the moment-curvature relationship of the exterior columns was carried out using Response-2000 [(12)]. The moments under these columns were measured with the help of the force transducers and the curvatures were calculated using measurements taken on both faces of the columns. The compressive concrete strengths of the two specimens and the axial loads that columns were subjected in the two tests were different than each other therefore different analyses were performed for each specimen. In Figures 21-24, moment curvature relationships of two exterior columns are given with the analytical prediction.

The axial load level of the columns showed almost 100% variation during the cyclic loading. Sometimes the columns were even under tensile axial force. For these reasons, several analyses were carried out with different axial force levels using the mentioned computer program. The axial forces at the bottom of the columns were taken from the force transducers and the corresponding curvatures were calculated from the strains measured on opposite faces. Using these data, some intervals were determined, where the peak of the axial load of the loading cycle was taken and assumed to be constant in that interval. The analytical results corresponding to those axial loads and the curvature intervals were used to obtain the overall analytical moment-curvature behavior. These intervals and corresponding axial forces are presented in Tables 5-6 for S1 and S2, respectively.

As can be seen from Figure 21, in specimen S1 on the positive curvature side, the left exterior column could not reach the capacity determined analytically. The reason for this might be the possible slip that could happen due to axial tensile force. Although the right exterior column reached the capacity predicted analytically, the column yielded under smaller curvature than predicted, Figure 22.

Some deviations from the analytical study can be seen from the figures. This might have occurred due to the change in the material property of the reinforcing bars because of the welding process. The readings for the curvature were taken just above the lap splice region where the reinforcing bars were connected to the plate of the force transducers. Welding L-shaped short bars to the reinforcing bars and to the plate provided this connection.

As already mentioned, exterior columns of the S2 did not yield during the test, Figure 23 and 24. Therefore the experimental graph is in the linear range and the analytical curves fit well with the test results.

5. CONCLUSIONS

Two-story, three-bay RC frames strengthened by infilling only the middle bay served for the purpose of the study. Additionally, test setup and instrumentation with specially produced force transducers were found to be satisfactory in order to investigate the seismic behavior of the specimen. In the light of the tests performed, and the results obtained, the following conclusions are drawn.

- The increase in lateral load capacity in both specimens was almost the same, and around 5 times greater than that of the bare frame.
- Lateral deformation capacity provided by the RC infill was very high. Because of the brittleness of the CFRP and because of the anchorage problem, the deformation capacity of the frame strengthened with CFRP was smaller.
- As compared to the RC infill, CFRP could be applied more easily and more rapidly. On the other hand, to choose the best strengthening technique, the cost of application must also be considered. In the cost analysis, the necessity of the evacuation of the occupants in the first technique should also be taken into account.
- While applying the second technique, namely strengthening the frame with CFRP, special care should be given and adequate anchorages should be supplied for a perfect bond between the structural element and CFRP.

ACKNOWLEDGEMENT

This study has been partially supported by NATO and The Scientific and Technical Research Council of Turkey.

REFERENCES

1. Sugano, S., "State - of - the - Art in Techniques for Rehabilitation of Buildings", Proc. of the 11th WCEE, Paper No. 2175, 1996.
2. Ersoy, U. and Uzsoy, S., "The Behavior and Strength of Infilled Frames", Report No. MAG-205 TUBITAK, Ankara, Turkey, 1971.
3. Altin, S., Ersoy, U., Tankut, T., "Hysteretic Response of RC Infilled Frames", ASCE Structural Journal, V.118, No.8, August 1992.
4. Sonuvar, M. O., "Hysteretic Response of Reinforced Concrete Frames Repaired by Means of Reinforced Concrete Infills", A Doctor of Philosophy Thesis in Civil Engineering, Middle East Technical University, June 2001.

5. Canbay, E., “Contribution of RC Infills to the Seismic Behavior of Structural Systems”, A Doctor of Philosophy Thesis in Civil Engineering, Middle East Technical University, December 2001.
6. Mertol, H. C., “Carbon Fiber Reinforced Masonry Infilled Reinforced Concrete Frame Behavior”, A Master of Sciences Thesis in Civil Engineering, Middle East Technical University, June 2002.
7. Keskin, R. S. O., “Behavior of Brick Infilled Reinforced Concrete Frames Strengthened by CFRP Reinforcement: Phase I”, A Master of Sciences Thesis in Civil Engineering, Middle East Technical University, July 2002.
8. Erduran, E., “Behavior of Brick Infilled Reinforced Concrete Frames Strengthened by CFRP Reinforcement: Phase II”, A Master of Sciences Thesis in Civil Engineering, Middle East Technical University, July 2002.
9. Ersoy, U., Ozcebe, G., Tankut, T., Erduran, E., Keskin, R. S. O., Mertol, H. C., “Strengthening of Brick Infilled RC Frames with CFRP”, SERU-Structural Engineering Research Unit, TUBITAK-METU, Report No. 2003/1, March 2003.
10. Triantafillou, T. C., “Strengthening of Masonry Structures Using: Epoxy-Bonded FRP Laminates”, ASCE-Journal of Composites for Construction, Vol. 2, No. 2, May 1998, pp.96-104.
11. Turkish Seismic Code, Ministry of Public Works and Settlement, Ankara, 1998.
12. Response-2000, Version 1.1, September 2001.

Table 1. Properties of Reinforcing Bars

Type	Bar Diameter (mm)	f_y (MPa)	f_{ult} (MPa)	Property
Column & Beam Longitudinal	8	430	622	Plain
Column & Beam Transversal	4	322	422	Plain
RC Infill Mesh *	6	378	484	Plain
Steel Anchor Dowels *	10	757	940	Deformed

- Used in the first specimen

Table 2. Concrete Strength

Strengthening Technique	Specimen	28 day f_c (MPa)	Test day f_c (MPa)
RC Infill	S1 1st Story	10.9	13.6
RC Infill	S1 2nd Story	10.7	11.8
RC Infill	S1 Infill Wall	30.4	32.2
CFRP	S2 1st Story	9.8	11.2
CFRP	S2 2nd Story	9.1	9.6

Table 3. Mechanical and Physical Properties of Carbon Fiber, C1-30

Property	Amount	Unit
Unit Weight	0.300	kg/m^2
Effective Thickness	0.165	mm
Characteristic Tensile Strength	3430	MPa
Characteristic Elasticity Modulus	230000	MPa
Ultimate Strain	0.015	mm/mm

Table 4. Comparisons of the Test Results

CHARACTERISTICS	Bare*	S1	S2
Lateral Strength (+) (kN)	14	70.6	64.7
Lateral Strength (-) (kN)	13	70.6	70.3
Initial Stiffness (kN/mm)	2.05	30.0	22.5
Stiffness at Max. Load (kN/mm)		4.97	6.21
Drift Ratio At Max. Load (%)	1.88	0.89	0.68
Max. Drift Ratio (%)	1.89	2.57	0.99

*Test performed by Canbay (5)

Table 5. Axial Forces and Corresponding Curvature Intervals for the Analytical Prediction of Moment-Curvature Variation of the S1

SPECIMEN S1			
Right Column		Left Column	
N(kN)	Range of K(rad/m)	N(kN)	Range of K(rad/m)
14	0 - (-0.01464)	6	0 - (-0.00636)
16	(-0.01461) - (-0.03452)	2	(-0.00636) - (-0.00909)
17	(-0.03452) - (-0.06723)	5	(-0.00909) - (-0.01388)
18	(-0.06723) -	0	(-0.01388) -
5.2	0 - 0.00455	13	0 - 0.00364
2	0.00455 - 0.00909	15	0.00364 - 0.00818
4	0.00909 - 0.02144	17	0.00818 - 0.01464
5	0.02144 -	18.5	0.01464 - 0.02144
		19.5	0.02144 - 0.03452
		21	0.03452 - 0.04177
		23	0.04177 - 0.05560
		24	0.05560 - 0.06116
		21	0.06116 -

Table 6. Axial Forces and Corresponding Curvature Intervals for the Analytical Prediction of Moment-Curvature Variation of the S2

SPECIMEN S2			
Right Column		Left Column	
N(kN)	Range of K(rad/m)	N(kN)	Range of K(rad/m)
6	0 - (-0.00364)	14.3	0 - (-0.00636)
4	(-0.00364) - (-0.00546)	15	(-0.00636) - (-0.01331)
2.5	(-0.00546) - (-0.00636)	17	(-0.01331) -
1.5	(-0.00636) - (-0.00840)		
0	(-0.00840) -		
12.5	0 - 0.00723	5	0.00364 - 0.00818
14	0.00723 - 0.01000	3	0.00818 - 0.01464
15	0.00100 -	2	0.01464 – 0.02144

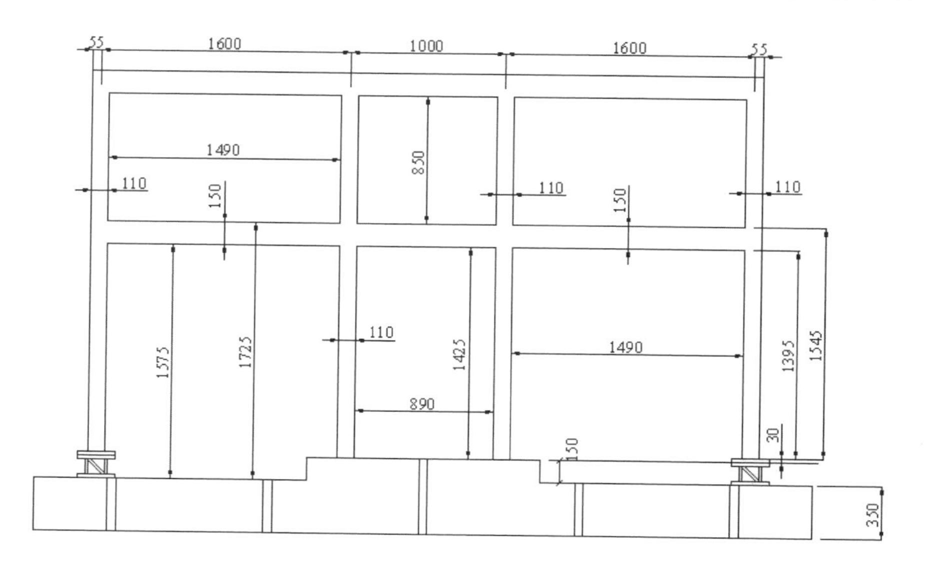

Figure 1. Dimensions of the test specimen (All dimensions are in mm)

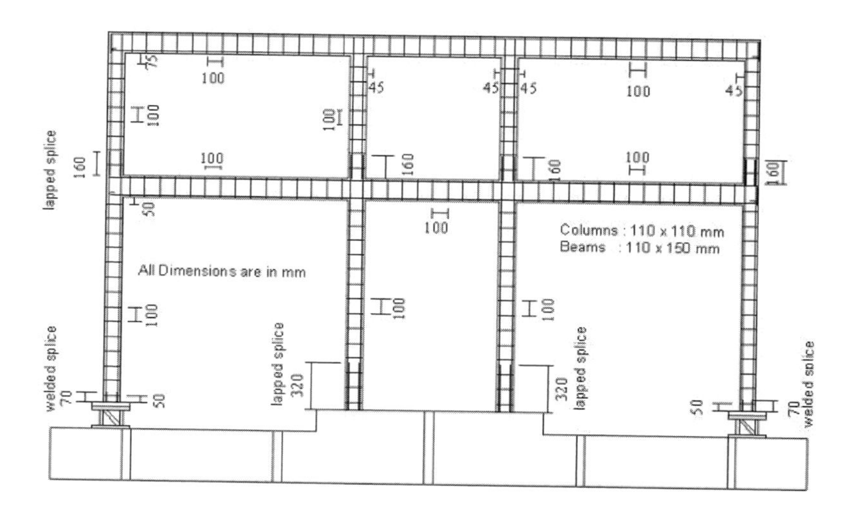

Figure 2. Reinforcement Detail of the RC Frame

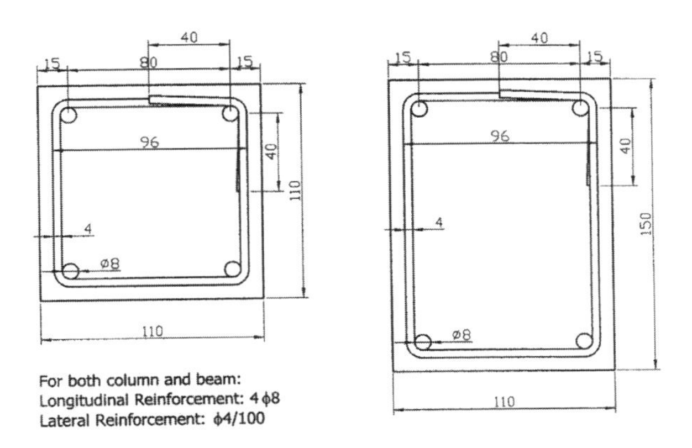

Figure 3. Reinforcement Detail of the Columns and the Beams

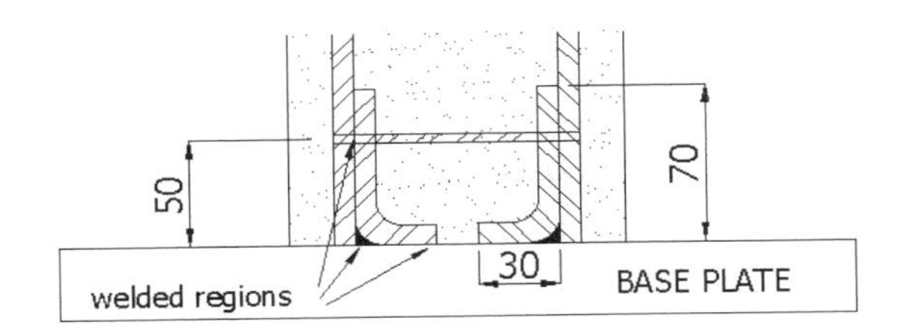

Figure 4. Detail of the Welded Splice Region at the bottom of the Outer Columns (All dimensions are in mm)

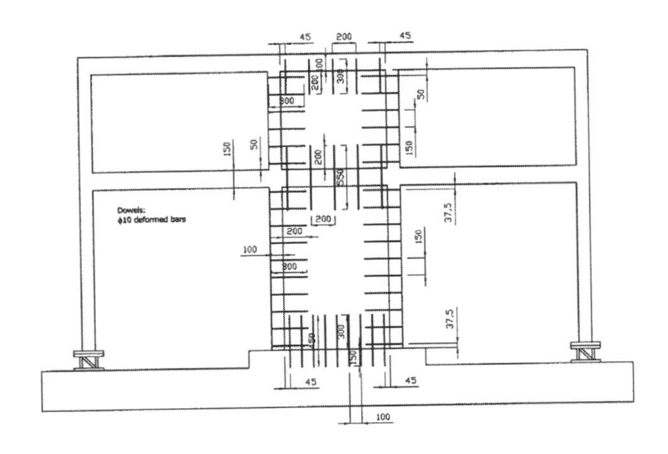

Figure 5. Infill dowels (All dimensions are in mm)

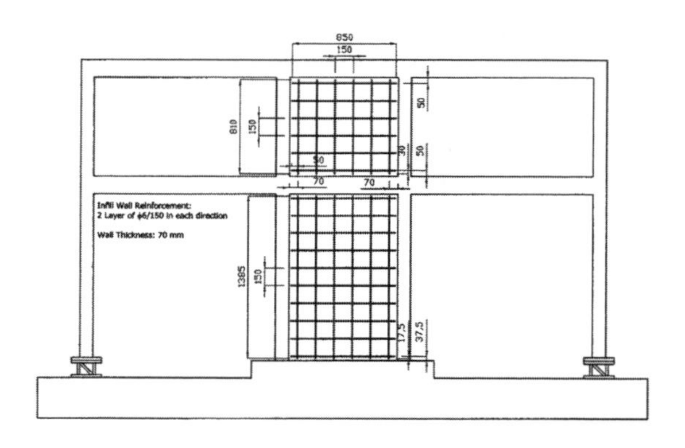

Figure 6. Infill Mesh Reinforcement (All dimensions are in mm)

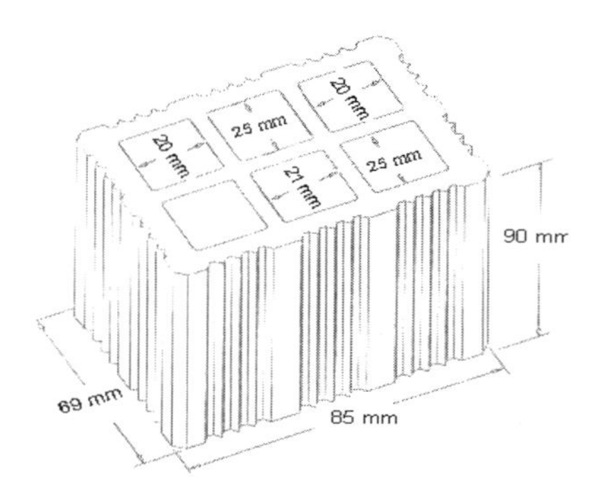

Figure 7. Dimensions of the Hollow Clay Tile

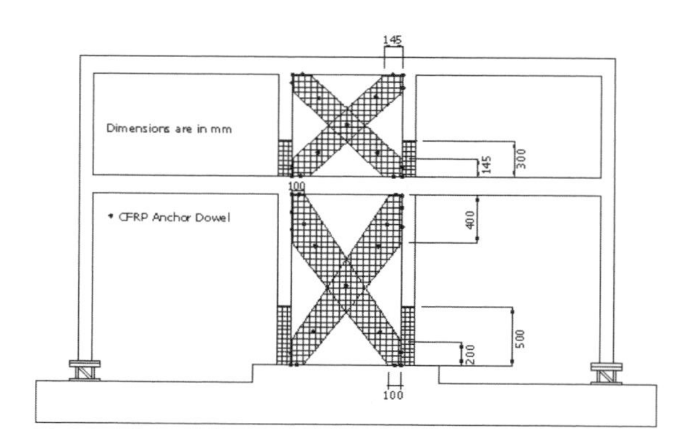

Figure 8. Configuration of the CFRP and Location of Anchor Dowels

Figure 9. Front View of the Specimen with CFRP

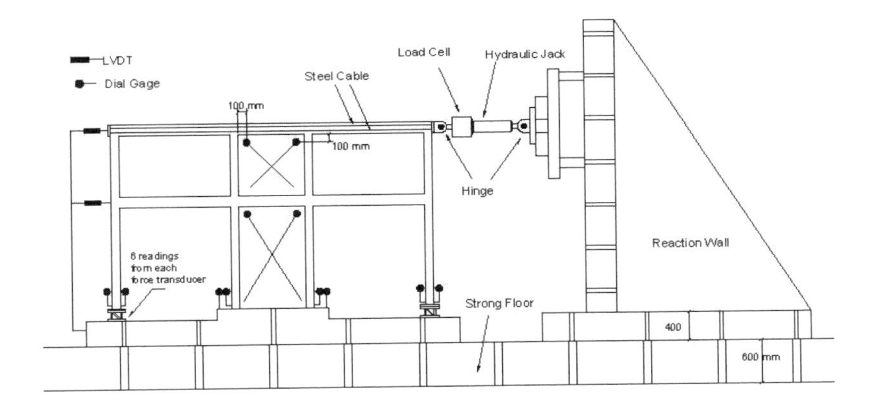

Figure 10. Instrumentation and Load Setup

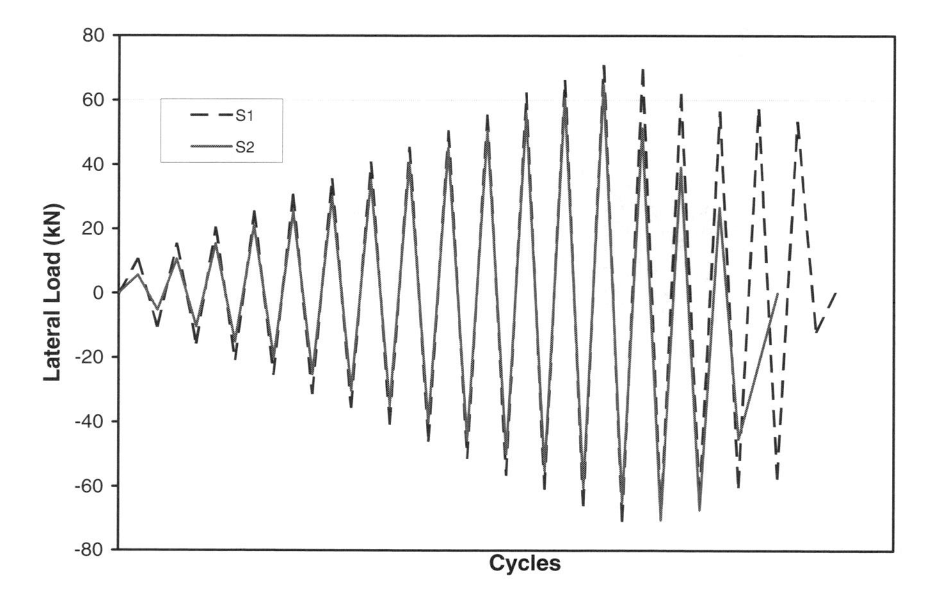

Figure 11. Loading History of the Specimens S1 and S2

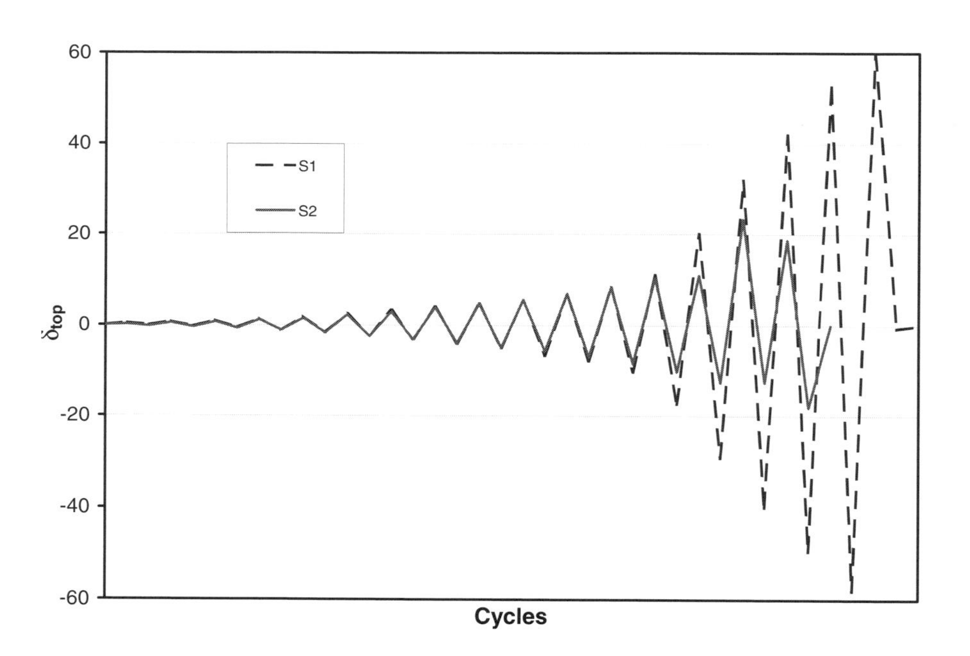

Figure 12. Displacement History of the Specimens S1 and S2

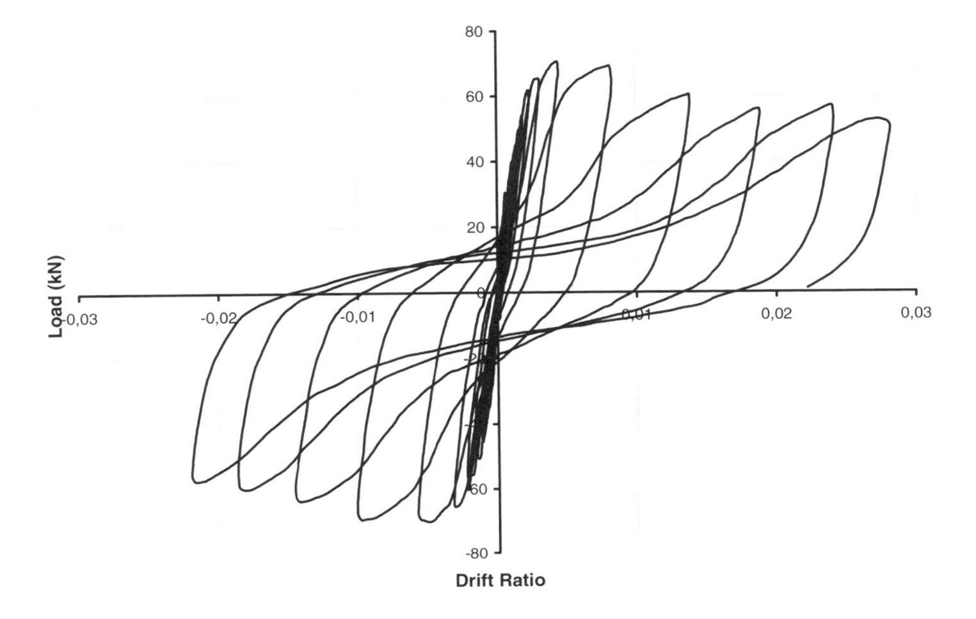

Figure 13. Variation of 1st Story Drift Ratio (Specimen S1)

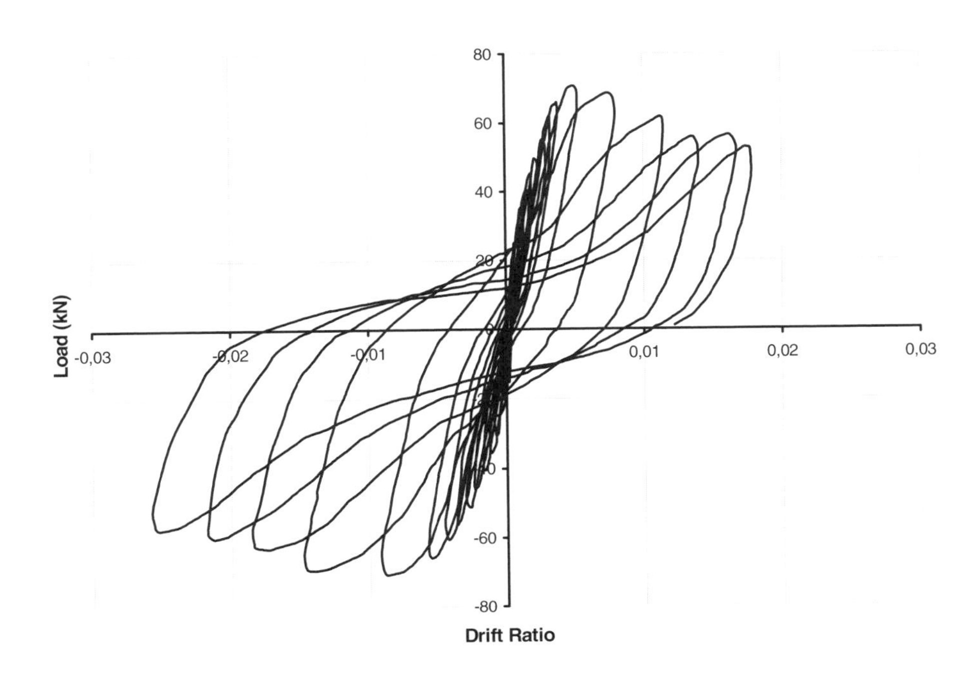

Figure 14. Variation of 2nd Story Drift Ratio (Specimen S1)

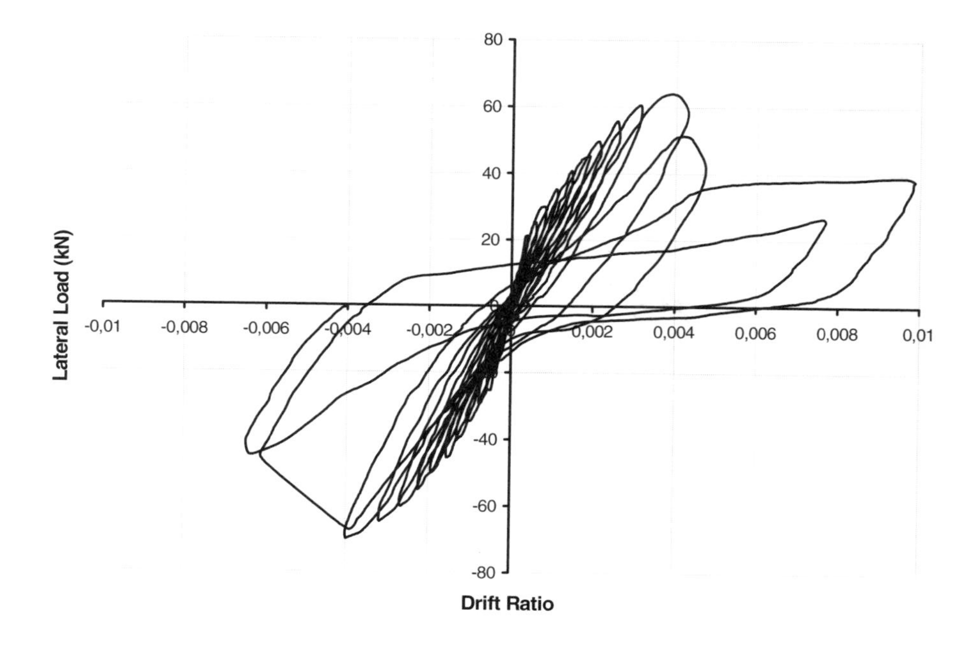

Figure 15. Variation of 1st Story Drift Ratio (Specimen S2)

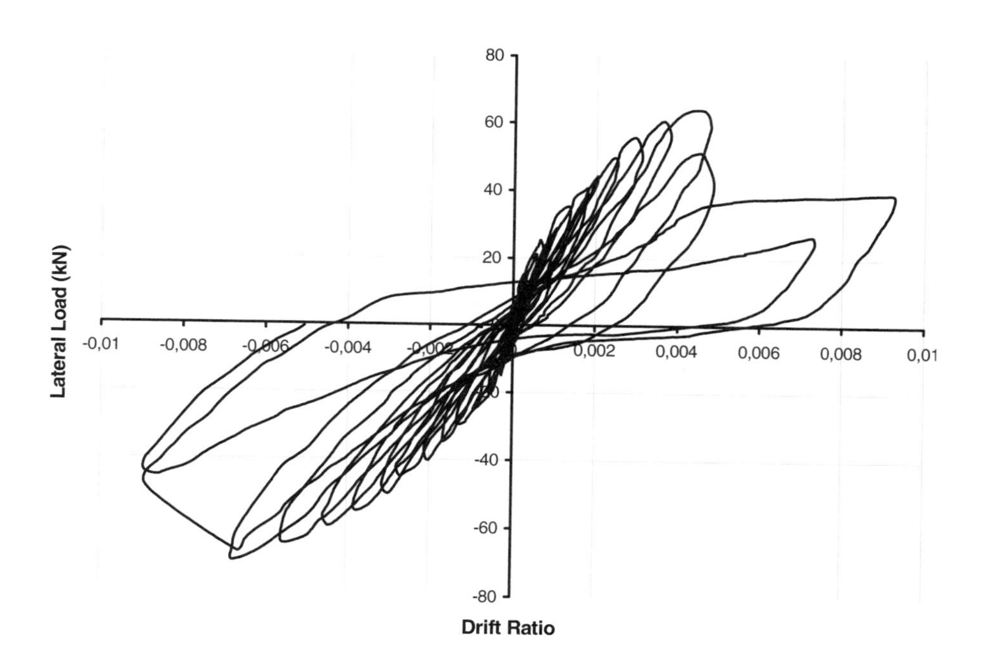

Figure 16. Variation of 2nd Story Drift Ratio (Specimen S2)

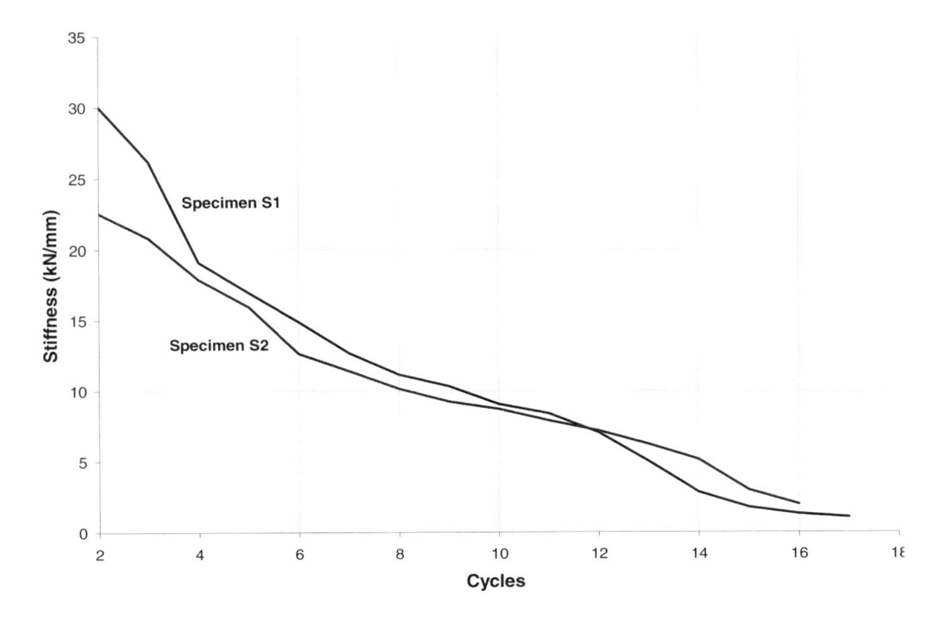

Figure 17. Stiffness change of the Specimens with the Cycles

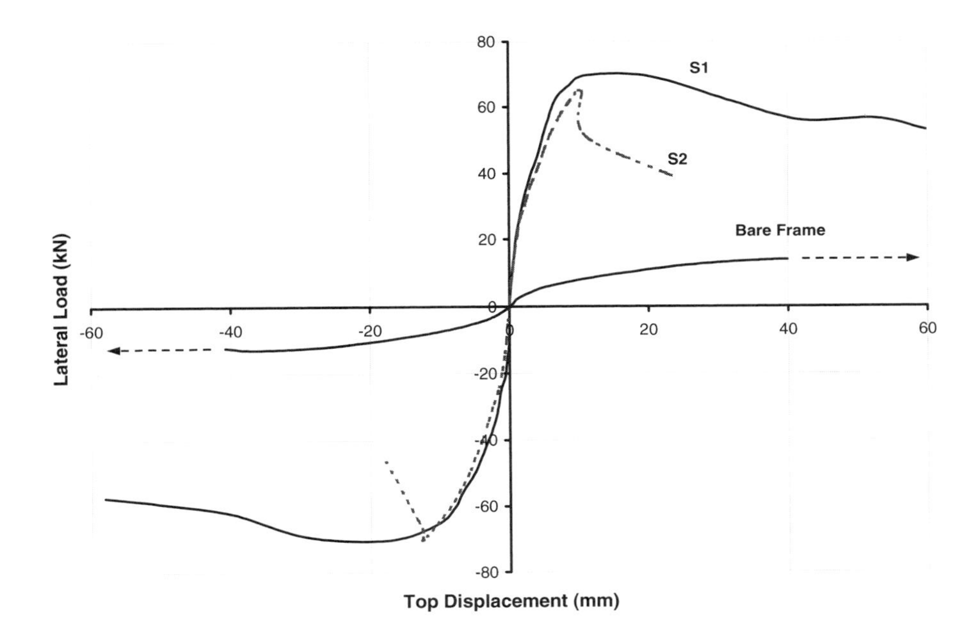

Figure 18. Envelope Curves of the Specimens (Bare Frame is taken from (5))

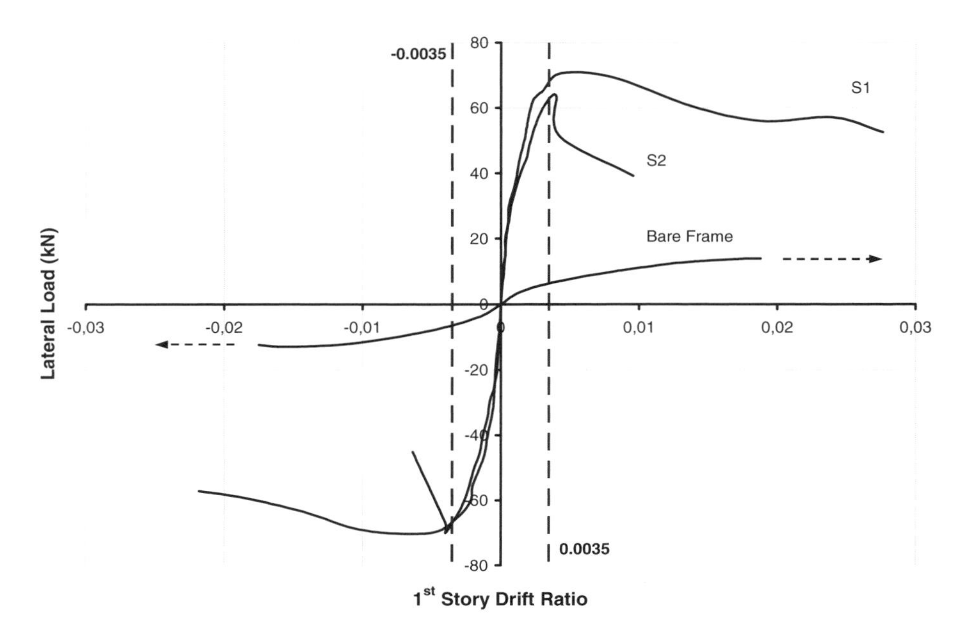

Figure 19. First Story Drift Ratio Envelopes of the Specimens

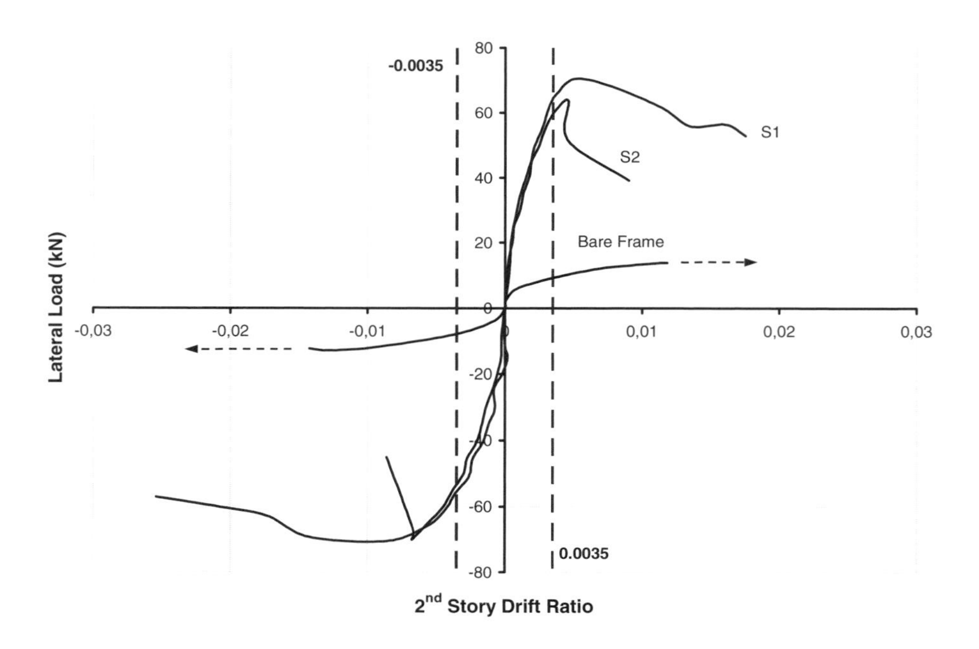

Figure 20. Second Story Drift Ratio Envelopes of the Specimens

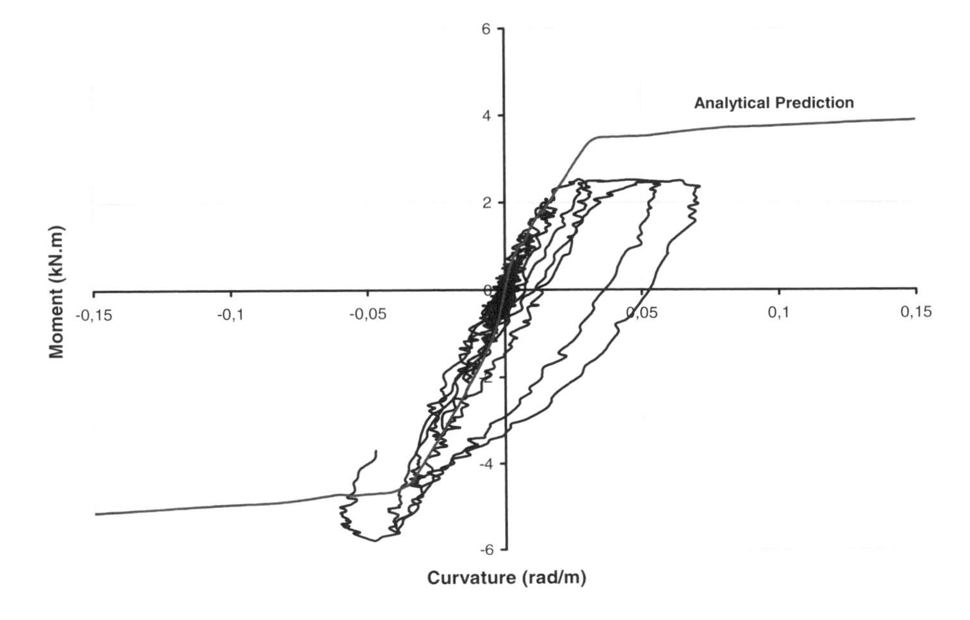

Figure 21. Analytical Prediction of Moment-Curvature and Test Results for the Left Exterior Column of Specimen S1

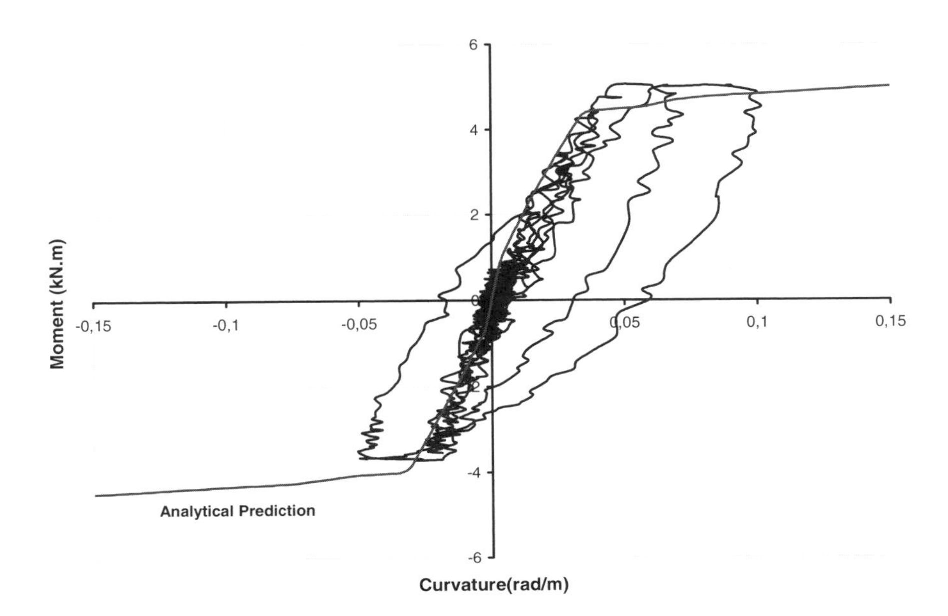

Figure 22. Analytical Prediction of Moment-Curvature and Test Results for the Right Exterior Column of Specimen S1

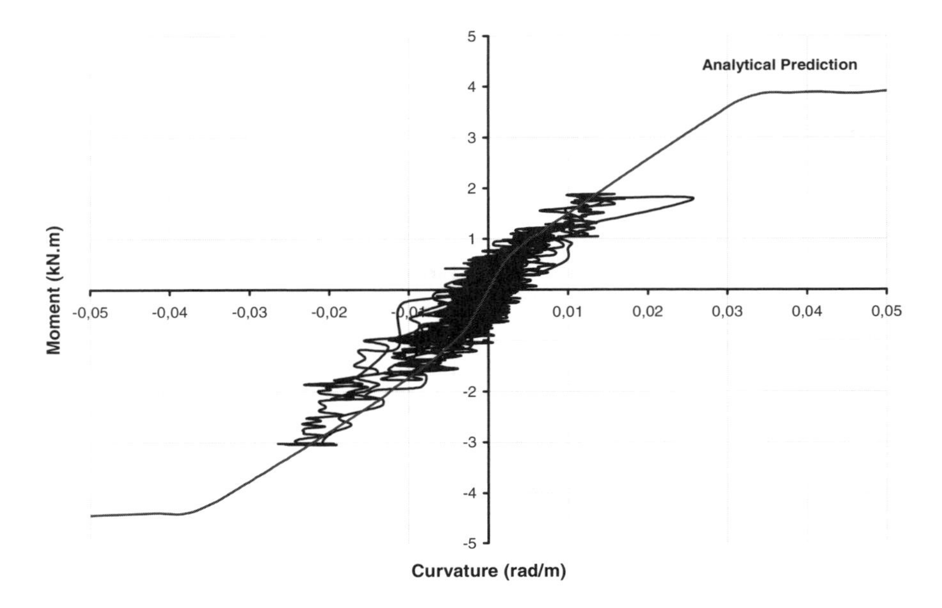

Figure 23. Analytical Prediction of Moment-Curvature and Test Results for the Left Exterior Column of Specimen S2

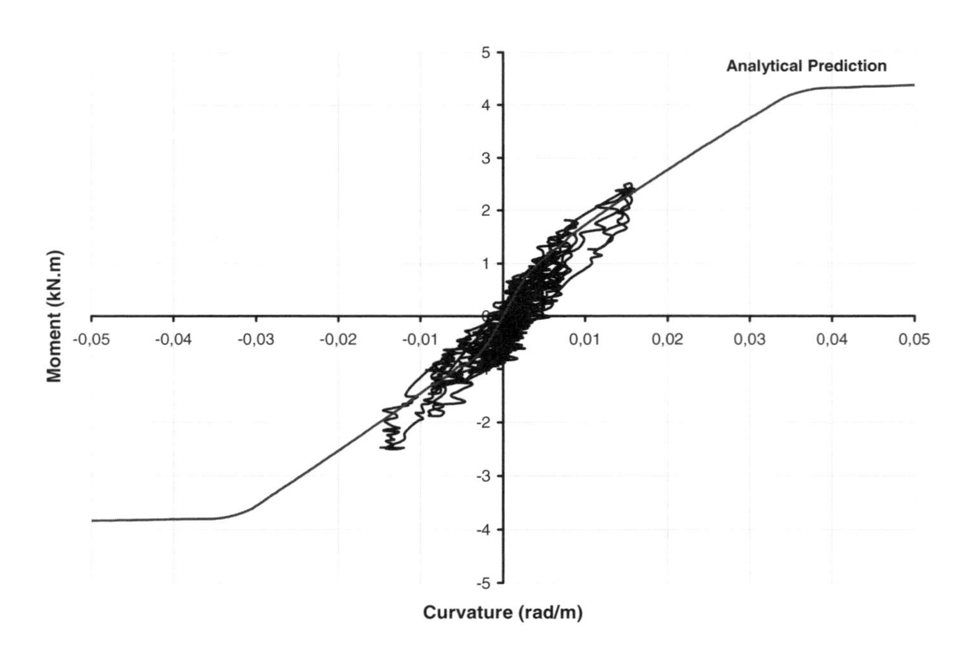

Figure 24. Analytical Prediction of Moment-Curvature and Test Results for the Right Exterior Column of Specimen S2

OCCUPANT FRIENDLY SEISMIC RETROFIT (OFR) OF RC FRAMED BUILDINGS

Mehmet Baran, Murat Duvarci, Tugrul Tankut, Ugur Ersoy, Guney Ozcebe
Structural Mechanics Division, Department of Civil Engineering
Middle East Technical University, Ankara, Turkey

Abstract: An innovative non-evacuation retrofitting technique is being developed for RC buildings. The introduction of cast-in-place RC infill walls is very effective, but involves messy construction work and requires evacuation. The proposed method transforms existing hollow masonry infill walls by reinforcing them with precast concrete panels epoxy glued to the wall and frame members. Three tests have so far been performed under reversed cyclic loading. The first specimen was an ordinary RC frame with hollow brick infill walls, to serve as a reference specimen. The infills of the other two specimens were strengthened with epoxy glued precast concrete panels. Both strengthened specimens exhibited superior behaviour and capacity, indicating the potential of the retrofitting technique.

Keywords: Seismic retrofit, seismic strengthening, infilled frames, lateral stiffness improvement

1. INTRODUCTION

A huge existing building stock awaits seismic vulnerability assessment followed by seismic retrofitting. Putting the complicated financial, legal, administrative aspects of the problems aside, one very challenging aspect is how to introduce a reliable yet economical structural strengthening intervention without evacuating the building, even without causing more

S.T. Wasti and G. Ozcebe (eds.), Seismic Assessment and Rehabilitation of Existing Buildings, 433–456.

disturbance to the occupants than a painting job. This challenging problem which definitely requires development of innovative retrofitting techniques is the subject matter of the ongoing comprehensive NATO project. The study reported in the present paper concerns the development of one such innovative retrofitting technique, suitable for the most common type of building structures in the region, namely, reinforced concrete frames with hollow brick masonry infill.

The seismic repair technique developed earlier for this kind of building structures has been successfully used in the post-quake repair of damaged buildings. This repair technique aiming at the improvement of the overall system behaviour, which in most cases suffer from the deficiency of insufficient lateral stiffness, consists of the introduction of cast-in-place reinforced concrete infill walls properly connected to the existing frame members by epoxy anchored dowels. However, this repair technique, which proved to be very effective in improving the overall seismic structural performance, involves messy construction work and definitely requires evacuation of the building for about six months.

2. RETROFITTING PRINCIPLE

Here, the idea is to create a similar strong infill by reinforcing the existing hollow brick infill wall with a comparatively thin layer of high strength precast concrete elements to be epoxy connected to the wall and the frame members. One single piece of precast concrete member would definitely be unmanageable, too large to go through doors and too heavy to be carried manually. Therefore it has to consist of individual panels of manageable size and weight, and has to be assembled on the wall by connecting the panels together.

The method of connecting precast concrete panels together is a critical issue, since these connections will have to be capable of transferring tension and compression as well as shear. This is why shear keys and welded connections on panel reinforcement were considered essential in the design of the first series of test specimens.

3. EXPERIMENTAL WORK

3.1 Test Specimens

3.1.1 Test Frames

One-third scale, one-bay two-storey reinforced concrete frames are being used as test units. These units are supposedly identical to the twin frames used in the other test series of the present sub-project. Intentionally, they have been poorly proportioned to have strong beam-weak column connections, poorly detailed to have insufficient confinement and poorly constructed using a low-grade concrete (C16) to represent the common local practice. These frames have their columns fixed to the rigid foundation beams. Dimensions and reinforcement of the test frames are illustrated in Figure 1.

Both the first and second floor frame bays are infilled with scaled brick walls covered with a scaled layer of plaster. Realising the critical consequences workmanship may have on the performance, ordinary workmanship was intentionally employed in wall construction and plaster application to reflect the ordinary practice.

3.1.2 Infill Wall Strengthening Panels

The factor dominating the design of precast panels is weight; each piece to be used in actual practice should not exceed 70~80 kg so that it can be handled by two workers. The other important factor is the panel thickness. Considering the relatively high strength of concrete (30~40 MPa) to be used in panels, 50~60 mm panel thickness can reasonably be proposed for the actual practice. Since the usual floor height is about 2.80~3.00 m and the usual beam depth is around 400~600 mm, a panel arrangement with three layers sounds rather sensible and leads to a panel size around 700~800 mm in vertical direction, horizontal size being around 600~700 mm. With all these considerations, Type 1 panel was designed to have the dimensions and reinforcement shown in Figure 2a. Type 2 reflects a different design approach, since it is a narrow and tall panel to cover the full floor height, Figure 2b. Two different precast panel arrangements so far used in the test units are illustrated in Figure 3.

Table 1 gives an idea about the panel properties in various test specimens. Extreme care and attention was initially paid to the panel connection, considering it to be the weakest link of the chain, and both shear keys and welded connections, fixing reinforcing steel bars to each other and to dowels epoxy anchored into the frame elements, were provided. However, the epoxy mortar used in the panel joints proved to be so successful in connecting the panels in PI1 and PI2 tests that both shear keys and welded connections appeared to be redundant. This is why no shear keys are planned for the future tests, and welded connections will probably be confined to the foundation level where maximum stresses occur.

3.1.3 Materials

A low strength concrete is intentionally used in the test frames to represent the concrete commonly used in existing building structures. On the other hand, relatively strong concrete should be preferred for the precast panels to provide the required load carrying capacity by using relatively thin layers of concrete, minimising the panel weight. Ordinary cement-lime mortar is used for the plaster, imitating usual practice. For the same reason, mild steel plain bars are used as reinforcement in both test frames and panels.

The average compressive strength values for concrete and plaster determined on the testing day, as well as the yield strength values for three different bar sizes used in frames and panels are listed in Table 2.

SIKADUR-31 epoxy repair mortar was used in panel joints and between the panels and the plaster on the wall. It is a relatively inexpensive two component mortar recommended for reinforced concrete repair works. Its rather viscous structure enables a practical use of this material on vertical surfaces. According to the manufacturer's manual, its setting time is about 45 minutes at room temperature and the tensile strength is nearly 20 MPa which is obviously beyond the strength required in concrete applications.

3.2 Test Set-up

Figure 4 gives a general view of the test set-up. As seen in this figure, tests are being performed in front of the reaction wall. Test units are subjected to reversed cyclic horizontal load resembling the seismic effects and the resulting deformations are measured at numerous locations to obtain data needed for a comprehensive analytical evaluation of the performance.

3.2.1 Loading and Supporting System

A reinforced concrete universal base, serving as a rigid foundation for the test unit and enabling various support configurations, has been prestressed to the strong testing floor of the Structural Mechanics Laboratory. Each test frame is cast together with a rigid foundation beam, which is suitably bolted down to the universal base as required. The quasi-static test loading consists of reversed cyclic horizontal load applied at floor levels, besides constant vertical load approximately equal to 20% of the column axial load capacity.

The vertical load application gear consists of a hydraulic jack and a load cell placed between a spreader beam and a cross-beam at the top, which is pulled down by two prestressing cables attached to the universal base on either side of the test unit. Having been supported as a simple beam with supports at the column heads as shown in Figure 4; the spreader beam divides the load developing in the ram and transfers the two equal components to the two columns. The load is continuously monitored and readjusted during the test. However, a significant readjustment is not normally needed, since the variation caused by sway displacements of the frame is not significant.

Reversed cyclic horizontal load is applied by using a double acting hydraulic jack bearing against the reaction wall as illustrated in Figure 4. The loading unit, consisting of the ram and load cell, has pin connections at both ends to eliminate any accidental eccentricity mainly in the vertical direction and tolerating a small rotation in the horizontal direction normal to the testing plane. The load is applied at one third span of the spreader beam to ensure that the load at the second floor level always remains twice as the load at the first floor level. At floor levels, clamps made of four steel bars connected to two loading plates at either end are loosely attached to the test frame to avoid from any unintended interference with the frame behaviour, which may possibly be caused by the external prestressing on the beams. Thus, in both pushing and pulling modes, a horizontal push is applied to the test unit through steel loading pads without inducing any undesirable axial load in the beam.

The infill wall, which has a considerably smaller thickness than the beam width, is placed eccentrically on the exterior side of the beam to reflect the common practice. Thus the contribution of the infill makes the frame behaviour somewhat unsymmetrical, and the load applied in the plane of symmetry creates warping, which may lead to significant undesirable out-of-plane deformations, especially towards the end of the test. A rather rigid external steel 'guide frame' attached to the universal base, is used to prevent any out-of-plane deformations. Four 'guide bars' two on each side, with roller ends, are attached to the guide frame, and they gently touch the test frame beam, smoothly allowing in-plane sway. The photograph given in Figure 5 shows the guide frame and the guide bars.

Typical load and displacement histories followed during the tests are illustrated in Figure 6. As can be deduced from the figure, ever-increasing load cycles are applied so far as the specimen is capable of resisting higher loads, and beyond that, deformation controlled loading is performed with ever-increasing displacement cycles, usually in steps of multiples of yield displacement.

3.2.2 Deformation Measurement System

All deformations are measured by displacement transducers; using either LVDTs or electronically recordable dial gauges as shown in Figure 7. Sway displacements are measured both at the first and second floor levels. Three measurements are normally taken at the top level to ensure a reliable collection of these very important data, even in the case of one or two transducers unexpectedly going wrong.

Infill wall shear deformations are determined on the basis of displacement measurements along the diagonals.

Displacement measurements taken at the column roots are meant for computation of rotations of the entire test unit, when the infill wall remains intact and the overall behaviour of the test unit resembles cantilever behaviour. However, they also provide data for monitoring the critical column section deformations; steel yielding in the tension side column, concrete crushing in the compression side column etc.

Although it is heavily prestressed to the strong testing floor, the rigid body rotations and displacements of the universal base are monitored using four dial gauges, during the tests to enable corrections in the critical measurements in the case of an unexpected movement.

3.3 Test Results

As already indicated in Table 1, three successful tests (PR, PI1 and PI2) have so far been performed, besides two preliminary tests.

3.3.1 Preliminary Tests

For decades, similar tests have been performed on horizontally tested twin frame specimens in the Structural Mechanics Laboratory. This set-up had been developed with the rather modest facilities then available in the laboratory, and required a lengthy and tedious testing process. After completing the last series of tests, the horizontal test set-up was put out of service by a retirement party, an excuse for the faculty to indulge in food and drinks.

The vertical test set-up used in the present test series was recently developed and employed, hoping that it would be identical to its sister lying down on the floor. Alas, the preliminary tests proved the contrary. In both of the preliminary tests, major damage took place in the second floor infill wall, despite the expectation of failure in the first floor, which has always been the case in the horizontally tested twin frames, Figure 8.

Application of the horizontal load at the second floor level was naturally an approximation to simplify the test procedure, and it had never caused a problem of this kind in the horizontal test set-up. After lengthy discussions, it was concluded that the unexpected type of failure was stemming from the reaction differently developing at the foundation level. The reaction was a concentrated force acting at the opposite end of the foundation beam of the twin frames, whereas it was distributed along the entire length of the foundation beam in the vertical test set-up, leading to a much wider compression strut development in the first floor infill wall. Since equal shear forces developed in both infill walls under the horizontal force applied at the top, the lower one had a higher chance to survive.

The complication could be remedied by using a much more realistic two-component horizontal load, applied at each of the two floor levels, the second floor receiving twice the load as the first floor.

Another complication was caused by unsymmetrically placed infill walls leading to eccentric loading, which created out-of-plane deformations. Since there was no measure to control warping, excessive out-of-plane deformations (beyond acceptable limits) could not be prevented in the preliminary tests. However, the guiding system, explained above in the 'Loading and Supporting System' section could solve the problem in the following tests.

3.3.2 Reference Test - PR

A specimen representing the present state of a typical existing building, an ordinary reinforced concrete frame with hollow brick infill walls plastered on both sides, was tested under reversed cyclic loading to serve as a reference for the behaviour and capacity of the strengthened specimens to be tested. The load history followed in this test is illustrated in Figure 9a.

The load-top displacement diagram given in Figure 10a displays clear indications of the hollow masonry infilled frame behaviour characterised by,

- Rather rigid and linearly elastic behaviour at the initial stages under relatively high loads;
- Relatively high capacity resulting from infill wall contribution;
- Rapid strength degradation and very rapid stiffness degradation upon infill wall crushing;
- Small ductility and relatively small energy dissipation.

This expected behaviour was concluded by a typical failure accompanied by excessive permanent first storey sway deformations as illustrated in

Figure 11a. The maximum lateral load level attained in this test was ~79 kN, probably 4~5 times the capacity of the same frame without the infill walls. The corresponding second storey lateral displacement was measured as 6.2 mm. The initial and the final lateral stiffness indicators of this test were computed, in terms of loading branch slope in each cycle, as ~42.3 kN/mm and ~0.6 kN/mm respectively.

3.3.3 Retrofitting by Type 1 Panels - PI1

Type 1 precast concrete panels were placed on the interior faces of the infill walls by using the epoxy mortar mentioned above in the 'Materials' paragraph. However, before the application of the epoxy mortar, bars extending out of the panel corners were welded to each other and to dowels epoxy anchored into the frame elements to ensure proper connections.

The load-top displacement diagram given in Figure 10b clearly illustrates the enormous improvement in behaviour and capacity, compared to the reference specimen. One can make the following observations at a glance:

- Significant increase in the load carrying capacity;
- Significant increase in the initial and final stiffness;
- Significantly delayed strength degradation;
- Considerably decelerated stiffness degradation;
- Improved ductility and significantly increased energy dissipation capacity.

Minor damage took place in the infill walls, indicating effectiveness of the wall strengthening panels. The overall behaviour, observed in this test, resembled monolithic shear wall behaviour rather than infilled frame behaviour. A major flexural crack separated the infill from the foundation beam; column reinforcement and anchorage bars remaining in the tension zone yielded, and finally, failure occurred by crushing of the column on the compression side. Figure 11b shows the general view of this specimen after failure. The maximum lateral load level attained in this test was measured as 193 kN and the corresponding second storey lateral displacement as 10.7 mm. The initial and final lateral stiffness indicators were computed as ~122.8 kN/mm and ~2.2 kN/mm respectively.

3.3.4 Retrofitting by Type 2 Panels - PI2

This specimen was very similar to PI1, except the Type 2 panels used as infill wall reinforcement. The other significant difference from the former strengthened specimen was the relatively low strength of concrete used in the test frame.

The load-top displacement diagrams given in Figures 10b and 10c clearly indicate the strong resemblance between the performances of the two strengthened specimens PI1 and PI2 apart from a few minor superiorities of the latter such as:

- A greater number of load cycles tolerated;
- Slightly higher load carrying capacity;
- A little slower strength degradation and better ductility;
- Somewhat higher energy dissipation as a result of the preceding items.

The above mentioned superiority of the specimen PI2 was, no doubt, caused by the larger number of epoxy anchored dowels (13ϕ6 versus 5ϕ6) connecting the panels to the foundation beam. The overall behaviour of the specimen PI2 was almost identical to that of the specimen PI1; namely, monolithic cantilever behaviour leading to flexural failure at the critical section where the frame and infill were connected to the foundation beam. The epoxy anchored dowels naturally served as additional reinforcement crossing this critical section.

Minor damage took place in the infill walls, and the precast panels remained almost intact apart from separation from the frame members. Figure 11c shows the general view of this specimen after failure. The maximum lateral load level attained in this test was measured as 198 kN and the corresponding second storey lateral displacement as 12.9 mm. The initial and final lateral stiffness indicators were computed as ~123.2 kN/mm and ~3.5 kN/mm respectively.

4. DISCUSSION OF TEST RESULTS

4.1 Interpretation of Test Results

The behaviour of the strengthened specimens, explained briefly in the section above, indicate a very satisfactory improvement in the seismic performance of the test frames by introduction of panel type strengthening. The load capacity as well as ductility and energy dissipation characteristics were significantly improved. Envelope load-lateral displacement curves given in Figure 12 indicate a significant increase in the load carrying capacity and significantly delayed strength degradation, leading to a much better ductility. Beside these, the comparison of the individual load-top displacement curves given in Figure 10 indicates a significantly increased energy dissipation capacity in the strengthened specimens, which is also evident from the energy dissipation curves given in Figure 13. Significantly increased initial and final stiffness values and considerably decelerated stiffness degradation can be observed in the stiffness variation diagrams in Figure 14.

Some critical values measured or computed for each of the three specimens tested are listed in Table 3 for ease of comparison. (Strengthened/ Reference) ratios are also given. These ratios speak for themselves and quantify the significant improvement achieved by the proposed strengthening intervention.

4.2 Future Tests

The panel connections were initially considered very critical, and both shear keys and welded connections, fixing reinforcing steel bars to each other and to dowels epoxy anchored into the frame elements, were provided in addition to the epoxy mortar used in the panel joints. However, the epoxy mortar proved to be so successful in connecting the panels in PI1 and PI2 tests that both shear keys and welded connections appeared to be redundant. To observe the performance in an extremely simplified case, specimens having precast panels with no shear keys and no welded connections will be studied in the following two tests. Improving measures will be introduced in the later tests on the basis of any possible deficiencies observed. However, epoxy anchored dowels at the foundation level where the most critical load effects take place, are considered essential and therefore unavoidable.

The proposed technique appears to be effective when the panels are placed on the interior face of the wall, in other words, when the precast concrete panel layer is confined all around by the frame members. However, the question remains, if the technique will still be effective when it is applied to the exterior face of the wall. It should be realised that the beam is not wider than the wall in many practical cases, and the concrete layer has to be connected to the outer faces of the frame members. This problem will also have to be investigated in the following stages of the work.

5. CONCLUDING REMARKS

The initial results obtained from the limited number of tests so far performed indicate good prospects for the non-evacuation retrofitting technique proposed for hollow masonry infilled reinforced concrete building structures, although there are still a number of problems to be investigated. Tests have shown that strengthening of the existing infill walls by epoxy connected precast concrete panels may lead to a significant improvement of the seismic performance to an extent comparable to that of the cast-in-place reinforced concrete infills.

Analytical investigations are obviously needed to verify the amount of infill walls to be reinforced with precast concrete panels in an existing structural system for a satisfactory seismic performance. It may intuitively be said at this stage that a certain percentage of the floor area can be specified as a rule of thumb to be used as a preliminary design guide. It is again the intuition of the authors that a very satisfactory performance can easily be achieved by doubling the number of ordinary cast-in-place reinforced concrete infills.

One major concern is the lap splice problems inherent in the columns of the existing frame and the poor quality concrete used in the existing structure. Recalling that minor damage took place in the infill walls reinforced with precast concrete panels and failure occurred by crushing of the column on the compression side, one can conclude that the overall capacity is controlled mainly by the properties of the existing columns which serve as boundary elements for the shear wall like infill walls. The weaknesses in the existing columns may therefore lead to premature failures. In such cases, strengthening of columns may be needed, which will obviously be very undesirable, since it involves messy work. However, this problem can be overcome by applying the proposed technique on a few more infill walls than what is absolutely necessary.

REFERENCES

1. Yuzugullu, O., “Repair of Reinforced Concrete Frames with Reinforced Concrete Precast Panels”, final report in Turkish of the TUBITAK sponsored research project MAG 494, Ankara, 1979.
2. Duvarci, M., “Seismic Strengthening of Reinforced Concrete Frames with Precast Concrete Panels”, MSc thesis, Middle East Technical University, Ankara, 2003.
3. Baran, M., “Precast Concrete Panel Reinforced Infill Walls for Seismic Strengthening of Reinforced Concrete Framed Structures”, PhD thesis in progress, Middle East Technical University, Ankara.

Table 1. Precast panel properties in the test units

Specimen	Wall side with panel	Panel pattern	Shear key	Welded connection	Remarks
PPR1	None	---	---	---	Preliminary test
PPR2	None	---	---	---	Preliminary test
PR	None	---	---	---	Reference test
PI1	Interior	Type 1	Yes	Yes	Tested successfully
PI2	Interior	Type 2	Yes	Yes	Tested successfully
PI3	Interior	Type 1	No	At foundation	Planned
PI4	Interior	Type 2	No	At foundation	Planned
PI5	Interior	Type 2	No	At top/bottom	If necessary
PE1	Exterior	Type 1	No	Mechanical	Intended
PE2	Exterior	Type 2	No	Mechanical	Intended

Table 2. Material strengths on the testing date

Specimen	Concrete/plaster strength (MPa)			Steel yield strength (MPa)		
	Test frame	Panel	Plaster	Φ4	Φ6	Φ8
PR	16.6	---	6.5	220	378	330
PI1	18.2	32.5	6.5	220	378	330
PI2	13.0	38.1	6.2	220	378	330

Table 3. Some critical test results

Specimen	At maximum load		Initial stiffness	0.85 displ	Energy dissp
	Load (kN)	Sway (mm)	(kN/mm)	ductility	($\times 10^6$ J)
PR	79	6.2	~42.3	~4	~6.9
PI1	193	10.7	~122.8	~6	~17.5
PI2	198	12.9	~123.2	~8	~22.1
PI1/PR	~2.5	~1.7	~3	~1.5	~2.5
PI2/PR	~2.5	~2.1	~3	~2	~3.2

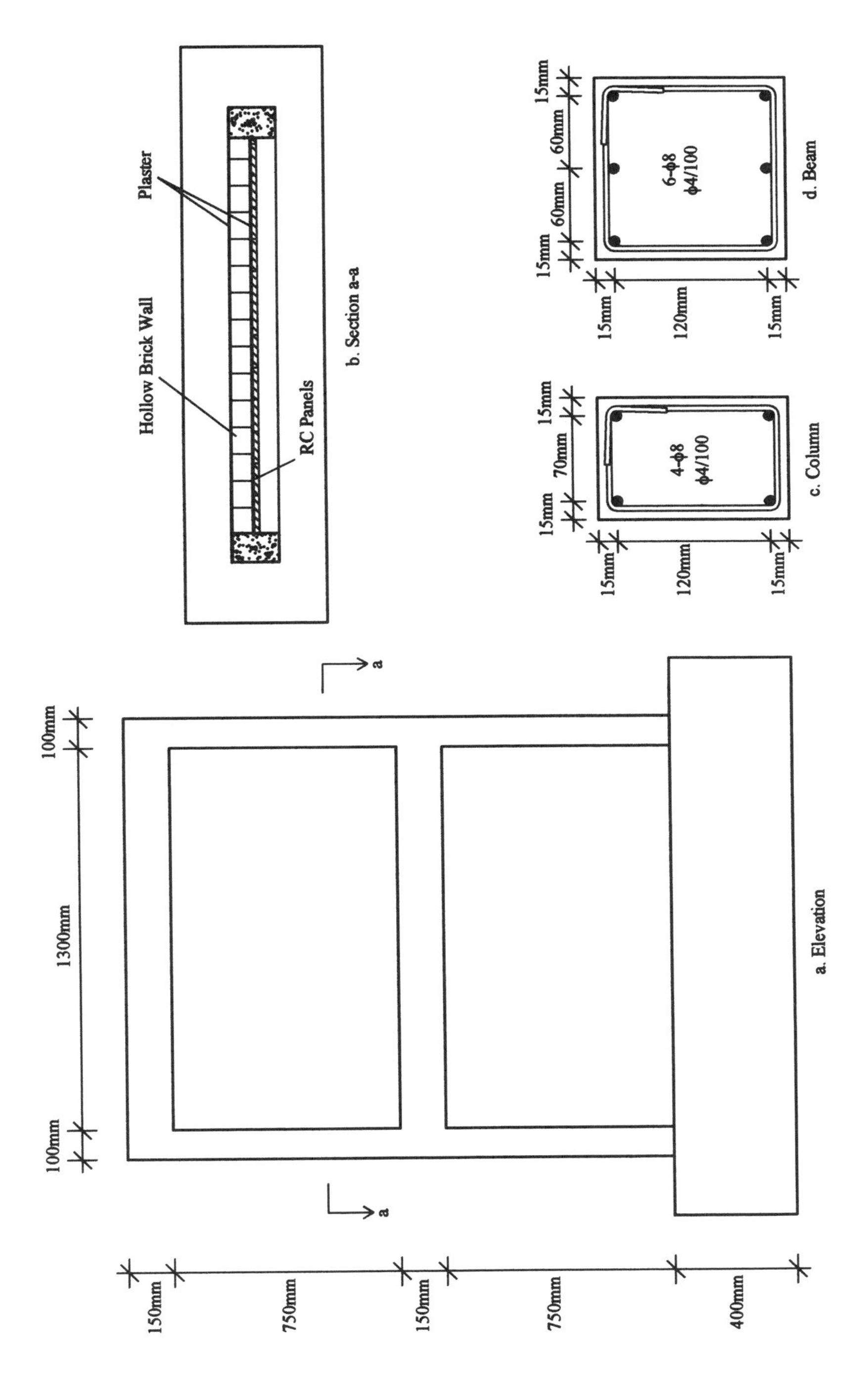

Figure 1. Test frame details

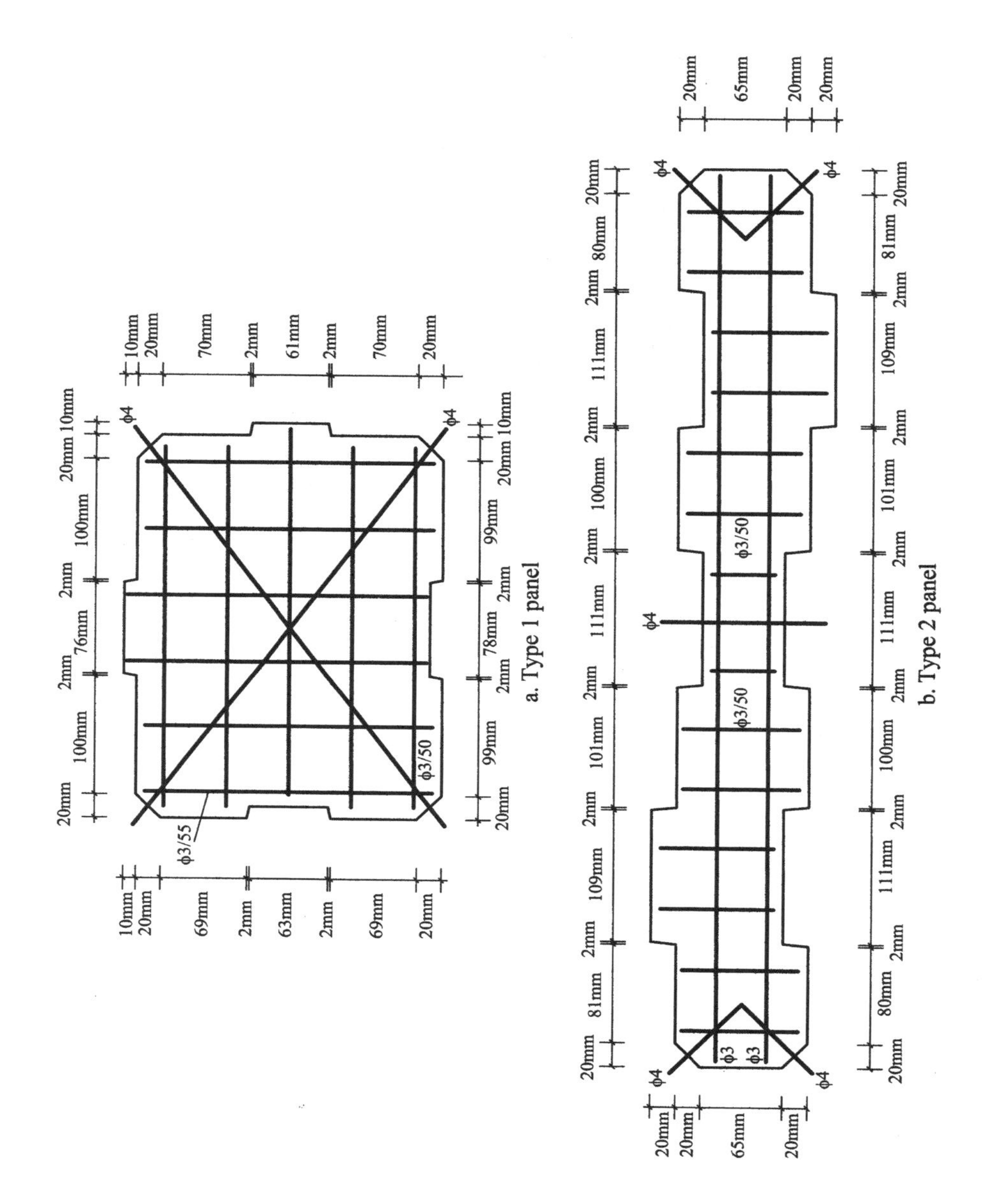

Figure 2. Dimensions and reinforcement of panel types

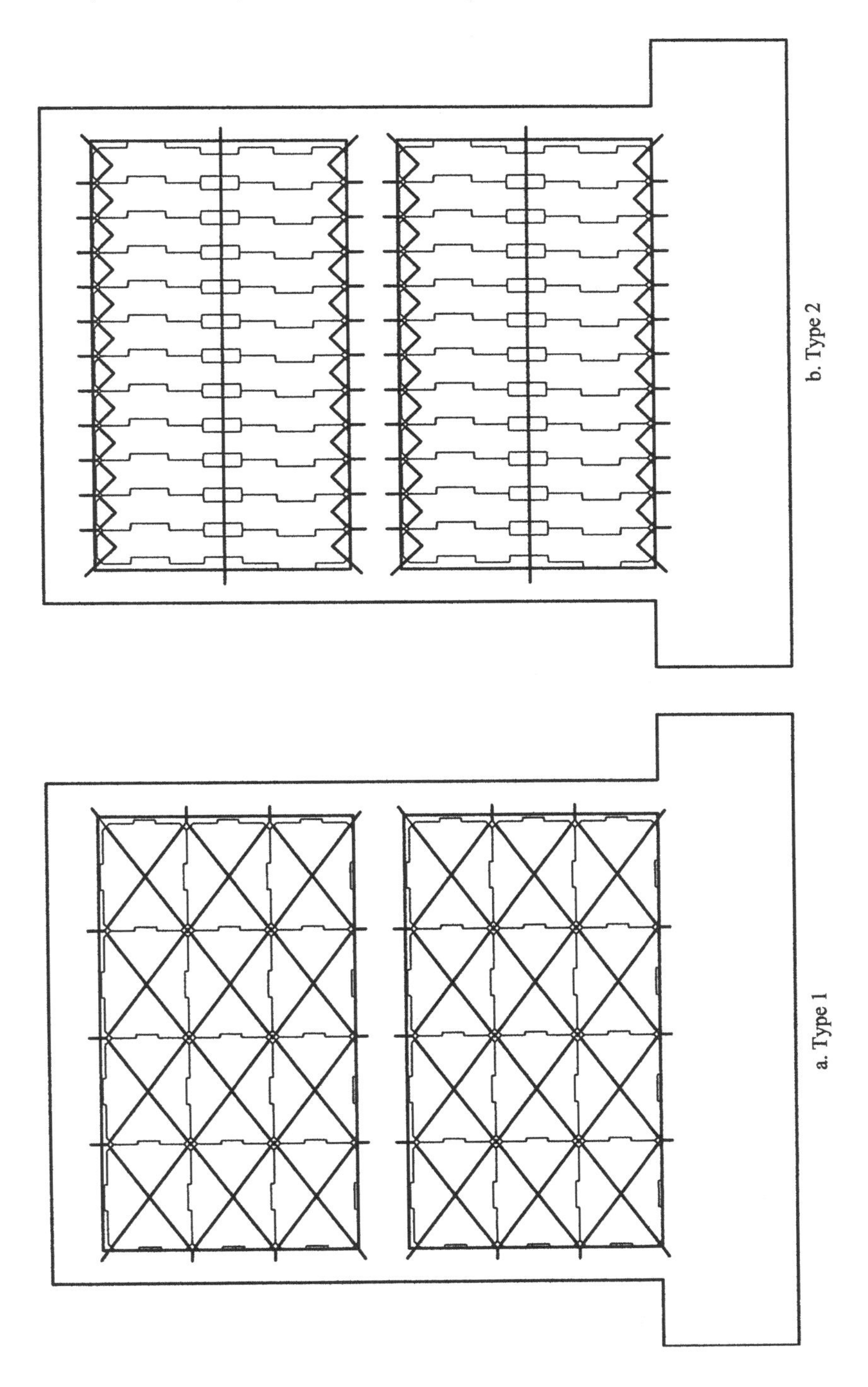

Figure 3. Panel arrangements

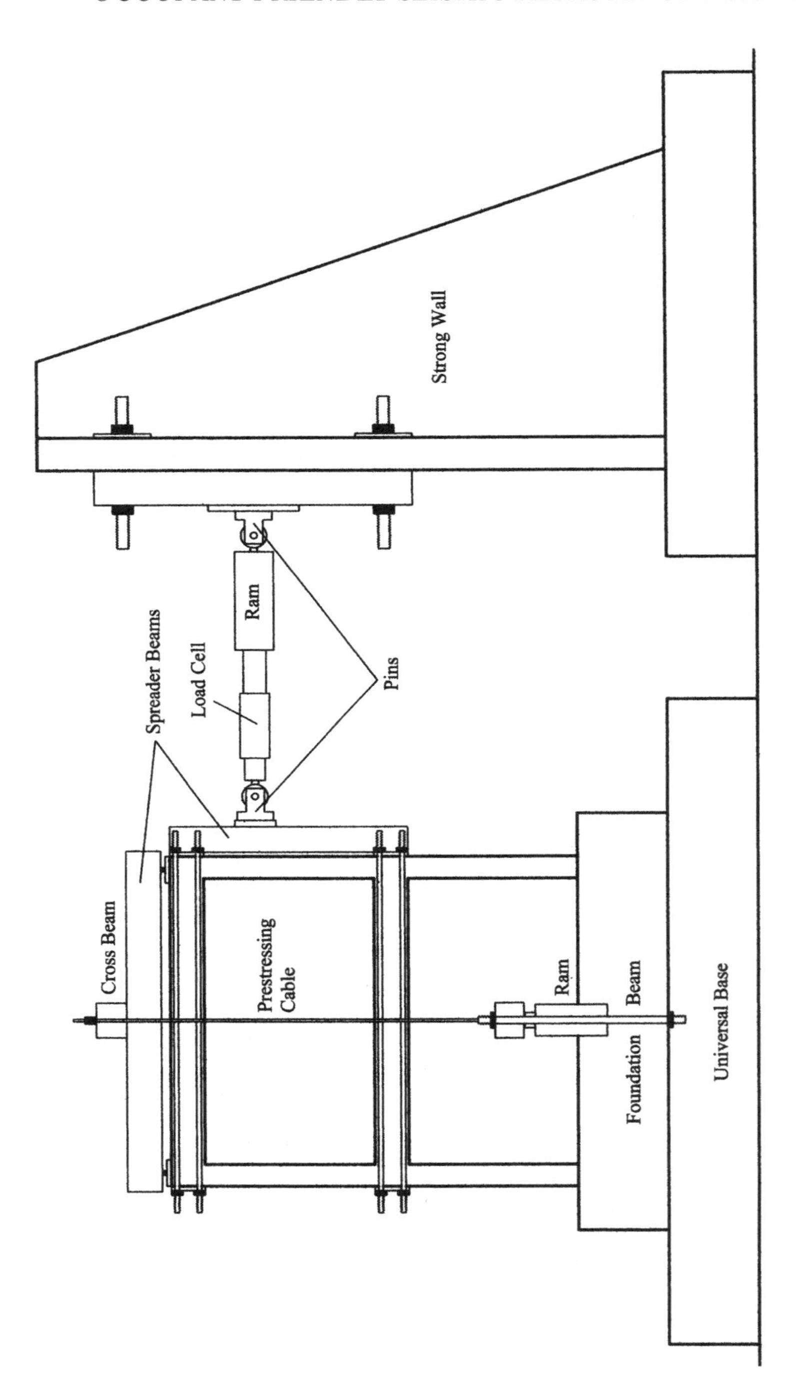

Figure 4. Loading and supporting system

Figure 5. Guide frame and guide bars

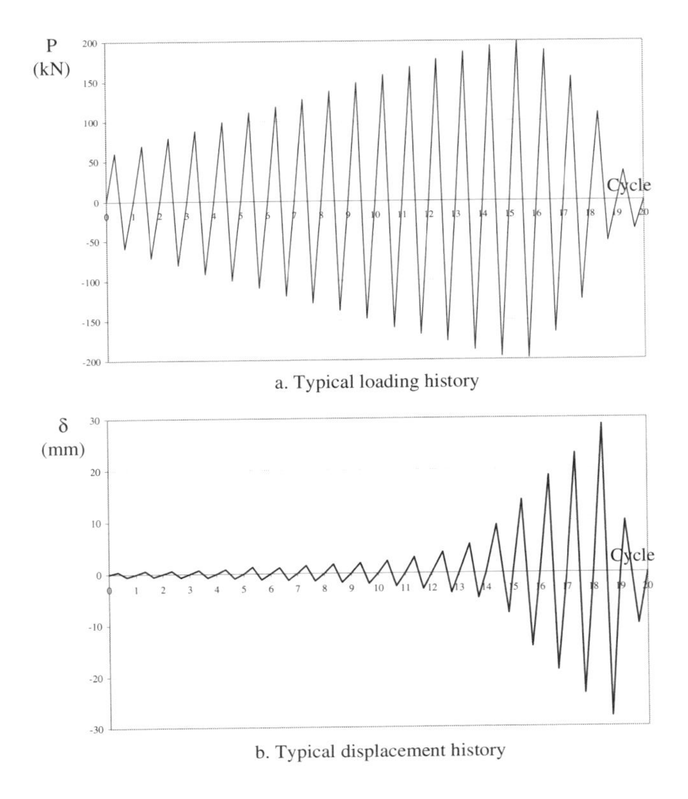

Figure 6. Typical loading and displacement history patterns

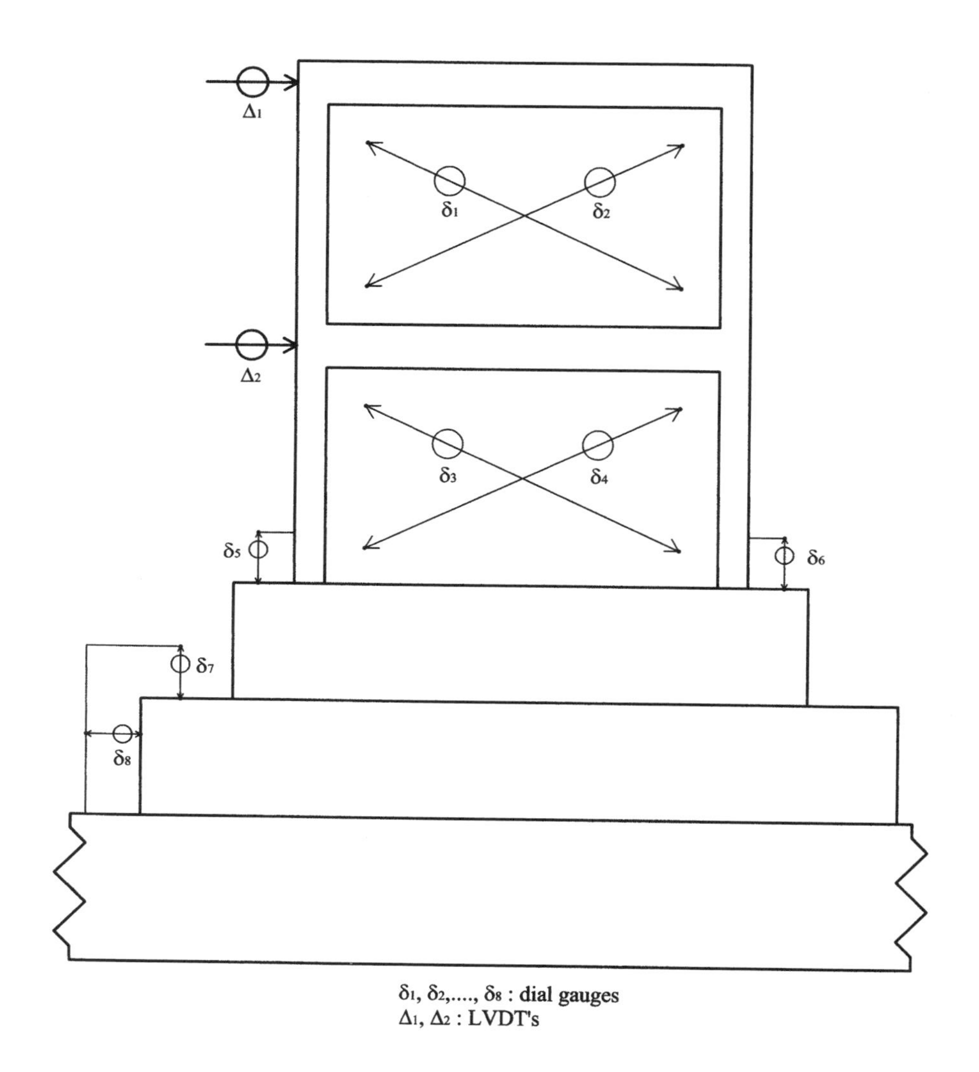

Figure 7. Instrumentation

b. Specimen PPR2

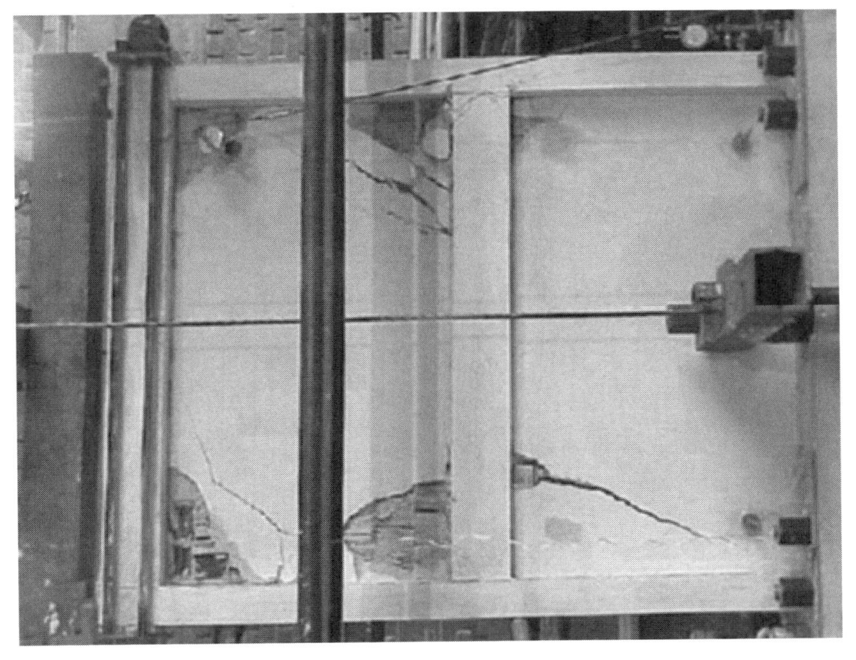

a. Specimen PPR1

Figure 8. Preliminary test specimens at the end of the test

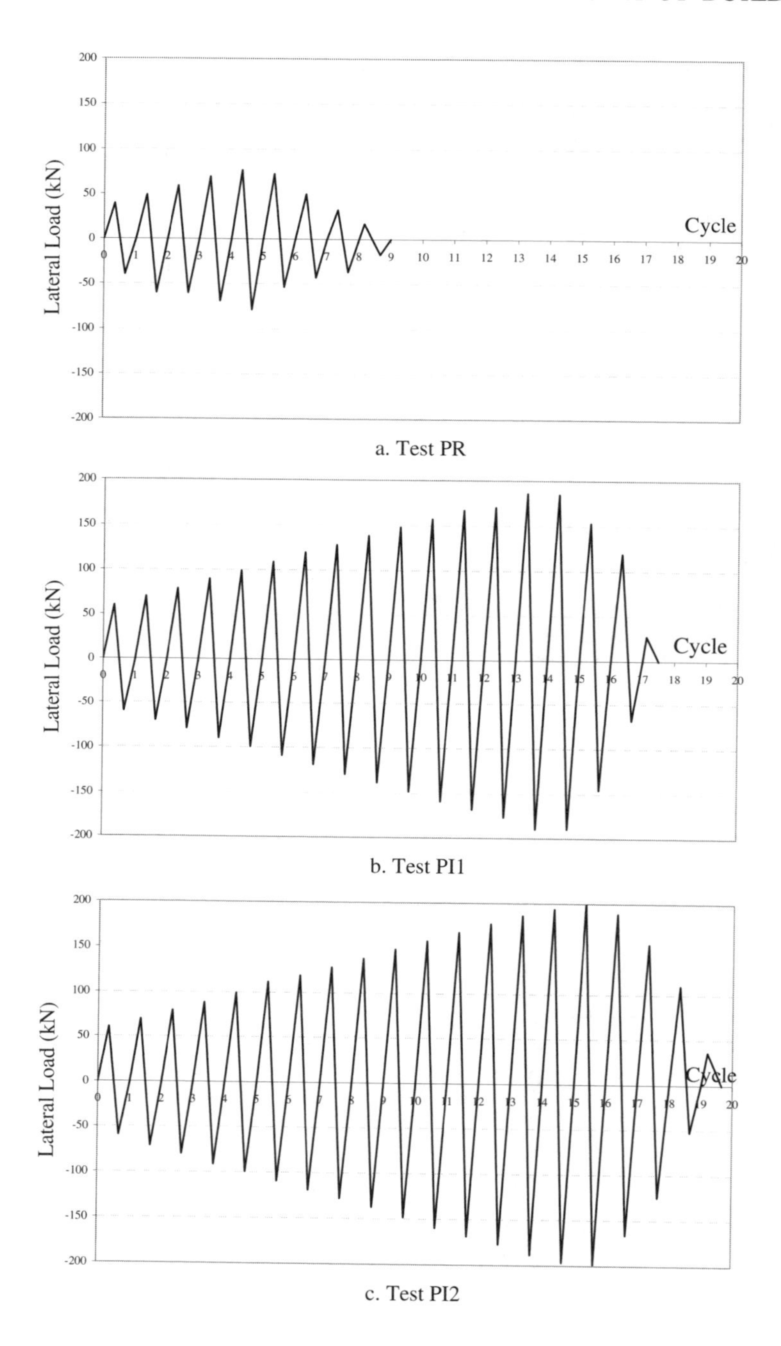

a. Test PR

b. Test PI1

c. Test PI2

Figure 9. Loading histories applied

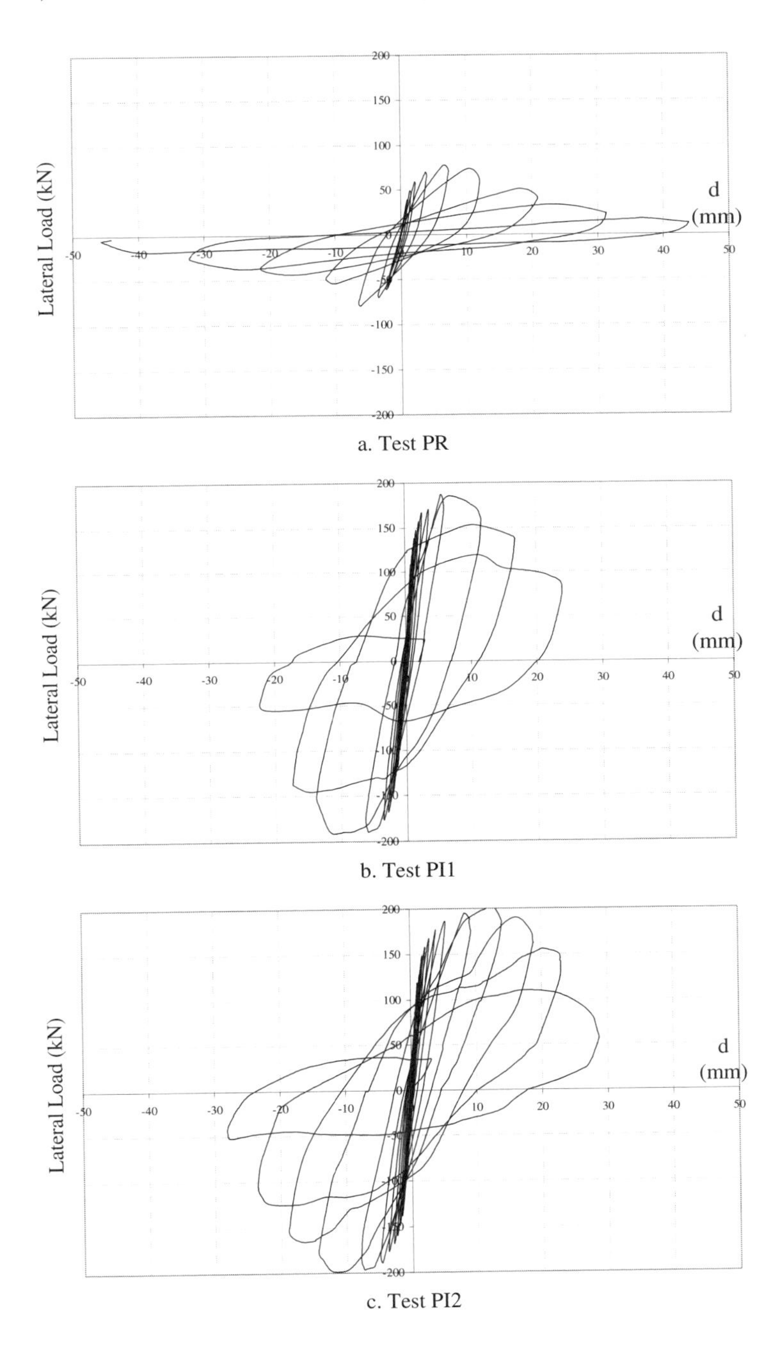

a. Test PR

b. Test PI1

c. Test PI2

Figure 10. Load-top displacement curves

c. Specimen PI2

b. Specimen PI1

a. Specimen PR

Figure 11. General views of the specimens at the end of the test

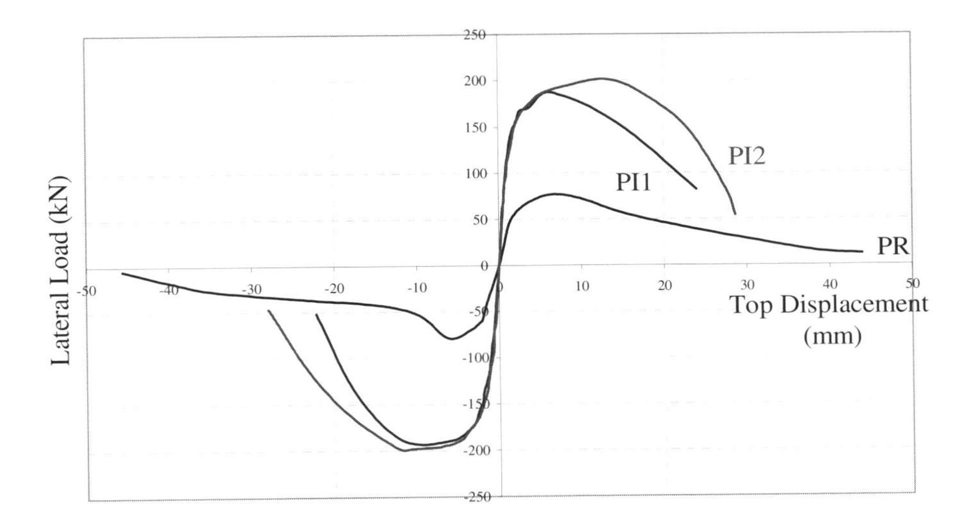

Figure 12. Envelope load-displacement curves

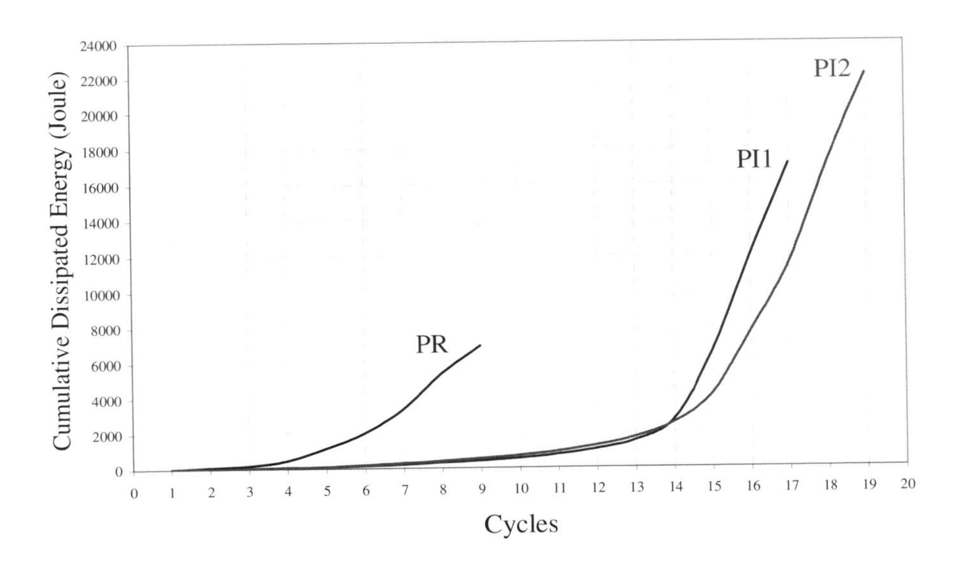

Figure 13. Energy dissipation curves

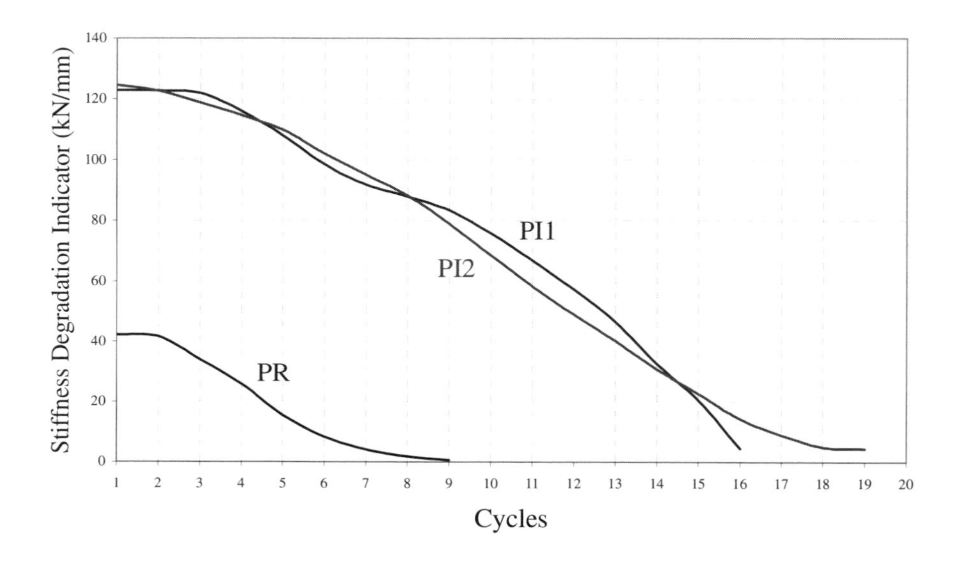

Figure 14. Stiffness degradation curves

SEISMIC RETROFIT OF REINFORCED CONCRETE STRUCTURES

Recent Research at the University of Ottawa

Murat Saatcioglu
Department of Civil Engineering, University of Ottawa, Ottawa, Canada

Abstract: A comprehensive research program is underway at the University of Ottawa on seismic retrofit methodologies for RC structures. The scope includes external transverse prestressing and fiber reinforced polymer (FRP) jacketing of concrete columns; lateral bracing of masonry infill walls by FRP sheets or steel strips; and the use of smart structure technology. The results indicate that the transverse prestressing of columns improves shear capacity, promoting flexural behavior. FRP jacketing of circular columns results in significant improvements in lateral deformability of columns. FRP sheets applied diagonally on infill walls control damage to masonry. Steel strips improve strength and deformability. Active control can reduce inelastic deformations and seismic forces.

Key words: Earthquake resistant design; FRP, masonry walls; reinforced concrete; seismic retrofit; shear walls

1. INTRODUCTION

Previous earthquakes have shown that the majority of structural failures and the resulting human and economic consequences can be blamed on seismically deficient buildings and bridges. These structures constitute a large part of the existing infrastructure. Buildings designed prior to the enactment of modern seismic design guidelines of 1970's, as well as those built more recently without the appropriate seismic design and detailing

S.T. Wasti and G. Ozcebe (eds.), Seismic Assessment and Rehabilitation of Existing Buildings, 457–486.

practices, possess particularly high seismic risk. Because it is not economically feasible to replace existing buildings and bridges with new and improved structures, seismic retrofitting remains to be an effective seismic risk mitigation strategy. Recent developments in new and innovative seismic retrofit methodologies offer great potentials for seismic risk reduction.

A comprehensive research program is currently underway at the University of Ottawa to develop improved seismic retrofit techniques. The research program consists of both experimental and analytical investigations. The experimental tasks involve tests of large scale structural components under simulated seismic loading while the analytical tasks involve development of analytical models, computer software and design procedures. This paper provides a summary of the research project.

The specific areas of research include; column retrofitting by either external transverse prestressing or FRP wrapping, retrofitting infill masonry walls in reinforced concrete frames with FRP sheets, retrofitting concrete and masonry shear walls with diagonal steel strips, and active control of frame buildings to minimize seismic deformations and forces.

2. TRANSVERSE PRESTRESSING OF CONCRETE COLUMNS

Existing concrete columns may suffer seismic damage primarily due to; i) insufficient shear strength, ii) lack of concrete confinement and iii) improper splicing of longitudinal reinforcement. A new method of retrofitting was developed for such columns, involving transverse prestressing. Transverse prestressing has been shown to be effective in overcoming all three deficiencies (Saatcioglu et al. 2000, Saatcioglu et al. 2002, Saatcioglu and Yalcin 2003). The technique, known as "Retrobelt," was developed through experimental research, which involved tests of full size concrete columns with different shear spans, geometry and reinforcement arrangement. Figure 1 illustrates the details of a shear-critical column specimen and typical cross-sectional geometry considered.

The columns were tested in pairs, one with and the other without the retrofit. External prestressing was done by placing 7-wire strands around a column in the form of hoops. The strand ends were tied using specially developed anchors. The anchors were first fastened to the column at required locations by means of concrete nails. The strands were then threaded through the anchors. One end of each strand was locked in to the anchor by means of a standard wedge. The strands were then prestressed to desired levels by means of a hollow hydraulic jack. The jacking ends were then locked in to the anchors using the same standard wedges. Figure 2 illustrates the hardware required for transverse column prestressing.

Circular hoops produce active uniform lateral pressure along the circumference, resulting from hoop tension. This is not possible in square columns. If the prestressing strands are placed on a square column and stressed, they would apply external pressures only at the corners, producing stress concentrations and the crushing of corner concrete. Furthermore, the lack of distribution of external pressure along the perimeter would not permit the development of proper mechanism of confinement, which would become important to protect compression concrete against premature crushing. Therefore, steel spreader frames were developed and used at each hoop location to distribute prestressing forces on column faces. The spreader frames were manufactured from small size hollow structural sections (HSS) with a metric designation of HSS 31.8 x 31.8 x 6.35, also specifying dimensions in mm. Semi-circular discs (pulleys) were welded on HSS sections at three locations per column side to raise the prestressing strands from the column surface so that the appropriate perpendicular force components could be developed. Steel pieces were cut from 50 mm diameter circular pulleys and were placed over column corners to avoid crushing due to stress concentration. The number and spacing of raisers were established to have similar effects as those caused by column cross ties.

The columns were prestressed with hoops placed at 150 mm spacing within their critical regions for shear and flexure. The level of stress was approximately 25% of the strand strength, which was 1860 MPa. However, this was found to be insufficient for improvements in bond between spliced longitudinal bars and concrete. Therefore, the hoops in this region were prestressed up to 50% of the strand strength, and were placed with a spacing of 100 mm. Figure 3 illustrates a flexure dominant circular and a square column, retrofitted with the Retrobelt system.

The columns were subjected to a constant axial compressive force of 15% of the computed concentric capacity, accompanied by incrementally increasing lateral deformation reversals. Experimentally recorded lateral force-drift ratio hysteretic relationships for two pairs of columns are shown in Figures 4 and 5. These include a pair of shear dominant circular columns and a pair of flexure dominant square columns with lap splices near the base. The results clearly indicate that Column BR-C1 and BR-S5, representing the as built conditions of shear and flexure/splice deficient columns, respectively, have a limited lateral drift capacity of approximately 1%. The companion columns, BR-C2 and BR-S6 retrofitted by external prestressing, exhibited significantly increased inelastic deformability, reaching lateral drift levels of in excess of 4%. Test data consistently showed similar improvements in other column tests when retrofitted with external transverse prestressing (Saatcioglu et al. 2000, Saatcioglu et al. 2002).

The retrofit method includes surface treatment, often in the form of shotcreting, to protect the strands against outside effects, such as corrosion, fire and vandalism. While this may be a recommended practice for bridge columns, the same protection can be attained in building structures with drywall enclosures.

3. FRP JACKETING OF CONCRETE COLUMNS

Full-scale reinforced concrete specimens, companion to those externally prestressed for seismic retrofitting, were retrofitted with carbon fiber polymer (CFRP) sheets and tested under simulated seismic loading to investigate the effectiveness of CFRP wrapping on their strength and deformability (Saatcioglu and Elnabelsy 2001). The columns were dominant in flexure, having a shear span of 2.0 m. They were reinforced with 12 # 20 (19.6 mm diameter) grade 400 MPa deformed reinforcement, equally distributed along the section perimeter. Each bar was spliced near the base with a splice length of 20 times the bar diameter, resulting in a 390 mm splice length. Each column had #10 (11.3 mm diameter) deformed transverse ties, spaced at 300 mm. Circular columns had circular hoops with overlapping ends, and the square columns had square hoops with 135° bends.

The specimens were fixed to the laboratory strong floor through heavily reinforced footings, which provided full fixity against rotation. Two servo-computer controlled MTS hydraulic actuators were used to apply constant axial compression, equal to 15% of column concentric capacity (P_0). A third MTS actuator was positioned horizontally and connected to the pin end of column to apply incrementally increasing deformation reversals.

The behavior of columns was investigated by comparing pairs of columns with identical geometric and material properties. While one of the columns in a pair was used as a reference column, reflecting as built conditions, the other column was retrofitted with MBrace CF 130 Carbon fibre reinforcement system (CFRP sheets). Both circular and square sections were considered. The circular columns had four plies, whereas the square columns had five plies of CFRP sheets, as per manufacturer's recommendations. Prior to the application of CFRP material, the column surface was first treated with an epoxy-based primer. The CFRP sheets were pre-cut to the required length and applied on the column with epoxy saturant. Figure 6 illustrates various stages of the application of CFRP jackets and the test setup. Coupons were cut from CFRP jackets and were tested to establish the actual stress-strain relationship of the CFRP material. Accordingly, the jackets had approximately 60,000 MPa elastic modulus and 700 MPa ultimate strength with linear elastic stress-strain relationship up to failure.

Column BR-C8 was a reference circular column without the retrofit. The column exhibited gradually increasing flexural cracks under deformation reversals during ½% and 1% drift cycles. The starter bars yielded at 2% drift causing vertical cracks of approximately 2 mm in the splice region. The peak load resistance was reached during the first cycle of 2% drift. The lateral load dropped to 57% of the peak during the second cycle. At this load stage, the column was considered to have failed because of the substantial drop in lateral load resistance. The failure occurred due to the slippage of longitudinal reinforcement within the splice region. The drift capacity was 1% prior to a significant strength decay caused by the slippage of longitudinal reinforcement.

Column BR-C8-1 was a companion retrofitted circular column as shown in Figures 6 (a) and 6 (d). The column did not show any sign of damage up to 1% lateral drift, at which load stage the yielding of longitudinal reinforcement was recorded. The maximum lateral load resistance was reached during the first cycle of 2% drift ratio. The column maintained a stable behaviour during the subsequent deformation reversals with little strength degradation up to about 8% lateral drift. Beyond this load stage the test was discontinued since the instruments started reaching their limits due to excessive deformations. The hysteretic moment-drift relationships shown in Figure 7 indicate that the CFRP jacket was highly effective in improving bond between the reinforcement and concrete within the splice region, while also improving the confinement of concrete.

Column BR-S5-1 was a retrofitted square column companion to BR-S5, whose hysteretic relationship is illustrated in Figure 5(a). The column was retrofitted with five plies of CFRP sheets, after rounding its corners to have a diameter of approximately 25 mm. The specimen did not show any sign of damage during initial cycles of loading. The yielding of longitudinal reinforcement was observed at the column-footing intersection during the first load excursion to 1% lateral drift. No sign of deterioration was observed up to 2% drift cycles, at which load stage the maximum load resistance was reached. The column was able to maintain 95% of its peak load resistance during the first cycle of 3% drift, but the strength dropped to about 80% of the peak load during the third cycle at 3% drift ratio.

The degradation of capacity continued during subsequent deformation cycles, exhibiting signs of de-bonding of lap splices. At 4% drift, the load resistance dropped to 75% of the peak load. The drop continued during subsequent deformation cycles, reaching down to 65% of the peak at 5%

drift ratio. The hysteretic relationship of Column BR-S5-1 is illustrated in Figure 8.

The hysteretic response shown in Figure 8 indicates that the CFRP jacket used in the square column resulted in limited improvements. The jacket could only provide lateral restraining action to concrete near the corners. The observations during testing indicated that the jacket opened up between the corners near the base, and was ineffective in improving bond between the reinforcement and concrete. The column could only sustain 3% drift with about 20% strength decay. It may be concluded that CFRP jackets for square and rectangular columns have limited effectiveness. Therefore, their effect should be critically examined before field applications.

4. MASONRY INFILL PANELS RETROFITTED WITH FRP SHEETS

A large number of buildings, particularly those constructed prior to the enforcement of ductile design philosophy of 1970's have utilized unreinforced masonry (URM) infilled frames as their structural system. These buildings were primarily designed and detailed to resist gravity loads. It has been a common practice to consider these infill walls as non-structural elements, though they often interact with the enclosing frame, sometimes creating undesirable effects. These elements may contribute to the increase in strength and stiffness of the enclosing frame, though they are not capable of improving inelastic deformability. Any seismic improvement to be expected from URM infills is limited to the elastic range of material. The elastic behavior may not be ensured during strong seismic excitations, and the subsequent brittle failure may lead to disastrous consequences. Buildings that were designed for higher seismic forces than those corresponding to the elastic threshold of URM need to be protected during seismic response. One technique to achieve this goal is to strengthen the URM by means of epoxy-bonded carbon fiber reinforced polymer (CFRP) sheets.

Two half-scale reinforced concrete frame-concrete block infill assemblies were tested under simulated seismic loading (Serrato, Saatcioglu and Foo 2003). The first specimen was built to reflect the majority of existing buildings constructed prior to the 1970's, with a gravity-load designed frame. Figure 9 illustrates the details of the specimens.

A professional contractor was hired to build the masonry infills to implement the actual practice in industry, as illustrated in Figure 10. The second specimen was retrofitted after the first one was tested so that the observed behavior would provide guidance as to how to best retrofit the specimen. Two aspects of the retrofit strategy employed were of concern; i) the amount and arrangement of CFRP sheets and ii) the possibility of delamination of sheets and measures against them. After carefully examining the results of the first test, it was decided to use one sheet per face parallel to each of the two diagonals, resulting in two sheets per wall face. The CFRP sheets were placed diagonally to increase their efficiency since their primary function was to resist diagonal tension. Specially designed CFRP anchors were used to minimize/eliminate the delamination of CFRP sheets from the surface of the wall. This was done by drilling holes in the frame members adjacent to the wall and inserting the anchors to be epoxy glued into the concrete. A hammer drill was used to make the holes (hole diameter = 13 mm) in columns and beams. The placement of CFRP sheets followed the recommendations of the supplier, as described below.

- Masonry substrate was prepared. Loose concrete and dust was removed for improved adhesion.
- A coat of epoxy primer was applied by a roller.
- Epoxy filler putty was applied to fill the mortar joints as much as possible, and any imperfections in concrete blocks as shown in Figure 11(a). This was done using a trowel.
- One coat of epoxy resin (saturant) was applied over the primed and puttied surface.
- The first layer of fibers (sheet) was applied. Fibers were saturated with epoxy prior to the placement over masonry surface.
- The second coat of saturant was applied over the first layer of sheets.
- The second saturated layer of sheets was applied.
- The final coat of saturant was applied, as illustrated in Figure 11(b).
- As the last step, the CFRP anchors were carefully embedded, through two layers of fibre sheets, into the pre-drilled holes in frame elements, and bonded to the laminates with epoxy resin as shown in Figure 12.

The frame-infill wall assemblies showed a significant participation of the walls in frame response. The walls were able to stiffen the frames significantly during the initial stages of loading. The wall without the retrofit experienced gradual stiffness degradation under reversed cyclic loading. Progressive cracking of masonry units and the mortar joints led to the dissipation of energy, without affecting the strength of frame. The lateral drift was controlled by the stiffening effect of the wall. Of particular interest was the simultaneous degradation of strength and stiffness of the wall and the frame, contrary to the belief that the masonry walls would disintegrate early in seismic response, leaving the frames as the only structural system to resist earthquakes. This may be the characteristic of the wall system tested, as the wall was constructed to be in full contact with the frame. However, it does indicate that the two materials, i.e., reinforced concrete and infill masonry are compatible in resisting lateral forces. The initial resistance was provided mostly by the wall. The load resistance was gradually transferred to the frame through progressive cracking and softening of the wall. The eventual failure of the unretrofitted specimen was caused by the hinging of columns within reinforcement splice regions near the ends, while a significant portion of the cracked infill wall remained intact. Figure 13 shows the hysteretic force-lateral drift relationship for the unretrofitted frame-wall assembly. The relationship indicates that the peak load of 273 kN was attained in the direction of first load excursion at approximately 0.25% lateral drift ratio and remained constant up to about 1% drift, though there was substantial stiffness degradation during each cycle of loading. Gradual strength decay was observed after 1% drift, and the assembly failed due to the failure of concrete columns in their reinforcement splice regions at about 2% lateral drift.

The retrofitted specimen showed a substantial increase in elastic rigidity and strength. The initial slope of the force-deformation relationship was high, corresponding to that of uncracked wall stiffness and remained at the uncracked stiffness level until the specimen approached its peak resistance. It was clear that the FRP sheets controlled cracking and helped improve the rigidity of wall. The specimen resisted a peak load of 784 kN in the direction of first load excursion, indicating an improvement of approximately a factor of 3. The peak load was attained at approximately 0.3% lateral drift ratio. The FRP sheets maintained their integrity until after the peak resistance was reached. There was no delamination observed throughout the test. However, the FRP sheets started to rupture gradually near the opposite corners in diagonal tension. This resulted in strength decay. By about 0.5% lateral drift, approximately 25% of the peak load resistance was lost. The load resistance continued to drop during subsequent deformation reversals and the resistance dropped to the level experienced by the unretrofitted specimen at approximately 1% drift ratio. The behaviour beyond this level was similar to that of the unretrofitted specimen, and the failure occurred at 2% drift ratio. The hysteretic relationship recorded during the test is illustrated in Figure 14. The comparison of two hysteretic relationships indicate that the CFRP sheets increased lateral load resistance of the entire subassembly by a factor of 3.0, with no significant effect on deformability.

5. RETROFITTING WALLS WITH STEEL STRIPS

Six large-scale walls with rectangular cross-sections were tested to develop a seismic retrofit strategy utilizing steel strips (Taghdi, Bruneau and Saatcioglu 2000). The specimens consisted of pairs of reinforced concrete, unreinforced masonry and reinforced masonry shear walls. Each pair was identical, with one wall in a pair representing as built conditions and the other representing retrofitted walls. All the walls had an aspect ratio of 1.0 and a wall height of 1.8 m. The specimens were subjected to constant axial compression, accompanied by incrementally increasing lateral deformation reversals. Two actuators were positioned vertically to apply the axial compression, while a third actuator was positioned horizontally to apply lateral deformations. Figure 15 illustrates the test setup.

Retrofit was accomplished by adding two 220 mm wide diagonal steel strips of gauge 9 thickness (3.81 mm) on each wall face. Steel strips were added on both sides of the wall to prevent eccentric stiffness and strength distribution that may cause twisting of the walls, while also providing retrofit against the out-of-plane failures of walls (although not tested in this investigation). The strip width was selected to ensure yielding of the gross-section prior to net section fracture at bolt locations. Diagonal strips were placed along wall diagonals to primarily resist diagonal tension caused by shear. The measured yield strength of the diagonal strips was 227 MPa. Two 80 x 3.81 mm vertical strips, with a 248 MPa yield strength, were added as boundary elements on each side of the walls. Through-thickness 9.5 mm and 15.9 mm diameter structural steel bolts were used to fasten the vertical and diagonal steel strips to the walls, respectively. The spacing between these bolts was chosen to prevent elastic buckling of the strip, and avoid interference with the vertical and horizontal reinforcement. These bolts were also expected to brace the steel strips and confine the concrete or masonry in between. The steel strips were welded together at the center of the wall, where they meet, as well as to 300 mm long and 150x150x16 mm steel angles anchored into the concrete footing and top beam using 400 mm long high-strength anchor bolts. The details of specimens, observed behavior and test results are summarized below for each pair of walls.

5.1 Unreinforced masonry walls

The geometry and overall view of unreinforced masonry walls are illustrated in Figure 16. The walls are labelled as Wall 9 and Wall 9R for unretrofitted and retrofitted specimens, respectively. Each wall consisted of 9 courses of concrete masonry units with effective dimensions of 200 mm by 400 mm, including 10 mm bed and head mortar joints. Prism tests were conducted to establish the strength of ungrouted and grouted masonry units to be 12.5 MPa and 8.2 MPa (reduced strength due to the lower strength of grout), respectively.

Wall 9 behaved in a combination of rocking and sliding, as evidenced by the asymmetric hysteresis loops shown in Figure 17(a). The sliding was developed in one direction at an ultimate force of 64.5 kN, while the rigid body rocking (with small sliding) developed in the other direction at an ultimate force of -58.5 kN. The wall had a limited strength but exhibited relatively large deformations with minor strength decay beyond failure. Rocking and sliding occurred as a result of a continuous crack in the bed joint above the second course of blocks. Despite the low strength, sliding and rocking dissipated some energy, as can be seen in the hysteretic relationship.

Wall 9R was the retrofitted wall and exhibited superior behavior, developing approximately 5 times the strength of companion specimen Wall 9. The wall developed gradually increasing uniform cracking. Yielding of the steel strips in tension produced permanent plastic elongations that could not be recovered in compression. The diagonal strips yielded shortly after the yielding of vertical steel strips. However, because the diagonal strips were wider and had a more favorable anchor bolt configuration, they exhibited only limited buckling. The hysteretic relationship shown in Figure 17(b) resembles to that of a tension only braced steel frame. The loops also showed noticeable pinching, which is attributed to bolt slippage and buckling of steel strips. After 1.0% drift, the hysteresis loops showed a 25% strength drop with further pinching due to the excessive crushing of the masonry and global buckling of the vertical steel strips. The comparison of hysteretic relationships shown in Figure 17 indicates that retrofitting unreinforced masonry walls with steel strips provides a substantial increase in strength and energy dissipation of walls. The retrofitted wall was able to sustain approximately 5 times the strength of unretrofitted wall under inelastic deformation reversals of up to approximately 1% lateral drift ratio without a significant loss of strength.

5.2 Reinforced masonry walls

The overall geometry of reinforced masonry walls are the same as those of unreinforced walls shown in Figure 16. The walls are labelled as Wall 10 and Wall 10R for unretrofitted and retrofitted specimens, respectively. Each wall consisted of 9 courses of concrete masonry units with effective dimensions of 200 mm by 400 mm, including 10 mm bed and head mortar joints. Prism tests were conducted to establish the strength of ungrouted and grouted masonry units to be 12.5 MPa and 8.2 MPa (reduced strength due to the lower strength of grout), respectively.

Three #15 (16.4 mm diameter) reinforcing bars with 400 MPa yield strength were placed in each wall, one at each of the end masonry unit and the other one in the central unit. These three units were filled with mortar and the bars continued into the footing and the top beam to attain sufficient anchorage. The use of three bars with a total steel area of 600 mm^2 resulted in 0.17% vertical reinforcement ratio based on gross masonry area. Joint reinforcement, usually used in masonry walls to ensure some continuity while controlling shrinkage cracking, was used in the reinforced masonry walls. A truss type bed joint reinforcement was placed in mortar joints at every two courses. The truss reinforcement consisted of 4.11 mm diameter steel wires (No. 8 gauge) interconnected by welded cross wires. This resulted in a wall horizontal reinforcement area of 106 mm^2 and a horizontal wall reinforcement ratio of 0.13%. Figure 18 illustrates the details of the reinforcement used in Walls 10 and 10R.

Wall 10 developed initial cracking under reversed cyclic loading at lateral loads of 68 kN and -50 kN in two directions. These cracks formed along horizontal and vertical mortar joints, resulting in a clear stair-step shaped diagonal pattern. A continuous horizontal crack formed along the mortar joint at the base. The first crack within a masonry block occurred at 0.2% lateral drift. The maximum lateral force was attained at the same deformation level, reaching 119 kN and -120 kN in two opposite directions, respectively. Diagonal cracks increased at the same deformation level. These cracks became wider at 0.4% drift cycles and a 23% strength drop was recorded. Further strength decay of about 45% was recorded at 0.6% drift level, accompanied by diagonal crushing of some of the masonry blocks. Further cycling at this deformation level resulted in the crushing of end blocks with the infilled mortar, exposing buckling of reinforcement. Severe crushing of masonry was observed in the corners as well as in the middle of the wall when lateral drift was increased to 0.8% drift. The hysteretic relationship recorded during the test is depicted in Figure 19(a).

Retrofitted Wall 10R initially developed well distributed horizontal and vertical cracks in mortar beds during 0.1% and 0.2% drift cycles. A long diagonal crack formed and propagated during the cycles at 0.6% drift. The horizontal crack along the wall base became wider and the lateral load resistance reached 397 kN during this deformation level. Buckling of diagonal steel strips near the base started to occur during the first cycle at the same deformation level of 0.6% drift. The buckling of vertical strips was observed during the second cycle of 0.6% drift. The peak load resistance was reached during 0.8% drift cycles, with the beginning of crushing in the masonry. At 1.0% lateral drift the weld that connected the diagonal strips failed and the test was stopped. It was observed that the weld had not been done properly. The joint was re-welded and the test resumed. The wall was cycled at 1.1% drift, resulting in the spalling of masonry shell around the vertical bar on the west end of the wall accompanied by bar buckling. During the subsequent deformation level of 1.5% drift, the steel strips continued showing yielding in tension and buckling in compression. One of the vertical strips ruptured due to low cycle fatigue associated with excessive buckling when lateral drift was increased to 1.6%.

The hysteretic relationship of Wall 10R is illustrated in Figure 19(b). It shows a marked improvement over the behaviour of companion Wall 10. The strength increased by approximately a factor of 4.0 and the wall was able to maintain its lateral deformability up to approximately 1% lateral drift without strength decay. When the behavior of Wall 10R is compared with that of Wall 9R, it can be seen that the presence of reinforcement in the reinforced masonry wall delayed the buckling of steel strips, resulted in slightly improved strength, inelastic deformability and energy dissipation characteristics.

5.3 Reinforced concrete walls

Two reinforced concrete walls with 100 mm thickness were designed and constructed as representatives of 1950's and 1960's practice. The walls were reinforced with 3 pairs of 9.5 mm diameter (No.3 U.S. size) vertical and horizontal reinforcement providing a total vertical and horizontal steel area of 426 mm^2 each, and a corresponding vertical and horizontal reinforcement ratio of 0.24%. Coupon tests revealed that the yield strength of reinforcement was 480 MPa. Although an effort was made to find reinforcement of lower grade steel, consistent with the 1950's and 1960's practice, this was not successful. Figure 20 illustrates the reinforcement details used in reinforced concrete walls, labelled as Wall 11 and 11R. The concrete strength used was 29 MPa.

Wall 11 exhibited a flexural crack along its base, during the first three cycles at 0.05% lateral drift. The crack extended up to 1680 mm during increased lateral drift of 0.1% at a lateral load resistance of 138 kN. The wall developed vertical splitting cracks near its compression ends, exhibiting signs of vertical bar buckling. During the third cycle at 0.1% lateral drift the east end bars buckled and cover spalled completely, exposing buckled reinforcement. The flexural crack at the base widened significantly during the first cycle to 0.2% lateral drift, extending throughout the entire length of wall. Cover spalling and bar buckling was observed on the west side of the wall. The resistance of the wall did not improve beyond this stage of loading and the wall started rocking in both directions without any sign of sliding. Further cycling the wall resulted in the deterioration of the wall footing interface, crushing the concrete near the base and exposing reinforcement. There was no appreciable damage to the rest of the wall and the web remained intact. Hysteretic relationship shown in Figure 21(a) illustrates the limited load resistance and significant pinching of loops observed.

Wall 11 was only damaged along its construction joint. The rest of the wall did not experience any damage, or even cracking. Therefore, it was decided to repair the wall to restore its flexural capacity to a level higher than that corresponding to the diagonal tension capacity of the wall. This was done by adding vertical steel strips at both ends on both sides of the wall. The strips were well anchored to the foundation by means of steel angles (Taghdi, Bruneau and Saatcioglu 2000). The resulting repaired wall specimen was labelled as Wall 11RP and was tested in the same manner as the previous wall specimens. The wall behaved in much the same manner as its original unrepaired behavior until 1% lateral drift, with increased load resistance. Some rocking of the wall at the base was observed due to the pullout of the anchor bolts that secured the steel strips to the foundation through steel angles. This pullout was measured to be limited to 12 mm. Increased lateral load resistance resulted in diagonal tension cracking of the wall at 1.16% lateral drift. The wall reached a peak lateral load resistance of 260 kN at 2% lateral drift. Subsequent cycles resulted in the buckling of diagonal strips and gradual strength degradation. The hysteretic force-displacement relationship of repaired reinforced concrete wall is shown in Figure 21(b), and shows a significant improvement over Wall 11.

Wall 11R is a companion reinforced concrete wall, retrofitted with vertical and diagonal steel strips. The wall showed well distributed and well controlled flexural cracks up to the deformation cycles of 0.2% drift. A diagonal crack was observed at this drift level, covering the entire wall height, at a lateral force resistance of 300 kN. The first yielding of vertical steel strip was detected at 0.5% drift, at a corresponding lateral load of 366 kN. The first buckling of the compression steel strip and diagonal steel strut occurred during the second and third cycles of 0.5% lateral drift, respectively. Severe buckling of these compression members followed during subsequent deformation cycles. The wall sustained improved lateral load resistance with increased inelastic deformability up to 2.7% lateral drift. The improvements associated with retrofitting can be observed in Figure 21(c), which illustrates the hysteretic relationship for Wall 11R.

6. ACTIVE STRUCTURAL CONTROL AGAINST SEISMIC FORCES

Like living organisms sensitive to subtle changes in the environment, buildings and bridges can be equipped to sense and react to their surroundings. The sensing can be achieved either by linear variable differential transducers capable of measuring seismic induced building deformations, or by hair-thin glass fiber sensors embedded in structural materials, capable of carrying information and measuring changes in stresses and deformations. The data sensed and collected is subsequently fed into a local computer for the assessment of corrective action to minimize seismic damage. The corrective action often involves active control of structural response by means of actuators.

Seismic control involves the computation of control forces during seismic excitations and their applications to structures to reduce or minimize the effects of earthquakes. Two seismic response quantities become important for structural control; i) lateral drift and ii) seismic forces. Both of these response quantities are affected significantly by inelasticity, which is unavoidable in reinforced concrete structures under strong earthquakes. Gradual cracking of concrete and yielding of reinforcement change strength and stiffness during response, producing inelastic behavior. Therefore, the consideration of inelastic seismic response gains importance in structural control. The current research involves the development of a control algorithm that enables the application of optimum corrective forces to minimize seismic induced deformations in inelastic concrete structures. Computer software developed for this purpose is equipped with a hysteretic force-deformation model that can trace inelastic seismic response, while introducing the stiffness changes to the equation of motion given below.

$$M\ddot{x} + C\dot{x} + Kx = E\ddot{x}_g(t) + Du(t) \tag{1}$$

where, M , C and K are mass, damping and stiffness matrices, respectively, $\ddot{x}_g$ and u are vectors which determine ground excitation and control force, and $\ddot{x}$, $\dot{x}$ and x are response acceleration, velocity and displacement, respectively. Matrix E represents the degree of freedom in which the ground excitation is applied, and matrix D shows the location of control forces. The equation of motion is solved by step-by-step linear integration technique, and the control force u(t) is established by the Instantaneous Optimal Algorithm.

When bending moment at the end of a flexural member reaches yield moment M_y, the behavior changes and the member develops an inelastic region. The properties of the plastic hinge can be represented by a rotational spring, placed at the end of an analytical model. The stiffness of the rotational spring is established through a hysteretic model that defines the stiffness values during various stages of loading, unloading and reloading. In this investigation, the hysteretic model proposed by Takeda et. al (1970) was incorporated into the computer software developed.

A 5-story reinforced concrete frame building was selected to investigate the characteristics of active control algorithm employed in this investigation.

The building was analyzed and designed on the basis of the National Building Code of Canada (NBC 1995) and CSA Standard A23.3, Design of Concrete Structures for Buildings (1994). An exterior frame of the building was modeled as illustrated in Figure 22, where the control force is applied at the middle bay of each floor, throughout the height of the structure, in the form of diagonal tension ties. The control forces are computed from horizontal forces required to satisfy the control algorithm, and applied at either corner of the bay so as to produce a diagonal tension force. No diagonal compression is applied.

The structure was analyzed under the 1940 El Centro Earthquake record. The results are summarized in Figures 23 and 24. Figure 23 shows the variation of first and fifth-story horizontal displacements with and without the control forces. The results indicate a substantial reduction in lateral drift when the control forces are applied. They also indicate that the ductility demands are reduced significantly. For the structure and earthquake ground motion considered, the maximum first-storey column ductility ratio decreased from 5.7 to 2.3 and the maximum beam ductility ratio recorded at the second floor level, decreased from 1.9 to 1.2. The reduction in inelastic column deformations can also be seen in Figure 24 where hysteretic relationships are plotted for a column member with and without active control.

7. CONCLUSIONS

The following conclusions can be made based on the research program reported in this paper:

- External prestressing concrete columns in transverse direction improves column shear capacity, concrete confinement and reinforcement bond stress in splice regions, resulting in improved inelastic deformability.
- CFRP jacketing of circular concrete columns results in significant ductility enhancements. However, the improvement obtained in square columns is limited. Therefore, caution should be exercised in applying the technique to square and rectangular columns.
- CFRP sheets can effectively be used to increase the strength of masonry infill walls in reinforced concrete frames. While this technique does not result in ductility enhancement, it does improve elastic seismic load resistance very significantly.
- Seismic resistance of reinforced concrete, reinforced masonry and unreinforced masonry shear walls improves significantly by retrofitting these elements with steel strips, placed vertically near extreme tension

and compression regions and diagonally to resist diagonal tension and compression caused by shear.

- Active control of reinforced concrete frame structures may result in significant reductions in seismic forces and deformations, and can be used as a viable seismic retrofit strategy.

REFERENCES

1. CSA. 1994. Design of concrete structures. CSA-A23.3-94, Canadian Standards Association, Rexdale, Ont., Canada, p.199.
2. NRCC. 1995. National building code of Canada 1995, 11th ed. Canadian Commission on Building and Fire Codes, National Research Council of Canada, Ottawa, Ont.
3. Saatcioglu, M., Chakrabarti, S., Selby, R. and Mes, D. 2002. "Improving Ductility and Shear Capacity of Reinforced Concrete Columns with the Retro-Belt Retrofitting System." Proceedings of the 7th U.S. National Conference on Earthquake Engineering, Earthquake Engineering Research Institute, 2002.
4. Saatcioglu, M. and Elnabelsy, G. "Seismic Retrofit of Bridge Columns with CFRP Jacket." FRP Composites in Civil Engineering. Elsevier Science Ltd. pp. 833- 838, 2001.
5. Saatcioglu, M. and Yalcin, C. "External Prestressing Concrete Columns for Improved Seismic Shear Resistance." ASCE Journal of Structural Engineering, Vol. 129, in print, 2003.
6. Saatcioglu, M., Yalcin, C., Mes, D., and Beausejour, P. "Seismic retrofitting concrete columns by external prestressing." OCEERC Research Report, Ottawa-Carleton Earthquake Engineering Research Centre, University of Ottawa, Ottawa, Canada, 2000.
7. Serrato, F., Saatcioglu, M. and Foo, S. "Seismic Retrofit of Masonry Infill Walls using CFRP Sheets." The Proceedings of the Natural Hazards Mitigation Workshop, Ottawa, Public Works & Government Services Canada, Hull, P.Q., Canada, 2002.
8. Taghdi, M., Bruneau, M., and Saatcioglu, M. "Seismic Retrofitting of Low-Rise Masonry and Concrete Walls Using Steel Strips," ASCE Journal of Structural Engineering, Vol. 126, No. 9, 2000, pp.1017-1025.
9. Takeda,T., Sozen, M.A. and Nielson,N.N. Reinforced Concrete Response to Simulated Earthquake. Journal of the Structural Division, ASCE, Vol.96, No.ST12, December 1970, pp.2557-2573.

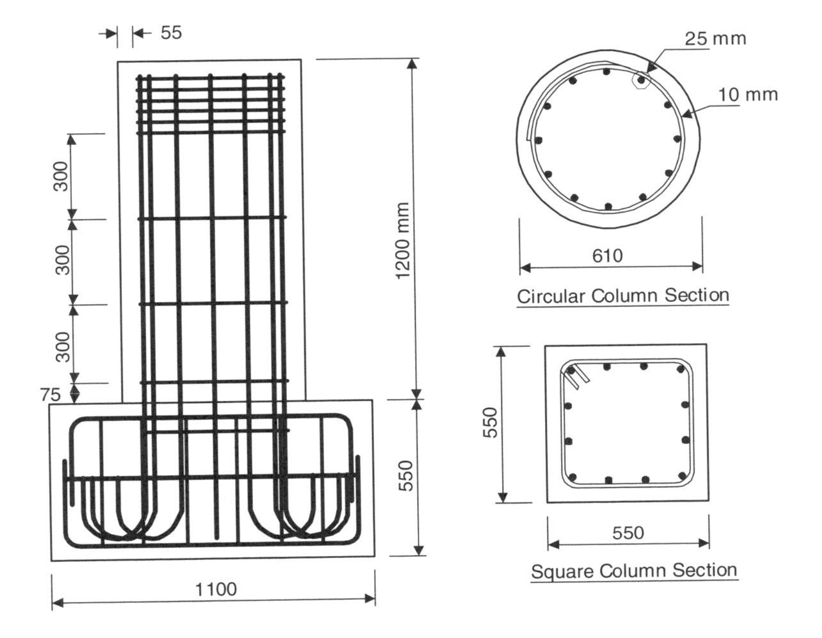

Figure 1. Geometry of typical specimens

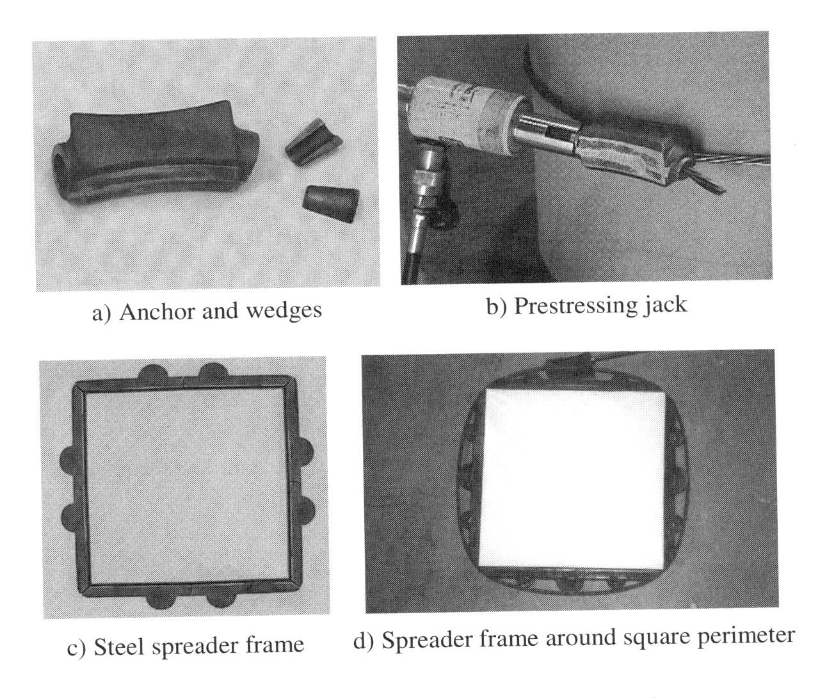

a) Anchor and wedges
b) Prestressing jack
c) Steel spreader frame
d) Spreader frame around square perimeter

Figure 2. Hardware used for external prestressing

a) Circular hoop

b) Square hoops

c) Circular column

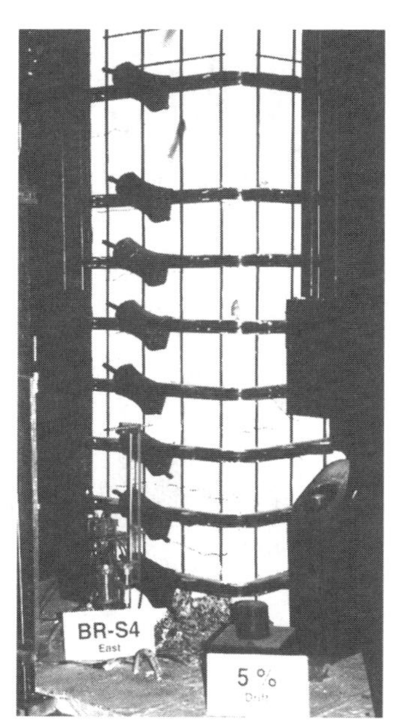

d) Square column

Figure 3. Circular and square columns retrofitted with Retrobelt System

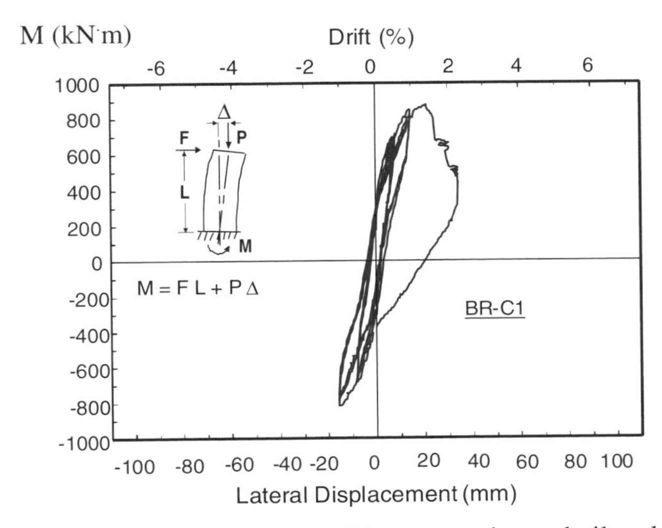

a) Reference specimen BR-C1, representing as built columns

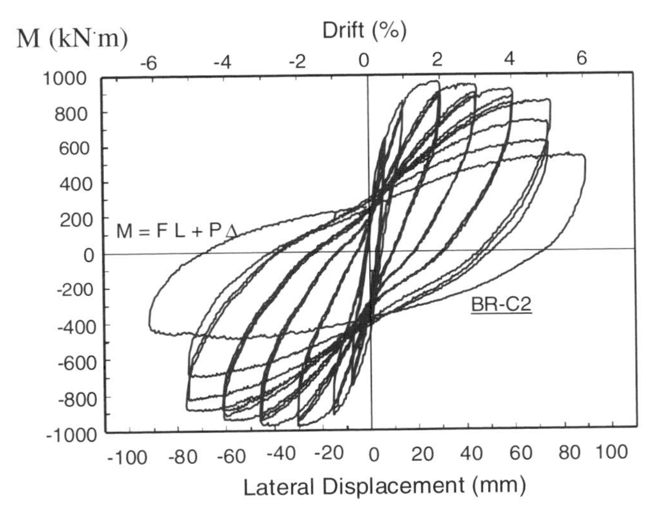

b) Column BR-C2, retrofitted with Retrobelt System

Figure 4. Shear critical circular columns

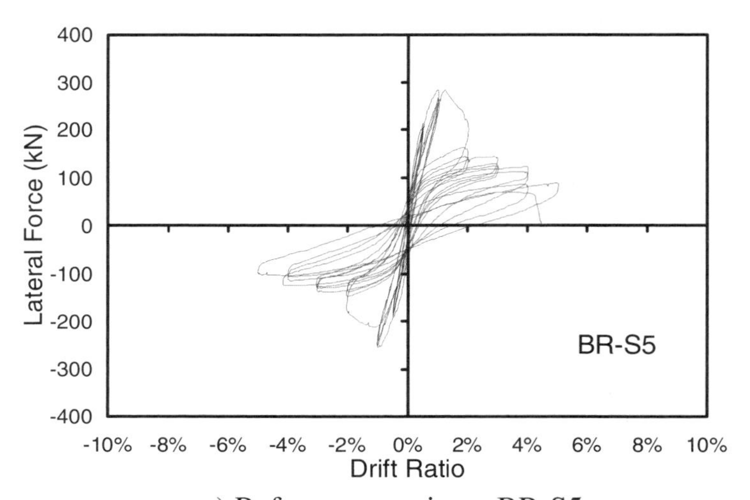

a) Reference specimen BR-S5,
representing as built columns

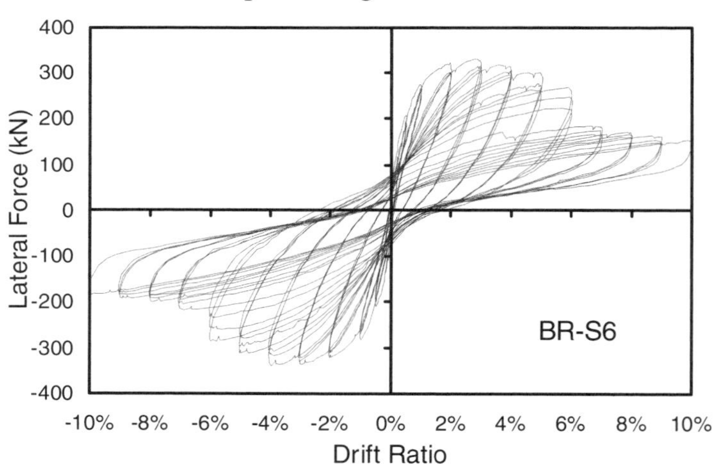

b) Square flexure/splice deficient column BR-S6,
retrofitted with Retrobelt System

Figure 5. Flexure-dominant square columns with spliced reinforcement

a) Circular column wrapping

b) Square column wrapping

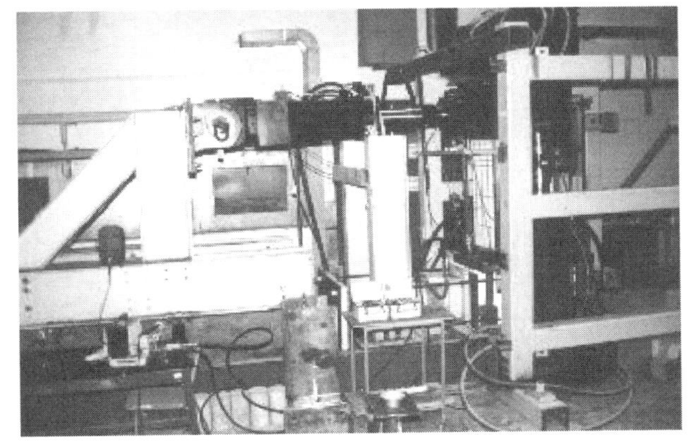

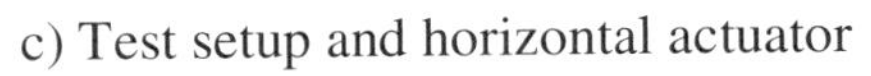

c) Test setup and horizontal actuator

d) Vertical actuators

Figure 6. Application of CFRP jacketing and test setup

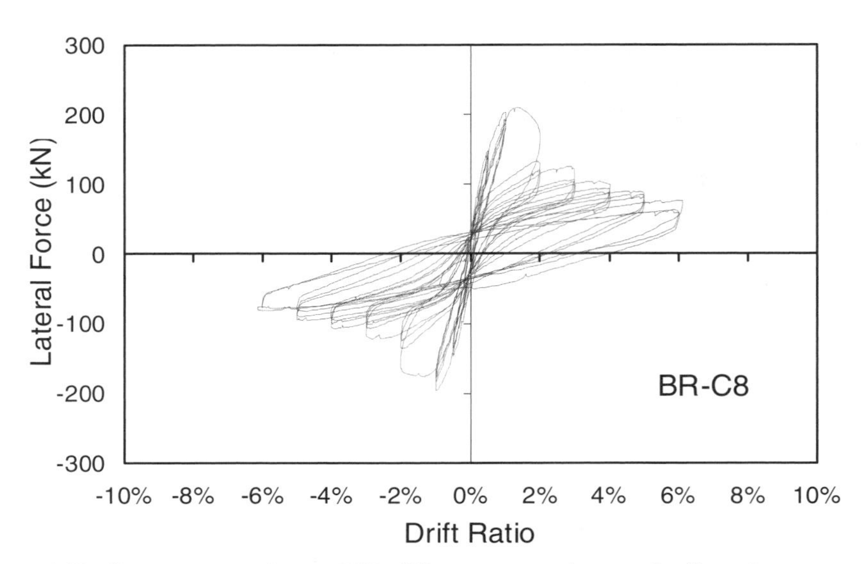

a) Reference specimen BR-C8, representing as built columns

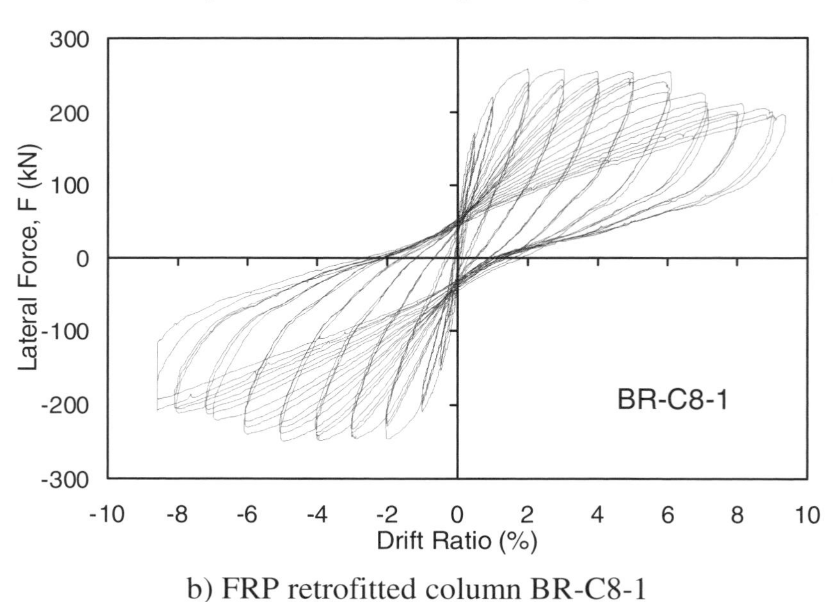

b) FRP retrofitted column BR-C8-1

Figure 7. Flexure dominant circular columns with spliced reinforcement

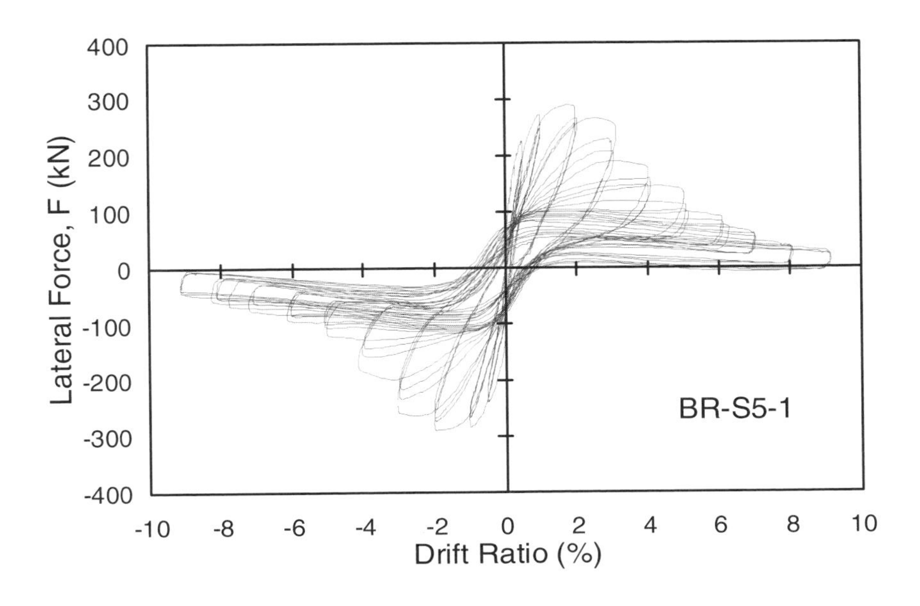

Figure 8. FRP retrofitted flexure-dominant square column with spliced reinforcement

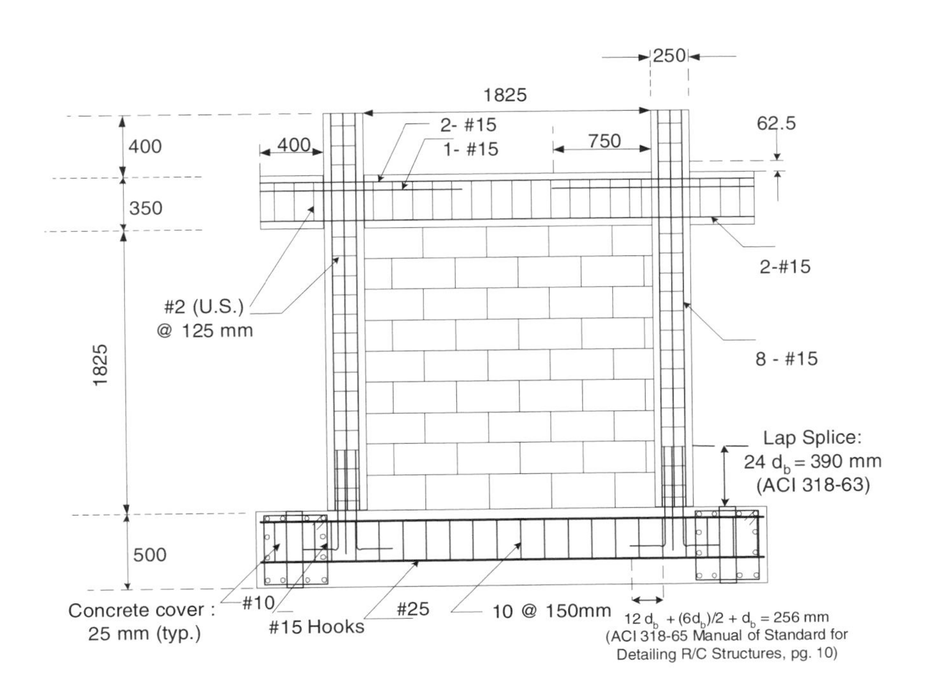

Figure 9. Details of reinforced concrete frames with infill masonry walls

Figure 10. Construction of frames and infill masonry walls

a) Application of epoxy filler

b) Application of FRP sheets

Figure 11. Retrofitting masonry infill wall with FRP sheets

a) FRP anchor

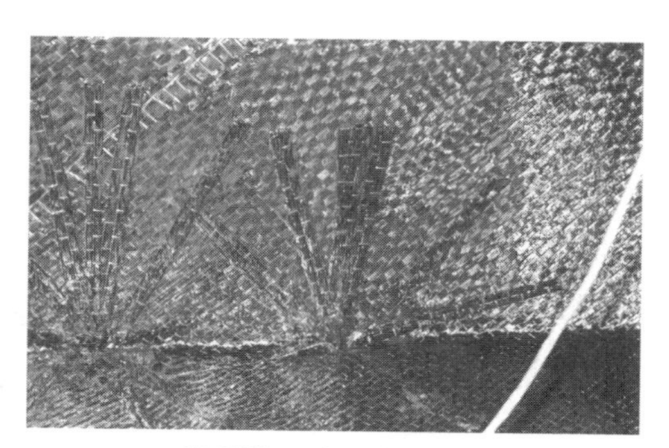

b) FRP anchors in place

Figure 12. Application of FRP anchors

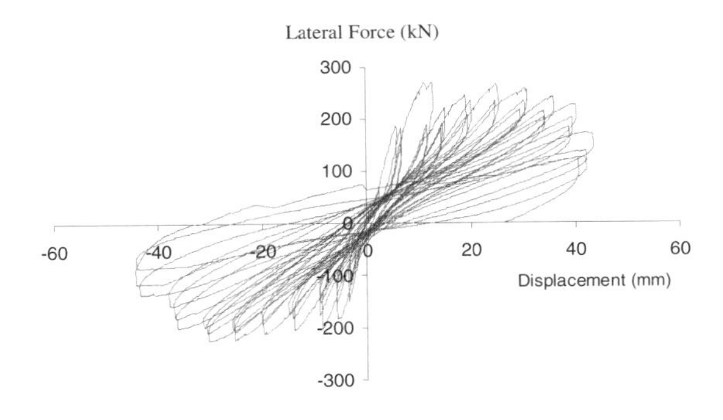

Figure 13. Hysteretic relationship of unretrofitted frame-wall assembly

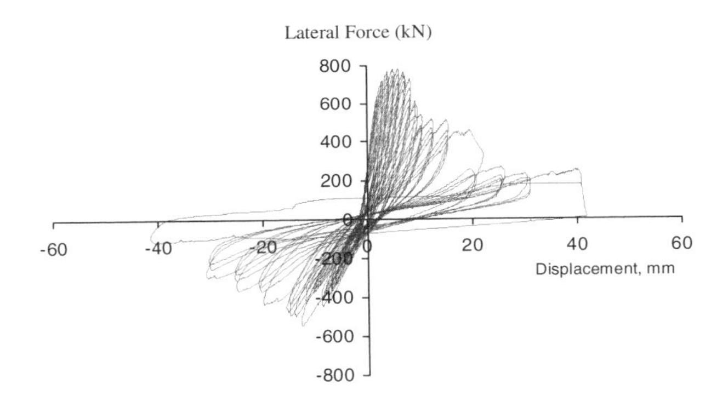

Figure 14. Hysteretic relationship of retrofitted frame-wall assembly

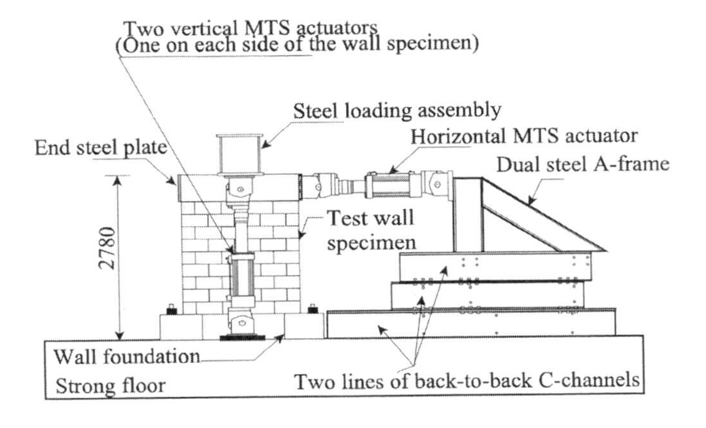

Figure 15. Setup for wall tests

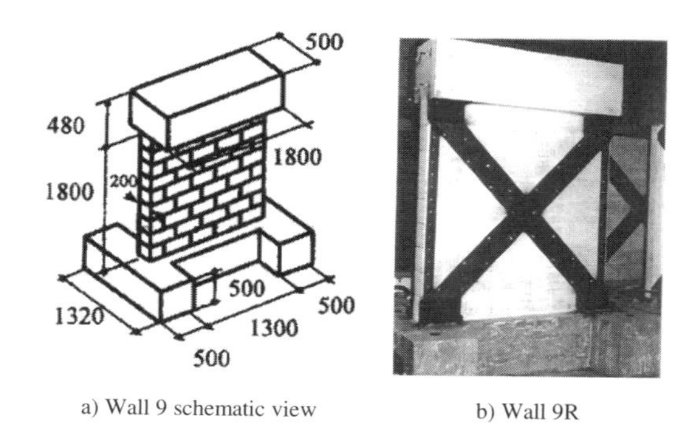

a) Wall 9 schematic view b) Wall 9R

Figure 16. Unreinforced masonry walls; Wall 9 unretrofitted and Wall 9R retrofitted

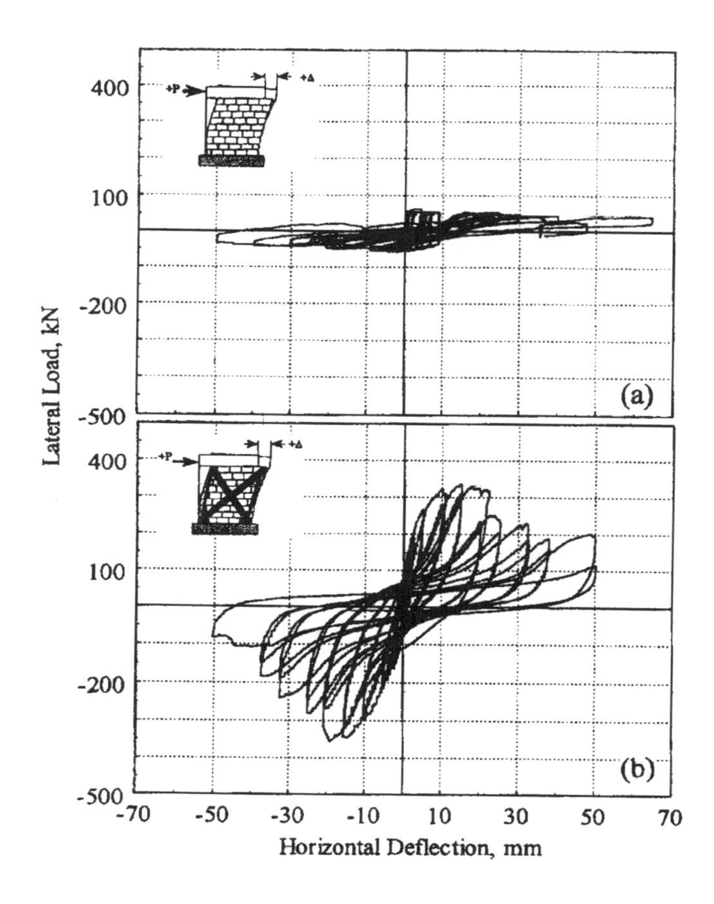

Figure 17. Hysteretic relationships of unreinforced masonry walls; (a) unretrofitted Wall 9; (b) retrofitted Wall 9-R

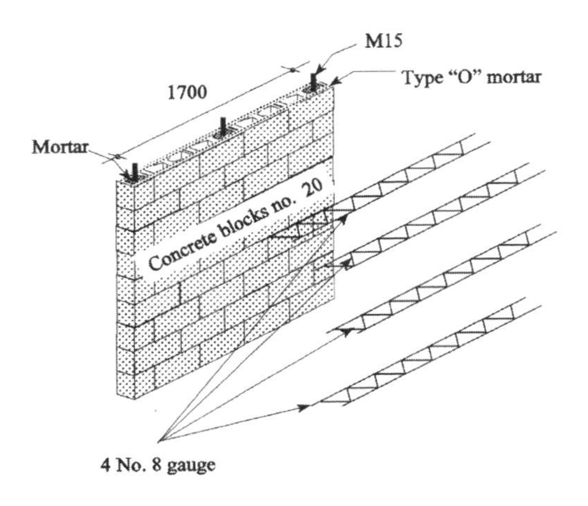

Figure 18. Reinforcement details in reinforced masonry walls

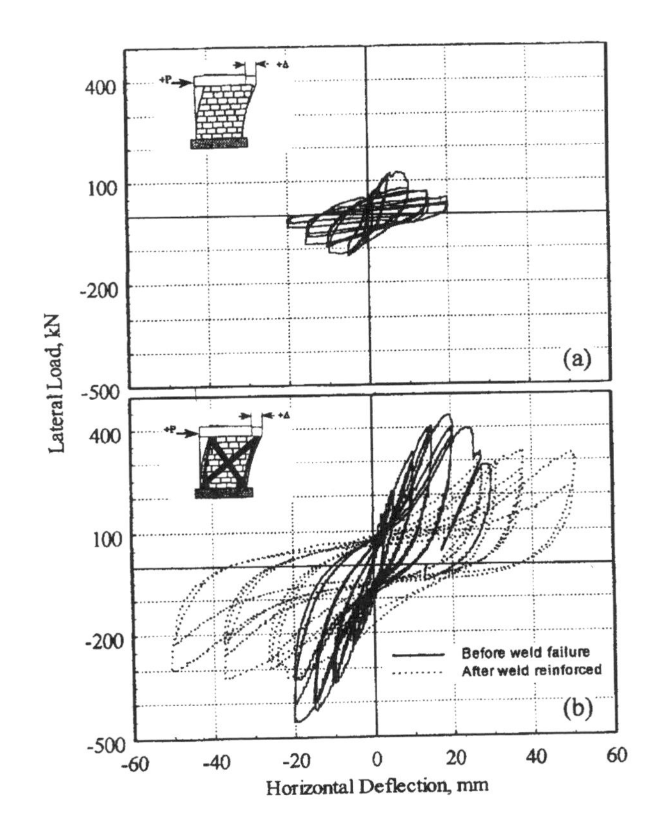

Figure 19. Hysteretic relationships of reinforced masonry walls; (a) unretrofitted Wall 10; (b) retrofitted Wall 10R

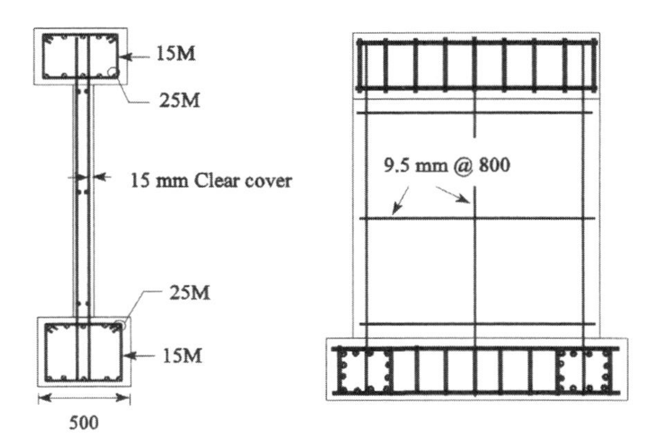

Figure 20. Reinforcement arrangement used in Wall 11 and Wall 11R

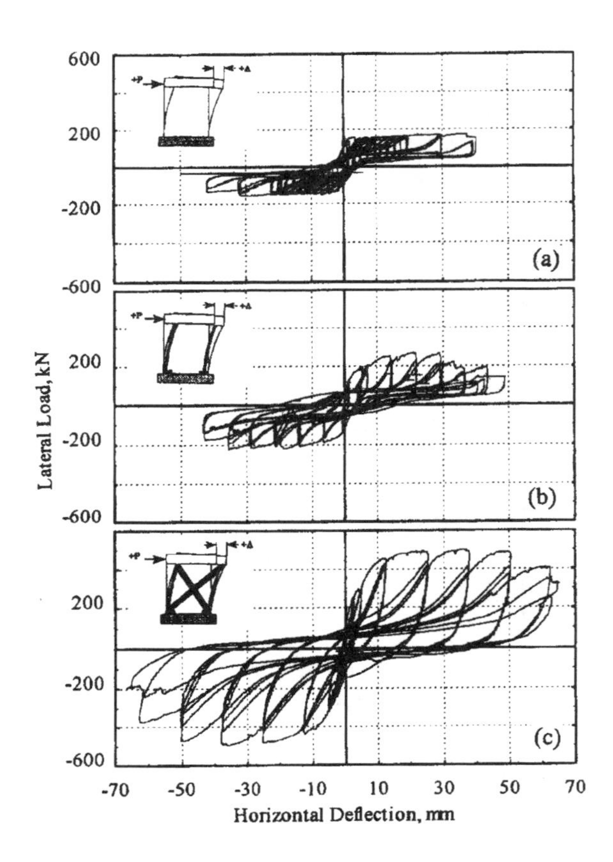

Figure 21. Hysteretic relationships for reinforced concrete walls; (a) reference Wall 11; (b) repaired Wall 11RP; (c) retrofitted Wall 11R

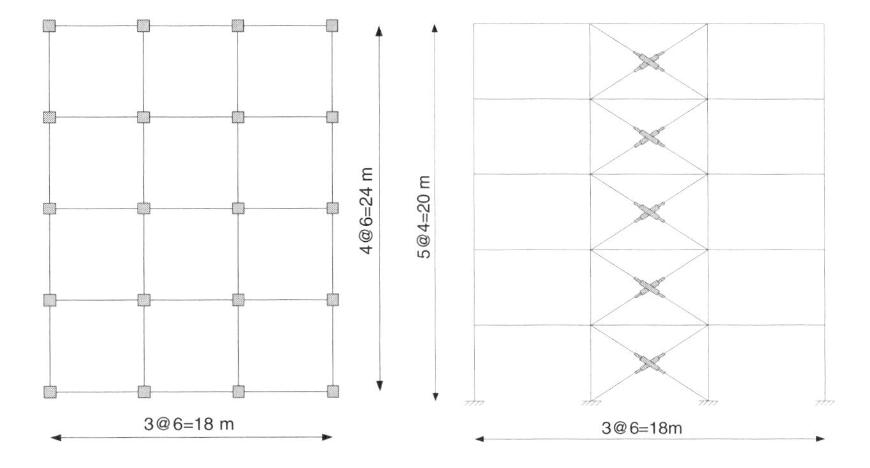

Figure 22. Plan and elevation of the structure selected for analysis

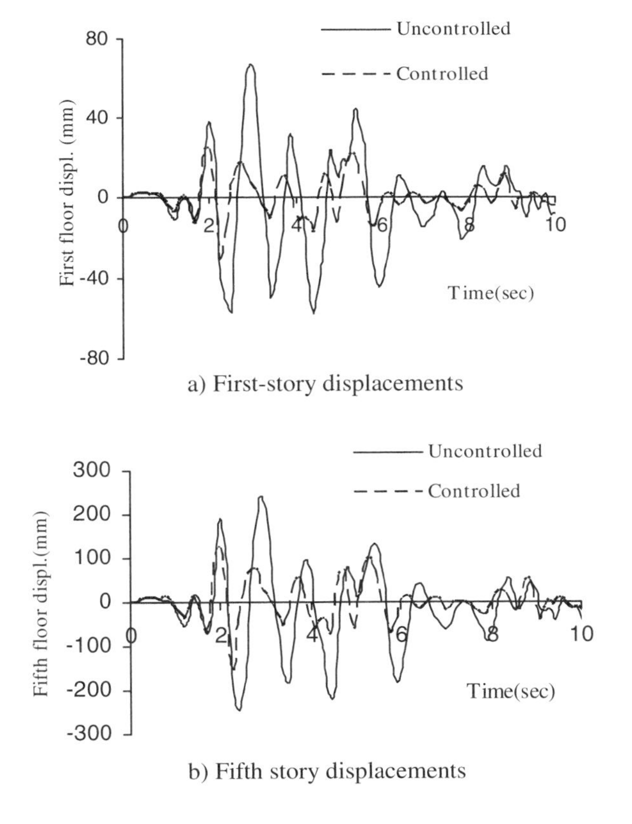

Figure 23. Comparisons of first and fifth story displacements with and without control

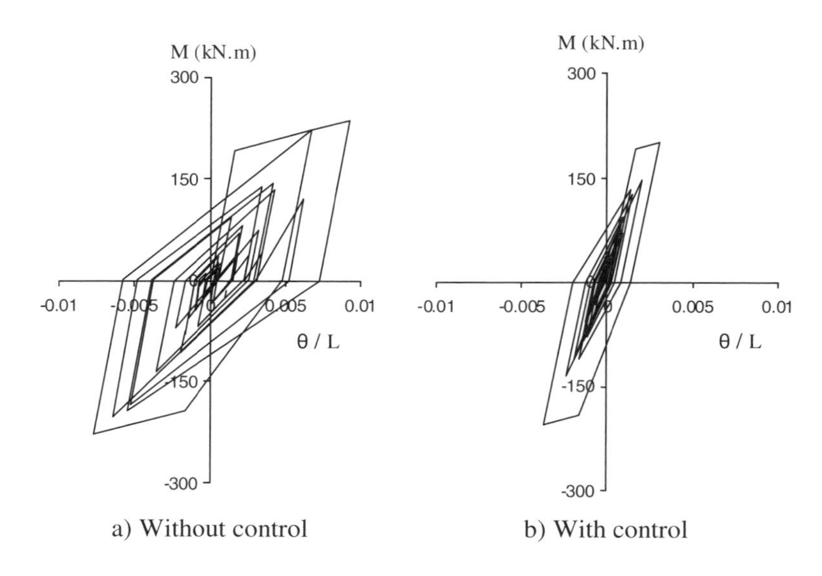

Figure 24. Comparisons of hysteretic relationships (moment vs chord angle per shear span) for a first-story column with and without active control

EXPERIMENTAL AND ANALYTICAL INVESTIGATION OF 1/3-MODEL R/C FRAME-WALL BUILDING STRUCTURES

PART I (Model Design and Analytical Evaluation of Dynamic Characteristics of the Model)

M. Garevski, A. Paskalov, K. Talaganov and V. Hristovski
Institute of Earthquake Engineering and Engineering Seismology, IZIIS, University "St. Cyril and Methodius", Skopje, Republic of Macedonia

Abstract: This study is a part of the research program that is being carried out within the NATO Science for Peace Project SfP977231 "Seismic Assessment and Rehabilitation of Existing Buildings" Within the framework of the research, a new retrofitting method is being experimentally investigated. The dynamical behavior of brick infilled reinforced concrete frames strengthened by Carbon Fiber Reinforced Polymers (CFRP) is also being studied. A 1/3- model of a R/C frame-wall structure with infill walls' behavior has been analytically investigated. The construction of the test models is underway.

Key words: model design, shaking table test, dynamic analysis, seismic input

1. INTRODUCTION

Three large scale test structures will be tested on the shaking table, under real earthquake excitations, with the aim of verifying the validity and applicability of the proposed Carbon Reinforced Fiber Polymers (CFRP) retrofitting technique [1]. The test specimens will consist of reinforced concrete frame structures with brick infill walls. The specimens will be

S.T. Wasti and G. Ozcebe (eds.), Seismic Assessment and Rehabilitation of Existing Buildings, 487–498.

scaled to 1/3 of the dimensions of a real structure, while keeping the same original material characteristics.

2. DESCRIPTION OF THE SHAKING TABLE

The shaking table on which structural models are to be tested is a prestressed reinforced concrete plate, 5.0 × 5.0 m in plan. Four vertical hydraulic actuators support the table with a total force capacity of 888 kN, located at the four corners at a distance of 3.5 m in both orthogonal directions. The table is controlled in the horizontal direction by two hydraulic actuators at a distance of 3.5 m with a total force capacity of 850 kN. The total weight of the shaking table is 330 kN. The natural frequency of the shaking table is 48 Hz for the maximum loading mass placed at the center of the table.

The maximum weight of the tested model should not exceed 400 kN with a total height of 6.0 m. The maximum applied accelerations are: vertical 0.50 g and horizontal 0.70 g, with a maximum displacement in the vertical direction ± 0.050 m and in the horizontal direction ± 0.125 m. The frequency range is 0 - 80Hz. The gravity load of the table and model is sustained by a static support nitrogen system with a bearing capacity of 720 kN. All actuators are supported by a rigid reinforced concrete structure. The analog control system controls the displacement, velocity, pressure and acceleration of all actuators. For analog control, a servo controlled closed loop system, a 4 channel A/D converter with analog output and 64 channel A/D converter for analog input and a 2 channel frequency analyzer HP 3582A are installed. A package of computer programs for control and acquisition is used. A function generator, a random noise generator or a computer (through D/A converter) can be used as input. Accelerometers, displacement transducers and strain gages can be used for measuring. The generated earthquake motion, accelerations and displacement time histories are stored in digital form (digital control and data acquisition) in a computer. Collection of data is provided by a fast data acquisition system.

3. MODEL DESIGN

The design of specimens for shaking-table tests poses a problem, due to the difficulty of satisfying all similitude requirements for structural models. In particular, it is difficult to find suitable substitutes for construction materials that satisfy similitude requirements in the inelastic range of behavior. In most cases, prototype materials must be used in the construction of the model, and ballast is added to increase the density of the model to satisfy similitude requirements for the materials. Therefore, most models used in shaking-table experiments are not true replicas, but they are adequate for predicting prototype behavior, provided that the requirements for geometrical similitude and earthquake scaling are satisfied. As the scale of the model is reduced, its fabrication becomes more difficult.

Similarity between a prototype structure and a small-scale model is maintained by proper scaling of the significant physical quantities that govern the structural behavior. The design of a model is initiated by performing a dimensional analysis. In a dimensional analysis, the first step is to identify the significant variables that affect the structure (Table 1).

The selection and production of materials is probably the most difficult step in a successful research investigation using models. Exact duplication of the prototype material properties is required if the model is to simulate elastic and nonlinear inelastic behavior of the structural system up to failure. Since the mass characteristics of the model are an important parameter, the distortion of the mass density can be minimized by using artificial mass simulation. In artificial mass simulation, mass is added to the structure in such a manner that it will not appreciably change its stiffness. Care must be taken to mount the additional mass adequately without affecting the structural system. The added mass should be distributed over locations where it would normally act. The proper distribution is often difficult, as geometrical restrictions often dictate where the additional mass can be placed. For this project the artificial mass simulation type was used, following the idea envisaged in the project of using the original prototype structure materials [5].

4. TESTING MODELS

Three separate tests are anticipated in this project (Figure 2).

The first specimen consists of a RC 3D frame structure with infill walls. The second specimen will have additional CFRP straps on the faces of the infill walls. After studying the behavior of the second specimen, the third test specimen will be determined. The dimensions of the test specimen in the longitudinal direction are: the side bays have a width of 1600 mm; the middle bay (infill brick wall) is 1000 mm wide; the height of the first story is 1500 mm and the second story 1000 mm. In the transversal direction, the frames are 1400 mm apart. The columns are 110×110 mm and beams are 110×150 mm (Figure 3) [2].

The model rests on a 250 mm thick foundation plate which is 5000 × 2200 mm in plan (Figure 4).

In the beams and columns 4ϕ8mm plain bars will be used as longitudinal reinforcement. The beam top reinforcement is extended into the exterior column and bent 90° downward. The beam bottom reinforcement is extended into the external column and its ends are hooked inward. The middle bay of the frame will have 9 stirrups and each side bay will have 15 stirrups. The length of a transversal beam is 1400 mm with 4ϕ8mm longitudinal reinforcement and 13 stirrups. The column reinforcement is spliced at the floor levels with a splice length of 20ϕ (160 mm). The longitudinal reinforcement of all columns is spliced at the foundation level. Four dowels (4ϕ8mm) will be cast with the foundation and will have a length of 40ϕ (320 mm). The longitudinal reinforcement of all columns will be lapped with these bars. Each column has 4ϕ8mm longitudinal plain reinforcing bars and 14 stirrups with ϕ4mm. The stirrup spacing is 100 mm both for columns and beams. The stirrup ends have 90° hooks. The slabs (50 mm thick) are reinforced with ϕ6/100 mm top and bottom reinforcement in both directions. The reinforcing details of the frames and the model display are given in figures 5-7.

The target compressive concrete strength for the frame and the foundation will be set at 15 MPa and 30 MPa, respectively. Standard cube and cylinder test specimens will be taken. The test cubes are 150 mm in all three dimensions. The test cylinders have 150 mm diameter and 300 mm height. They will all be kept under the same conditions as the test model specimens [6].

5. ANALYTICAL MODEL

Analytical studies were undertaken to evaluate the linear response of the model using the computer programs COSMOS/M and TOWER. The mass of the model frame with infill walls and foundation plate is 9.44 t. An additional mass of 8.0 tons is added to the model to satisfy the similitude requirements. Twelve ingots (i.e. 4.8 t) are placed at the second floor slab and eight ingots (i.e. 3.2 t) at the first floor slab. A 3D FE model consisting of 849 nodes and 928 elements (BEAM3D; SHELL4) is made. The material characteristics (Table 2) are as follows:

Figures 8-9 show the orthogonal view of the 3D model, as well as, the 3D FE model used for analyses.

After defining the model for analysis the first step is to check its static deformations at some points. The deformed shape of the model and three selected points for checking are shown (Figure 10 and Table 3).

The next stage is calculation of the mode shapes (Table 4 and Figures 11-12).

5.1 Ground acceleration input excitation

The Marmara region earthquake in Izmit (August 17, 1999; E-W direction) will be applied in the longitudinal direction of the model. The original record consists of 6000 points with a time step $\Delta t=0.005$ sec. To follow the similitude requirements it is necessary to change the time step increment to 0.00288 sec, i.e., the total span of the earthquake record that will be applied on the shaking table is 17.32 sec. The acceleration values of the record will not be altered (Figure 13)

6. INSTRUMENTATION

The following instrumentation will be used for testing of the specimens: LVDTs (Linear Variable Displacement Transducer), accelerometers, LPs (Linear Potentiometers), clip gages and strain gages for reinforcement and concrete. Longitudinal deformations will be measured by three LPs. Two LPs are to be placed on the second floor slab of the frame model in each corner and the third LP on the first floor slab in the transversal mid span. Eight LVDTs for measuring shear deformations of the infill walls and two clip gages for measuring the curvature of the column bases will be used. A total of eight accelerometers will be utilized in the dynamic tests (one accelerometer placed directly on the shaking table to check the input motion; one accelerometer will be placed on the foundation plate to see whether slip motion between the shaking table and the test specimen exists, i.e. whether the test specimen is ideally fixed; three accelerometers will be mounted on each floor slab- two for longitudinal motion and one for transversal motion). The intention is to check whether any torsion appears. Twenty-eight steel reinforcement strain gages will be placed (sixteen on the reinforcement on the column bases-two strain gages in each column base and twelve in the longitudinal beams-mid spans and on the beam-column connection) (Figure 14).

6.1 Test procedure

Following the curing period of the concrete, the specimen without infill walls will be placed on the shaking table. The first step will be to load the model with the ballast to measure the static deflection of the beams and the slabs, in order to compare it with the analytical calculations. Afterwards, a small shaking table will be placed on the top slab to excite the model structure, in order to measure the mode shapes and the corresponding eigen frequencies. Checking of the mode shapes will also be done by using the ambient vibration technique. The same three experimental steps will be undertaken after construction of the infill walls, to measure the dynamic characteristics of the complete model. The next step will be application of a low-level sinusoidal loading time history for measuring the linear behavior of the model. This step is needed for checking the integrity of the model and its linear elastic response. Next, real earthquake record scaled appropriately according to the similitude requirements will be applied. The experimental results from the model study will compared to the analytical calculations of the model. The same test procedure will be applied to the second and the third test models.

ACKNOWLEDGMENT

The financial support from NATO Science for Peace Project SfP977231 is gratefully acknowledged.

REFERENCES

1. Ozcebe G. et al., Seismic Assessment and Rehabilitation of Existing Buildings, NATO SfP977231, 2000.
2. Canbay E., *Contribution of RC Infills to the Seismic Behavior of Structural Systems*, Ph.D. Thesis, Department of Civil Engineering, METU, Ankara, Turkey, 2001.
3. *COSMOS/M, Finite Element Analysis System;* User Guides, Volumes 1~4, Version 2.0, SRAC Inc, California, USA, 1994.
4. *TOWER, 3D Model Builder.* User's Manuals, Version 4.1, Radimpex, Belgrade, Yugoslavia., 2001.
5. Caccese V. and Harris Harry G., *Earthquake Simulation Testing of Small-Scale Reinforced Concrete Structures*, ACI Structural Journal, V. 87, No.1, January-February 1990.
6. NATO SfP977231 International Advanced Research Workshop, *Seismic Assessment and Rehabilitation of Existing Buildings*, Unofficial Draft Restricted Circulation, Izmir, Turkey 13-14 May 2003.
7. Aktan A.E., Bertero V.V., Chowdhury A.A., Nagashima T., Experimental and Analytical Predictions of the Mechanical Characteristics of a 1/5-Scale Model of a 7-Story R/C Frame-Wall Building Structure, EERC Report No. UCB/EERC-83/13, University of California, Berkeley, USA, June 1983.

Table 1. Model similitude requirements

Scaling Parameters	Model Type Scale Factors		
	True Replica Models	Artificial Mass Simulation	Gravity Forces Neglected
Length, l_r	l_r	l_r	l_r
Time, t_r	$l_r^{1/2}$	$l_r^{1/2}$	$l_r(E/\rho)_r^{1/2}$
Frequency, f_r	$l_r^{-1/2}$	$l_r^{-1/2}$	$l_r^{-1}(E/\rho)_r^{1/2}$
Velocity, v_r	$l_r^{1/2}$	$l_r^{1/2}$	$(E/\rho)_r^{1/2}$
Acceleration, a_r	1	1	$l_r^{-1}(E/\rho)_r$
Strain, ε_r	1	1	1
Stress, σ_r	E_r	E_r	E_r
Modulus of Elasticity, E_r	E_r	E_r	E_r
Displacement, δ_r	l_r	l_r	l_r

Table 2. Material characteristics

Material	Modulus of Elasticity E (kPa)	Poisson ratio ν	Density ρ (t/m^3)
concrete 15MPa	2.55×10^7	0.2	2.5
concrete 30MPa	3.15×10^7	0.2	2.5
brick	0.70×10^7	0.2	1.8

Table 3. Vertical displacements

Displacements [m]	COSMOS/M	TOWER
Node 9	-0.09×10^{-3}	-0.07×10^{-3}
Node 5	-0.382×10^{-3}	-0.33×10^{-3}
Node 181	-0.28×10^{-3}	-0.30×10^{-3}

Table 4. Mode shape analysis

Periods of vibrations [sec]	COSMOS/M	TOWER
T_1	0.2361	0.2388
T_2	0.1113	0.1093
T_3	0.0667	0.0657

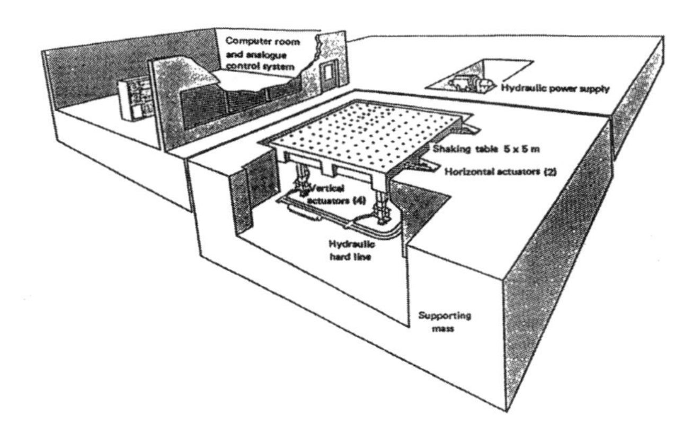

Figure 1. Display of the shaking table

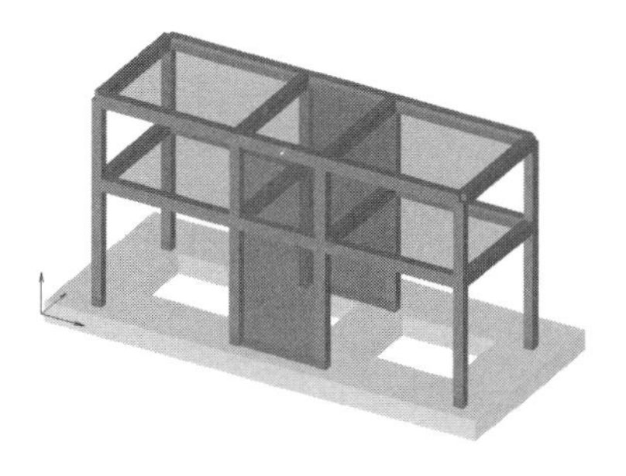

Figure 2. 3D display of the testing model specimen

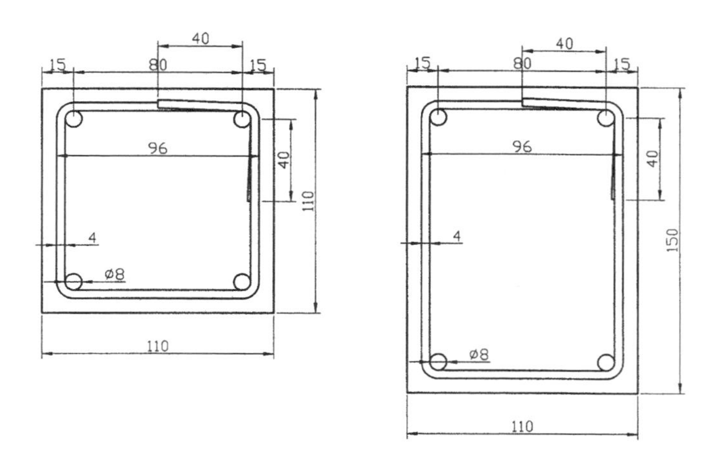

Figure 3. Cross sections of columns and beams

Figure 4. Display of the foundation plate

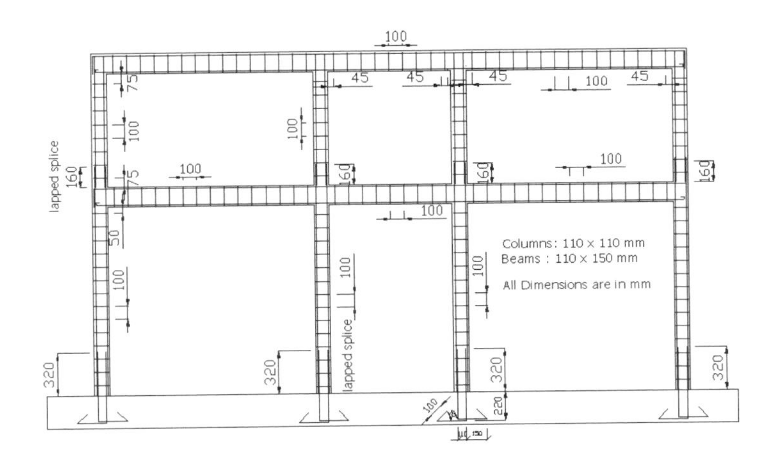

Figure 5. Reinforcing details of the longitudinal frame

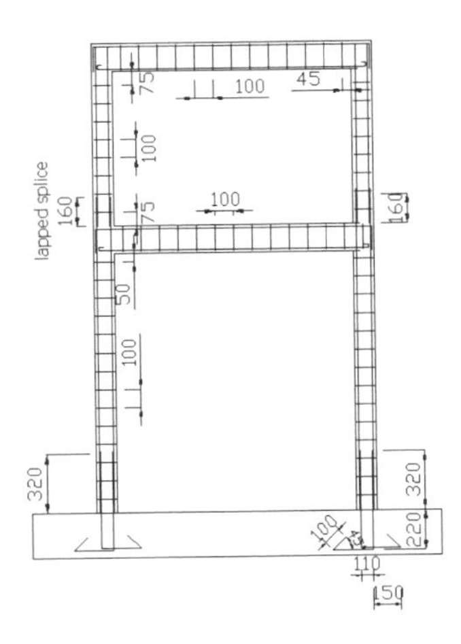

Figure 6. Reinforcing details of the transversal frame

Figure 7. Display of the test specimen

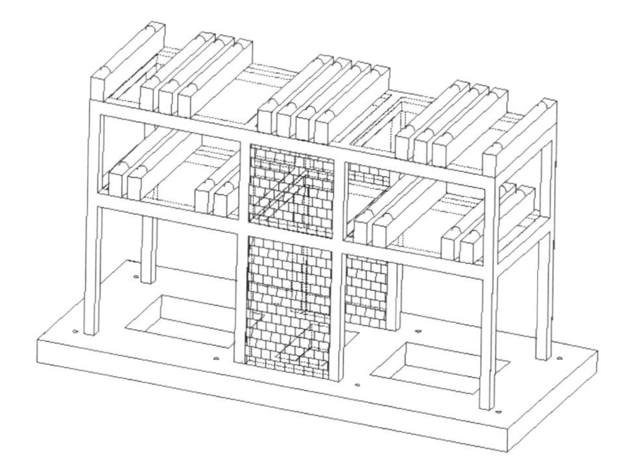

Figure 8. 3D view of the applied ballast to the model

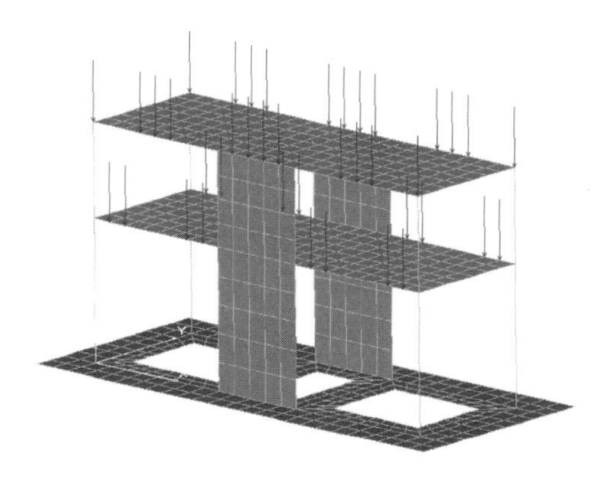

Figure 9. 3D FE model with application points for the ballast

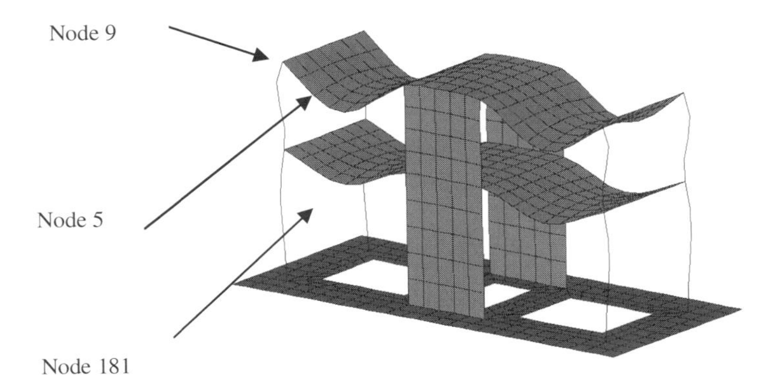

Figure 10. 3D deformed model

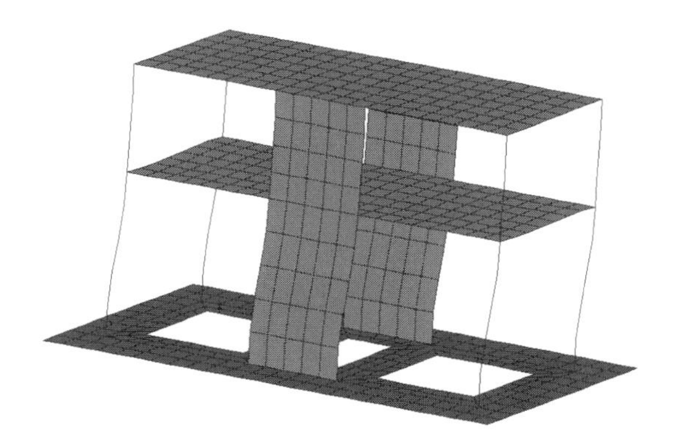

Figure 11. Mode shape 1 [T1=0.2361 sec]

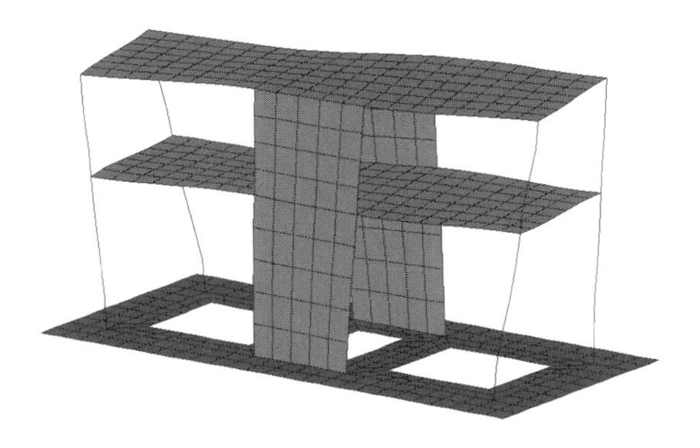

Figure 12. Mode shape 2 [T2=0.111 sec]

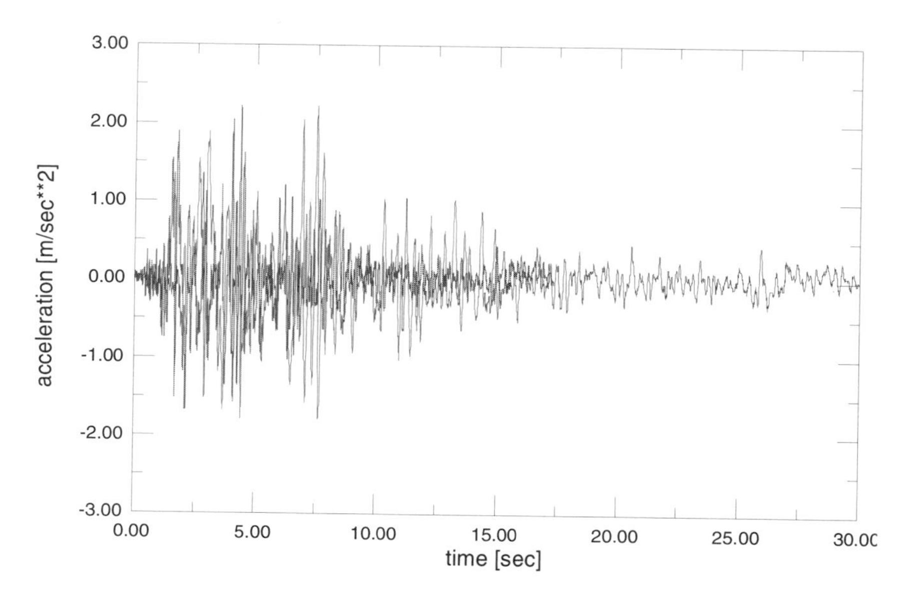

Figure 13. Original and altered Izmit E-W earthquake record (a_{max}=2.151 m/sec^2)

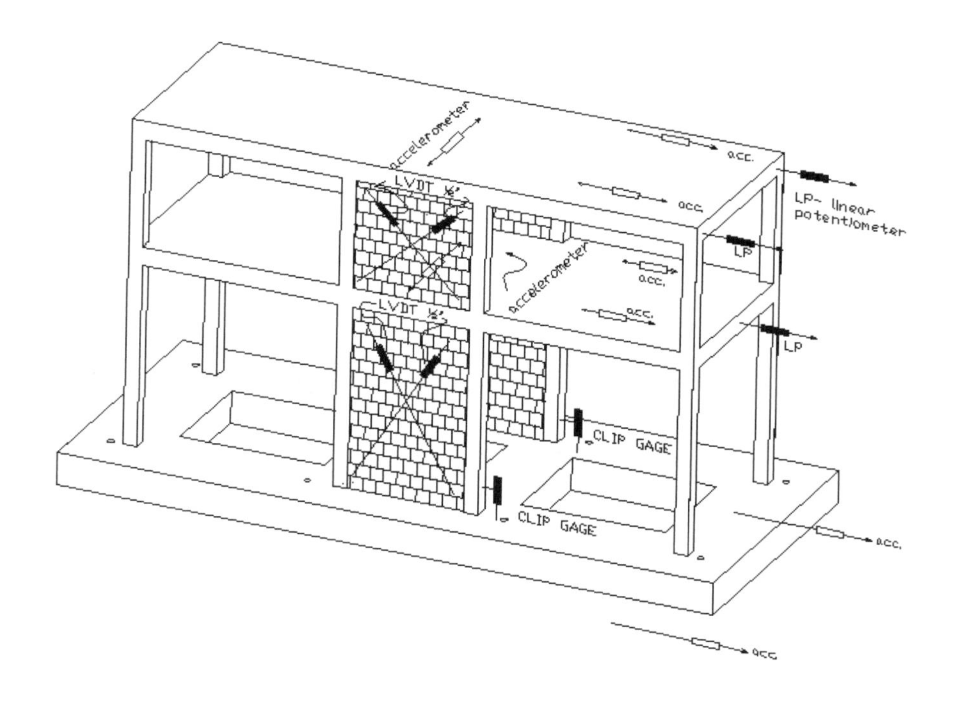

Figure 14. View of part of the instrumentation

EXPERIMENTAL AND ANALYTICAL INVESTIGATION OF 1/3-MODEL R/C FRAME-WALL BUILDING STRUCTURES

Part II (Nonlinear Analytical Prediction of Structural Behavior)

M. Garevski, V. Hristovski, A. Paskalov and K. Talaganov
Institute of Earthquake Engineeing and Engineering Seismology, IZIIS, University "St. Cyril and Methodius", Skopje, Republic of Macedonia

Abstract: The paper presents the results referring to the nonlinear analytical prediction of the behavior of a specimen to be tested at IZIIS (a 3D reinforced concrete frame-structure with infill walls) within the scope of the NATO-Science for Peace project SfP977231 "Seismic Assessment and Rehabilitation of Existing Buildings". The push-over capacity analysis as well as dynamic time-history analyses for a given history of ground accelerations have been performed, emphasizing the influence of the infill walls on the overall structural response. This analytically predicted structural response shall serve as a basis for the experimental investigations to be conducted at IZIIS.

Key words: push-over analysis, nonlinear time-history analysis, finite element method

1. INTRODUCTION

Based on the information presented in Part I of this paper regarding the shaking-table testing of a 2-story specimen of a frame structure with infill walls (the first test specimen), under real earthquake excitations with the purpose of verifying the validity and applicability of the proposed retrofitting techniques, a so called "blind" analytical prediction of the

S.T. Wasti and G. Ozcebe (eds.), Seismic Assessment and Rehabilitation of Existing Buildings, 499–516.

structural behavior has been performed. A push-over analysis for the evaluation of the lateral capacity and ductility of the structure has first of all been performed followed by dynamic time history response analyses under given ground acceleration records. The information obtained by these simulations has been used for the final detailed model design of the first test specimen, as well as for studying the influence of the infill walls on the overall structural response.

2. ANALYTICAL PREDICTION OF THE LATERAL CAPACITY AND DUCTILITY OF THE DESIGNED SPECIMEN

The analytical prediction of the lateral capacity and ductility of the designed specimen – RC frame with masonry infill, i.e. the push-over analysis has been performed using the software package FELISA/3M [1], that has originally been developed at IZIIS and is a general purpose program for linear and nonlinear static and dynamic structural analysis. Using a bi-axial hypo-elastic fracture constitutive model for concrete and elastic-plastic model for steel, 2D finite element analysis has been performed using the symmetry of the structure in plan, so that only one frame has been analyzed subjected to in-plane lateral and normal forces (Figure 1). Cracking has been modeled using the smeared approach, without emphasizing any particular crack. Steel has also been treated in a smeared, layered manner. The adopted uni-axial constitutive relation for concrete in compression is based on Saenz's relation [2]. The effect of confinement due to lateral reinforcement has not been taken into account in the analytical model. However, the influence of tension-stiffening during cracking has been included using the Shirai's-function [3]. The Kupfer yield curve has been adopted as a failure criterion; however, the influence of concrete strength degradation in compression due to cracking has been taken into account [4]. The bond between the steel and concrete has been assumed as perfect.

The concrete and masonry properties adopted in the analysis have been shown in Table 2.1.

In Table 1, E is Young's modulus of elasticity, fc is the compressive strength, ft is the tensile strength, ν is Poisson's ratio and ρ is density. The steel is modeled as smeared, with E=200,000 MPa and fy=240 MPa. The analytical model consists of 298 2D iso-parametric plane-stress elements with 8 nodes (Figure 1), using 3×3 Gaussian integration quadrature scheme. The total number of nodes is 1097.

During the push-over analysis, first the gravity load due to own weight (columns, girders and slab) has first been applied, together with the additional mass of 80 kN (40 kN per frame), which has been added because of similitude requirements. This additional mass has been distributed in such a way that 3/5 has been placed on the upper floor and 2/5 on the first floor. The analysis due to total gravity load has shown that some slight cracks in girders are likely to occur (Figure 2a). However, is it assumed that some part of the tension stresses in concrete will be undertaken by the slab which has not been taken into account in the analytical model, so that, in reality, these cracks will be less apparent.

Once the full gravity load and load due to additional masses have been applied incrementally, the upper girder has been subjected to a lateral load from zero to the maximum calculated value of F_{max}=43.67 kN. This lateral load has also been incrementally applied. The progressive failure process until the maximum applied lateral force is shown in Figure 2. For the characteristic analysis steps, the crack paths and state of deformations have been simulated. The maximum horizontal displacement of 14.976 mm has been calculated, corresponding to the maximum applied lateral force. As can be seen from the figures, because of the presence of the central masonry infill wall, the concentration of stiffness has caused the middle columns and infill to be first overstressed and then cracked. As the lateral force increases, the cracked zone spreads, especially at the contact between the wall and the foundation structure. In reality, the contact between the infill wall and the foundation structure (as well as the RC frame) will not be specially detailed, so that it is assumed that this zone will be responsible for the failure in the wall. However, since this zone has not been specially modeled in the analyses, a refined response on the contact has not been obtained. Anyway, despite the lack of a more sophisticated model, the distribution of the cracks shows that such a tendency exists, and failure during the experimental test will probably occur due to shear-slip between the infill wall and the foundation slab. After the tests, it will probably be shown that the analytically calculated deformations are underestimated, because of this phenomenon, which has not been taken into account in the model.

The total force-displacement response during the progressive failure process is shown in Figure 3 with denoted stages, according to the characteristic ones shown in the previous Figure 2,. From the diagram, the achieved ductility regarding the displacements can be calculated and it is approximately du/dy = 6.

Regarding the reliability of the obtained results, due to the inaccuracy related to the material characteristics (they were not known in advance), it is likely that the experimental results could eventually differ from the analytical ones, depending on the differences between the assumed and the real properties of the materials achieved during the experiment.

This push-over analysis has been performed up to the maximum applied load, so that the descending branch of the force-displacement diagram has not been obtained. However, it is expected that it will be possible to obtain larger displacements during the shaking-table test for higher level of input ground-motion excitations. Since larger deformations would mobilize mechanisms like the dowel-action of reinforcement, bond-slip deterioration between steel and concrete and aggregate-interlock between cracked surfaces, and these phenomena could strongly influence the post-failure behavior. However, the applied model has not taken into account these mentioned phenomena concentrating on response only up to the maximum capacity which is considered to be a sufficient parameter for getting an insight into the eventual behavior of the specimen.

3. NONLINEAR DYNAMIC TIME-HISTORY PREDICTION

Analytical prediction of the dynamic time-history response due to base excitation in terms of given acceleration time history record has been performed using the computer program DRAIN-2D [3] for 2D nonlinear dynamic analysis of structures. Since the constitutive model for reinforced concrete in DRAIN-2D is based on a sectional M-N interaction ultimate curve, it was necessary to calculate these interaction diagrams for column and beam sections. The software package referred to as SECAP/3M, that has originally been developed at IZIIS [4] was used for that purpose. The results from the sectional Mu-Nu analyses are shown in Figure 4 and Table 2.

The time step of the applied ground acceleration time-history record, (Izmit EQ with peak acceleration a_{max}=2.151 m/sec^2) consisting of 6000 points, has been set to Δt=0.00288 sec in order to satisfy the similitude requirements (see Part I of this paper), without alteration of the acceleration values. This time-scaled excitation (which is to be applied on the shaking table, too) has been applied in the longitudinal direction of the analytical model. The analytical model of the analyzed structure consists of frame and panel elements and it is shown in Figure 5, together with the static loading conditions.

In the above Table, Nc is the ultimate axial force for pure compression, Mb and Nb are bending moments and axial forces respectively to balance point and Mo are bending moments for pure bending. The values for positive and negative diagrams are identical since the sections are symmetrically reinforced.

In order to estimate the influence of the masonry infill on the overall structural response, an analysis of the pure frame system without infill walls has first of all been performed. For the given dynamic ground excitation mentioned above (with factor f=1.0), the maximum displacement of 9 mm has been obtained. However, the strength properties of the masonry infill were not known prior to these analyses, so that it has been assumed that the shear modulus has a value of G=980 kPa and that the shear strength varies between τ_m=100 kPa to τ_m =200 kPa. Performing parametric analyses regarding the shear strength of masonry for values of 100, 150 and 200 kPa, as well as scaling the accelerations applying factors f=1.0, 1.1, 1.2, 1.3 and 1.4, some information has been gained about the possible behavior of the specimen structure to be tested under the shaking-table excitation. In addition, the phenomena of sliding and separation of the infill wall relative to the reinforced concrete frame elements (i.e., relative to the columns and foundation) have not been taken into account in the analytical models, so that it is supposed that the real response should be slightly softer (with greater displacements) than the analytically obtained one. The results from the performed parametric analyses, showing the possible plastic hinge formation, are shown in Figure 6. In the same figure, we can see the time histories of the obtained displacements on the top of the structures. It is characteristic that, for the pure frame without masonry infill, the accumulated plastic displacements are about 5 mm. The contribution of the infill to the seismic resistance is obvious, as can be seen further in this figure. So, even for very small shear strength of masonry (100 kPa), the maximum absolute displacement dropped from 9 (for pure frame) to 5 mm, with accumulated plastic deformations of 2 mm.

According to these analyses, if the shear strength of the built masonry infill varies between 100 and 200 kPa, and using the peak acceleration of 2.151 m/sec^2 (factor f=1.0), the maximum absolute displacements ranging between 3 to 5 mm are expected at the top. On the other hand, for increased intensity of the accelerations (from f=1.0 to f=1.4) and shear strength between 150 and 200 kPa, maximum absolute displacements ranging between 4 and 6 mm are expected on the top.

4. INFLUENCE OF MASONRY INFILL ON THE OVERALL BEHAVIOR OF THE STRUCTURE

The masonry infill of the actual test-specimen will be constructed after hardening of the reinforced concrete frames, in contact with them, but without any special connection to them. This type of masonry infill is very common in southern European countries where the seismicity is very high.

Generally, although the masonry infill is considered as a non-structural part of the structure, it has significant effects on the overall structural response. The presence of masonry infill contributes to the increased structural stiffness. Consequently, the fundamental period of natural vibrations is decreased and the base shear seismic force is increased. Also, the distribution of the lateral stiffness of the structure in elevation is modified. According to the performed time history analyses, it has been shown that a shear-type behavior has been observed. The great part of the seismic forces is carried by the infill, and also, which is very important, the ability of the structure with respect to dissipation of energy is significantly increased.

Since masonry infill in general is a rigid and non-ductile component compared to the RC frame, the more flexible the structural system (as in our case), the greater the above effects of the infill. Therefore, because of its inability to follow the large displacements developed in ductile frames, this structural component fails first, as has also been shown in the analyses. Generally, the failure modes in infill members are due to separation from the frame and very often x-cracks. In the push-over analysis performed for the purposes of this research, the separation from the base structure was clearly emphasized with developed large shear cracks. In any case, these components absorb and dissipate large quantities of seismic energy, acting as the first line of seismic defense of the structure.

5. CONCLUSIONS

From these analytical investigations, the following conclusions can be drawn:

1. For the purpose of prediction of the behavior of a 2-story specimen of frame structure with masonry infill walls to be tested on the shaking-table at IZIIS under NATO Science for Peace Project SfP977231, a push-over analysis for evaluation the capacity and ductility as well as nonlinear dynamic time-history analyses have been performed.
2. The push-over analysis has been performed using a sophisticated constitutive model for reinforced concrete and FEM based analytical model. This prediction has shown that the infill walls would act as the first line of structural resistance. The progressive failure process of the structure has been simulated up to the maximum calculated lateral force of F_{max}=43.67 kN with corresponding maximum displacement on the top of the frame u_{max}=14.976 mm, achieving a ductility of u_{max}/u_y=6.
3. The dynamic time-history analyses have been performed parametrically using different values for shear strength of the masonry infill since this parameter is not known in advance. Also, different factors for the input

acceleration time-history have been applied. According to these analyses, for shear strength of the infill between 100 and 200 kPa and intensity of the input acceleration record between f=1.0 to f=1.4, maximum displacements ranging between 3 and 6 mm have been calculated.

4. The results from the dynamic time-history simulations have been compared to the results obtained for the pure frame without infill walls. The influence of the infill walls has proved to be significant.
5. Since the exact properties of the materials have not been known prior to the analyses, it is expected that the real behavior of the specimen could vary, depending on the deviation of these parameters.
6. Since the phenomenon of slippage of the infill wall along the foundation structure has not been taken into account in the models, it is also possible that the real test of the structure give greater values for displacements under corresponding ground motion levels.
7. However, despite the described limitation of the applied analytical models for simulation of the behavior of the specimen to be tested, it is expected that the results from these analyses will offer a reasonable qualitative and quantitative insight into the mechanism of behavior of the designed specimen.

REFERENCES

1. Hristovski V. and Noguchi H., Comparative Study of FEM Based Reinforced Concrete Analytical Models and Their Numerical Implementation: Software Package FELISA/3M, Proceedings of the First fib Congress, Session 13: Failure Mechanism and Nonlinear Analysis for Practice, pp. 403-410, Osaka, Japan, October 2002
2. Saenz, L.P., Discussion of "Equation for the Stress-Strain Curve of Concrete" by Desayi and Krishnan, ACI Journal Vol. 61, September 1964, pp., 1229-1235
3. Noguchi, H., Ohkubo, M. and Hamada S., *Basic Experiments on the Degradation of Cracked Concrete under Biaxial Tension and Compression,* Proc. of JCI, Vol. 11, No.2, 1989, pp.323-326 (in Japanese)
4. Shirai N and Sato T., *Bond Cracking Model for Reinforced Concrete*, Trans. of the JCI Vol. 6, pp. 457-468, 1984
5. Powell G. H., DRAIN-2D Users Guide, Finite Element Analysis System; Report No. EERC 73-22, October 1973
6. Hristovski V., *Diagnosis of the States and Functionalities of Bridge Structures,* Ph. D. Thesis, October 1998, IZIIS, Skopje, Macedonia

Table 1. Material properties adopted in the analysis

Component	E (MPa)	fc (MPa)	ft (MPa)	ν	ρ (t/m^3)
INFILL	2,352	5.0	0.6	0.2	2.5
FRAME	25,500	15.0	1.5	0.2	2.5
FOUND.	31,500	30.0	2.4	0.2	1.8

Table 2. Obtained characteristic values of Mu-Nu diagrams

	Column	Beam
Nc [kN]	229.5	295.5
Mb (+) [kNm]	4.31	7.34
Nb (+) [kN]	87.31	119.58
Mb (-) [kNm]	-4.31	-7.34
Nb (-) [kN]	87.31	119.58
Mo (+) [kNm]	2.09	3.03
Mo (-) [kNm]	-2.09	-3.03
Nt [kN]	-48.00	-48.00

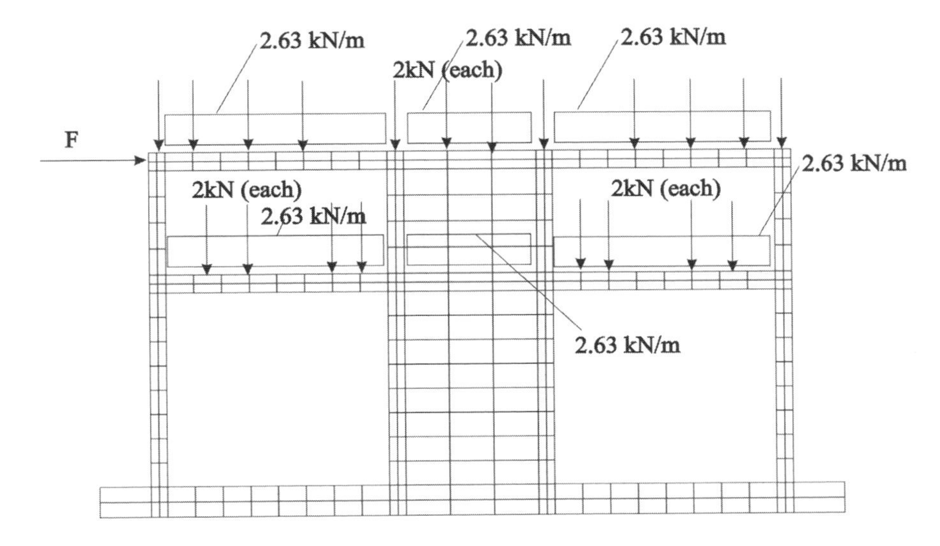

Figure 1. 2D display of the analytical model and loading

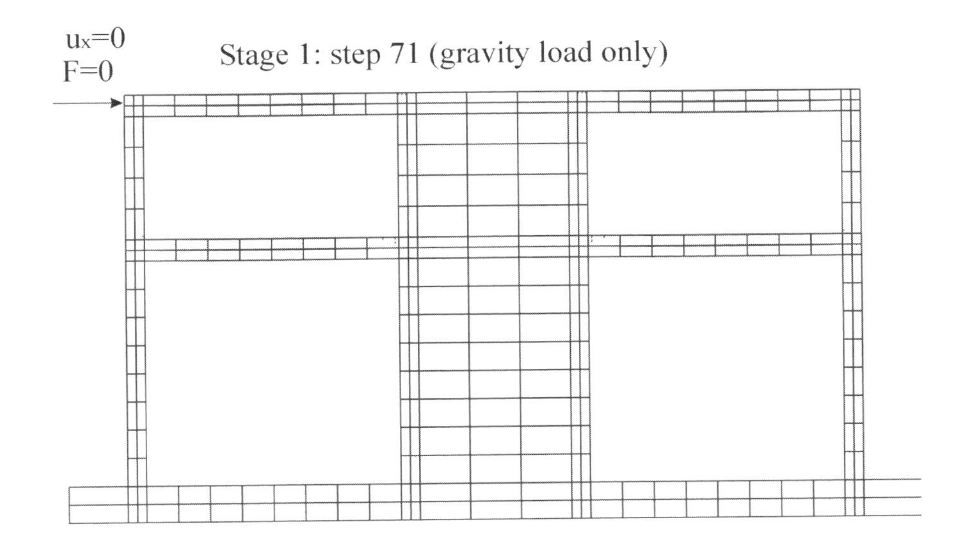

Figure 2a. State of deformations and cracks after applying the total gravity load

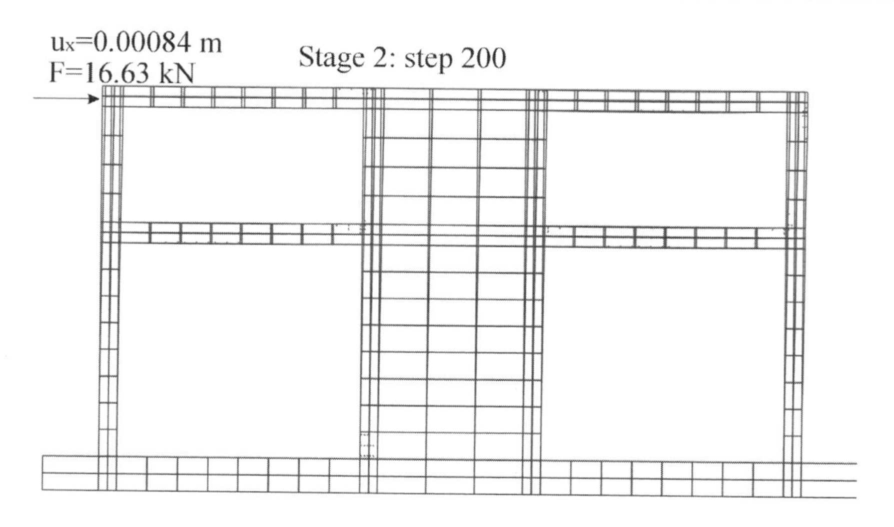

Figure 2b. State of deformations and cracks after applying lateral force F=16.63 kN

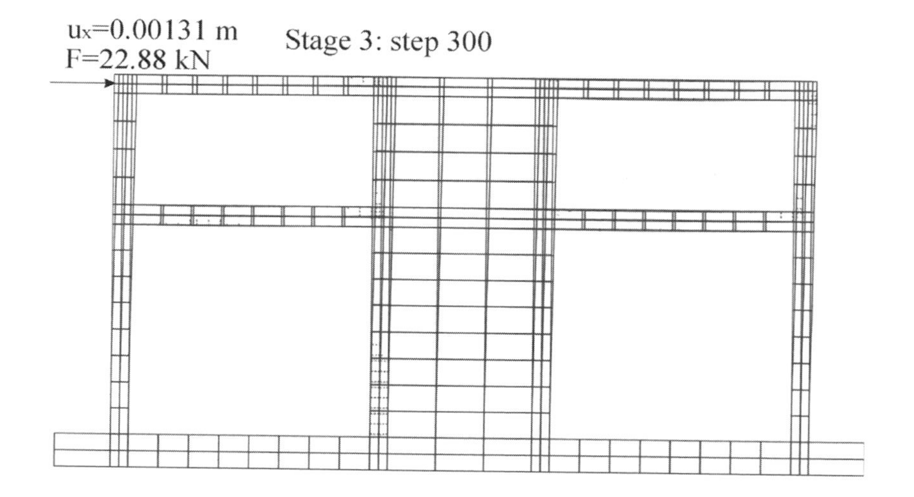

Figure 2c. State of deformations and cracks after applying lateral force F=22.88 kN

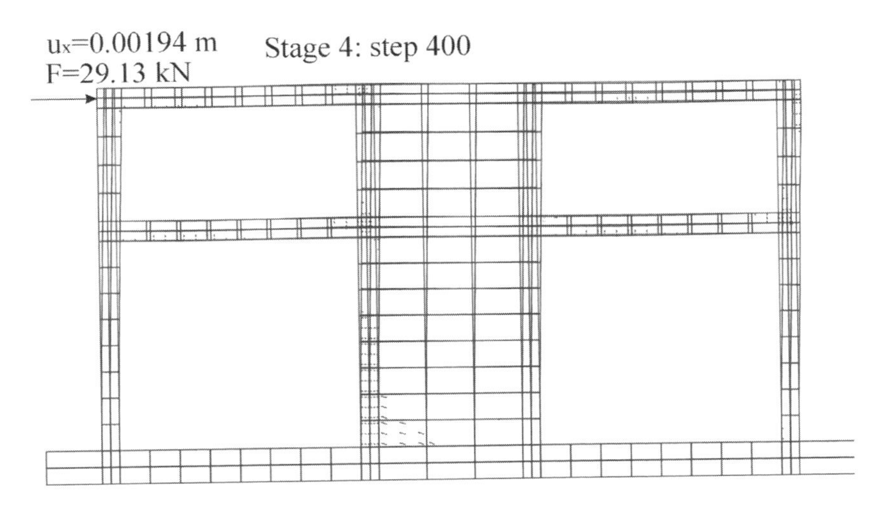

Figure 2d. State of deformations and cracks after applying lateral force F=29.13 kN

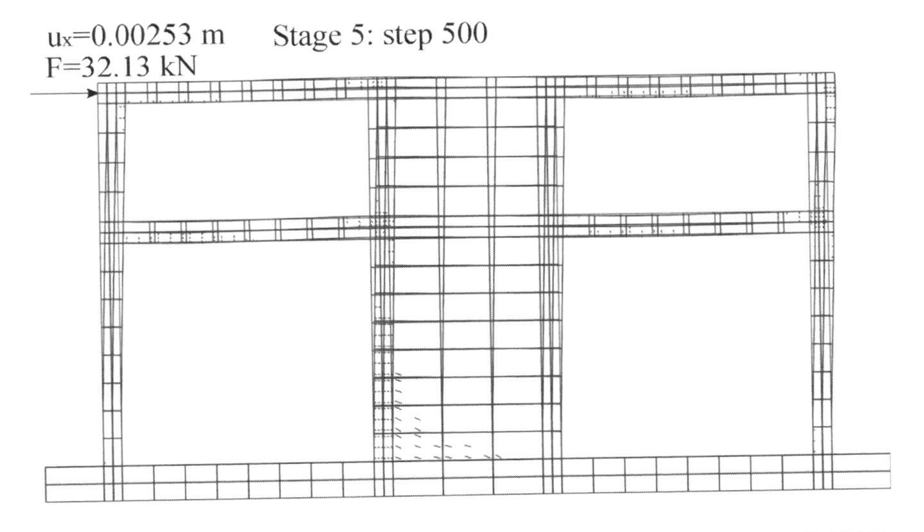

Figure 2e. State of deformations and cracks after applying lateral force F=32.13 kN

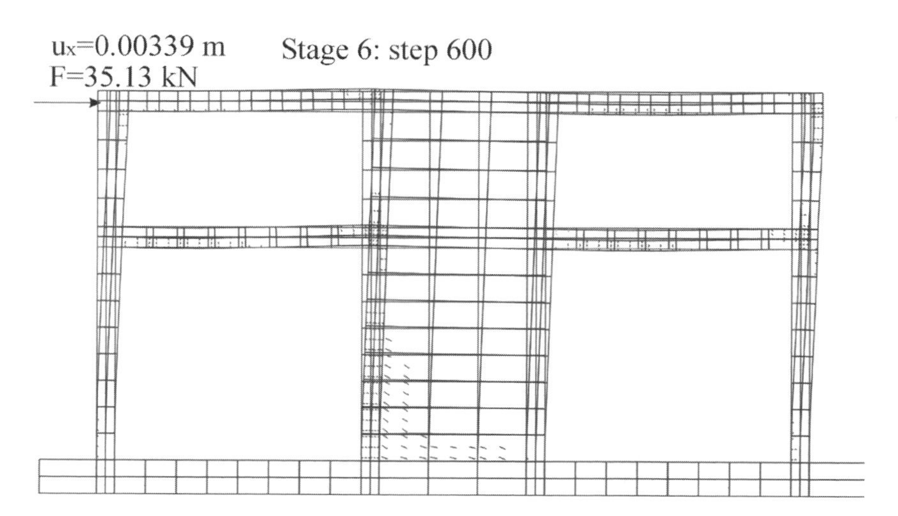

Figure 2f. State of deformations and cracks after applying lateral force F=35.13 kN

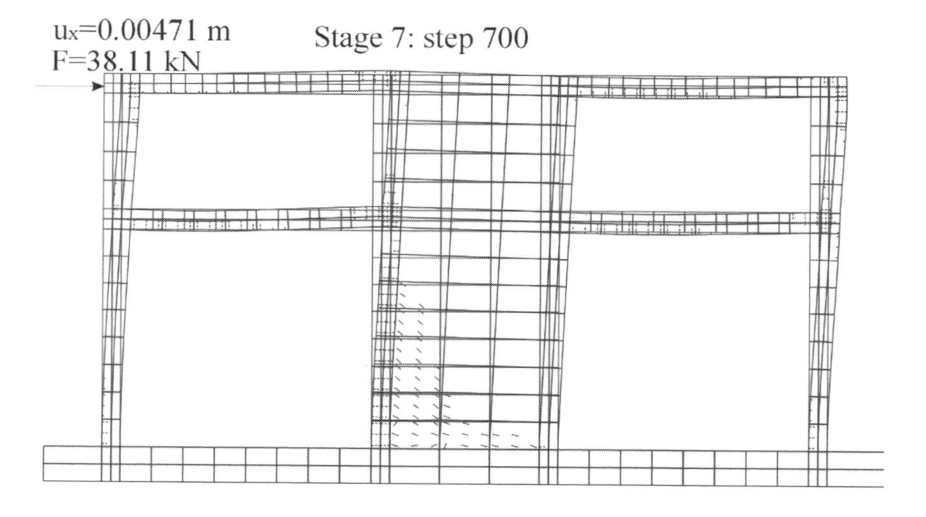

Figure 2g. State of deformations and cracks after applying lateral force F=38.11 kN

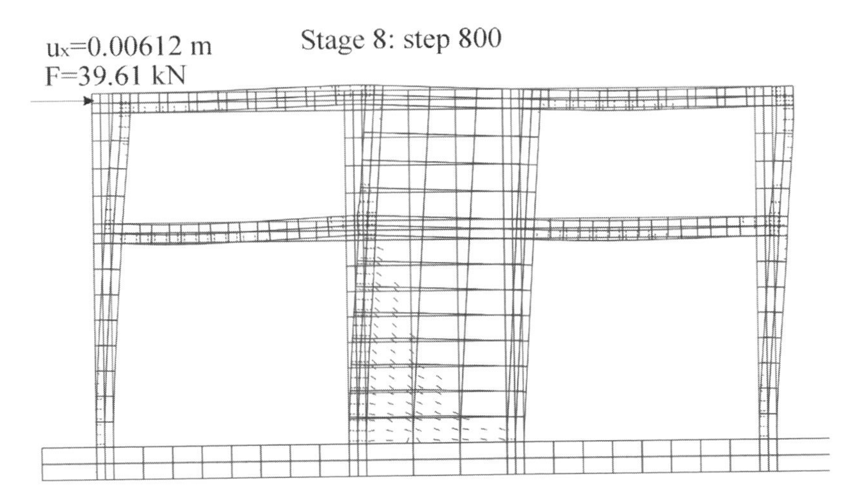

Figure 2h. State of deformations and cracks after applying lateral force F=39.61 kN

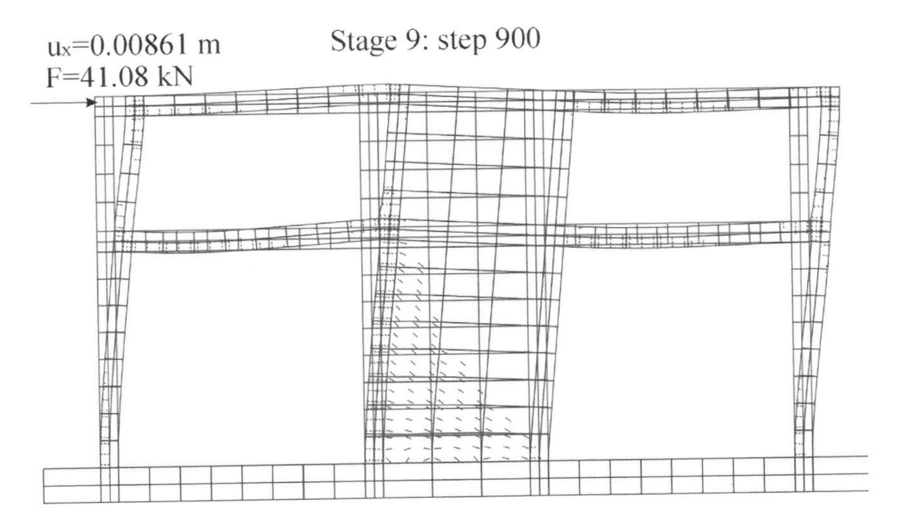

Figure 2i. State of deformations and cracks after applying lateral force F=41.08 kN

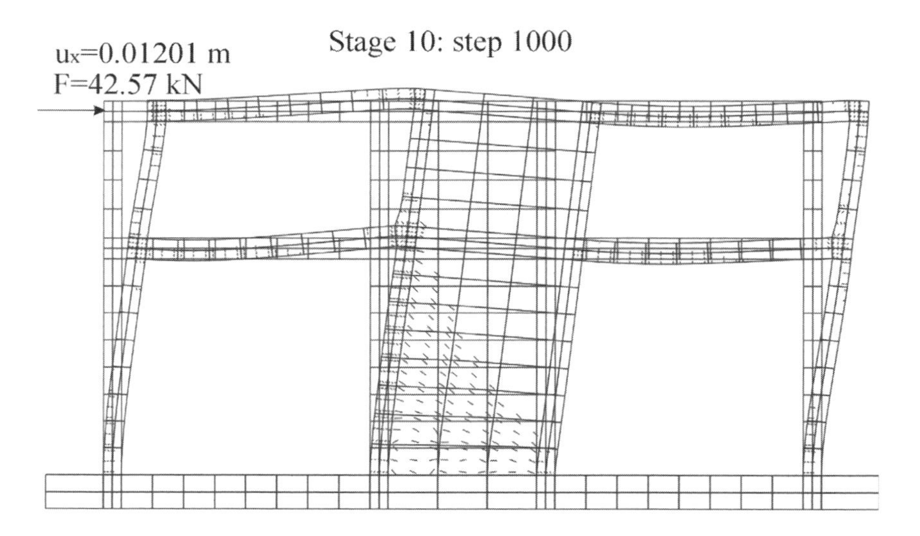

Figure 2j. State of deformations and cracks after applying lateral force F=42.57 kN

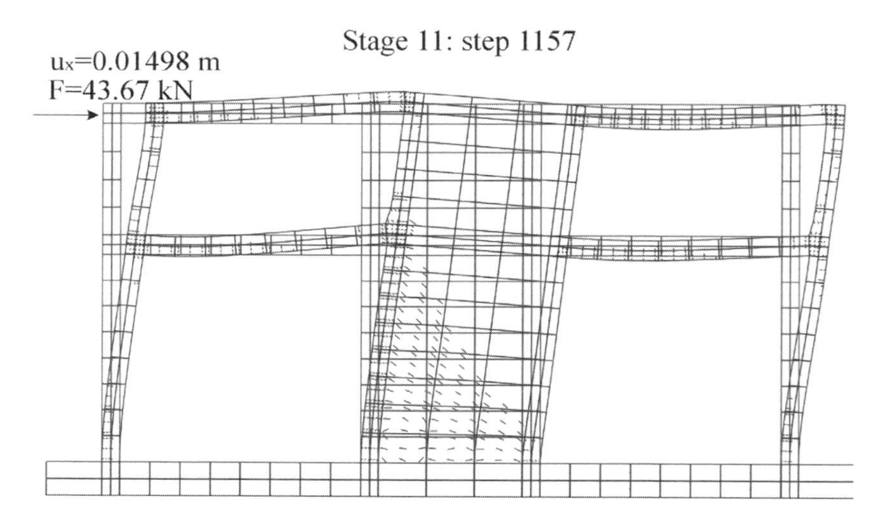

Figure 2k. State of deformations and cracks after applying lateral force Fmax=43.67 kN

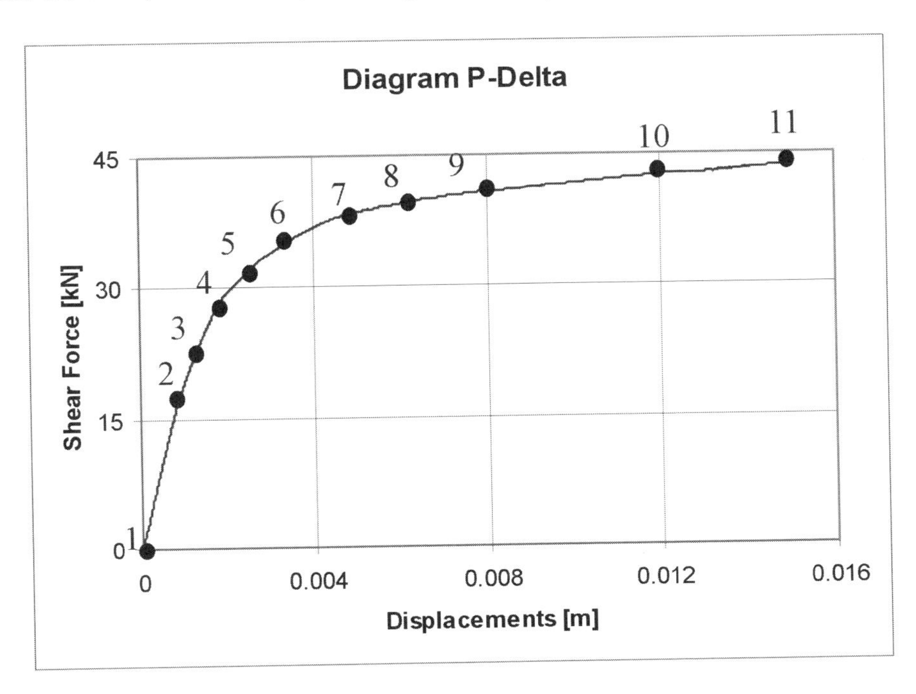

Figure 3. Predicted force-displacement diagram from the push-over analysis with denoted stages corresponding to Figure 2

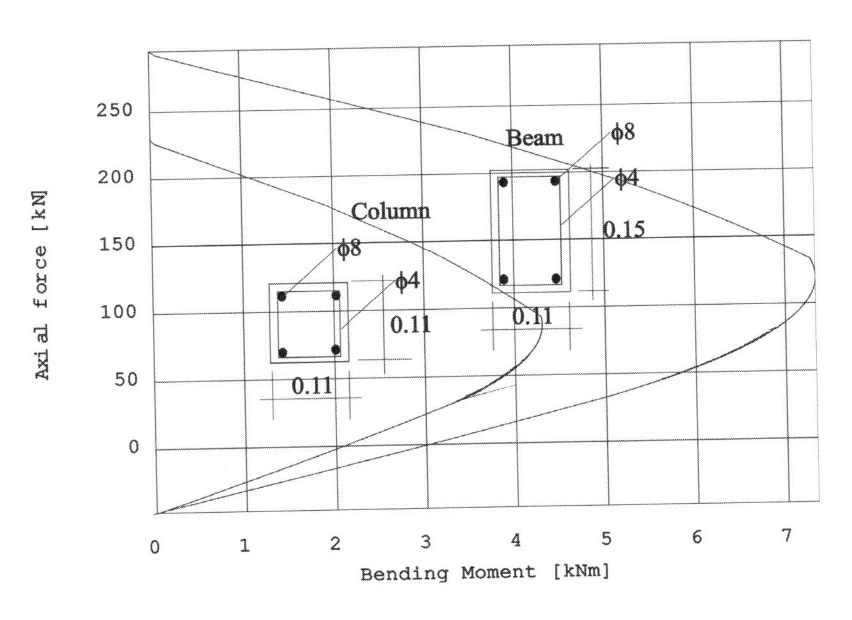

Figure 4. Interaction Mu-Nu diagrams for column and beam sections

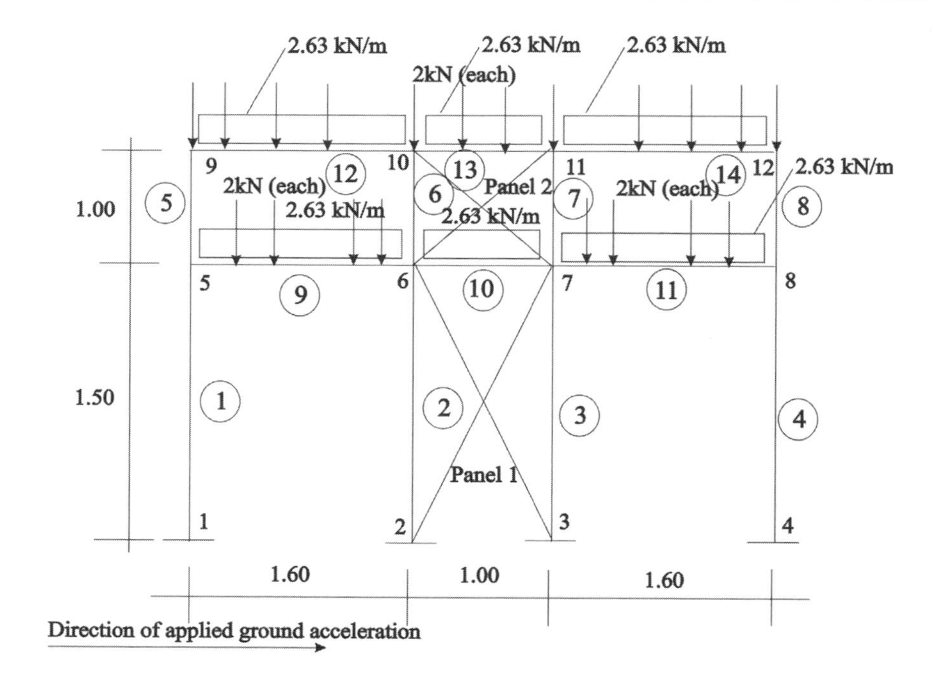

Figure 5. Analytical model used for nonlinear dynamic time-history analyses

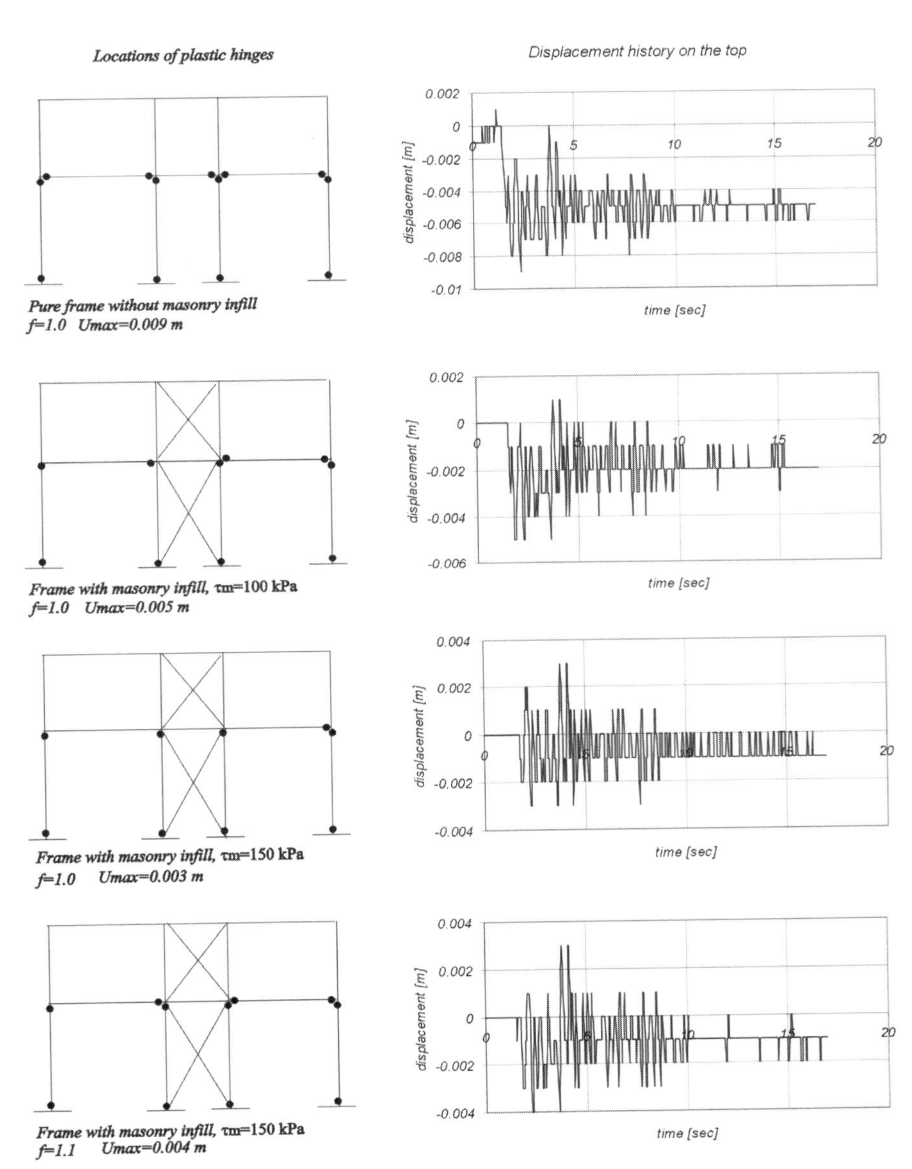

Figure 6. Results from the performed parametric analyses

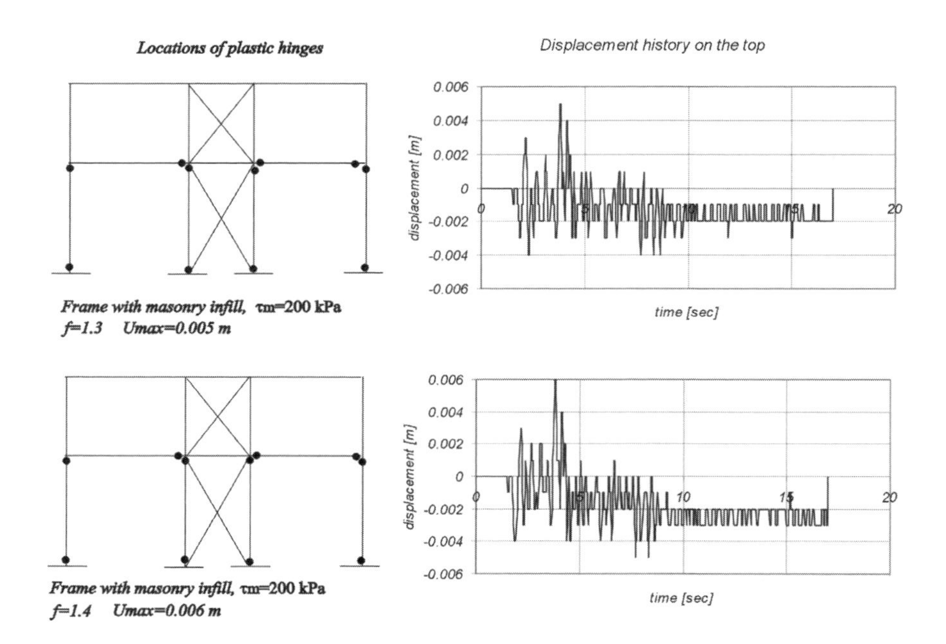

Figure 7. Results from the performed parametric analyses (contd.)

A BUILDING CODE IS OF VALUE ONLY IF IT IS ENFORCED

Sukru M. Uzumeri and Yaman Uzumeri
Middle East Technical University Ankara, Turkey
Chief Building Official (Ret.) City of Toronto, Canada

Abstract: Recent Earthquakes have shown that, to reduce the life and property loss, it is necessary to have good building codes and to have the provisions of these codes enforced. This paper discusses the issues related to enforcement of Building Codes and, using Canadian experience, provides a road map for an effective system to enforce the provisions of the Building Codes.

Keywords: Building Code Enforcement, Inspection, Supervision of Construction, Earthquake Damage Reduction, Regulatory Agencies

1. PURPOSE

The purpose of this paper is to underline the importance of enforcement of Building Codes. It is clear that in many countries there are very up-to-date building codes in effect. Yet in the same countries there are great losses of life and property due to earthquakes. The obvious reason for this is the fact that the presence of an up-to-date Building Code offers no protection, unless it is enforced. This paper discusses the issues related to enforcement of Building Codes and, using Canadian experience, provides a road map for an effective system to enforce the provisions of the Building Codes.

S.T. Wasti and G. Ozcebe (eds.), Seismic Assessment and Rehabilitation of Existing Buildings, 517–526.

2. ROLE OF BUILDING CODES

Buildings for most locations in the world are designed to have the strength to resist the applied gravity and wind loads. In the case of earthquake resistant design for normal buildings, achieving the deformability of the structure under earthquake action without failure is an essential additional requirement. In earthquake prone areas, the designer and the builder need to revise their standard procedures. They need to apply special rules and procedures during both the design and the construction stages. These special rules and procedures are provided in the *Building Codes*.

In a country, authorities having the jurisdiction on safety of the public will need to focus on having in place an up-to-date and implementable Building Code. An equal responsibility of the authorities is the enforcement of their building codes. Just as a Building Code, with provisions that cannot be enforced is not of much use; an excellent Building Code, if it is not enforced, provides no protection at all.

A concrete building appropriately reinforced with steel is a very versatile as well as an economical structural type. However, there may be some construction problems that never surface until a failure. For a builder whose intentions are less than honourable and/or whose knowledge is less than adequate, a reinforced concrete structure is ideal, because, once the concrete is cast and forms are removed, unless the structure is subjected to a real life test, like an earthquake, builder/designer's all faults are hidden from the eye.

When a major earthquake occurs and some buildings collapse, and there is life and property loss, the first priority is to help in the recovery of the victims. Subsequently, the rubble is removed and the site is cleared. After that, the focus may become the determination of the "guilty party". Accusations and court cases follow. On the one side are the victims and their families as well as the public prosecutor and their accusations and demands and on the other side is the need of accused persons to have a fair trial.

Unless the design and construction of the structure in question has been well documented, it is impossible to come to a fair assessment of guilt as well as the assignment of responsibility among parties. The good thing is, whatever steps are necessary for "documentation" are the same steps to ensure a quality construction. Therefore, emphasis must be placed on monitoring and documenting the quality of the product during design and construction. If this is done, in the rare cases that a well designed and well documented building fails, it would be an easier task to apportion responsibility.

3. NEED FOR CODE ENFORCEMENT

Recent earthquakes in Turkey have clearly demonstrated the need for immediate action to put in place an effective building code enforcement process. The Turkish Building Code is a good code but it is obvious that the presence of a good design code is not sufficient. The Building Code must be enforced. Enforcement of a good Building Code would not only help to improve earthquake resistance of buildings but would also improve resistance of buildings against other equally destructive events such as fires and floods. Governments must act on the lessons provided by the disaster, to provide their citizens with a reasonable protection against natural disasters. In all civilized countries regulation and control of building construction is a natural outcome of many years of experiences, resulting from fire and collapse of buildings due to substandard building codes and inadequate construction practices. In spite of many years of destruction of buildings even today conditions still exist that will result in needless deaths and injuries because of inadequate building regulations and more importantly absence of effective enforcement, even when the regulations are adequate.

The design and construction of buildings require up to date building codes and well-qualified designers, builders and inspectors who understand and appreciate the behaviour of structures subjected to events such as earthquakes. Building construction is a team effort, and each team member plays an important role for the completion of a successful project. Designers such as engineers and architects must understand the intricacies of building codes and must have a good knowledge of local construction practices. Builders, including contractors and material suppliers must appreciate the importance of executing the design elements shown on construction drawings prepared by the engineers and architects. Building inspectors who may be government employees or working for private companies provide another but very important independent pair of eyes to oversee the design and construction of projects. A chain is no stronger than its weakest link.

4. ADMINISTRATION OF BUILDING CODE ENFORCEMENT

In order to effectively protect the health and safety of people in buildings, a meaningful building code administration and enforcement system must be developed. The system should take into account many complex and interrelated issues. In general terms these issues may be categorized as:

- up-to-date building codes and reference standards,
- appropriate skill sets and knowledgeable practitioners, and

- realistic systems approach to validate and verify compliance

In this context, the Building Code is meant to include all the local laws regulating construction of buildings including structural, electrical, plumbing and mechanical components. Generally speaking, the building department or another similarly identified entity within a government should be responsible for the enforcement of the building code. Even though some of the above components may be administered by different agencies, building departments should have the overall responsibility to satisfy itself that the component regulations are being adhered to.

5. BUILDING DEPARTMENTS OR PRIVATE AGENCIES

In modern building codes, life safety of building occupants has become the paramount objective. Because of their importance for the public good, code provisions related to structural integrity and fire safety are normally kept within the domain of building departments. Within the past 5-10 years a small number of countries around the world have started using private agencies for code enforcement. This is an area that should be explored further, giving careful consideration to the factors such as:

- overall business culture within the country,
- general attitude to privatization,
- degree of readiness of the construction sector,
- insurance schemes for the practitioners,
- liability of the design professionals,
- certification of private agencies (who, what, when),
- resolution and appeal mechanism for dispute resolution,
- support of the legal system

6. ROLE OF THE DESIGN PROFESSIONALS

Role of the design professionals in the design and construction phases of buildings can not be overstated. With the possible exception of relatively small buildings, it is important that engineers and architects be involved in all phases of projects, until the project is completed and ready for occupancy.

Qualifications of designers and their areas of expertise play a major role in the execution of a successful project. Normally, responsibility to monitor and control the conduct of design professionals rests with governmental agencies or professional associations empowered to take punitive measures when the professional's conduct is suspected to be questionable.

7. BUILDING CODE ENFORCEMENT BY THE BUILDING DEPARTMENTS

The mission of a building department is to police the acts of individuals to insure compliance with statutes and regulations. The principle behind this is that in order to protect the health, safety and welfare of the many, governments must limit the acts of the few.

Similar to the majority of the western countries, in Canada, enforcement of building regulations which have been formulated and approved by National expert groups, is the responsibility of local municipal governments. Perceived and real independence of governments, knowledge of local conditions, proximity to job sites, and the necessity for a quick response to emerging problems are the main reasons local municipalities are preferred delivery agents for enforcement of construction regulations.

8. TOOLS FOR BUILDING CODE ENFORCEMENT

Building codes are normally enforced through a building permit process. This process also provides for collection of fees to recover costs associated with the enforcement activities. The typical building permit process may be divided into the following five distinct but related activities:

- application for a building permit ,
- plan examination,
- permit issuance,
- inspection of construction,
- occupancy certification.

8.1 Application for a building permit

Application for a building permit is the first step towards effective administration and enforcement of building regulations.

The owner or the authorized representative of the owner should make the application to the municipality, on the forms to be obtained from the Building Department. The applicant should give clearly and fully the information required to complete the forms and should certify the correctness of the information.

Application should also contain information regarding professional designers who are involved in the design of the project. For significant buildings, it should be mandatory that structural design and field review of construction be carried out by qualified structural engineers.

The permit application should be accompanied by:

- Site plan and survey, prepared by a certified surveyor, showing the location and siting of the building on the property and its proximity to the property lines, other structures or futures that may have an impact on the building's performance.
- Soil report, prepared by a geotechnical engineer, providing geological information, showing the capacity of the soil and groundwater conditions based on bore holes, recommended foundation design and lateral earth pressures,
- Architectural plans, prepared by an architect, showing the interior layout and exterior elevations of the building as well as types of occupancies and intended use of spaces within the building.
- Structural plans, prepared by a structural engineer, with sufficient detail, including design assumptions, lateral and vertical loads to establish the structural capacity and integrity of the structure under design forces established by the building code,
- Commitment from the owner that the design engineer or his/her authorized representative will be retained for field review of construction.
- Acknowledgement from the design engineer that he/she has been retained for the field review of construction.
- Payment of appropriate permit fee. The fee may be based on the value of the project. It should be sufficient to cover the expenses of the agency.

8.2 Plan examination

Plan examination begins after plans and supporting documents are submitted to the building department. The plan examiner reviews the plans to determine compliance with the code requirements. The extent of the review usually depends on resources, and complexity of the project.

The structural review does not mean the re-design of the entire structure. By carrying out random checks or structural calculations, and reviewing the structural details specified, it is expected that the plan examiner would be able to confirm that the requirements of the building code are complied with. During this process missing information or inaccurate calculations, assumptions, structural details etc. are brought to the attention of the design professional for correction. After completion of the review the plan examiner would clear the project for permit issuance and obtain counter signing of the project from his/her supervisor.

8.3 Permit issuance

After the plan examination is complete the building permit may be issued. At this point and just before the permit issuance, documents are reviewed to insure that all relevant approvals from other agencies are in place and that the building permit fee paid at the time of application is appropriate. If everything is in order, the permit would be issued to the person who made the initial application.

If the information submitted with the application indicates that the work proposed would not comply in all respects with the provisions of the building code or other applicable governmental regulations, the permit should be refused until the drawings and specifications are appropriately amended or corrected.

8.4 Inspection

Inspection of buildings under construction is a very important phase of building code enforcement. The mission of a code enforcement agency such as a building department is to police the acts of individuals to insure compliance with statutes, rules and regulations related to construction of buildings. Such compliance is established through inspection of all new buildings, structures and installation of equipment during the course of construction.

Available resources also dictate frequency and extent of inspections. Using specially trained inspectors for inspection of concrete high rise structures or welded steel frames is another alternative that should be considered. This of course would depend on the size of the municipality as well as the type of building activity within it.

9. RECORD KEEPING AND THE ROLE OF COMPUTERS

Effective building code enforcement will fail without an efficient record keeping, plan storage and retrieval system. Updating of records on a daily basis, tracking of permit applications and inspection results are fundamental to enforcement activities. Records that are kept must be up-to-date, accurate and easily retrievable.

With an ever increasing number and complexity of buildings, manual record keeping, except in relatively small jurisdictions will not produce the desired results. Therefore the use of computers and other electronic devices should form part of a comprehensive and integrated building permit system. Many products are available in the market to establish such a system.

10. TRAINING, EDUCATION AND QUALIFICATION

Training and education of practitioners involved in the design and construction of buildings, such as building designers, building officials and others is another important and integral part of an effective building code enforcement process. Professionals involved in the design and construction must be well qualified for their tasks. In many parts of the world, qualification requirements for technical persons include certification such as a "Professional Engineer" and/or "Specialist Engineer" in addition to the individual's graduation from an accredited program. For an effective design and construction process in a country, clear qualification steps must be established so that design professionals are given goals and encouraged to improve themselves. The most significant benefit of such a process is to ensure that the technical personnel involved in design and in construction are adequately equipped and qualified.

Building officials are required to possess technical knowledge and they must have opportunities to upgrade themselves to remain current with the changes being made in the regulations. Most of the time problems encountered by the building officials are people problems, political problems and procedural problems. To dwell at length on the technical aspects, at the expense of sociological, political and legal aspects of code enforcement may distort requisite attributes of a building official. Training of officials to deal with such problems is essential and would have a long lasting and positive effect on the achievement of enforcement objectives.

11. GENERAL RECOMMENDATIONS

The following is a list of general recommendations for effective code enforcement. It is not an exhaustive list but it includes items for review of plans prior to issuance of permits and inspection activities that take place during construction. Implementation of these recommendations will result in substantially improved building code enforcement in Turkey.

1. Require submission of plans with building permit applications that are sufficiently detailed to insure substantial code compliance.
2. Establish a procedure for processing of plans through all governmental agencies having a legitimate interest in the project. It is important to be selective in this process so that permit issuance is not delayed

unnecessarily. Such delays may easily result in premature start of construction without a permit.

3. Check all legal description against ownership information given on the application forms.
4. Establish a central computerized filing system for all open permits.
5. Establish a systematic and uniform procedure for the processing of code violations, such as by issuing *order to comply* and if necessary *stop work orders*
6. Establish fees for permit issuance and inspections. The fees should be collected at the time of permit issuance and should be sufficient to fully cover expenses associated with all phases of code enforcement.
7. Make periodic inspections of all buildings under construction. Frequency of inspections to be determined on the basis of complexity of the project, capacity of the contractor and the critical stages of construction. This inspection includes the review of the results of all material tests carried out by independent laboratories or firms.
8. Require inspectors to document all inspections and complete a written report of any violation. Each violation report should be referenced to a specific section number of the code that has been violated.
9. Require every notice of violation to be approved by the supervisor.
10. Use multi-copy notices for follow-up and to secure compliance. Keep one copy in the office until compliance is achieved.
11. Do not issue certificates of occupancy until all violations are corrected and any special conditions have been fulfilled.
12. Require inspectors to prepare route sheets in the office.
13. When work is done without a permit, issue violation notice to the owner as well as to the contractor.
14. Require utility companies to secure a written release from the Building Department before turning on utilities.
15. Establish a date file for open violations with expected compliance dates and reinspect on the agreed on compliance date.
16. When compliance is not achieved require inspectors to contact responsible parties immediately.
17. Require inspectors to recommend an appropriate course of action to the supervisor.
18. Require the supervisor to review and approve or modify the course of action recommended by the inspector.
19. Repeat the re-inspection process until compliance is achieved.
20. If compliance is not achieved, issue a *stop work order*.
21. If work is not stopped or compliance still not achieved, prepare the necessary documents and submit to the courts for prosecution.
22. Introduce an electronic building permit control system.

23. Improve training and education opportunities for those involved in building code enforcement.

12. CONCLUSIONS

In this paper an attempt is made to outline a framework, to improve the building code enforcement process. The need for such improvement to Turkish practice has been made abundantly clear in view of the devastation caused by the recent earthquakes.

However above all else, the Government of Turkey must seize this opportunity, and embrace the need for action and begin to introduce legislation as well as structural changes within the system that will result in much improved building code enforcement throughout the country.

ACKNOWLEDGEMENT

During the preparation of this paper, critical comments and suggestions made by J. W. Wright Director and Chief Building Official of the Town of Markham, Ontario, Canada are gratefully acknowledged.

EDUCATIONAL ASPECTS OF SFP977231 – SPREADING THE INSIGHT

Syed Tanvir Wasti
Structural Mechnaics Division, Department of Civil Engineering, Middle East Technical University, Ankara, Turkey

Abstract: Continuing technical education updates knowledge so that the engineer can face the challenge of technical change. Engineering research always provides spin-off for application in training and distance learning. However, the absence of a suitable nexus between producers and consumers of know-how results in vast areas of expertise remaining unexploited. Timely transmission of research results in digestible form to engineers must be ensured. Information from NATO Project SfP977231 is being processed and disseminated to the technical community in user – friendly formats to increase preparedness against future disasters.

Keywords: Dissemination of technical information, continuing education, seismic technology transfer, specialized distance learning

1. INTRODUCTION

Priority fields in continuing civil engineering education would include relatively specialized applied disciplines that have not permeated the regular curriculum. The standard engineering curriculum is like a human skeleton in that one can do little to change its composition or make it look different. Training, professional experience and / or graduate study are needed in order

S.T. Wasti and G. Ozcebe (eds.), Seismic Assessment and Rehabilitation of Existing Buildings, 527–538.

to put flesh on the skeleton and give it some life. The contents of earthquake engineering, stochastic analysis, neural networks, biomechanics and health monitoring of structures are examples of developments that have not filtered down to the level of the competent civil engineer in many countries. However, all continuing education need not be highbrow or possess a primarily research – oriented content. Extremely valuable if somewhat mundane technical and professional activities would include the formulation and development of engineering codes, standards and specifications, the compilation of suitable design handbooks and even the preparation of technical dictionaries for use in both developed and developing countries. The simple academic process of analysis and application does not always prepare the engineer for the semi-legal intricacies of, say, the earthquake code. Commentaries on code clauses, complete with numerical illustrations, are a requirement for the practising engineer today. Furthermore, in earthquake – prone countries, damage assessment, structural rehabilitation and disaster management are important areas in which the latest knowledge, techniques and expertise need to be disseminated to engineers, builders, technicians and even managers.

In recent years the pace of civil engineering development has considerably increased worldwide. The swift and successful completion of ambitious construction projects such as taller buildings, longer bridges and safer structures requires 'know-how' in planning, in design and in actual execution on site. The know-how is needed not only in the use of local materials and skills but also in the acquisition of the latest design and construction techniques. Continuing education, both on the formal and informal levels, can provide the necessary means of supplementing as well as building on the knowledge acquired at the university, so that the engineer can augment his store of knowledge and thereby successfully face technological challenge. As Dr Samuel Johnson has sagely remarked: "**Knowledge is of two kinds. We know a subject ourselves, or we know where we can find information upon it**."

2. IMPORTANCE OF CONTINUING EDUCATION

According to Smerdon[1], a group of experts in the mid 1980's estimated how long it would take for half the knowledge an engineer was given in his field to become obsolete, and came up with a figure of 7.5 years for a mechanical engineer and 2.5 years for a software engineer. Admittedly, although we cannot say that earth-shaking events do not occur in civil engineering – this very Workshop is testimony to the fact that earth-shaking events occur frequently – we may admit that civil engineering is a conservative discipline and suggest a half-life of a dozen years. Since a typical working career can easily span 30 years, it becomes incumbent to provide engineers with the facilities for professional enhancement. The engineer must be encouraged to keep fully abreast of developments in his special area. He must either shape up or ship out or, as Smerdon puts it: "...engineers must be prepared to switch nimbly to a new field when the old one peters out."

This lifelong learning process that seems to have caught the present generation unawares is not confined to engineers alone. All professions are mutating, but it is the swiftness of technological change that is responsible for the formal induction of continuing education into engineering disciplines. In the words of Bowman: "A concerted effort is needed to...promote the mind-set that the initial degree is only the first step in the pursuit of engineering competency."[2] As mentioned above, advancements in the field of civil engineering are so rapid that it has become difficult for educationists to develop suitable long–term standard curricula for undergraduate courses. In drawing up such a curriculum the major objective is to ensure a balance between the imparting of theoretical material on the one hand and the inculcation of practical skills on the other. In a sense, engineers are the victims of their own success. Time limitations do not allow course lectures to adequately cover specialized and interdisciplinary areas such as engineering management, adverse environmental effects, advanced computer applications, statistical methods, new materials and earthquake engineering. It becomes, therefore, essential that the engineers working in such specialized fields should move towards acquiring a postgraduate degree. However, it is not possible for most practising engineers, especially in developing countries, to undertake postgraduate studies due to various reasons. Firstly, the facilities for postgraduate studies to be conducted side by side with one's routine job are not presently available in most such countries, although the introduction of distance learning techniques has begun to make inroads into this problem. Secondly, the socio-economic condition of a civil engineer may not allow him to leave his job and take up specialized studies even on a non-full-time basis. The dissemination of new engineering knowledge and application techniques, summarized under the general rubric of Continuing Education, as envisaged and implemented in most contemporary universities, would provide a stimulus to practising engineers, architects and contractors to improve their detailed or in-depth knowledge of certain subjects. While it is not proposed that continuing technical education should provide a method for the atonement of academic neglect during one's undergraduate career, there is no doubt that such aspects will probably heighten its appeal.[3]

Continuing education can also be effectively used for bridging the ever-increasing gap of engineering know-how between developed and developing countries. However, the greater importance of continuing education lies in the early dissemination of the results of research work being carried out in any one institution, whereby engineers all over the world working in the same area may reap the benefits. Another advantage of continuing education is that there is usually an adult audience with greater motivation and fewer distractions. No miracles can be expected in education, and continuing education cannot lead to "instant technology". Rather, it is the exploitation of new methods and learning systems in an effort to prolong the process of education beyond the confines of the classroom. It is, in fact, an illustration of the old saying, which states: "**give a man a fish and you have fed him for a day – teach him how to fish and you have him fed for life**".

3. PRIORITY FIELDS IN CONTINUING EDUCATION

Priority fields in continuing education in the area of civil engineering will mainly depend on the specific needs of a particular country. Many countries adopt or adapt specifications and codes for the design and construction of structures from available 'advanced' codes. However, continuing advancements in the knowledge of structural engineering are not readily available in many countries and therefore it becomes difficult for practising engineers to upgrade or even follow the philosophy of 'advanced' codes and specifications. Short 'deficiency' courses on the philosophy of design or even the latest techniques of analysis therefore need to be offered regularly in all countries. Design handbooks, charts and technical manuals should also be prepared for use by freshly graduating engineers. The inculcation of technical expertise by means of special programmes on public television should be encouraged.

Reinforced concrete is the most important material of construction in a country like Turkey. In order to ensure good quality concrete, the recommended methods of mix design batching, placing, curing and testing should be publicized. Short courses and training programmes should be organized wherein experts teach suitable practices of hot and cold-weather concreting.

A big knowledge-gap in many countries located in seismic regions is that of the design and construction of seismic-resistant low-cost buildings including houses, schools and industrial buildings. Continuing education should be used for training the engineers and architects in the design and construction of low cost buildings, with emphasis on materials and techniques suitable for a particular region. There is occasionally a tendency to ignore the seismic behaviour of masonry structures and to emphasize only reinforced concrete structures. Particularly in less urbanized regions, it would be desirable to introduce reinforced masonry construction for buildings of unto two or three storeys in height. As public buildings often possess standard architectural features and may be produced in the form of prestressed and prefabricated structures, thereby introducing a certain degree of industrial construction into the design, attempts must be made to provide engineers with information on the seismic behaviour of such buildings. Furthermore, bearing in mind the vulnerability of the building stock in many countries, a more urgent need is the availability of knowledge and techniques for swift and economically feasible seismic rehabilitation of structures, whether pre- or post-disaster.

4. EDUCATIONAL ASPECTS OF NATO PROJECT SFP977231

NATO Project SfP977231 is a comprehensive project with analytical as well as experimental phases in the general area of Seismic Assessment and Rehabilitation of Existing Buildings. Site applications and monitoring also form part of the project goals. As such, the project area of interest spreads over a very wide range.

The primary objectives of the completed project are to develop simple methodologies for the seismic vulnerability assessment and sound, practical and economical rehabilitation techniques [including rapid retrofitting methods] for seismically vulnerable buildings in Turkey and the Balkans. Seismic evaluation and rehabilitation work for existing buildings necessitate the participation of a large number of practicing engineers. However, these tasks require special expertise. Therefore, a major thrust of this project is to develop engineer-training programs to enhance the knowledge of the practising engineer and to disseminate the findings and the end-results of the research phase of the proposed research program. However, in order to enable research findings to be understood, appreciated and put into use by practising engineers, it is necessary to provide such basic information in the areas of earthquake engineering, structural dynamics, reinforced concrete behaviour, computer modelling, codes and specifications.

Technical universities have a great responsibility in arranging the continuing education of engineers. These institutions, while preferring to deal with advancing the frontiers of technical knowledge must also, by virtue of being oriented towards the applied sciences, undertake less cerebral research work on the properties of locally available materials, the establishment of design standards and specifications and the development of new materials and methods of construction. Not surprisingly, it is the results of this second-tier research work which may have a direct impact on the development and wealth of a country. Hence, technical universities should be encouraged to disseminate such results to practising engineers through frequent long-term and short-term extension courses. In its current website, the American Society for Engineering Education [ASEE] correctly states that "...**the quality of our engineers affects the quality of our lives.**" In severely active seismic environments, it is partly the obligation of technical universities to ensure that proper levels of specialized education are imparted to engineers so that careless and dangerous building design does not affect the quality of our deaths.

In many countries, it is expected that technical universities will in future be able to set up courses leading to postgraduate degrees in specialized areas e.g. earthquake engineering, water resources engineering, environmental engineering etc. To these specialized fields a postgraduate course in 'Disaster Engineering' could also be introduced. Such a course would enable the engineer to broaden his knowledge of the role of civil engineering in disaster relief implementation. The course may include such areas as management, economics, lifeline engineering, and the design, construction and maintenance of seismic-resistant structures. The seismic design of utilitarian structures such as storage tanks, silos, sheds, small bridges and small dams could also feature in such a course. Depending upon the needs of a particular country similar courses in 'National Development Engineering' may also implemented, to include such areas as intermediate technology, water resources and irrigation engineering, transportation systems, lifeline engineering and industrialized construction. Engineers who attend such courses should be encouraged to work out alternate solutions of the actual problems of a particular region, and the practical application of these solutions should be stressed. The idea behind such interdisciplinary programmes is not just to make engineering look fashionable, but to allow and indeed enable those dedicated engineers who have not chosen to make academic research a full time career utilize at first hand the practical results obtained from contemporary research to enhance the quality of the structures they build. A study of the long and short term academic offerings being advertised by several renowned universities in the United States and Western Europe indicates that subjects from specialized fields hitherto considered almost incompatible, e.g. Structural Design of Health Care Facilities, Environmental Effects of Multi-storey Car Parks, Seismic Risk and Zonal Planning, etc., are now being jointly proposed as courses both in the regular as well as continuing education programmes.

5. TECHNIQUES FOR DISSEMINATION OF INFORMATION

If continuing education is to have the right impact on practising engineers in an earthquake-prone country it must be imparted in an efficient and attractive manner. Knowledge must be well–packaged and even well–advertised. In the colourful words of Stewart Henderson Britt: "**Doing business without advertising is like winking at a girl in the dark. You know what you're doing, but nobody else does.**" These days, I suppose, the same is true of education, training and other didactic activities. Delivering a message is not enough – the message must be seen to be conveyed.

The chalk and blackboard are becoming increasingly outmoded instruments of instruction. Filmstrips, slides and transparencies often combined with special manuals for self-study are taking their place. The use of television for both classroom as well as home teaching is extensive. Most universities in the U.S.A. now have departments of visual instruction. Automated learning by computer now provides means of step-by-step teaching as well as performance evaluation.

In the teaching of civil engineering, the use of models, photo-elastic methods, video-cassettes of important and complicated experiments e.g. soil tests, pseudo-dynamic lateral load tests as well as shaking table experiments, etc. should be encouraged. Construction know-how, incorporating all aspects of seismic-resistant construction from the excavation stage to the finished structure, e.g. soil improvement, in-situ placing of reinforcement, design of joints, RC shear wall detailing, masonry infill walls, should be made available in order to enable engineers to apply their knowledge to practical problems.

"Packaging" of education in the form of video-cassettes, films, slides and records with supplementary books and booklets will relieve what is probably the biggest problem of technical education in several countries - the acute shortage of qualified and talented teachers. New educational methods enable a teacher to reach a potentially unlimited audience and also liberate him from the less attractive facets of regimented teaching - the reluctant docility of students, payment on the basis of hours taught and the carrot and stick approach in the class room.

Possibilities of home study, as with the courses of the Open University in Great Britain and also numerous other institutions of distance learning have now become completely feasible. The ability to obtain a formal certificate on completion of a set of distance learning courses will provide an added incentive to such programmes.

6. ACHIEVEMENTS UNDER THE SUBPROJECT

Subproject 4 has been planned to include a program of extensive publication of research and professional results suitable for engineering application, two national/international workshops and engineer training sessions including distance learning techniques. For this purpose the Coordination Workshop held in Antalya between 25 and 28 May, 2001 provided excellent opportunities for brain-storming. As a result of the decisions at Antalya, a summary of recent progress may be given under different headings below.

It is suggested that the dissemination of research results from the present project be conducted in the following formats:

- Individual lectures, seminars, short refresher courses and training workshops on the latest methodologies and techniques for the seismic assessment and rehabilitation of buildings
- The preparation of desktop-published course material for the personal and professional development of engineers, explaining and illustrating salient items of the Turkish and other contemporary earthquake codes; structural behaviour during earthquakes; damage assessment for urban and rural structures; rehabilitation techniques as developed from experimentation and research, etc.
- Broadcasting of systematized knowledge relating to the seismic amelioration of structures and buildings of different types by means of compact discs, video films, the Internet and television channels [if possible]

It is expected that after some trial runs, the continuing education activities could develop into formal programs leading to evaluation of student performance and possible award of certificates of proficiency.

6.1 Implementation of the structural engineering research unit [SERU] website and net

It will be appreciated that an important aspect of knowledge dissemination deals with the provision of updated and readily accessible information about the NATO Project itself, in addition to suitable links with related persons and projects. A comprehensive web-site in both English and Turkish is now more and less complete, under the address below: http://www.seru.metu.edu.tr/

A short introduction to the function and capabilities of the SERU net designed and programmed for the sub-project by Mr Ufuk Yazgan is as follows:

SERU will work as both as a Bulletin Board and a Clearing house for research information connected with NATO Project SfP977231. SERU net is basically an Internet platform developed for SERU members to share their computer files over the World Wide Web. SERU net is an Active Server Pages(ASP) application running on a Microsoft Windows® 2000 Server using Internet Information Services®(IIS) 5.0 technology.

SERU net provides its users the ability to upload their files on the server where they are shared on a password protected environment and can be downloaded or viewed by other SERU members from anywhere on the world as long as there is a World Wide Web connection available. Users may also send messages to other members using the interface provided in SERU net. SERU net has the capability of maintaining about 100 simultaneous connections to the server.

Users can upload their files to SERU net server using the interface provided by the web page or using the File Transfer Protocol(FTP). All requests sent to SERU net are automatically filtered by the Microsoft URL Scan® utility for possible misuse. It is planned that SERU net is going to be further improved with the feedback coming from users.

6.2 Implementation of the excel- based sorting and information retrieval facility

The dissemination of knowledge and technical information from research publications requires the organization and management of vast amounts of data. A large database of information on seismic analysis and design, structural assessment and rehabilitation has been compiled as part of the subproject. Asst. Professor Dr Ahmet Turer has developed a user – friendly EXCEL program called **SUPER**, an acronym from **S**ortable **U**serfriendly **P**rogram: **E**arthquake **R**eferences. SUPER.XLS has been produced as part of the Dissemination of Information subproject of the NATO Project No.. SfP977231. The program runs under a regular workbook taking full advantage of the tabular format provided by Excel. The SUPER.XLS file has two alternative input windows which are identical in nature but provided in the English and Turkish languages with an attached database which is currently in use for author and keyword retrieval and search of relevant documents, as well as the sorting of similar new input.

6.3 Development of a computer – based earthquake slide archive

As of November 2002, Assistants Baris Yalim and Melih Susoy have begun work on a process for the scanning, identification, labelling and conversion into electronic format of available 36 mm diapositive colour slides dealing with all aspects of earthquake engineering. It is intended to develop a series of presentations for teaching and information dissemination purposes incorporating various aspects of seismic damage, structural assessment and rehabilitation techniques.

6.4 English translation of TS500 – Building code requirements for reinforced concrete

Work has been completed in recent months on the technical translation into English of the recently published edition of the Turkish Standard TS500 entitled Requirements for Design and Construction of Reinforced Concrete Structures. The standard serves as a national code for the gravity and seismic design of RC members and structures. The Turkish Standards Institution gave its approval for the publication and educational utilization of the translation, and a grant from the Scientific and Technical Research Council of Turkey [TUBITAK] made publication possible in the middle of March, 2003. The publication of the English version of this Standard will enable it to be used for specific teaching purposes. However, it is also intended to expose the contents of the Standard to some international scrutiny and discussion, and thereby to collect feedback regarding the clauses of this Standard, especially those that deal directly with seismic analysis, design and detailing.

7. NATO INTERNATIONAL WORKSHOP IN IZMIR

In order to enable exchange of information, present intermediate analytical / experimental / other results obtained and to provide a forum for technical discussions related to continuing work in each subproject, a "Work In Progress" NATO International Workshop under Subproject 4 of NATO Project SfP977231 was organized. The proposed dates for such a Workshop to be held in Izmir, Turkey were chosen as Tuesday 13 May 2003 and Wednesday 14 May 2003. It is gratifying that many project members were able to participate in the Workshop, the dates for which had been conveniently been chosen to follow those of the 2003 FIB Symposium in

Athens on Concrete in Seismic Regions. Apart from participants in NATO Project SfP977231 a few personal invitations were also issued to prominent researchers working in the areas covered by the project whose contributions have significantly enhanced the deliberations of the Workshop.

8. OTHER ACADEMIC ACTIVITIES RELATED TO THE SUBPROJECT

Publications as well as seminars on the national level dealing with various aspects of the NATO Project are continuing as part of the educational fall-out of the subproject. Such publications include research reports stemming from the NATO Project, the first of which has been published under the SERU logo.[4] Turkish team members have also contributed papers in project-related areas, e.g. to the Ergin Çitipitioglu Memorial Symposium on Structural and Earthquake Engineering [held in Middle East Technical University in October 2002].

As components of the academic training aspects of the project, it should be mentioned that various young faculty members have been and currently are being supported for short periods of training and academic experience as well as for participation in international conferences.

9. OVERALL CONCLUSIONS

Over 40 years ago, as pointed out by Johnson,[5] engineering graduates could be split up on the basis of future careers into the following categories:

- Engineer-Scientists. .. 5 – 10%
- Creative Design Engineers. .. 10 – 20%
- Functional Engineers. .. 20 – 40%
- Engineer Technicians. .. 10 – 20%
- Non-Engineering work. .. 5 – 10%

In several countries these percentages may change radically, and while precise data are not available, observation and experience suggest that engineer technicians engaged in routine engineering operations and engineers in non-engineering work are often over-represented. While such lateral career modification may sometimes be to the detriment of the profession and the job satisfaction of the engineers themselves, occupational mobility is also becoming a feature of modern professional life. In a recent study, Rosen and Paul[6] have mentioned that in North America, over 12 million people change their occupations every year. The important matter is to make available those tools to the engineer that will continue to allow him to remain as a qualified, productive and up-to-the-minute analyst and designer.

Continuing technical education and dissemination of research information should have as its goal the continual upgrading of the engineer's task by providing him with the knowledge and expertise that will enable him to progress up the ladder towards becoming a functional or a creative design engineer if he so wishes.

Funds allocated to education in many countries often fall far short of requirements. It is in the best interest of all countries to acquire and then consolidate the technical know-how that is a by-product of applied research. One manner in which such knowledge can be spread is through the support of creative comprehensive programmes of continuing technical education.

REFERENCES

1. Ernest T. Smerdon, "Lifelong Learning for Engineers: Riding the Whirlwind", *The Bridge*, Vol. 26, Nos. 1 & 2, Washington D.C., 1996.
2. C. W. Bowman, "Lifelong Learning for Professional Engineers", *Engineering Issues* [Canadian Academy of Engineering], No. 6, September 1997, pp. 1-2.
3. K. Mahmood and S. T. Wasti, "Continuing Education in Developing Countries", unpublished report, Lahore, Pakistan, 1978.
4. G. Ozcebe, U. Ersoy, T. Tankut, E. Erduran, R.S.O. Keskin and H.C. Mertol, " Brick-Infilled Frames with CFRP", SERU Report No. 2003/1, Middle East Technical University, Ankara, March 2003.
5. J. S. Johnson, "A Philosophy of Engineering Education", *Journal of Engineering Education*, Washington, D.C., U. S. A., March 1959.
6. S. Rosen and C. Paul, *Career Renewal*, Academic Press, New York, October 1997.

Acknowledgments

We wish to express thanks to many organizations and individuals in connection with the Workshop and publication of the book:

- The NATO Science for Peace Office, Brussels, represented most ably by Dr Chris de Wispelaere and Dr Susanne Michaelis, for their continued positive sponsorship. Dr Michaelis personally participated in the deliberations of the Workshop in Izmir and was responsible for liaison with the NATO Science for Peace Office.
- The Scientific and Technical Research Council of Turkey [TUBITAK] for matching funds and support
- The Publishers, Messrs Kluwer of the Netherlands and their energetic and helpful representative Wil Bruins
- METU Civil Engineering Department Research Assistants Baris Yalim and Melih Susoy for efficient infrastructural support
- Bakhus Travel and Tours, Ankara for their professional management of local hospitality

Syed Tanvir Wasti
Guney Ozcebe

Index

Note: Very frequently occurring individual words like assessment, building, code, concrete, deformation, earthquake, frame, reinforced, response, structure, system(s), etc., have not been included in the index.

A

B

C

D

E

F

G

H

I

J

L

M

N

O

P

Q

R

S

T

U

V